PRISON LAW

Prison Law

TEXT AND MATERIALS

STEPHEN LIVINGSTONE

AND

TIM OWEN

CLARENDON PRESS · OXFORD

1993

Oxford University Press, Walton Street, Oxford OX2 6DP
Oxford New York Toronto
Delhi Bombay Calcutta Madras Karachi
Kuala Lumpur Singapore Hong Kong Tokyo
Nairobi Dar es Salaam Cape Town
Melbourne Auckland Madrid
and associated companies in
Berlin Ibadan

Oxford is a trade mark of Oxford University Press

Published in the United States
by Oxford University Press Inc., New York

British Library Cataloguing in Publication Data
Data available

Library of Congress Cataloging in Publication Data
Owen, Tim
Prison law: text and materials | Stephen Livingstone and Tim Owen.
p. cm.
Includes index.
1. Prisons—Law and legislation—Great Britain.
I. Livingstone, Stephen, 1961- . II. Title.
KD8450.094 1993 344.41'035—dc20 [344.10435] 93–14381
ISBN 0–19–876265–8
ISBN 0–19–876264–X (p)

Set by Hope Services (Abingdon) Ltd.
Printed in Great Britain
on acid free paper by
Biddles Ltd
Guildford & King's Lynn

Foreword

Frequently enough to carry conviction, a prisoner will recount how somebody in authority has said 'I'm the law here'. The sense of impotence and isolation the phrase creates is designedly chilling. In too many instances, moreover, the officer is right: he or she is in sole control and there is no recourse to any legal authority.

Such absolute power is the antithesis of the rule of law. Yet for an age the courts were either afraid or unwilling to take responsibility for the protection of citizens behind bars from unlawful treatment. It was not said, of course, that prisoners had no rights; only that the assurance of such rights as they had could safely be entrusted to their custodians. Behind this lay, I think, an assumption that in any contest the custodian would be stolidly in the right and the prisoner a mendacious troublemaker, so that nothing was to be gained by giving a prisoner a day in court.

The acquiescent attitude of the courts was not confined to prisons. With unimpressive exceptions it ran across the whole range of public administration. But when in the 1960s and 70s the judicial review of administrative action began to wake from its long sleep, the problem remained that prisoners who alleged that the power of their custodians over them had been abused were simply not going to be believed. It was the Hull prison riots which changed the story, aided by one of the peculiarities of the laborious bureaucracy of prison administration—the practice of recording disciplinary proceedings word for word in longhand. Departures from natural justice which no judge would have believed merely on a prisoner's say-so turned out to have been faithfully verified in manuscript and routinely filed. In consequence, in the landmark case of *St Germain* at the end of the 1970s, the Court of Appeal placed a judicial foot in the prison door, holding that judicial review lay to Boards of Visitors in their (then) disciplinary capacity.

The Court of Appeal's decision is hallmarked by the judgment of Lord Justice Shaw, who as Sebag Shaw QC had been not a public law practitioner but one of the country's leading criminal advocates. In a judgment which contains one of the handsomest prose passages in the modern law reports as well as one of the most far-sighted, Shaw spelt out the status of the prisoner as a citizen behind bars whose entitlement to the court's vigilance for those rights he or she retained was as great as any other citizen's.

I do not believe that it was a coincidence that this breakthrough occurred when and where it did. The Hull prison riots, like others since, were the

product of a suppressed sense of grievance at unjust and sometimes inhuman treatment in our prisons. Although public law had independently
begun to rediscover the doctrines and to develop the self-confidence that
were necessary to bring prison administration within the law, it was the
Hull riots which afforded not only the opportunity but—more important—
the incentive to carry it through. The more recent syndrome of the 1990
prison riots followed by the radical and reforming Woolf Report is in many
ways a replication of the Hull–*St Germain* sequence. This is not an argument for violent disorder: it is the case for proactive and vigilant courts of
public law at the elbow of a conscientious and law-abiding public administration.

The other important dimension of these early developments lay in
Strasbourg. Sidney Golder, a prisoner who was unable to obtain redress
through the English courts for an injustice done to him in prison, established his claim under the European Convention on Human Rights. Those
who have since practised in this field have been in little doubt that, while it
was until recently impossible to cite the Convention as even an indirect
source of law, the courts have consciously taken decisions on prisoners'
rights designed at least in part to rescue the UK government from the
prospect of further embarrassment in Strasbourg.

It should not be forgotten that it takes not only awareness but a degree
of courage on a prisoner's part to take his or her custodians or their
departmental superiors to court. The occasionally expressed judicial perception of disgruntled and devious prisoners sitting in their cells devising fresh
ways of making life difficult for the authorities, even if partly true, has to
yield to the fact that prisoners' challenges to the prison administration have
been few in number but proportionately pretty successful by comparison
with public law challenges in other fields.

Although there are now some useful books in this important field of law,
none does the job that the present volume does. Stephen Livingstone and
Tim Owen are unusually well equipped and qualified for the task. Stephen
Livingstone, an already distinguished young academic lawyer in Belfast, has
the invaluable touchstone of knowledge of the American penal system and
of the rights litigation which it has generated. Tim Owen, one of the handful of barristers specializing in prison and inquest law, has in a succession
of important cases moved outward the frontiers of legal protection for prisoners. The book they have written does something which even today few
legal books accomplish: it combines an excellent and compendious account
of the state of the law on prisons and prisoners with a historical and political analysis of where the present system has come from and where it is
headed. They do not place a narrow meaning on 'law': they recognize that
administrative rules and practices and non-legal avenues of redress are as
important as substantive law and recourse to the courts, and they provide a

full treatment of the European dimension which, acknowledged or not, is unquestionably influential.

The concluding chapter is an essay of distinct importance. It attempts an appraisal of the now historic meeting of the prison system and the courts. Legal scholarship is traditionally deficient in this vital appraisal of law not as a thing in itself but as an element in a many-faceted social process. The authors delineate the unevenness of judicial intervention, which has been powerful in relation to procedural norms of discipline but hesitant in relation to standards of treatment. Debate is needed as to whether the latter are best left to respected public invigilators such as Judge Stephen Tumim and Lord Woolf, whose effect has been palpable, or whether at some point minimum standards must become justifiable too. Before lawyers acclaim the latter, a sober look is needed at the actual effect that the courts' pronouncements of legal principle have had on the day-to-day treatment of prisoners. Attention has also to be given to the view that it is from the Prison Service that concrete reform comes. My own view, and I think that of the authors, is that these are all moving parts of a constantly changing whole. Just as prison discontent has plainly catalysed judicial and administrative responses, so the setting of legal standards has concentrated administrative minds; and just as the Prison Service does the detailed work of devising and implementing reforms, so the public pressure generated by independent reports furnishes the political impetus under which many of its policy choices are actually made.

The rule of law will not, however, have finally permeated the prison system until the vocabulary of administration and discipline no longer contains the sentence 'I'm the law here', and that—as Stephen Livingstone and Tim Owen demonstrate—is still some distance from realization.

LONDON   Stephen Sedley
May 1993

Preface

For many years it has been clear that British prisons are in a most unsatisfactory state. Academics and journalists have produced a stream of accounts stressing how overcrowding, staffing problems, lack of resources, and confusion over the objectives of imprisonment have produced prisons that are often dirty, tedious, unhealthy, and sometimes dangerous. Such accounts have been given impressive official confirmation by the reports of Her Majesty's Chief Inspector of Prisons, Lord Woolf and Judge Tumim's Inquiry into the 1990 prison disturbances and the initial report of the European Committee on the Prevention of Torture published in November 1991. Where other institutions fall into disrepute, lawyers often become involved as attempts are made to set new standards and find ways to ensure their achievement. Yet when it comes to prisons, lawyers do not seem to be particularly vocal as to what should be done. Is this because the law has little to say about the prison?

At one time this might have been true but with the extension of judicial review and the growing influence of the European Convention of Human Rights since the 1970s the law has had a lot to say about prisons. Yet most lawyers, even those whose work in criminal practice frequently involves them in decisions about who should go to prison, know comparatively little about the law relating to prisons. In large part this is no doubt due to the haphazard development of prison law in a variety of different contexts such as judicial review, civil actions, European cases and even some criminal law. Keeping track of all these developments is likely to test even the most assiduous lawyer. That is why we, an academic lawyer and a practising barrister, have sought to bring together these developments and offer what we hope is a fairly comprehensive account of the law relating to prisons. Though our primary emphasis is on summarizing the legal developments which have occurred over the past twenty years, we also take the view that in many respects the rule of law is a long way from being achieved in Britain's prisons. Throughout the text we point out what we believe to be failures to achieve acceptable standards and we hope that in addition to the information we provide the views we express will stimulate other lawyers to become involved in what is already a rapidly developing area of law. It is also an area, like any other area of law, which is too important to be left solely to lawyers. As the reforms initiated by the Woolf Report are gradually put in place, the law is likely to have an increasing effect on the lives of all those who live and work in prisons.

The book is certainly not aimed exclusively at lawyers and law students. We hope and intend that it will also be of value to prisoners, prison officers, prison administrators, and others involved with the prison system, all of whom have a role to play if the rule of law is to be established and respected in prisons.

In writing *Prison Law* we have incurred many debts of gratitude. Richard Hart and Jane Williams of OUP have been helpful and sympathetic editors. Their confidence in us throughout this project has been greatly appreciated. Milica Djuradjevic's copy editing redeemed us from many errors. Stephen Sedley QC (as he used to be), Kate Akester, Peter Madden, Padrigin Drinan, Seamus Treacey and Edward Fitzgerald have been imaginative, trail-blazing 'prison lawyers' who have provided a wealth of ideas and encouragement over the years. David Harris QC and Nick Blake have also been valued devil's advocates to tease out individual points of law. The former Director General of the Prison Service, Joe Pilling, and Brian Caffarey at the Home Office were models of open government and, by ensuring us access to Standing Orders and Circular Instructions, gave us the opportunity to make this a more comprehensive work than it might have been. Angela Francis assisted us with some valuable background research. Among those who provided information or read parts of the text we would especially like to thank Geoff Coggan, Kieran McEvoy, Steve Nathan, Martin O'Brien, Adam Sampson and Stephen Shaw who, as Director of the Prison Reform Trust, has a limitless fund of knowledge of developments in the prison system. Last, but by no means least, a number of prisoners (past and present) have contributed ideas as well as the courage to take a stand on issues of general importance to all prisoners. It would be impossible to list them all but Mark Leech, Christopher Hague, John Bowden, Wadi Williams, Michael Hone, Ben Wilson, Roger Payne and John Hirst deserve particular mention.

We are sure that there are many things we have missed and points where others would disagree with our views. We look forward to hearing about them.

Recent developments in the law since completing the main text of the book testify to the rapid pace of change in prison law.

The first is the Court of Appeal's decision of 19th May 1993 in the case of *R v Home Secretary ex parte Leech*. Therein a unanimous court overturned the decision of Webster J in the High Court which we have discussed, and criticized, in Chapter Six. Steyn LJ indicated that opening a prisoner's correspondence with a lawyer, even if that prisoner had not yet become a party to legal proceedings, constituted a hindrance on a prisoner's right of access to the courts. Such a right he saw as a constitutional right, even in a country without a written constitution, and he was extremely doubtful whether it could be infringed by secondary legislation, let alone an regarded as permitting the reading or stopping of legal correspondence it was *ultra vires* the 1952 Prison Act. The decision brings UK law into line with the decision of the European Court of Human Rights in *Campbell* v. *United Kingdom*, also discussed in Chapter Six. Arguably the Court of Appeal has gone further. While the Strasbourg decision is based on the right of privacy and seems to envisage some circumstances where legal correspondence may be read, the *Leech* ruling would appear to give unqualified confidence to correspondence between prisoners and their lawyers.

The second development in the courts occurred on 24 June 1993 when the House of Lords gave judgment in the appeals and cross-appeals in *R* v. *Home Secretary ex parte Doody & others*, an important decision affecting the rights of mandatory life sentence prisoners. The speech of Lord Mustill (speaking for a unanimous House) is an authoritative analysis of the development of the life sentence and points to a growing convergence between the rights of discretionary and mandatory lifers, though differences remain which only Parliament can resolve. The decision largely confirms the decision reached by the Court of Appeal (set out in Chapter 9, 'Release from Prison') and declaratory relief was granted as follows:

(i) The Secretary of State is required to afford a prisoner serving a mandatory life sentence the opportunity to submit in writing representations as to the period he should serve for the purposes of retribution and deterrence before the Secretary of State sets the date of the first review of the prisoner's sentence.

(ii) Before giving the prisoner the opportunity to make such representations, the Secretary of State is required to inform him of the period recommended by the judiciary as the period he should serve for the purposes of retribution and deterrence, and of any other opinion expressed by the judiciary which is relevant to the Secretary of State's decision as to the appropriate period to be served for these purposes.

(iii) The Secretary of State is obliged to give reasons for departing from the period recommended by the judiciary as the period which he should serve for the purposes of retribution and deterrence.

In reaching its conclusion on (iii) above, the House of Lords also finally overruled the decision of the Court of Appeal in *Payne* v. *Lord Harris of Greenwich*, holding that 'because of the continuing momentum in administrative law towards the openness of decision-making . . . [its] reasoning . . . cannot be sustained today.'

On the administrative front, since the Prison Service became an Executive Agency on 1st April there has been a change in the way in which staff are updated on changes in Service Policy. Instead of Circular Instructions, there are now 2 types of document issued above the personal signature of the new Director General, Derek Lewis—'Advice to Governors' and 'Instruction to Governors', the different titles reflecting a difference in the discretion each document allows to reside with prison governors.

We have attempted to present the law as at 1 July 1993. As will be clear the primary focus of this book is on the law affecting prisons in England and Wales, though we also draw upon the experience of one of the authors in relation to prison law in Northern Ireland at a number of points in the book.

1 July 1993   *Stephen Livingstone*
                                               Faculty of Law
                                               The Queen's University of Belfast
                                               Belfast BT7 1NN

                                               Tim Owen
                                               2 Garden Court
                                               Middle Temple
                                               London EC4Y 9BL

Contents

Table of Cases

Principal references printed in bold

DECISIONS OF THE EUROPEAN COURT OF HUMAN RIGHTS

DECISIONS OF THE EUROPEAN COMMISSION ON HUMAN RIGHTS

UNITED STATES OF AMERICA

AUSTRALIA

CANADA

SOUTH AFRICA

UNITED NATIONS HUMAN RIGHTS CASES

THE INTER-AMERICAN COURT OF HUMAN RIGHTS

Table of National Legislation

(c)Non-Statutory Sources

(1) Circular Instructions

USA

Table of Treaties
and Other International Instruments

1
Introduction to Prison Law

Until recently a textbook devoted to the prison law of England and Wales would have been an unlikely undertaking. Whereas a number of, now, well-recognized areas of law have developed over the past twenty years (immigration, social security, discrimination and individual employment law), mainly in response to legislative initiatives, there has been uncertainty as to whether 'prison law' exists as a discrete entity rather than as a mere branch (or twig) of administrative law or simply as a specific example of the law of tort in action. The fact that no primary legislation affecting the totality of the prison system had been passed since 1952 undoubtedly influenced the perception that nothing fundamentally new could be said about the law of prison and, more particularly, the legal position of prisoners.

Such a perception is no longer valid today. While it remains true that the prison system of England and Wales continues to exist within the static legislative framework of the Prison Act 1952 and the Prison Rules 1964, the last fifteen years have witnessed a significant increase in legal intervention in prison life. And recent legislative and administrative initiatives presage further alterations to the status quo. Four factors have combined to bring about these changes.

The first and most significant of these has been the expansion of judicial review into all areas of prison life. Until the late 1970s, prison discipline was considered to be a matter solely within the discretion of governors and Boards of Visitors. This meant, in effect, that prisoners could be dealt with unfairly and arbitrarily but had no means of recourse via the courts to complain about their treatment. The decision of the Court of Appeal in R v. *Board of Visitors of Hull Prison ex parte St Germain*[1] that the disciplinary adjudications of Boards of Visitors are susceptible to judicial review began a process of judicial intervention which culminated in 1991 with an acceptance by the House of Lords that all decisions taken in pursuance of the Prison Rules, including non-disciplinary decisions such as segregation under Rule 43, were amenable to judicial review.[2]

The second factor has been the influence of the European Court of

[1] [1979] QB 425.
[2] R v. *Deputy Governor of Parkhurst Prison ex parte Hague* [1992] 1 AC 58. The expansion of public law remedies for prisoners has coincided with a reduction in both the scope and effectiveness of private law remedies. See for example the outcome of *Ex parte Hague* and the decisions of the Court of Appeal in *H* v. *Home Office* the Independent 6 May 1992 and *Racz* v. *Home Office* The Times 17 Dec. 1992.

Human Rights ('ECHR'). As a result of a series of judgments beginning in the mid-1970s, the ECHR has acted as a spur to reform of the UK prison system in areas such as access to legal advice and to the courts,[3] the disciplinary system[4] and, most recently, release procedures for discretionary life sentence prisoners.[5] Even though the English courts continue to resist the idea that any administrative decision which conflicts with the jurisprudence of the ECHR and the European Convention on Human Rights is necessarily unlawful,[6] the UK government has not hitherto refused to accept the legitimacy of specific decisions in so far as they impact on domestic prison law. The ECHR's influence has been buttressed more recently by the Council of Europe Committee for the Prevention of Torture and the Prevention of Inhuman or Degrading Treatment or Punishment which published a report in December 1991 on conditions in three English prisons which it found to be 'inhuman and degrading'.[7]

The third factor was the series of prison disturbances in April 1990 which began in Strangeways Prison and triggered five other serious confrontations between prisoners and staff.[8] Far more serious than anything previously witnessed in the history of the British prison system, the significance of the 1990 disturbances lay in the reaction which it prompted. While a massive police inquiry led to a long sequence of trials of those prisoners who were charged with criminal offences (trials which continued into 1993), the government appointed Lord Justice Woolf (as he then was) to head a major inquiry into the events of April 1990. As well as examining in detail what happened during the six most serious disturbances and how they were handled, Woolf's task was to examine the causes of rioting and what should be done to prevent it from happening in the future. With the assistance of the Chief Inspector of Prisons, Judge Stephen Tumim, Lord Justice Woolf published a 600 page report in February 1991 which presented a comprehensive package of reforms.[9] The Woolf Report is a major document whose influence will be seen throughout this book. Although the

---

[3] Golder v. United Kingdom (1975) 1 EHRR 524, Series A No. 19, and *Silver* v. *United Kingdom* (1983) 5-EHRR 347, Series A No. 61.

[4] *Campbell and Fell* v. *United Kingdom* (1984) 7 EHRR 165, Series A No. 80.

[5] *Weeks* v. *United Kingdom* (1987) 10 EHRR 293 and *Thynne, Wilson and Gunnell* v. *United Kingdom* (1990) 13 EHRR 666.

[6] R v. *Secretary of State for the Home Dept. ex parte Brind* [1991] 2 WLR 588.

[7] *Report to the United Kingdom Government on the Visit to the United Kingdom Carried out by the European Committee for the Prevention of Torture and Inhuman or Degrading Treatment or Punishment (CPT)*, Council of Europe, CPT/Inf (91) 15. The three prisons identified by the Report were Brixton, Leeds, and Wandsworth. The UK government's response has also been published (CPT/Inf (91) 16) and, not surprisingly, it rejected the CPT's assessment that conditions at the three prisons amounted to 'inhuman and degrading treatment'. It agreed, however, that they needed 'considerable improvement'.

[8] The other five prison disturbances took place in Glen Parva YOI, Dartmoor, Cardiff, Bristol and Pucklechurch Remand Centre.

[9] *Prison Disturbances April 1990* (Woolf Report), Cm. 1456 (London, 1991).

government continues to be selective in its implementation of the twelve key recommendations of the Woolf Report, there can be no doubt about its achievement in defining the agenda for change in the prison system into the next century.

The fourth factor, whose influence is as yet unclear, may properly be described as the government's commitment to ending the status of the Prison Service as a traditional, public sector monopoly closely supervised by Ministers on a day to day basis. This commitment led to the opening in 1992 of the first private sector ('contracted-out') prison and the achievement of 'agency status' for the Prison Service from 1 April 1993. It is far too soon to say to what extent these 'administrative' reforms will lead to a fundamental change in the quality of life for prisoners in England and Wales but they indicate a determination to bring about change after decades of inertia. The appointment of a new Director-General of the Prison Service from the private sector world of television management has coincided with a new vocabulary for the Service.[10] Both new and existing prisons are to be 'market tested' with a view to being 'contracted out'. Individual prison services such as court escorts, laundry and education services have been subjected to competitive tendering with contracts being won by private sector firms. The Prison Service has a Framework Document which identifies key performance indicators (the number of escapes and assaults on staff, time out of cell for prisoners, etc.) as the basis for examining whether each prison establishment is achieving its targets.

Against this background of radical change, we attempt to paint a comprehensive picture of the law and practice regulating prisons in England and Wales. In the process we will also analyse the legal status of prisoners, the rights which they can be said to enjoy and the remedies which exist to correct abuses of power. As well as explaining the state of domestic law we also compare and contrast the UK prison system with the law as developed by the European Court of Human Rights as well as that of the United States of America, where a very different legal system has produced a massive body of prison law but, arguably, no better penal system.

In Chapter 2 we summarize the legislative history of the prison system during the nineteenth century and examine in detail the current legislative framework of the Prison Act 1952 and the Prison Rules 1964 together with the implications of the provisions of the Criminal Justice Act 1991 which deal with 'contracted-out prisons'. The legal status of Standing Orders and Circular Instructions is also explained as well as the administrative structure of the Prison Service now that it has become an executive agency.

---

[10] The new Director-General, Derek Lewis, used to be the Chief Executive of the Granada Group and UK Gold Television. Until his appointment he had never set foot in a prison and in an interview in the *Independent*, 4 Apr. 1993, was quoted as saying, 'I had thought about prisons very little—probably no more than the average man in the street.'

In Chapter 3 we look at the various avenues of redress which it is open to prisoners to pursue in respect of issues arising from their confinement. These include the internal complaints procedure and the newly created Prisons Ombudsman as well as the options of an application for judicial review or a private law claim for damages in the courts. We also summarize the kind of documents to look for in prison law actions and the special problems which can arise when preparing to mount a prison claim. In Chapter 4 we examine the avenues of redress available to prisoners via the European Convention of Human Rights and look at other international human rights mechanisms relevant to the maintenance of minimum standards for the treatment of prisoners.

In Chapters 5–8 we focus on the four aspects of prison law which have the greatest impact on the day-to-day lives of prisoners. Chapter 5 deals with the broad issue of prison conditions, a term which embraces not just the physical environment in which prisoners must live but also the other aspects of the prison regime such as food, work, clothing, facilities (or 'privileges'), education, and health care. In this latter category we look in particular at the problems of HIV positive prisoners. We also analyse the Home Office's duty to take reasonable care for the safety of prisoners and the special problems of women in prison.

In Chapter 6, under the general heading 'Access to the Outside World and Maintenance of Family Contacts', we look at the non-physical barriers which regulate prisoners' contacts with the world beyond the prison walls. This includes contacts with family and friends by means of telephone, letters, and visits as well as access to legal advice and the courts. We also describe the new provisions governing home leave and the possibility of prison transfers both within the domestic system of England and Wales and from one jurisdiction to another. The possibility of such transfer is a vital aspect of maintaining contact with family or friends outside.

In Chapter 7 we describe the operation of the prison disciplinary system, the area of prison law which has received most attention from the courts over the past fifteen years and which has recently been fundamentally altered by the removal from Boards of Visitors of their disciplinary role. The internal disciplinary code provides the official means whereby good order and discipline are maintained in a prison but it is not the only weapon in the armoury of the prison authorities. In Chapter 8 we examine the non-disciplinary powers available to enforce good order and discipline, including the power to categorize, transfer and segregate prisoners according to their perceived threat to the prison regime and their potential threat to the outside world should they escape. We also look at the provisions in the Rules which authorize the use of non-medical restraints for particularly violent/disturbed prisoners.

In Chapter 9 we turn to the important question of early release from prison via the new machinery introduced by the Criminal Justice Act 1991 as well as the law governing the release of life sentence prisoners, which has been heavily influenced by decisions of the ECHR. The statutory provisions and the case law have combined to produce a complex set of provisions and before analysing them we place the current system in context by describing the development of the scheme of remission and parole from the nineteenth century onwards.

Finally, in Chapter 10, we move away from a descriptive approach and seek to draw together some conclusions about the relationship between law and prisons in the light of the changes brought about by legal interventions over the past two decades. In the process, we try to answer the 'big' question in prison law, namely to what extent has the courts' growing willingness to intervene in prison life led to improvements in prison conditions and the relationship between prisoners and the prison authorities?

It will quickly be observed that the focus of this book is on male adult prisoners, who make up the majority of the prison population. Where different legal principles or administrative policies apply to women and young prisoners, we seek to make this clear.

2

The Legal Framework and Administrative Structure of the Prison System

INTRODUCTION

Since 1877 prisons and prisoners in England and Wales have been subject to the exclusive authority and jurisdiction of the Home Secretary. Until 1963 this was exercised via a managerial body of five Prison Commissioners but the passage of the Prison Commissioners Dissolution Order 1963 resulted in the absorption by the Home Office of all responsibility for administering the penal estate.[1] Thirty years later in 1993, in a move intended to give greater autonomy to the administrators and greater freedom from day-to-day ministerial interference in the running of prisons, the Prison Service was made an Executive Agency.

Prior to 1877 there were two separate prison systems governed by separate administrative authorities—the local prison system and a much smaller number of prisons run by central government known as 'convict' prisons. Local prisons were run independently, with little or no central control, by local Justices of the Peace. The description 'local prison' embraced a variety of institutions—bridewells and houses of correction (which existed to encourage vagabonds, beggars, and the 'idle poor' in the ways of work but which also housed minor offenders) as well as small local gaols. The local gaol housed felons, misdemeanants, convicted or unconvicted, civil debtors, and, until 1867, those awaiting transportation to Australia.[2]

Towards the end of the eighteenth century, central government began to acquire its own penal institutions. So long as transportation was the normal method of dealing with the long-term prisoner, it was only necessary to use the prisons to hold offenders awaiting trial or the next transport, and local prisons were able to carry out this function adequately. The American War of Independence brought a temporary end to transportation in 1776 and so

---

[1] SI 1963/597. The Prison Service was and remains part of the Home Office. Agency status has not changed this and accordingly Prison Service staff remain civil servants employed by the Home Office.

[2] Felony was the common law term employed to describe graver crimes below high treason and petty treason. Until 1827 all felonies were liable to be punished by death. Misdemeanours applied to all offences below felonies whether punishable on indictment or on summary conviction.

the government invented the Hulks as an alternative. These were old ships taken out of commission, moored in rivers or estuaries and adapted to house convicts. In 1779 an Act was passed to authorize the punishment of 'hard labour' for offenders who would normally have been transported, the hard labour in question being cleansing and improving the navigation of the River Thames and any other navigable port.[2a]

The 1779 Act helped to establish imprisonment with hard labour as a standard punishment for felony. The 'discovery' of Australia by Cook led to the use of transportation to Australia which began in 1787 and continued to be the mainstay of penal treatment for convicts up to 1857. But the 1779 Act resulted in an increasing growth of prisons for long sentence prisoners. In addition to the Hulks, a national penitentiary was built at Millbank in 1816 under the control of the Home Secretary. There followed a prison for juvenile convicts at Parkhurst (1839) and Pentonville (1842), which was intended to be a model prison for local prison authorities to emulate. Portland (1849) and Dartmoor (1850) were the precursors of what became an outbreak of government prison building following the creation of the Board of Directors of Convict Prisons in 1850. The new Board was soon able to abolish the use of the Hulks and the way was clear for the introduction of penal servitude which, by 1857, superseded transportation.

The nineteenth century saw a progressive movement towards greater central government intervention in the local prison system, a movement which culminated in the unified, centralized system achieved by the 1877 Act. The 1791 Gaols Act had authorized Justices to appoint a governor for gaols and houses of correction within their jurisdiction. It also authorized and required them:- 'to make such Rules and Orders for receiving, separating, classing, dieting, clothing, maintaining, employing, reforming, governing, managing, treating and watching all offenders during their respective confinement'. The Act was very much a response to the prison conditions described by John Howard's *State of the Prisons* (1776), which advocated that prisons should be sanitary and secure, the sexes should be effectively separated, the keeper should be a paid and responsible servant of the Justice, who, in turn, should exercise effective supervision.

The 1823 Gaols Act was a further attempt at reform. It consolidated the twenty-three pre-existing statutes on the subject of gaols and houses of correction. The aim was to provide for the safety of prisoners but also 'more effectually to preserve the health and to improve the morals of . . . prisoners'.[3] The means to achieve this end was to enforce a strict system of 'separation, superintendence, employment and instruction of prisoners'.[4] Section 10 set out

---

[2a] Chap 74, Geo III. For an interesting summary of the development of the prison system in the late 18th and 19th centuries see *The Struggle for Penal Reform* Gordon Rose, (London, 1961) to which the current analysis is indebted.

[3] Gaols Act 1823, s. 1.   [4] Ibid.

a uniform code of twenty-four rules and regulations to be observed in every prison and gaol within the jurisdiction of Justices. Sidney and Beatrice Webb commented in their authoritative work *English Prisons under Local Government* that the code was 'the first measure of general prison reform to be framed and enacted on the responsibility of the national executive'.[5] The Act provided for regular reports to be made by Justices to the Home Secretary and for systematic inspections to be carried out by the Justices. But it had only a limited impact in that it only applied to a minority of prisons in the country—the London debtors' prisons were excluded as were 150 gaols in the boroughs. Furthermore, there was no system for enforcement or any national inspectorate to investigate whether standards were being maintained.

The 1835 Prisons Act changed this and empowered the Home Secretary to appoint prison inspectors to inspect on his behalf and report to him. It also required Justices to submit such rules as they had drafted to the Home Secretary for his approval and possible amendment.

The legislative reforms of the 1820s and 1830s did not, however, produce the desired results. The 1863 inquiry conducted by the Select Committee of the House of Lords on Prison Discipline revealed widespread discrepancies and anomalies in the administration of the local prison system. Of greatest concern was the fact that most prisons were failing to enforce the 'separate system' which had been introduced in 1838 following the recommendation of the original Inspectors of Prisons. The 1865 Prison Act was a consolidation of thirteen other Acts of Parliament. It amalgamated the gaols and houses of correction into what were henceforth to be called local prisons. It also strengthened the Home Secretary's powers of control by incorporating in a Schedule to the Act detailed 'Regulations for Government of Prisons' which were binding on all prisons authorities charged with administering the various penal institutions.[6]

The regulations placed a statutory duty on the governor to visit the whole prison every day and see each prisoner. Further specific duties were cast upon the prison doctor and chaplain. The Act also gave power to the Home Secretary to close any prison which failed to meet the requirements of the Act for four successive years. Although the 1865 Act succeeded in securing the closure of a large number of smaller prisons, too many local prisons remained. They were already expensive to run and to put them in a state where they could comply with the law and satisfy the Inspectors placed an intolerable burden on the county rates. Further, there remained wide variations in the method of enforcing the statutory code. As Sir Louis Blom-Cooper, QC, has commented, 'the situation was ripe for a national take-over'.[7]

---

[5] Quoted in Sir Louis Blom Cooper QC's stimulating article 'The Centralisation of Government Control of National Prison Services with Special Reference to the Prison Act 1877', contained in J. Freeman (ed.), *Prisons Past and Future* (London, 1978).

[6] Prison Act 1865, s. 20.   [7] 'The Centralisation of Government Control', 64.

The 1877 Prison Act led to the control of all local prisons vesting in the Home Secretary and the cost of their maintenance was transferred to public funds. Their general superintendence, subject to the overall political control of the Home Secretary, was vested in a Board of Prison Commissioners assisted by Inspectors, appointed by the Home Secretary, and a departmental staff. The rule-making power of Justices having passed to the Home Secretary, a new code of rules was issued in 1878 and from 1 April of that year all prisons, local and convict, fell under one central control and single set of rules. For the first time, the local community ceased to have any statutory administrative function in respect of the prisons located in its midst.

The Prison Act 1898 and the 1899 Rules made thereunder, together with the (largely) unrepealed portions of the 1865 and 1877 Acts, formed the legal basis for the administration of prisons until the passage of the Prison Act 1952.[8] And despite the numerous inquiries into the state of the prisons, culminating in the publication of the Woolf Report in 1991, the 1952 Act remains in force as the primary source of legislative authority governing prisons and imprisonment. The detailed provisions of the Act and the Rules will be analysed in later chapters. Before descending into the mire of statute, case-law, and administrative guidance, however, it will be helpful to stand back and examine the legislative framework which regulates life in a prison. Having done that we will look at the administrative structure of the Prison Service before, finally, we turn to the implications in both law and administrative practice of the introduction of 'contracted-out' prisons.

THE LEGISLATIVE FRAMEWORK

The Prison Act 1952 is a relatively brief statute and is expressed in remarkably general terms given its subject-matter.[9] As originally enacted it vested powers of management of prisons in the Prison Commissioners subject to the overall control of the Home Secretary. In addition, an adjudicative power in relation to prison disciplinary offences was divided between the Commissioners and visiting Justices. The 1963 Order 'dissolving' the Commissioners meant that their functions of oversight became merged with those of the Home Secretary while the adjudicative powers were distributed between the Home Secretary and the Boards of Visitors.

Apart from identifying the Home Secretary as the source of power and authority and providing for a central administration, a Chief Inspector of Prisons, a Board of Visitors for each prison, as well as 'officers' to run the prisons, the Act is little more than a series of enabling and deeming

---

[8] The full text of the Prison Act 1952 is to be found at App. 1 below.
[9] Its fifty-five sections, with only forty-one still in force, are six fewer than the 1877 Act.

provisions designed to give the Home Secretary maximum discretion in the organization of the prison system.[10] In effect the Act simply calls upon the Home Secretary to create and police an internal regime for prisons. With one exception in relation to disciplinary proceedings against a prisoner, the Act creates no clear statutory rights for prisoners.[11] The details of prison life are left to be expressed in the Rules, which, by the rule-making power in s. 47 (1), require the Home Secretary to provide for the 'regulation and management of prisons and for the classification, treatment, employment, discipline and control of persons required to be detained therein'. Precisely because the Rules contain the 'meat' of prison law, most judicial attention has focused on their implications and effect, but certain aspects of the Act require analysis.

Section 4 is of some significance, particularly sub-sections (1) and (2) which state:

(1) The Secretary of State shall have the general superintendence of prisons and shall make the contracts and do the other acts necessary for the maintenance of prisons and the maintenance of prisoners.

(2) Officers of the Secretary of State duly authorised in that behalf shall visit all prisons and examine the state of the buildings, the conduct of officers, the treatment and conduct of prisoners and all other matters concerning the management of prisons and shall ensure that the provisions of this Act and of any Rules made under this Act are duly complied with.

Section 4 (1) thus creates a general duty of 'superintendence' over prisons which is cast upon the Home Secretary. (Its reference to a power 'to make the contracts' necessary for the maintenance of prisons and prisoners is to be distinguished from the power now contained in the Criminal Justice Act 1991 to 'contract out' the running of any prison.) Section 4 (2) imposes a duty upon the Home Secretary, expressed in broad terms, to ensure compliance with both the 1952 Act and the Prison Rules. For several years prior to the decisions of the House of Lords in R. v. *Deputy Governor of Parkhurst ex parte Leech*[12] and R. v. *Deputy Governor of Parkhurst ex parte Hague*,[13] s. 4 (2) was relied upon by the Home Secretary, and accepted by the courts, as a form of ouster clause which prevented the High Court from exercising its supervisory jurisdiction to grant judicial review in respect of disciplinary and administrative decisions taken by prison governors. Underlying this approach was a powerful aversion to the courts becoming embroiled in the day-to-day decisions of prison life. Accordingly, it was sought to rationalize this concern by reference to the public policy principle

---

[10] The personal responsibility of the Home Secretary for the running of the Prison Service does not of course deny his wide powers of delegation—see e.g. R. v. *Home Secretary ex parte Oladehinde* [1991] 1 AC 254, in which the Home Secretary's ability to delegate his powers under the Immigration Act 1971 was considered.

[11] See s.47(2).   [12] [1988] 1 AC 533.   [13] [1992] 1 AC 58.

enshrined in s. 4 (2) whereby Parliament, it was argued, had made the Secretary of State uniquely responsible for ensuring that the prison regime was carried out according to law. Only his acts or omissions in carrying out that task were susceptible to judicial review. Governors were immune from direct challenge by way of judicial review in order, so it was believed, to ensure the smooth administration of the prison system.

This argument was accepted by the Court of Appeal in *R. v. Deputy Governor of Camphill Prison ex parte King*,[14] in which a prisoner sought to quash a disciplinary adjudication conducted by a governor on the basis that he had misconstrued his powers in relation to a specific disciplinary charge. While accepting that, logically, no proper distinction could be drawn under the legislative scheme between the disciplinary powers of Boards of Visitors (which were subject to review[15]) and those of governors, the majority of the Court of Appeal in *King* pointed to s. 4 (2) as the justification for a different approach. Lord Justice Griffiths said the section meant that:

the court should, in the first instance, be prepared to assume that the Secretary of State will discharge the duty placed upon him by Parliament to ensure that the prison governor is doing his job properly. If it is shown that the Minister is not discharging this duty and allowing a prison governor to disregard the prison rules then judicial review will go to correct that situation by requiring the Minister to perform his statutory duty.

A different decision was reached by the Northern Ireland Court of Appeal in *R. v. Governor of the Maze Prison ex parte McKiernan*,[16] where it was accepted that jurisdiction to review governors' disciplinary decisions existed. But the Northern Ireland legislation had no equivalent to s. 4 (2).[17] In *Leech* however the House of Lords decisively rejected the Home Secretary's argument that s. 4 (2) had any relevance to the issue of jurisdiction. Dismissing as 'fundamentally fallacious' the argument that the court had jurisdiction to entertain an application for judicial review only where that jurisdiction was shown to be necessary in order to secure compliance with the law, Lord Bridge said that to invoke the Home Secretary's general duty to ensure compliance with prison legislation in order to oust the jurisdiction of the court in relation to governors' disciplinary powers 'is to stand the doctrine by which the limits of jurisdiction in this field are determined on its head'.[18] Furthermore, the mere fact that a prisoner could petition the Home Secretary under Rule 56 asking him to remit a disciplinary award imposed by a governor and that the Home Secretary was obliged pursuant to s. 4 (2) to consider whether the governor had acted in

[14] [1985] QB 735
[15] See *R. v. Board of Visitors of Hull Prison ex parte St Germain* [1979] QB 425
[16] (1985) 6 NIJB 6.   [17] See the Prison Act (Northern Ireland) 1953.
[18] [1988] 1 AC 533 at 562 B–D.

accordance with the Rules did not constitute an alternative remedy which prevented an immediate challenge to the governor's decision.

Notwithstanding the decision in *Leech*, a further attempt to rely on s. 4 (2) as an ouster provision preventing the court from directly reviewing a governor's decision was made in *Hague*, which concerned a decision to segregate a prisoner under Rule 43. Applying the reasoning in *Leech*, the Divisional Court rejected this argument and it was not renewed on appeal. Accordingly s. 4 (2) in no way inhibits the exercise of the High Court's supervisory jurisdiction in relation to either the disciplinary or administrative decisions of prison governors or their delegates.[19]

Section 5A of the 1952 Act provides for the appointment of a Chief Inspector of Prisons whose duty it is to inspect prisons in England and Wales and to report to the Home Secretary on his findings. He is required in particular to report on prison conditions and the treatment of prisoners.[20] The Home Secretary may refer specific matters connected with prisons and prisoners to the Chief Inspector and direct him to report on them.[21] An annual report must also be submitted to the Home Secretary to be laid before Parliament. The current Chief Inspector, Judge Stephen Tumim, has pursued a vigorous, well-publicized campaign over the impoverished regimes, poor conditions, and depressing, insanitary buildings which characterize prisons in England and Wales.[22] He has also produced authoritative reports on specific issues, most significantly an investigation of suicide prevention policies in prisons in the wake of an alarming series of suicides particularly of young prisoners.[23] He has declined however to inspect the conditions 'enjoyed' by prisoners in police cells who are detained there pursuant to the Imprisonment (Temporary Provisions) Act 1980. In his 1992 Annual Report, Judge Tumim stated that police cells do not come within his 'lawful remit' and that 'it would be impracticable to inspect police cells without inspecting police stations, which would be wholly outside [my] duties'. This is strictly correct since the terms of s. 6 of the 1980 Act expressly contrast detention in a 'prison, remand centre . . . or detention

---

[19] In *Hague* at 116A Taylor LJ, was content to accept that the courts should approach the exercise of discretion in matters relating to prison administration/management 'with great caution', an observation which was perhaps unnecessary in the light of the great reluctance displayed by English judges to intervene with reckless enthusiasm in prison cases.

[20] Prison Act 1952, s. 5A (3). The term 'prison' is negatively defined in s. 53 (1) as not including a naval, military, or air force prison. Until the publication of the May Committee Report (*Report of the Committee of Inquiry into the United Kingdom Prison Services*, Cmnd. 7673 (London, 1980)) the Inspectorate was an internal Prison Service body and did not lay a report before Parliament—see 983 *House of Commons Report* (5th series), 30 Apr. 1980, col. 1395.

[21] s. 5A (4).

[22] Judge Tumim was co-author with Lord Justice Woolf of the Woolf Report, *Prison Disturbances: April 1990*, Cm. 1456 (London, 1991).

[23] *Suicide and Self-Harm in Prison Service Establishments in England and Wales'* Cm. 1383 (London, 1990).

centre' with detention 'in the custody of a constable'. The Chief Inspector's statutory duty is to inspect 'prisons' and to report on the treatment of 'prisoners . . . in prisons'.[24] This technical obstacle to the assumption of jurisdiction to inspect conditions in police cells is extremely unfortunate since it prevents a complete assessment of the treatment of prisoners in England and Wales. Though the 1980 Act was, as its name suggests, intended to be a temporary measure to cope with difficulties caused by industrial action by the Prison Officers' Association, it has become a permanent feature of the penal landscape. Conditions for detainees in police cells are significantly worse than for other prisoners and, by not subjecting them to the same process of inspection as other prisons, a false impression is given of prison conditions in general.[25] The Home Office point out that police cells are subject to inspection by the inspectorate of constabularies but these reports are not published. If police cell detention is to be permanent, an amendment to the 1980 Act is surely desirable so that the Chief Inspector is able to inspect the entire prison system.

Sections 6–9 of the 1952 Act provide for the personnel necessary for the running of every prison. Section 6 requires the Home Secretary to appoint a Board of Visitors for each prison and for rules to be made requiring individual Board members to pay frequent visits to the prison and to hear any complaints. The Board of Visitors is to be distinguished from individual prison visitors, whose role is to visit and befriend individual prisoners. The existence of official prison visitors is not dependent on any discrete statutory provisions and they have no statutory functions. Until April 1992, Boards of Visitors had two distinct functions. First, they are supposed to act as watchdogs of the daily life and regime in an individual prison. To assist them in this task, Boards have a statutory right of free access to every part of the prison and to every prisoner at any time.[26] Their second function was to act in an adjudicatory capacity within the prison disciplinary system. Their remit extended to those disciplinary cases which were regarded as too serious to be dealt with by the governor.

Since April 1992 Boards no longer have any role in the formal disciplinary process. This was a direct result of the Woolf Report which confirmed what many critics of the system had long felt, namely that the Boards' adjudicatory role impinged upon and undermined their watchdog functions. In particular, it was widely felt by prisoners that Boards were not truly

---

[24] s. 5A (2).

[25] See e.g. *R. v. Commissioner of Police for the Metropolis ex parte Nahar* The Times 28 May 1983, a case in which two prisoners detained in police cells sought writs of habeas corpus on the basis of the conditions of their confinement. They failed, but the court accepted that minimum standards of confinement had to be complied with under the 1980 Act. For a strong criticism of the Chief Inspector's refusal to inspect police cells, see Rod Morgan, 'Prisons Accountability Revisited', *Public Law* [1993].

[26] s. 6 (3). See also rules 88–96 Prison Rules 1964.

independent of the prison authorities and were too easily persuaded to uphold the views and interests of staff in preference to those of prisoners. This partiality was especially obvious in disciplinary hearings, where too many Boards accepted the word of prison officers rather than that of the accused and handed out severe penalties for relatively minor disciplinary infractions.[27] The abolition of the disciplinary jurisdiction of Boards does not mean that they have no role to play in the informal disciplinary process. For example, it is still necessary for a governor who wishes to extend a prisoner's segregation on Rule 43 beyond three days to obtain the approval of a member of the Board of Visitors.[28] It remains to be seen whether, freed from their disciplinary role, members of Boards of Visitors are able to develop a recognized role as fearless, independent, public critics of the numerous aspects of the prison system which deserve such attention. With some notable exceptions, most Boards have hitherto appeared to be content to pursue a low profile and somewhat cosy relationship with the prison authorities. So long as Boards have no statutory powers to change the decisions of prison staff, still less to alter policy at a higher level, it is difficult to see how they will be able to do more than create the appearance of being independent scrutineers of the system in action.[29]

Section 7 demands that 'every prison shall have a governor, a chaplain and a medical officer and such other officers as may be necessary'. It follows that the term 'officer', which is not defined in the Act, embraces not merely uniformed prison officers (or discipline staff as they are described) but all the statutory officers who work in a prison and are, administratively, members of the Prison Service. They are to be distinguished from the

[27] The best example of the partiality of Boards of Visitors concerned their approach to the granting of legal representation to prisoners appearing before them in disciplinary hearings. After the extraordinary decision of the Court of Appeal in R. v. *Risley Board of Visitors ex parte Draper* The Times 24 May 1988, in which the Risley Board's refusal to grant legal representation to all prisoners who appeared before them following a two-day roof-top protest was upheld, the grant of legal representation slowed to a trickle. It could hardly be said that standards of justice had improved to the extent that prisoners no longer required legal representation before Boards. Rather it seemed to be the case that Boards were unduly discomfited by the presence of lawyers, regarding their presence as productive of delay and unnecessary 'legality' in what they liked to think of as an inquisitorial rather than adversarial process. The lawyer acting for a prisoner in such hearings tended to be regarded as a rather unwelcome guest who had to be tolerated through gritted teeth before being shown the door.

[28] See Rule 43 (2). The decision to authorize continued segregation requires, as was pointed out by the Court of Appeal in R. v. *Deputy Governor of Parkhurst Prison ex parte Hague* [1992] 1 AC 58, a reasoned approach as to both the need for continued segregation at all and the period for which it might be necessary. There is very little hard evidence but the impression is that Board members rarely disagree with the views of prison governors. An obvious weakness in the system is that if a governor knows that one particular Board member is too independently minded for his liking, he can simply approach another 'tame' member of the Board for the authorization needed to extend segregation.

[29] A Prison Service working party is due to report in 1993 on whether or not to appoint a President of the Boards of Visitors who would be responsible for recruiting and training new Board members, thus distancing the Home Secretary from the process of appointment.

Home Secretary's 'officers', whose duties are defined in s. 4.[30] Since the pay and productivity agreement known as Fresh Start the number of grades of officer between the basic grade officer and the 'number one' governor has been reduced to eight, of whom there can, in any one prison, be seven.[31] Consistent with the sketchy nature of the Act, there is no detailed exposition of the powers and duties of the various prison 'officers'. This is left to the Prison Rules save in one respect. Section 8 states that 'every prison officer while acting as such shall have the powers, authority, protection and privileges of a constable'.[32] The words 'while acting as such' limit the effect of the section to the actual discharge of a prison officer's duties. Accordingly, prison officers do not have the status of 'constable' at all times. It means, however, that prison officers, while acting as such, have the same powers of arrest as police officers and that an assault on a prison officer constitutes an assault on a constable in the execution of his duty.[33]

The office of prison governor, it can be seen, is a creation of statute and although the exercise of his powers is governed by the Rules made pursuant to s. 47, the powers themselves are original and are not derived from or exercised on behalf of the Home Secretary. The mere fact that content is given to the governor's functions and duties by the Rules drafted on behalf of the Home Secretary, who is thus able to decide what the extent of the governor's powers should be, does not derogate from the governor's discrete status under the Act. Accordingly in *Leech* Lord Bridge dismissed the argument that, in exercising his disciplinary powers, the governor could somehow be distinguished from the Board of Visitors:

A prison governor may in general terms be aptly described as the servant of the Secretary of State, but he is not acting as such when adjudicating upon a charge of a disciplinary offence. He is then exercising the independent power conferred on him by the rules. The Secretary of State has no authority to direct the governor, any more than the Board of Visitors, as to how to adjudicate on a particular charge or what punishment should be awarded. If a Home Office official sought to stand behind the governor at a disciplinary hearing and tell him what to do, the governor would probably send him packing.[34]

For similar reasons the power given to 'the governor' in Rule 43 to segregate a prisoner is confined to the governor of the prison in which a prisoner is, for the time being, detained. It follows that neither the Home

---

[30] See e.g. Rule 46 (4).

[31] The Lygo Report on *The Management of the Prison Service*' (London, 1991) commented that there were too many grades which 'made for a very long chain of command and inhibits career progress for the most able'.

[32] *Home Office* v. *Robinson* EAT The Times, 20 Oct. 1981.

[33] See the Police Act 1964, s. 51. The same applied to the offence of obstructing an officer in the execution of his duty.

[34] [1988] 1 AC 533, at 563 D. Boards of Visitors no longer have any adjudicatory role but the principle remains applicable to governors.

Secretary, acting through his officers, nor the governor of any other prison may initiate the prisoner's segregation under Rule 43.[35]

Sections 12 and 13 of the 1952 Act deal with the 'place' in which a prisoner may be lawfully confined and the legal custody of prisoners. Section 12 (1) states that prisoners, whether convicted or on remand, 'may be lawfully confined in any prison'. In *Ex parte Hague* the House of Lords held that this provision was not subject to any implied term making lawful confinement dependent on compliance with the Prison Rules.[36] Even gross or 'fundamental' breaches of the Rules could not therefore affect the lawfulness of a prisoner's confinement such that he could sue in false imprisonment. Both Lords Bridge and Jauncey in *Ex parte Hague* concluded that the broad terms of s. 12 (1) would always provide a complete answer to any claim for false imprisonment against the governor or anyone acting on his authority.[37]

Section 12 (2) gives the Home Secretary the power to transfer a prisoner from one prison to another during the course of his imprisonment. This power can be delegated to a governor who may wish, for example, to transfer a prisoner in the interests of good order or discipline pursuant to Circular Instruction 37/1990.[38] While the power to transfer is a broad one it is susceptible to challenge by means of an application for judicial review. In *R. v. Home Secretary ex parte McAvoy* the Divisional Court said that a transfer decision reached without reference to a remand prisoner's 'right' to receive legal and family visits could 'in principle' be quashed as an unlawful exercise of discretion.[39]

Section 13 deals with the formalities of the legal custody of a prisoner and established two principles. First, a prisoner is deemed to be in the legal custody of 'the governor of the prison'.[40] This simply defines the legal responsibility or liability of the governor of the prison to which a prisoner has been allocated under s. 12. In the event of a challenge to the legality of a prisoner's detention by means of an application for habeas corpus, the governor, as legal custodian, is the proper respondent to the writ. Secondly, s. 13 (2) extends the concept of legal custody to include any period of time when the prisoner is being taken from one prison to another or is for any reason outside the prison in the custody or control of an officer of the prison.

Section 14 of the 1952 Act is an important provision since it is the only statutory provision which deals with the necessity for the Home Secretary

---

[35] *Ex parte Hague* [1992] 1 AC 58, at 105–8.   [36] Ibid., at 163 F.

[37] In reaching this conclusion both Lord Bridge and Lord Jauncey distinguished 19th-century cases involving claims for false imprisonment (such as *Cobbett* v. *Grey* (1850) 4 Exch. 729 and *Osborne* v. *Millman* (1887) 18 QBD 471) on the basis that there was no 19th-century equivalent of s. 12. This is not correct—see Prison Act 1877, s. 25.

[38] See App. 4 below for the full text of CI 37/1990. See also per Taylor, LJ, in *Ex parte Hague* [1992] 1 AC 58, at 107 H.

[39] [1984] 1 WLR 1408.   [40] s. 13 (1).

to satisfy himself that 'in every prison sufficient accommodation is provided for all prisoners'.[41] The means by which the Home Secretary is to satisfy himself on matters such as the size, lighting, heating, ventilation, and fittings in cells is by a system of certification by an inspector of every cell. The certificate may specify the period of time during which a prisoner may be separately confined in the cell.[42] Since the Prison Commissioners Dissolution Order 1963, the reference to an 'inspector' is to be read as a reference to 'an officer (not being an officer of a prison) acting on behalf of the Secretary of State'.[43] It can readily be seen that, since s. 14 permits no independent means of certifying the sufficiency of prison accommodation, the Act cannot itself act as a brake on overcrowding. If overcrowding places a strain on the system, making it impossible for the available cells to comply with their certificates, the Home Secretary merely has to alter the criteria for certification.[44] Since the one thing that all observers of the prison system agree upon is that for at least the last fifteen years local prisons in England and Wales have been grossly overcrowded, so that 'sufficient accommodation' has not been provided for all prisoners, it is hard to resist the conclusion that s. 14 of the Prison Act 1952 is probably the most ineffective statutory provision on the statute book.

Aside from the power to pass rules for the management of prisons, which we examine below, the rest of the 1952 Act deals with a number of unconnected and largely uncontroversial matters which have rarely been the subject of litigation. These include provisions dealing with the temporary discharge of prisoners on grounds of ill health,[45] the maintenance and closure of prisons,[46] criminal offences such as assisting a prisoner to escape,[47] and provision for the payment of all expenses incurred under the Prison Act from money provided by Parliament.[48]

THE PRISON RULES

The important rule-making power is contained in s. 47 of the Act:

(1) The Secretary of State may make rules for the regulation and management of prisons . . . and for the classification, treatment, employment, discipline and control of persons required to be detained therein.
(2) Rules made under this section shall make provision for ensuring that a person who is charged with any offence under the rules shall be given a proper opportunity of presenting his case.

---

[41] s. 14 (1).   [42] s. 14 (3).
[43] SI 1963/597, Art. 3 (2). Such officers are appointed pursuant to Prison Act 1952, s. 3 (1).
[44] See Rule 23 (2).   [45] s. 28.   [46] ss.33–7.   [47] ss.39–42.
[48] s. 51 and see also *Becker* v. *Home Office* [1972] 2 QB 407.

Section 52 (1) of the Act requires that any power to make rules shall be exercisable by statutory instrument. Any rules made pursuant to s. 47 are subject to the negative resolution procedure.[49] The Prison Rules may therefore be distinguished from, for example, the Immigration Rules, which are issued as House of Commons papers rather than by statutory instrument. In *Alexander* v. *Immigration Appeal Tribunal* the House of Lords held that the Immigration Rules are not to be construed with all the strictness applicable to the construction of a statutory instrument.[50] Lord Roskill said that the rules 'give guidance to the various officers concerned and contain statements of general policy regarding the operation of the relevant immigration legislation'. The Prison Rules by contrast must be regarded as delegated legislation in the strict sense.

The Rules are divided into five parts.[51] And in an exemplary analysis of their content, Professor Graham Zellick has described the Rules as falling into five broad categories.[52] Category one is 'rules of general policy objectives' which include the 'mission statement' of the Prison Service set out in Rule 1, namely that 'The purpose of the training and treatment of convicted prisoners shall be to encourage and assist them to lead a good and useful life.'[53] Rule 2 sets out a broad principle in relation to the way in which good order and discipline is to be maintained ('with firmness, but with no more restriction than is required for safe custody and well ordered community life').

Category two is 'rules of a discretionary nature' which expressly leave a number of important matters within the discretion of the prison authorities. Examples include temporary release[54] and communications with the outside world.[55] Many rules use the words 'so far as reasonably practicable'[56] and 'subject to any directions of the Secretary of State',[57] which have a similar effect. Category three is the 'rules of general protection' which deal, broadly speaking, with health and welfare 'standards'. Thus Rule 21 (4) specifies that 'food provided shall be wholesome, nutritious, well prepared and served, reasonably varied and sufficient in quantity' and others deal with the provision of bedding and 'toilet articles necessary for . . . health and cleanliness'.[58]

Professor Zellick's fourth category was the 'rules as to institutional struc-

---

[49] Criminal Justice Act 1967, s. 66 (4).　　　　[50] [1982] 1 WLR 1076.

[51] The Rules subdivide into 'Prisoners', 'Officers of Prisons', 'Persons Having Access To Prisons', 'Boards of Visitors', and 'Supplemental'.

[52] 'The Prison Rules and the Courts', [1981] *Crim. LR* 602. See also 'The Prison Rules and the Courts: A Postscript' [1982] *Crim. LR* 575.

[53] To continue to describe Rule 1 as the Prison Service's mission statement is slightly misleading. History, and a 'new realism' about the reforming potential of prisons, has led to the Prison Service's Statement of Purpose which recognizes three tasks for the Service. First, to keep secure those the courts put in its custody. Second, to treat those in custody with humanity. Third, to help prisoners lead 'law abiding and useful lives' in custody and after release.

[54] Rule 6.　　　　[55] Rules 33–4.　　　　[56] Rule 16.　　　　[57] Rule 30.

[58] Rules 24 and 26.

ture and administrative functions'. These he described as plainly designed to benefit prisoners but framed in such a way as to cast duties upon particular persons or bodies including the governor, medical officer, chaplain and the board of visitors. Thus the governor is required to hear requests and complaints by prisoners every day[59] and the chaplain is required to make daily visits to prisoners belonging to the Church of England (and any others willing to see him) who are 'sick, under restraint or undergoing cellular confinement'.[60] The various duties of Boards of Visitors are set out in rules 88–97 and include an obligation to inquire into any report that a prisoner's health, mental or physical, is likely to be injuriously affected by any conditions of his imprisonment.[61] The fifth and final category covers 'rules of specific individual protection'. These include the disciplinary code to be found in Rules 47–55 as well as non-disciplinary rules relating to good order or discipline such as Rules 43, 45, and 46.

Professor Zellick's article was written at a time when a breach of the Rules was not regarded as giving rise to a remedy in either public or private law. Lord Denning's description of the Rules in *Becker* v. *Home Office*[62] as 'regulatory directions only' seemed to suggest that they were not amenable to any form of judicial supervision. The purpose of the fivefold division was to support an argument that the courts had hitherto adopted an erroneous 'all or nothing' approach towards the Rules which failed to distinguish those rules which were plainly not intended to be 'actionable' (category one) from those in category five where the specific protective purpose required strict compliance by the prison authorities. As Professor Zellick pointed out, 'these provisions were hardly included in the rules so that the prison authorities could obey them or not as they preferred'. In his view 'strict compliance goes hand in hand with judicial oversight'.

Since Professor Zellick's article was published judicial attitudes to the status of the Prison Rules have progressed slightly. There is no doubt but that individual rules are 'justiciable' in the sense that breaches may give rise to a public law remedy.[63] But in *Hague* the House of Lords finally ruled that they are not 'actionable' in that they cannot in any circumstance give rise to a claim for breach of statutory duty.[64] The Woolf Report did not recommend any change in this area of prison law. So, for example, the Code of Accredited Standards, which was recommended as the means whereby an individual prison could move gradually towards 'accredited status' based on its ability to offer identified regime standards, would

---

[59] Rule 8 (2).   [60] Rule 11 (2).   [61] Rule 95 (3).   [62] [1972] 2 QB 407.

[63] See *Ex parte McAvoy* [1984] 1 WLR 1408 R. v. *Home Secretary ex parte Hickling & JH (a minor)* [1986] 1 FLR 543 and most recently *Ex parte Hague* [1992] IAC 58.

[64] But see the different approaches of Lord Bridge and Lord Jauncey on whether the rule-making power on s. 47 is broad enough to cover the enactment of rules which could create private law right. Lord Bridge said it was. Lord Jauncey disagreed. The remainder of their Lordships expressed no view on the matter.

eventually be the subject of a Prison Rule 'and so enforceable by judicial review'.[65] Woolf did not favour the creation of private law rights in prisoners which would entitle them to sue for damages, an approach consistent with the prevalent mood of judicial distaste for such a remedy.[66]

STANDING ORDERS AND CIRCULAR INSTRUCTIONS

The broad canvas created by the Prison Act 1952 and the Prison Rules is 'filled in' by a plethora of administrative guidance and directions contained in Standing Orders which, in turn, are updated and amended by Circular Instructions. These documents, which emanate from Prison Service Headquarters, have no legal status whatsoever despite the fact that they contain massive detail relevant to the conduct of daily life in prison. Though several prison rules refer to the Secretary of State's power to make directions in regard to a particular matter,[67] the directions themselves have no legislative authority and are never debated by Parliament. Rather like the circulars issued by various government departments responsible for town and country planning, they are no more than non-statutory guidance to those charged with managing the prison system. Standing Order 5, which deals with prisoners' letters and visits, does not define the Secretary of State's power to interfere with a prisoner's right to communicate with the outside world. Thus in *Raymond* v. *Honey* the House of Lords considered the legality of Standing Orders issued in purported compliance with rules governing correspondence and access to legal advisers. The effect of the Standing Orders was to restrict access to the courts by a prisoner in respect of any complaint against a prison officer which had not been ventilated through the internal complaints system (the now abolished 'prior ventilation' rule). The House of Lords held that there was nothing in the Prison Act 1952 which was capable of conferring a power to make regulations which would deny, or interfere with, the right of a prisoner to have unimpeded access to a court. Section 47 of the Act was wholly insufficient to authorize hindrance or interference with so basic a right. Accordingly, neither Rule 34 nor Rule 37A and certainly not the Standing Orders could sanction a decision by a prison governor to stop a prisoner from communicating with the courts. As Lord Wilberforce said:

The standing orders, if they have any legislative force at all, cannot confer any greater powers than the [rules] which, as stated, must themselves be construed in accordance with the statutory power to make them.[68]

[65] Woolf Report, para 12.117.
[66] See the trilogy of cases *Hague* [1992] 1 AC 58, *H* v. *Home Office* the Independent 6 May 1992, and *Racz* v. *Home Office* The Times 17 Dec. 1992.
[67] See e.g. Rule 34 (1).

Of course where advice contained in a Standing Order or Circular Instruction leads to unlawful administrative action, there will be no difficulty in obtaining declaratory relief via an application for judicial review to correct the error. In *Gillick* v. *West Norfolk AHA* Lord Bridge said that the question whether advice tendered in a non-statutory form such as a departmental circular is good or bad, reasonable or unreasonable, cannot, as a general rule, be subject to any form of judicial review.[69] But he concluded that there were exceptions and that:

We must now say that if a government department, in a field of administration in which it exercises responsibility, promulgates in a public document, albeit non-statutory in form, advice which is erroneous in law, then the court, in proceedings in appropriate form commenced by an applicant or plaintiff who possesses the necessary locus standi, has jurisdiction to correct the error of law by an appropriate declaration.

Exactly such a result was achieved in *Ex parte Hague* when the court ruled that Circular Instruction 10/1974 was *ultra vires* Rule 43.

Merely because Standing Orders and Circular Instructions do not have the force of law does not mean that they are irrelevant to a consideration of what 'rights' prisoners enjoy. The issuing of guidance and advice may give rise to a legitimate expectation on the part of a prisoner that a particular administrative procedure will be followed whose denial will give rise to an application for judicial review.[70] Thus Circular Instruction 37/1990, which deals with transfers in the interests of good order or discipline, states that 'inmates must be told the reasons for their transfer as far as is practicable, and as soon as possible'.[71] A refusal to give such reasons would, arguably, be susceptible to judicial review although before CI 37/1990 was issued it was held that a prisoner had no legal right to demand them.[72]

Access to and awareness of the content of Standing Orders and Circular Instructions is a problem for both prisoners and prison staff. Prison Rule 7 requires each prisoner within twenty-four hours of reception to be provided with written information about the Rules 'and other matters which it is necessary that he should know'. The Prison Service does not regard this as including the provision of copies of all Standing Orders and current Circular Instructions, though a refusal to do so in the face of a direct request by a prisoner could be challenged by means of an application of judicial review.

[68] [1983] 1 AC 1. The Court of Appeal adopted a similar approach in *Racz* v. *Home Office*.
[69] [1986] 1 AC 112, at 193 A.
[70] See for example *R.* v. *Secretary of State for the Home Dept. ex parte Khan* [1985] 1 WLR 1337, where the setting out of criteria in a letter of guidance concerning the exercise of discretion to permit trans-national adoptions was held to give rise to a legitimate expectation that entry clearance would be granted where those criteria were satisfied. See also *R.* v. *Home Secretary ex parte Ruddock* [1987] 1 WLR 1482.
[71] See Appendix 4 below.   [72] See per Taylor, LJ, in *Hague* [1992] 1 AC 58, at 112.

Whereas Standing Orders are now published and generally available in prison libraries, Circular Instructions tend to be sent to the governor, who decides how best to disseminate the new material to the appropriate recipients. The Chief Inspector has regularly criticized the poor level of communication of relevant information to staff, stating in his 1991 Report that 'our impression is that prison staff do not absorb much of the written information supplied for them'.[73] The authors of the Prison Reform Trust book *Prison Rules: A Working Guide* discovered during their research in 1992 that none of the prison libraries which they visited, including the Prison Service College in Wakefield, had a complete set of current Circular Instructions![74]

THE ADMINISTRATIVE FRAMEWORK

While the Home Secretary remains personally accountable to Parliament for the running of the Prison Service, the responsibility for administering the penal estate on a day-to-day basis rests with the Prison Service. Since 1990 the Service has been subject to almost continuous administrative restructuring and reform, culminating in April 1993 with the Prison Service becoming an Executive Agency. In a book devoted to explaining the nature of prison law, it is important to avoid becoming obsessed with the minutiae of these administrative changes. It is also important to stress that none of the radical alterations to the administrative structure has been accompanied by any changes to the legislative framework set out in the 1952 Act and the Prison Rules 1964. Agency status does not mean that the administrators are no longer subject to the rule of law or that the basic legal ground rules regulating prisoners' lives have changed, either for worse or better. Nevertheless some understanding of the administrative structure is essential to a complete understanding of the operation of the prison system.

The current, reformed administrative system is the result of the recommendations of the Woolf Report, which dealt in some detail with the managerial lessons to be drawn from the prison disturbances of spring 1990, and Sir Raymond Lygo's *Review of the Management of the Prison Service*. Woolf commented on 'a most remarkable dichotomy within the Prison Service'

---

[73] *Report of Her Majesty's Chief Inspector of Prisons January 1990–March 1991* (London, 1991).

[74] In the same work a prison governor is quoted as describing the present state of Standing Orders and Circular Instruction as 'a management system out of control'. The authors comment that 'they extend back over a long period of time, do not differentiate between minor administrative matters and important policy decisions; they have not been kept up to date and they are not effectively indexed, making it very difficult to find major topics (for example the Index to Standing Orders has no entry for "segregation")'. Nancy Loucks and Joyce Plotnikoff, *Prison Rules: A Working Guide* (Prison Reform Trust, London, 1993), 17.

between 'the high calibre and deep commitment of the majority of Prison Service staff at all levels' and 'the dissension, division and distrust which exists between all levels of Prison Service staff. They labour under a blanket of depression. They lack confidence in the value of what they do. They harbour a deep sense of frustration that the effort which they are devoting to the Service is not appreciated.'[75] Woolf's answer was three-fold. First, there was a need for more visible leadership of the Prison Service with the appointment of a Director-General who is, and is seen to be, in day-to-day charge of the Service. Secondly, the relationship between the Prison Service and Home Office Ministers needed to be more clearly structured to allow the Director-General to exercise the leadership needed to run the Service. Ministers should establish policies, priorities and resources. The Director-General should then be allowed to 'get on with the job'. Woolf recommended that the mechanism for achieving the new relationship should be a new document drawn up by the Prison Service and approved by Ministers which sets out the tasks and objectives for the Service for the coming year and the available resources. The document would be laid before Parliament and would be, in effect, a 'compact' or 'contract' between the Director-General and the Secretary of State. Finally, the relationship between Prison Service Headquarters and governors needed to be reformed so that the latter had greater discretion in the implementation of Prison Service policy. Each prison should be able to decide, as far as possible, how its budget is spent. The governor's contract with the Area Manager should set out specifically what his prison intends to provide for each prisoner as well as laying down the Service's obligation to provide the necessary resources.

The Lygo Report, commissioned by the Home Secretary in 1991 to provide a detailed managerial assessment of the working of the Service, endorsed many of Woolf's recommendations and concluded that agency status was the means whereby the necessary reforms could be achieved. Lygo commented that, since the absorption of the Prison Commissioners by the Home Office, numerous reports on the Prison Service had called for greater managerial independence for the Service from day-to-day Home Office control. Yet such independence remained illusory. While recognizing that the Home Secretary must remain politically accountable to Parliament for the Prison Service, Lygo said that:

the need for greater managerial independence suggest a move away from the present arrangements and towards a much more independent Prison Service, clearly separate from the rest of the Home Office but having responsibility to the Home Secretary. In current Civil Service thinking, this would equate to 'agency status'.

He concluded that the critical factor in the success or failure of the new arrangement would be the ability of Ministers to allow the Prison Service

---

[75] Woolf Report, para 12.1.

to operate in an almost autonomous mode while retaining their responsibility to Parliament for the overall policy and conduct.

The government accepted Lygo's basic recommendation and since 1 April 1993 the Prison Service has acquired agency status.[76] The main features of an executive agency are that it is headed by a Chief Executive[77] directly accountable to the Secretary of State, that the agency has clearly defined aims and objectives agreed with Ministers and set out in a Framework Document, and that, while the agency remains part of its parent government department, it has a great deal of autonomy, particularly in financial and personnel matters. The Secretary of State decides on the agency's resources, is responsible for policy, and remains ultimately accountable to Parliament. The Chief Executive is responsible for day-to-day management of the agency and, in the case of the Prison Service, will be the Home Secretary's main policy adviser. In December 1992 a draft Framework Document for the Prison Service was issued for consultation and a copy of this appears at Appendix 7.

The administrative structure of the Prison Service is based on reforms carried out in September 1990. Beneath the Director-General is a nine-member Prisons Board chaired by the Director-General and comprising two part-time, non-executive members plus the Heads of the six Directorates, namely Building and Services; Healthcare; Personnel and Finance; Inmate Administration; Custody; Inmate Programmes. The latter three Directorates have operational as well as policy responsibilities. In other words they are responsible for one or more policy divisions and, in addition, they have 'line management' responsibility for the fifteen Area Managers who form the next tier in the system. Each Area Manager is responsible for approximately nine penal establishments based on geographical divisions of England and Wales.

CONTRACTED-OUT PRISONS

The Criminal Justice Act 1991 introduced a new dimension to the prison system by giving the Home Secretary the power to 'contract-out' the running of a prison.[78] In effect, the Act provides for the privatization of prisons.

The contracting-out provisions potentially extend to any kind of prison although the draftsman of the 1991 Act has achieved this by a curiously cir-

---

[76] Agency status is an key feature of the government's Next Steps programme. Since 1988 some seventy-six Executive Agencies have been set up and another twenty-nine are in the pipeline.
[77] In fact the 'Chief Executive' of the Prison Service has retained the old title of Director-General.
[78] See App. 7 below: Criminal Justice Act 1991, ss. 84–7.

cuitous route. Section 84 (1) confines the power to contract out the running of a prison solely to remand prisons established after the commencement date of the section. Section 84 (2) then enables the Home Secretary by statutory instrument (approved by both Houses of Parliament) to apply the provisions of s. 84 (1) to any prison—whether or not established after commencement and regardless of whether it holds remand or convicted prisoners or both. In July 1992 a statutory instrument was approved which enables the Home Secretary to contract out any prison, holding remand or sentenced prisoners, which is established after the commencement date of s. 84.[79] A subsequent Order removes even this limitation, enabling the Home Secretary to contract out all prisons regardless of their age.[80] The first private prison opened in April 1992. The Wolds was a new prison built to hold remand prisoners and is run by the private security firm Group 4 Remand Services. The first 'old' prison to be 'market tested' was Strangeways, which indicates that the government does not intend to restrict the power to contract out to small, modern prisons holding low security prisoners.[81]

The legal implications of contracting out a prison are less drastic than the term suggests. Section 84 (1) states that while any contract between the Home Secretary and 'the contractor' who has contracted to run a prison is in force, 'the prison to which it relates shall be run subject to and in accordance with sections 85 and 86 below, the 1952 Act (as modified by section 87 below) and prison rules'. Section 85 creates three new officers to manage a contracted-out prison. Instead of a governor, each prison must have a 'director' who will be a 'prisoner custody officer appointed by the contractor and specially approved . . . by the Secretary of State'.[82] In addition to the director there must also be a 'controller' for each prison who is a Crown servant appointed by the Home Secretary.[83] Finally, instead of prison officers, 'every officer . . . who performs custodial duties shall be a prisoner custody officer who is authorised to perform such duties'.[84]

The director has the same duties and functions as the governor of a state prison save that he may not in any circumstance inquire into a disciplinary charge against a prisoner under Rules 47–50 nor may he conduct a disciplinary hearing or make, remit, or mitigate any disciplinary award.[85] Furthermore, except in cases of urgency, the director has no power to segregate a prisoner under Rule 43 nor any power to order confinement in a

---

[79] Criminal Justice Act 1991 (Contracted Out Prisons) Order 1992, SI 1656.

[80] Criminal Justice Act 1991 (Contracted Out Prisons) (No 2) Order 1992, which came into effect in Mar. 1993.

[81] The decision to 'market test' Strangeways sent a clear signal to the Prison Officers' Association that if it wished to maintain its position within a partly privatized prison system it would have to abandon existing policies on manning and resources in order to compete with private sector firms in the running of existing prison establishments.

[82] s. 85 (1) (*a*).   [83] s. 85 (1) (*b*).   [84] s. 85 (1).   [85] s. 85 (3) (*a*).

special cell under Rule 45 nor power to apply any restraint to a prisoner under Rule 46. All these powers are to be exercised by the controller by virtue of a new Rule 98A of the Prison Rules.[86] In addition the controller is under a duty to keep under review and report to the Home Secretary on the running of the prison by or on behalf of the director and must investigate and report on any allegations made against prisoner custody officers performing custodial duties at the prison.[87]

The powers and duties of prisoner custody officers are defined in s. 86 of the 1991 Act. They include the power to search prisoners in accordance with the Prison Rules[88] and any other person seeking to enter the prison.[89] This latter power is confined, however, to requiring a visitor to remove an outer coat, jacket or gloves. No intimate searching is authorized. Section 86 (3) defines the duties of prisoner custody officers as being to prevent the escape of prisoners, to prevent, detect, and report on other unlawful acts by prisoners (including prison disciplinary offences presumably), to ensure good order and discipline and to attend to prisoners' well-being. In carrying out these duties, prisoner custody officers are entitled to use reasonable force.[90] They do not, however, possess the powers of a constable under s. 8 of the 1952 Act.[91] All prisoner custody officers must be approved and certified accordingly by the Home Secretary to carry out both custodial duties and escort functions.[92]

Section 88 of the 1991 Act provides for the Home Secretary to intervene in a contracted-out prison where it appears to him that the director has lost or is likely to lose effective control of the prison or any part of it and that it is necessary to appoint a Crown servant to act as governor of the prison in the interests of preserving the safety of any person or to prevent serious damage to property. Where such an appointment is made, the governor assumes the powers and functions of both the director and the controller. This emergency provision is plainly intended to reassure the public that, in the event of a major prison disturbance, the entire resources of the state will be available to deal with it.

The key document which sets the standards for the running of a contracted-out prison is the contract concluded between the Home Secretary and the contractor. The Home Office refuses to publish any of the final contracts on the grounds of commercial confidentiality but the tender documents for The Wolds and other prisons subjected to market testing have been published. It seems clear that they require regime standards far better than those provided by state prisons. For example, the tender documents

---

[86] See s. 85 (4).   [87] Ibid.   [88] See Rule 39.

[89] Rule 86. An abuse of the power to search, including a failure to conduct a search in a reasonably seemly and decent manner, may give rise to an action for assault—see *Bayliss and Barton* v. *Governor of HM Prison Frankland*, Liverpool County Court, HHJ Marshall Evans, Legal Action Feb. 1993, 16.

[90] s. 86 (4).   [91] s. 87 (3).   [92] s. 89 and Schedule 10.

for The Wolds laid down a minimum entitlement of twelve hours out of cell for each prisoner each weekday and ten and a half hours at weekends. In marked contrast to the dreary regime of the average state local prison, the tender document opines that 'Ideally, prisoners should be free to leave their cells at the first unlocking in the morning and remain cell-free until final lock-up prior to the commencement of the night shift.' The documents identify 'performance criteria', for example an obligation to provide twenty-four hour on-call arrangements by a qualified doctor and daily medical visits. In the event of a failure to meet the agreed targets, stiff financial penalties are envisaged.[93] It remains to be seen whether the resource implications of a cell-free regime will be practical in a large, overcrowded prison such as Strangeways.

The tender documents also makes clear that, while the rules governing the management of The Wolds will be the Prison Rules 1964, Prison Service policy as expressed in Standing Orders and Circular Instructions will not automatically apply. The contractor will however be obliged to comply with any policy and procedural decisions which the Home Secretary deems applicable to contracted-out prisons. An example of such a requirement is in the area of suicide prevention and other healthcare matters. The Wolds' tender documents state that 'procedures for the identification of suicide risks and their prevention, and of known or identifiable mental or psychiatric disorder, and for the handling of AIDS and hepatitis cases must be carried out with standards at least consistent with those in the Prison Service'. Each contracted out prison is to have a Board of Visitors appointed by the Home Secretary and is subject to inspection by the Chief Inspector of Prisons.

The existence of a contract between the Home Secretary and a private contractor does not of course give any contractual rights to individual prisoners *vis-à-vis* the contractor. Basic principles of privity of contract would obviously prevent such claims. Prisoners' remedies in respect of unlawful treatment in a contracted-out prison will remain the same as in a state prison.[94] But the better regime standards required by the contract should provide a basis for a prisoner to argue that the contract creates a legitimate expectation of their provision. A failure by the contractor to meet these standards, or the Home Secretary's failure to enforce them, could therefore lead to a successful challenge by means of an application for judicial review or, possibly, by a private law claim for negligence based on a standard of care defined by the contract.

The proper respondent or defendant in such proceedings would be the Home Secretary (in an application for judicial review) or the Home Office,

---

[93] For a description of life at The Wolds see the article by Adam Sampson 'Private Tutors' in *Prison Report* (Summer 1992), PRT.

[94] See Ch. 3 for the scope of such remedies.

which remains departmentally responsible for prisons under the provisions of the Crown Proceedings Act 1947.[95] Merely because the 1991 Act enables the Home Secretary to contract out the running of a prison does not absolve him from his ultimate responsibility for all prisons under the terms of the Prison Act 1952.

[95] Crown Proceedings Act 1947, s. 17 (3).

3

Legal and Non-legal Avenues of Redress in the UK

INTRODUCTION

The fact of imprisonment invariably means that prisoners cease to have responsibility for and control over their environment. Living in a world governed almost entirely by the exercise of discretion naturally generates a wide range of grievances. The Woolf Report recognized this, pointing out that 'a prisoner, as a result of being in prison, is peculiarly vulnerable to arbitrary and unlawful action'.[1] Accordingly it is essential that prisoners have a number of avenues of redress open to them whereby the illegal exercise of power may be challenged and by which compensation can be recovered for the infringement of such civil rights as survive in all prisoners notwithstanding their imprisonment.[2]

Not all these avenues necessarily involve recourse to law. Indeed, as with life outside, by no means every problem which arises in prison is capable of successful resolution by legal means. In this chapter therefore we set out the range of legal and non-legal options which are available to prisoners to pursue in order to redress the various grievances and complaints which arise most frequently in prison.

The non-legal options are largely those which are 'internal' to the prison system itself, though they include recourse to the Prisons Ombudsman, Members of Parliament, and the Parliamentary Commissioner for Administration ('the Ombudsman'). These will be dealt with first under the general heading 'complaints procedures'. The legal options include two different forms of action—a civil action usually involving a claim for damages for a civil wrong (a tort) and an application for judicial review in the High Court. A third option, namely an application to the European Commission for Human Rights seeking a remedy from the European Court of Human

---

[1] *Prison Disturbances: April 1990* (Woolf Report), Cm. 1456 (London, 1991), para. 14.293.

[2] A 1977 American Report claimed that 'elementary psychology and fundamental justice both dictate that, wherever large numbers of human beings are confined involuntarily in close quarters, there must be effective credible machinery to provide an outlet for their complaints and dissatisfaction' (J. Keating, *Grievance Mechanisms in Correctional Institutions* (National Institute of Law Enforcement and Criminal Justice, Law Enforcement Assistance Administration, Washington, DC, US Government Printing Office, 1977).

Rights (ECHR) in Strasbourg, is dealt with in the following chapter.

We do not intend in this chapter to describe and analyse the detailed provisions of tort or administrative law as it applies in the prison context. Nor do we pretend that this book can be a substitute for specialist textbooks dealing with the substantive law and procedure for both civil actions and applications for judicial review. Rather our aim is to identify the appropriate means whereby the wide range of decisions and events which may affect prisoners can be challenged and the particular features which distinguish legal actions by prisoners from other forms of litigation.

As will, in due course, be explained, the division between non-legal and legal avenues of redress is not a rigid one, because it will frequently be the case that pursuing a complaint via the internal complaints mechanism will give rise to an application for judicial review. Indeed in many cases such internal recourse will be a wise, or indeed essential, prerequisite of mounting a successful legal challenge in the High Court.

COMPLAINTS PROCEDURES

Under the Prison Rules, prisoners have the right to pursue a request or complaint connected to or arising from their imprisonment with the governor of the prison or the Board of Visitors.[3] Though there is no longer any express reference to this in the Rules, it has long been accepted that in addition to these two avenues of redress, prisoners also have the right to complain to or petition the Secretary of State as the Minister with overall responsibility for the Prison Service and the care of prisoners.[4] In practice of course such petitions are dealt with by civil servants who act on behalf of the Home Secretary.[5]

---

[3] Rule 8 was brought into force by the Prison (Amendment) Rules 1990 and commenced on 25 Sept. 1990. The effect of the amendment was to delete from the old Rule 7 (1) any reference to the need to provide prisoners with information about how to petition the Secretary of State and to institute an entirely new machinery for dealing with prisoners' requests and complaints based on the recommendations in the Prison Service Working Group Report 'An Improved System of Grievance Procedures for Prisoners' Complaints and Requests' (P3 Division, July 1989). The Working Group's Report was itself a response to the Chief Inspector's 1987 Report *Prisoners' Complaints: A Review of the Procedures Used by Prisoners in Making Requests and Complaints* (Home Office, 1987). Both these reports provide a detailed and exhaustive analysis of the merits and demerits of various different methods of complaint/grievance resolution.

[4] The 'right' to petition the Secretary of State is not based on any express provision of the Prison Act or the Rules though SO 5C acknowledges its existence in para. 1: 'Inmates have the right to make requests or complaints to the governor, the Board of Visitors or to the Secretary of State'. Art. 5 of the Bill of Rights 1688 states that 'it is the right of the subjects to petition the King and all commitments and prosecutions for such petitioning are illegal'. On the *Raymond* v. *Honey* principle prisoners are no worse off than ordinary 'subjects'.

[5] Such civil servants were described as 'a faceless authority in Whitehall' by Lord Bridge in *Leech* [1988] 1 AC 533, at 568 C.

In September 1990 a new system of dealing with prisoners' complaints/ requests was introduced by the Prison Service. The previous scheme, developed piecemeal over the past 100 years, was generally regarded as unsatisfactory for a number of reasons. It was inefficient, slow, and lacking in coherence. The replies which were sent to prisoners, sometimes months after a complaint had been lodged, were almost always brief and wholly uninformative, thereby generating a complete lack of confidence in prisoners that their requests were taken seriously.[6] Furthermore, the system lacked any safeguards to ensure that complaints were not blocked by staff, contrary to the requirement in the European Prison Rules (EPR) that 'every prisoner shall be allowed to make a request or complaint, under confidential cover, to the central prison administration, the judicial authority or other proper authorities'.[7] Above all the old system made no provision for independent review of action taken by the prison authorities. The sole power to remedy any complaint lay with the governor or the Home Secretary, who was, either personally or through an agent, the author of the matter giving rise to complaint. Boards of Visitors, while nominally independent, had no power to take action in the absence of the agreement of the prison authorities and were in any event not regarded by prisoners as truly independent, compromised as they were by their disciplinary functions.[8]

Before analysing the extent to which the new scheme meets the criticisms levelled at the old one, we summarize the current provisions as they appear in the Rules, Standing Orders, and Circular Instructions.

[6] It was normal under the old complaints system for a prisoner to wait six months for a reply which, when it arrived, amounted to no more than a six-line statement informing the prisoner that his submissions had been considered carefully but that the original decision was upheld! This approach did not encourage confidence that a fearless, critical analysis had been conducted by the Whitehall bureaucracy. In her book *Bricks of Shame* (London, 1987), the Director of NACRO, Vivien Stern, commented that 'in this country the only appeal against . . . decisions is by petition to the Home Secretary whose officials made the decision in the first place. So if you came to prison with the feeling that it is an unfair world, and with a certain amount of anger and bitterness, your stay may well increase it.'

[7] EPR Rule 42 (3).

[8] The Woolf Report stated (para. 12.171) that 'We have no doubt whatsoever that the public and the Prison Service have cause to be indebted to the Boards of Visitors. The Boards should be preserved. We are conscious, however, that their endeavours on behalf of prisoners are inhibited by the fact that a substantial part of the prison population does not recognise the Boards' members as being as impartial as they in fact are'. Woolf recommended that the Boards' adjudicatory function be abolished and that organizational improvements be made to the Boards' structure including the appointment of a president, who would develop 'more effective methods for recruiting new members to Boards'. In May 1993, the Home Secretary announced that, following a period of consultation, it had been decided to reject this recommendation.

The statutory requirement to provide some mechanism for dealing with prisoners' complaints and requests is to be found in Rules 8 and 95, which state:

8 (1) A request or complaint to the governor or board of visitors relating to a prisoner's imprisonment shall be made orally or in writing by the prisoner.
(2) On every day the governor shall hear any requests and complaints that are made to him under paragraph (1) above.
(3) A written request or complaint under paragraph (1) above may be made in confidence.

95 (1) The board of visitors for a prison and any member of the board shall hear any complaint or request which a prisoner wishes to make to them or him.

There are a number of points to be noted about these provisions. First, Rule 8 (which was amended in September 1990 at the time the new complaints system was implemented) now requires 'the governor' to be available to hear any requests or complaints every day. Rule 98 enables the governor to delegate any of his powers under the Rules to another officer of the prison where the leave of the Secretary of State is obtained and Standing Order 5C confirms that the Secretary of State has authorized governors to delegate to any of their staff the task of hearing complaints pursuant to Rule 8. Accordingly a prisoner cannot demand to make his request/complaint to the 'Number 1' governor in charge of the prison and, so long as a system is arranged for a member of staff to be available to hear complaints or requests every day, then the requirements of Rule 8 can be fulfilled.

Secondly, prisoners can make their requests/complaints either orally or in writing but if they wish to do so in confidence then the request/complaint must be made in writing. Rule 8 says nothing about the circumstances in which such a confidential complaint may be made and there is nothing in the Rules which fully equates with the requirement in EPR 42 (2) that 'it shall be possible to make requests or complaints to an inspector of prisons during his inspection. The prisoner shall have the opportunity to talk to the inspector or to any other duly constituted authority entitled to visit the prison without the director or other members of staff being present.' Rule 96 (2) states that members of Boards of Visitors 'shall have access at any time to every part of the prison and to every prisoner and he may interview any prisoner out of the sight and hearing of the officers', but this is not the same as a right in the prisoner to demand that he be allowed to see the Board member in private and in confidence.

Thirdly, the requirement of daily availability to hear prisoners' requests/complaints does not apply to Boards of Visitors as it does to the governor. Rule 96 (1) requires the members of the Board to visit the prison frequently and to arrange a rota whereby at least one of the members visits the prison between meetings of the full Board. In practice most Boards ensure that one of their number visits at least three or four times each week. A prisoner who is dissatisfied with only seeing a single member of the Board can reasonably ask to have access to the full Board at one of their regular meetings.[9]

Finally, the Rules say nothing about time limits for responding to a prisoner's request/complaint although guidance is given in Standing Order 5C, which complies with the EPR 42 (4) requirement that prison authorities deal with complaints 'promptly' and 'without undue delay' unless they are obviously frivolous or groundless.[10] A failure to deal promptly with prisoners' complaints could therefore be challenged by an application for judicial review.

THE SCHEME IN PRACTICE

The internal complaints scheme envisages a two-tier process for dealing with prisoners' requests/complaints. The first tier seeks to resolve the matter within the prison if it can properly be dealt with at local level. The second tier involves taking the matter to the Area Manager or Headquarters either as an appeal against the decision reached at local level or because the subject-matter of the complaint is a 'reserved subject' which can only be dealt with by Headquarters/Area Manager. At either level the prisoner can take advantage of the right to make a request/complaint in confidence.[11]

In its document 'Notes for Prisoners: How to Make a Request or Complaint—The Stages Involved', the Prison Department suggests a five-step process for dealing with a request/complaint within the prison, that is, the first tier of the system. First, talk to a member of staff or personal officer. This is intended to be an informal means of resolving problems which falls short of a formal application. Secondly, make an oral application to the landing officer or wing manager. Thirdly, if the prisoner is still unhappy, he is advised to make a governor's application (in practice this is likely to be dealt with by a senior member of staff). Fourthly, the prisoner may make a formal written application using a request/complaint form and addressed to the governor.

Fifthly, there is the option of applying to the Board of Visitors. This advice may be sensible in many cases but it is only advice, and prisoners

---

[9] The practice adopted by Boards for hearing prisoners' complaints varies from prison to prison. The three principal methods are (1) an individual member seeing prisoners in the course of a routine visit, (2) one or more members holding a pre-arranged 'clinic' and (3) the Board (in whole or in part) seeing prisoners when they convene for their regular meetings.

[10] SO 5C, para. 9.

[11] Rule 8 (3).

cannot be required to exhaust the various options in the sequence suggested. Accordingly it would be unlawful for the Board of Visitors to require a prisoner first to have raised a complaint with the wing officer before approaching them and equally the governor can make no such requirement before responding within seven days to a written request/complaint form addressed to him.

Standing Order 5C states that in the case of a written request or complaint to the governor 'a reply shall be provided within 7 days of the form being received by the request/complaints clerk'.[12] If a full reply cannot be given within that time limit, an interim reply must be provided, save that in the case of allegations against members of staff the relevant time limit shall be fourteen days.

Prisoners dissatisfied with the reply received at local level or who wish to complain about a 'reserved subject' must move to the second tier of the complaints system and write to Prison Headquarters or the Area Manager of the Prison Service. 'Reserved subjects' include complaints about disciplinary adjudications, parole, any matter which directly challenges policies determined by Headquarters, and all complaints where, respectively, category A or life sentence status is relevant to the subject-matter of the complaint. SO 5C states that the time limit for dealing with appeals from local level or reserved subject complaints by Headquarters/Area Managers shall be six weeks from the receipt of the application/appeal. If a full reply cannot be given, an interim reply shall be provided.[13]

An important feature of the new complaints procedure is that all written requests or complaints to the governor or to Headquarters/Area Manager 'shall normally receive a written reply, save that in exceptional circumstances, the reply may be given orally' and 'if a request is refused or a complaint rejected the reply shall give reasons'.[14]

What constitutes 'reasons' for these purposes? Although the Court of Appeal has held that, as a matter of legal entitlement, a prisoner has no right to be given reasons for a decision to segregate him under Rule 43, that case arose before the new complaints system was in force and before the issuing of CI 26/1990, both of which state that written reasons shall be given for such decisions where requested by the prisoner.[15] Though the point has not as yet been tested in the courts, it is strongly arguable that the effect of the new commitment by the Prison Service to providing written reasons in response to a written request/complaint is that prisoners have a legitimate expectation that they will indeed be provided with reasons in writing unless there is some overriding reason for not doing so.[16] In

---

[12] SO 5C, para. 10.   [13] SO 5C, para. 14.   [14] SO 5C, para. 12.

[15] R. v. *Deputy Governor of Parkhurst Prison ex parte Hague* [1992] 1 AC 58, at p.112 A, per Taylor, LJ.

[16] The doctrine of a legitimate expectation whose frustration may provide the basis for an

what circumstances then will a failure to give any, or any sufficient, reasons be susceptible to a challenge in the courts by means of an application for judicial review?

There is clear authority that where a body is subject to a statutory duty to give reasons for its decisions, it must satisfy a minimum standard of clarity and explanation. It must also address the basic issues in dispute. In *Save Britain's Heritage* v. *Secretary of State for the Environment*,[17] Lord Bridge expanded on some earlier dicta that reasons must be proper, intelligible and adequate, stating that:

if the reasons given are improper they will reveal some flaw in the decision-making process which will be open to challenge on some ground other than failure to give reasons. If the reasons given are unintelligible, this will be equivalent to giving no reasons at all. The difficulty arises in determining whether the reasons given are adequate, whether they deal with the substantial points that have been raised or enable the reader to know what conclusion the decision-maker has reached on the principal controversial issues. What degree of particularity is required? I do not think that one can safely say more in general terms than that the degree of particularity required will depend entirely on the issues falling for decision

Lord Bridge was speaking there in the context of a planning case, where the legislative framework required that reasons be given by the Secretary of State for his decision. He concluded that the reasons for a planning inspector's or the Secretary of State's decision in a planning question should enable a person who is entitled to contest the decision to make a proper assessment whether it should be challenged but that the onus always lay upon the applicant to satisfy the court that he had been substantially prejudiced by the deficiency of the reasons. It is hard to see why a lesser standard should be applied to decisions taken in the context of prison life against an administrative framework which promises written reasons in response to all written requests/complaints. The more drastically a

application for judicial review must be kept within proper limits. It has been used primarily to identify these situations where there exists a right to be consulted and to make representations (in the absence of a clear statutory entitlement) before a public body changes its position (see e.g. *O'Reilly* v. *Mackman* [1983] 2 AC 237 and *R.* v. *Assistant Commissioner of Police ex parte Howell* [1986] RTR 52). There is a separate line of authority where the courts have held that, in the absence of a contrary public interest, a public authority may be obliged to exercise its discretion in accordance with its published policy (see e.g. *R.* v. *Home Secretary ex parte Khan* [1984] 1 WLR 1337; cf. *Re Findlay* [1985] AC 318; *R.* v. *Home Secretary ex parte Ruddock* [1987] 1 WLR 1482 and *R.* v. *Torbay BC ex parte Cleasby* [1991] COD 142). A published policy which requires reasons for a decision to be given to the person affected by it will obviously assist in establishing that 'fairness in action' (to use the words of Lord Donaldson, MR, in *R.* v. *Civil Service Appeal Board ex parte Cunningham* [1991] 4 All ER 310) demands that reasons be given. For a helpful analysis of the case-law on legitimate expectation see Richard Clayton and Hugh Tomlinson, 'Preventing Public Authorities from Backtracking', *Legal Action* (Feb. 1993), 17.

[17] [1991] 1 WLR 153.

decision affects a prisoner's civil rights or liberties, the greater should be the obligation to explain or justify it.[18]

If, in the absence of a clear statutory duty to provide reasons for an administrative decision, the common law will not impose such a duty the breach of which will be, by itself, a public law wrong, it is certainly true that in some cases an administrative authority will be unable to show it has acted lawfully without explaining itself.[19] There are numerous authorities which support this analysis but they all rest upon a broader basis than a mere failure to give reasons.[20] The cases reveal that there existed some separate basis for concluding that the decision under attack may be unlawful and the failure to give reasons then becomes part of the evidence which enables the court to conclude that the decision maker has erred in law or reached a perverse conclusion. More recent authority suggests that the courts may be moving towards a doctrine of minimum-reasons as part of the general duty to act fairly. In *R. v. Civil Service Appeal Board ex parte Cunningham*,[21] the Court of Appeal held that the Board, a creature of the prerogative, should explain why it had awarded the applicant such a low level of compensation for his unfair dismissal from the Prison Service. Lord

[18] In *Payne* v. *Lord Harris of Greenwich* [1981] 1 WLR 754 the Court of Appeal held that the duty to act fairly which was imposed on all administrative bodies by the rules of natural justice did not require that the Parole Board should provide reasons to a mandatory life sentence prisoner for refusing to recommend his release on parole. This was relied upon by Taylor, LJ, in *Ex parte Hague* to deny a right to reasons for a decision to segregate a prisoner under Rule 43. Interestingly, only two years later, Taylor, LJ was to refer in *R. v. Parole Board ex parte Wilson* [1992] QB 740 to *Payne* as a case 'decided in 1981 when established views on prisoners were very different from those of today' and ruled that it would be unjust to apply it to deny a discretionary lifer access to reports and other materials before the Parole Board considering his release on licence. Of course there is a difference between a right to access to adverse material to be presented to a body deciding on your freedom in advance of its decision and a right to reasons for its decision not to recommend your release. *Payne* is still, regrettably, good law on the latter issue and it is a case which has dogged the development of better standards of procedural fairness in prisons.

[19] *Padfield* v. *Minister of Agriculture, Fisheries and Food* [1968] AC 997 and more recently *Ex parte Chahal* unreported Nov. 1991, Popplewell, J., a political asylum case. For the principle that there is no general duty to give reasons see *Public Service Board of New South Wales* v. *Osmond* (1986) 60 ALJ 209; *Crake* v. *Supplementary Benefits Commission* [1982] 1 All ER 498. See more generally G. Richardson, 'The Duty to Give Reasons: Potential and Practice' [1986] *Public Law* 437–69 and the JUSTICE/All Souls Review, *Administrative Justice: Some Necessary Reforms* (1988), Ch. 3.

[20] See *Lonrho PLC* v. *Secretary of State for Trade and Industry* [1989] 1 WLR 525, R. v. *Secretary of State for the Home Dept ex parte Sinclair* (1992) The Times 5 Feb. See also *R. v. Lancashire County Council ex parte Huddleston* [1986] 2 All ER 941 where Lord Donaldson, MR, contrasted a local authority's obligation to provide reasons for refusing to make a discretionary grant to a student at the time of the decision and once leave to move for judicial review had been granted. No duty to provide reasons could be implied at the former stage but after leave had been granted it was the duty of the authority 'to make full and fair disclosure' and 'to explain fully what has occurred and why'. Once leave for judicial review is given the decision maker will have to disclose his hand 'with all cards face upwards on the table.'

[21] [1991] 4 All ER 310.

Donaldson, MR, in words which echo Lord Bridge's in the planning case, said that:

there was no general rule of common law or principle of natural justice that a public law authority should always or usually give reasons for its decisions . . . fairness required that the Board should give outline reasons sufficient to show what it was directing its mind to and thereby indirectly showing not whether its decision was right or wrong which was solely a matter for it, but whether its decision was lawful.

It has been said that 'no single factor has inhibited the development of English administrative law as seriously as the absence of any general obligation upon public authorities to give reasons for their decisions'.[22] This assessment is certainly confirmed by the history of prison law, where the absence of a clear duty to provide reasons for decisions as fundamental to liberty as release on parole has undermined a general progression towards better standards of fairness in prison life.

The early signs are that the 'new' requests and complaints system is not working well. The Chief Inspector of Prisons commented in his Report for 1991–2 that

'our initial impressions were that the system was reasonably well established and accepted in most prisons but that requests and complaints which required action in Prison Service headquarters did not receive prompt replies. The Prison Service's target that inmates should receive a reply within 6 weeks of submitting a request or complaint to Prison Service headquarters was rarely met. We did not detect any measurable improvement in 1991/92.[23]

The Chief Inspector also commented that the quality of replies had attracted adverse comment, with 'excessive legality' characterizing those responses dealing with disciplinary adjudications. He concluded that the system needed close monitoring and delays had to be reduced if it was to have any credibility.[24]

THE PRISONS OMBUDSMAN

An important recommendation in the Woolf Report was the creation of an independent Complaints Adjudicator.[25] According to Woolf, the existing

---

[22] JUSTICE, *Administration under Law*, 23.   [23] (HMSO, 1992).

[24] The Prison Reform Trust reported in its Winter 1992 journal *Prison Report* that the Home Office had confirmed that almost one in three of prisoners' requests and complaints are not answered within the six weeks deadline. In the London South area (comprising Albany, Parkhurst, Belmarsh, Wandsworth, and Brixton prisons) no less than 61% of prisoners were waiting for a reply after six weeks. The PRT comment that 'the complaints procedure introduced two years ago is in danger of falling into the same disrepute as the old system of petitions'. Governors' replies fare little better with only 53% of prisoners' requests/complaints being dealt with within the seven-day deadline.

[25] Woolf Report, paras. 14.342–14.363.

'internal' avenues for complaints lacked a key feature, namely recourse to an independent person/body with power to review the decisions of the prison authorities. While the Association of Members of Boards of Visitors and the Parliamentary All-Party Penal Affairs Group supported the idea of a 'Prisons Ombudsman', the Prison Service in its evidence to Woolf said that it was too soon to move to create such a post. It argued that until the new internal grievance procedures had had time to bed in, it was impossible to make a measured judgement on the need for a Prisons Ombudsman. The Woolf Report rejected this excessively cautious approach, concluding that:

the presence of an independent element within the Grievance Procedure is more than just an 'optional extra'. The case for some form of independent person or body to consider grievances is incontrovertible. There is no possibility of the present system satisfactorily meeting this point even once it has bedded down. A system without an independent element is not a system which accords with proper standards of justice.

Woolf recommended that the Home Secretary should appoint a Complaints Adjudicator who would have two distinct roles. In relation to the grievance procedure, his role would be to recommend, advise and conciliate at the final stage of the procedure. In relation to disciplinary proceedings, his role would be to act as a final tribunal of appeal. The existence of such an independent body at the top of the complaints system would, it was argued, improve the decision-making process as a whole. In order to underline his independence, Woolf recommended that the Adjudicator should be a person eligible for judicial office, i.e. a barrister or solicitor of not less than seven years standing. The Adjudicator's jurisdiction, as envisaged by Woolf, would arise once a prisoner had exhausted the various levels of the internal request/complaint system. He would be free to determine his own procedure and would be concerned both with the merits of the complaint and with the manner in which the decision giving rise to the grievance had been considered. He could receive a report from the Area Manager, the governor, or the Board of Visitors, where necessary. His powers, however, would not extend to the overruling of Prison Service decisions though he would be able to make recommendations to the governor, the Area Manager, Headquarters, or the Director-General. If the Adjudicator were unhappy with the response from the Prison Service, then he could make a report to the Home Secretary. The Adjudicator would not be concerned with the policy of the Prison Service generally in its dealings with and treatment of prisoners. Rather Woolf saw his role as being confined to individual grievances and how they were handled by the Prison Service.

Acknowledging that this might present problems in relation to complaints concerning the Prison Rules and Regulations and 'Reserved subjects', Woolf commented that:

So far as these matters are concerned, we would prefer there to be no special restriction placed upon the jurisdiction of the Complaints Adjudicator. However in practice, we would not expect the Complaints Adjudicator to make recommendations which conflicted with decisions on the merits by Headquarters. He would regard his role in such cases as being a supervisory role. It would be similar to that performed by the High Court in judicial review, a role which primarily involves scrutinising the decision-making process.[26]

The Woolf Report recognized that it was vital for the Complaints Adjudicator to establish time limits within which he should respond to the prisoner making an application to him. It recommended that a substantive reply should always be available within twelve weeks. Each prisoner should receive a written reply to his complaint with reasons for the Complaints Adjudicator's response and for any recommendations made.

The government has accepted the general thrust of the Woolf Report in relation to the need for an independent Complaints Adjudicator although, contrary to Woolf, it prefers the title Prisons Ombudsman.[27] It is intended that whoever is to be appointed to the new post will be in place by the end of 1993. No primary legislation is envisaged in order to give 'teeth' to the Ombudsman and this means that, initially, he will have to rely on powers of recommendation. He will not be able to make awards of compensation though he can recommend the making of an *ex gratia* payment to an aggrieved prisoner.

In its consultation paper published in March 1992, 'Independent Complaints Adjudicator for Prisons', the Prison Service identified five areas 'where it did not consider that it would be right for the [Complaints Adjudicator] to consider grievances'. These were parole decisions, decisions concerning the release of lifers, decisions involving the clinical judgement of doctors, the actions of Boards of Visitors other than in relation to authorizing continued segregation or the use of restraints, and cases which are already the subject of civil litigation or criminal proceedings in the courts. The paper also implied that, on security grounds, the Adjudicator should be excluded from considering the merits of some decisions relating to category A prisoners. These 'might include issues of categorisation, allocation, transfer and approved visitors where decisions are frequently based on confidential reports from the police, customs officials and others', though it was conceded that 'it would clearly be proper for the [Complaints Adjudicator] to consider whether proper procedures were followed in reaching decisions'. Professor Rod Morgan, one of the assessors who assisted Lord Justice Woolf and Judge Tumim in writing the Woolf Report, strongly criticized this approach. He pointed out that the areas listed in relation to category A prisoners embraced 'most of the decisions which

[26] Ibid., para. 14.362.
[27] At the time of going to press the identity of the new Ombudsman had not been announced.

high security prisoners might reasonably wish to challenge and which most fundamentally affect their quality of life.'[28] In the Home Office's latest pronouncements on the powers of the Prisons Ombudsman it seems that the criticisms of Professor Morgan and others have been listened to as matters relating to Category A prisoners are now to be included in his remit.

The appointment of a Prisons Ombudsman will not replace the jurisdiction of the Parliamentary Commissioner for Administration. Such a move would require amending legislation and accordingly prisoners are still free to approach the PCA via the statutory 'filter' of a Member of Parliament when they believe the Prison Service has been guilty of maladministration. In practice such complaints are rare and take many months to resolve. The statutory limits on the PCA's jurisdiction mean that he may not inquire into maladministration of the complaints procedure itself.

REDRESS THROUGH THE COURTS

In many cases resort to the internal complaints system will fail to resolve a particular problem either because it cannot lead to an award of financial compensation or because the response to the complaint is itself unsatisfactory, giving rise to the possibility of a legal challenge. In that event consideration must be given to litigation and a basic question then arises: what is the proper forum in which to commence proceedings? The answer to this question necessarily involves consideration of the difference between private law claims for damages and public law challenges to administrative decisions by means of an application for judicial review.

CIVIL ACTIONS

The decision of the House of Lords in *Ex parte Hague/Weldon*[29] clarified the scope of the private law remedies which a prisoner may pursue in respect of any complaint which might arise from his treatment by the prison authorities. There are essentially three tortious claims potentially available to cover the entire range of such complaints, namely actions for negligence, assault and battery, and misfeasance in public office.

Excluded from this list are potential claims for breach of statutory duty and false imprisonment. It is perhaps surprising that it was not until *Ex parte Hague/Weldon* reached the House of Lords in 1991 that it was conclusively decided that neither of these actions could be pursued by a prisoner (at least, in the case of false imprisonment, not against the prison

---

[28] *Public Law* [1993] (forthcoming).   [29] [1992] 1 AC 58.

authorities). It is important to understand the reasoning behind this decision since it has widespread implications for the potential of the law to remedy apparently wrongful conduct by the prison authorities.

Christopher Hague, who pursued an application for judicial review and included a claim for damages for false imprisonment and breach of statutory duty in his Form 86A,[30] had been held for twenty-eight days on solitary confinement at Wormwood Scrubs in breach of Rule 43 under the administrative procedures set out in CI 10/74, the predecessor to CI 37/90. Known by prisoners as the 'ghost train', a '10/74' was designed to transfer and segregate a supposedly 'subversive' prisoner. The Court of Appeal held this to be unlawful and, though the Home Office subsequently issued CI 37/90, which incorporated virtually all the procedural safeguards for which Hague had been arguing, it maintained the line in the House of Lords that he was not entitled to damages for his admittedly 'unlawful' (i.e. *ultra vires*) confinement at Wormwood Scrubs in substantially worse conditions than he would have enjoyed had he remained on normal location at Parkhurst.[31]

Kenneth Weldon had pursued a private law claim for assault and false imprisonment in which he alleged that he had been beaten up by prison staff at Armley Prison and then dumped without authority inside a strip cell, where he was left naked overnight. The appeal reached the House of Lords as a striking-out application by the Home Office, who claimed that no cause of action for false imprisonment arose at the suit of a prisoner as against the prison authorities. The Court of Appeal had dismissed the Home Office's application, holding that the facts disclosed a 'clearly arguable case' of false imprisonment.[32] Accordingly, the two conjoined appeals raised two key questions. First, whether a breach of the Prison Rules could in any circumstance give rise to a claim for breach of statutory duty. Secondly, whether a prisoner who was otherwise lawfully in prison could sue for false imprisonment in respect of an allegedly unlawful confinement.

Breach of statutory duty

Before *Ex parte Hague* reached the House of Lords, a prisoner seeking to pursue a private law claim action for breach of the Prison Rules faced two Court of Appeal decisions which held that no such claim could be entertained by the courts. *Arbon* v. *Anderson*[33] and *Becker* v. *Home Office*[34]

---

[30] Form 86A is the formal document to be used to make an application for judicial review.
[31] The judgment of the Court of Appeal in *Ex parte Hague* is at [1992] 1 AC 102.
[32] The judgment of the Court of Appeal in *Weldon* is at [1992] 1 AC 130.
[33] [1943] KB 252.
[34] [1972] 2 QB 407. For an excellent critique of this decision see A. M. Tettenborn, 'Prisoners' Rights' [1980] *Public Law* 74. Note however that the author wrongly states that the plaintiff in *Cobbett* v. *Grey* (1850) 4 Exch. 729 succeeded in his action and that the judgment of Denman, J. was reversed on appeal in *Osborne* v. *Milman* (1886) 17 QBD 514.

were both cases which reflected a judicial approach to prisoners' rights which today is almost unrecognizable. The common theme in the judgments of Lord Goddard in *Arbon* and Lord Denning in *Becker* was the fear that prison discipline would be fatally undermined if governors and prison officers had to perform their duties with the fear of legal action hanging over their heads.

The subsequent evolution of administrative law and decisions in the prison context such as *Ex parte St Germain*,[35] *Ex parte Tarrant*,[36] and *Ex parte Leech*[37] destroyed the validity of this approach, and so the fact that both *Arbon* and *Becker* were approved by the House of Lords in *Ex parte Hague* seems, at first sight, to mark a reversal of the *Raymond* v. *Honey*[38] principle that prisoners retain all civil rights save those which are taken away by express statutory provision or by necessary implication from the fact of imprisonment. In fact no such reversal has occurred. Lords Bridge and Jauncey (who gave the main speeches in *Ex parte Hague/Weldon*) approached the question of whether existing prison legislation gave rise to a cause of action for breach of statutory duty as, conventionally, one of legislative intention. Reviewing the leading authorities from *Groves* v. *Lord Wimborne*[39] to *P* v. *Liverpool Daily Post and Echo Newspapers PLC*[40] in the light of the provisions in Rule 43 designed to protect prisoners from unwarranted solitary confinement, Lord Jauncey concluded that the fact that a particular statutory provision was intended to protect certain individuals was not of itself sufficient to confer private law rights of action upon them. Something more was required to show that the legislature intended such conferment, namely some express reference enabling regulations to be made enforceable by a private law cause of action such as is to be found in s. 76 (2) of the Factories Act 1961.[41]

Lord Bridge, in similar vein, rejected the argument advanced on behalf of Hague that the 'ground rule' for ascertaining whether a plaintiff has a cause of action for breach of statutory duty is whether he belongs to a class of persons which the particular provision was intended to protect and has suffered damage in consequence of a breach of the duty of the kind from which the provision was intended to protect him—in Hague's case the requirement that only the governor of the prison in which he is located (and not the transferring governor) may decide to place him in Rule 43 segregation in the interests of good order or discipline. In Lord Bridge's view, such an approach was fallacious in the context of existing prison legislation, which differed fundamentally from, for example, the Factories Acts,

---

[35] [1979] QB 425.   [36] [1985] QB 251.   [37] [1988] 1 AC 533.   [38] [1983] 1 AC 1.
[39] [1898] 2 QB 402.   [40] [1991] 2 AC 370.
[41] In his stimulating article 'Liability in Tort for Breach of Statutory Duty' [1984] *LQR* 204, R. A. Buckley concluded that 'The fiction that liability depends exclusively on legislative intention should finally be abandoned. The courts should recognise that the decision whether or not to grant a civil action is ultimately one of policy for them.'

where the sole purpose of the statutory duties imposed was to protect various persons from the risk of personal injury. The Prison Act 1952 was concerned with the management and administration of both prisons and prisoners and covered a wide range of matters wholly unconnected with the protection of prisoners from personal injury. Accordingly it was impossible to conclude that Parliament had intended to confer private law rights of action in respect of any of the matters covered by the Prison Rules.[42]

In one potentially important respect Lord Bridge and Lord Jauncey differed in their analysis. Lord Jauncey accepted the Home Office submission that the rule-making power contained in s. 47 of the Prison Act did not authorize the creation of any private law rights in prisoners and thus any rule which expressly sought to do so would be *ultra vires*. Lord Bridge disagreed. In his view the power conferred by s. 47, with its express reference to the creation of rules for the 'treatment' of prisoners, was broad enough to cover the enactment of rules providing a cause of action for breach of statutory duty. None of the existing rules, including Rule 43, did so provide but Lord Bridge's analysis leaves open the possibility of a amendment to the Rules without the need for fresh primary legislation. The *vires* point remains moot, however, since Lords Ackner, Goff, and Lowry stated that they agreed with both Lords Bridge and Jauncey without expressing any view on the specific issue which divided them.

False imprisonment

Before *Hague/Weldon* reached the House of Lords there was no modern authority in which a prisoner had successfully sued the Home Office for damages for false imprisonment.[43] Indeed there were several which held

[42] In an interesting article 'The Prison Rules and the Courts' [1981] *Crim LR* 602, Graham Zellick advocated an approach to the Prison Rules which distinguished them into five categories. rules of general policy objectives (Rules 1 and 2); rules of a discretionary nature (Rules 33–4); rules of general protection (Rules 21 (4), 24 and 26 (1); rules as to institutional structure and administrative functions (Rules 95–6); and, finally, rules of specific individual protection including those dealing with discipline and segregation. Prof Zellick was writing before the 'new approach' signalled by *Raymond* v. *Honey* and he was concerned to identify those rules whose breach should *ipso facto* entitle a prisoner to redress in the courts in the wake of Lord Denning's assertion of non-justiciability in *Becker* v. *Home Office*. It was argued before the House of Lords in *Ex parte Hague* that Prof. Zellick's fifth category was a sensible basis for permitting a claim for damages for breach of statutory duty in some case of breaches of the Prison Rules. As he said, these rules 'are so concrete and precise in nature that it is inconceivable that any latitude should be left to the authorities in their implementation. Strict compliance goes hand in hand with judicial oversight.'

[43] A number of 19th-century cases had suggested that a false imprisonment claim could be sustained by an otherwise lawfully imprisoned prisoner—see *Osborne* v. *Angle* (1835) 2 Scott 500; *Yorke* v. *Chapman* (1839) 10 Ad & E 210; *Cobbett* v. *Grey* (1850) 4 Exch; *Osborne* v. *Milman* (1887) 18 QBD 471. In *Ex parte Hague/Weldon* these were described as cases which 'depended upon the strict classification of prisoners at the time and the statutory requirements as to where they should be confined dependent upon their classification' (per Lord Jauncey). The terms of s. 12 (1) of the Prison Act 1952 with its reference to the Home Secretary's power to confine a prisoner 'in any prison' demonstrated, according to Lord Jauncey, 'how different

that no such claim could be pursued.[44] In this sense *Hague/Weldon* did not represent a step backwards from the point of view of prisoners.

The House of Lords did however close the door on one potential basis for a false imprisonment claim stemming from the decision of the Court of Appeal in *Middleweek* v. *Chief Constable of Merseyside*,[45] namely one based on an allegation that the conditions of confinement were 'intolerable'. Prison conditions in late twentieth-century Britain do not make such a claim remote or fanciful and since false imprisonment carries with it the right to trial by jury, the ability to pursue such a remedy was potentially of some significance.

The facts of *Hague* and *Weldon* neatly raised two possible bases for a false imprisonment claim on the part of an otherwise lawfully detained prisoner. In *Hague* it was alleged that his segregation at Wormwood Scrubs, in breach of Rule 43 and in conditions identical to those on punishment, deprived him of his 'residual liberty' within the prison to associate with fellow prisoners and to have access to his possessions which he was entitled to retain unless deprived of it by a lawful order. He alleged that

is the position today to that which prevailed in the 19th century'. It is submitted, with great respect, that this supposed difference is not soundly based. *Cobbett,* for example, did not turn on a simple statutory provision requiring prisoners in a particular class to be confined and only confined within a particular defined place. The 1842 Act in question simply required prisoners in the Queen's Prison to be divided into seven classes and empowered the Secretary of State to make separate rules for each class of prisoners and 'as far as the construction of the prison will allow thereof, prisoners of each class shall be separated from each other. Debtors could be and were throughout the 19th century (and into the 20th) lawfully detained in prisons which also housed criminals—the Queen's Prison housed criminal prisoners as well as debtors. The issue in *Cobbett* was: on the assumption that Cobbett was confined in the 'right place', i.e. within a prison to which he could lawfully be sent, was there proof that he had been treated lawfully within the 'place'? The majority of the Court of Exchequer accepted that a breach of the Rules (i.e. a failure to satisfy the court that Cobbett had been dealt with in accordance with the Rules applicable to him) would give rise to an action for false imprisonment.

[44] See *Williams* v. *Home Office* (No 2) [1981] 1 All ER 1211 and *R.* v. *Board of Visitors of Gartree Prison ex parte Sears* The Times 20 Mar. 1985.

[45] Now reported as a Note at [1992] 1 AC 179. The 'intolerable conditions' analysis had been foreshadowed in an earlier decision, *R.* v. *Commissioner of Police for the Metropolis ex parte Nahar* The Times 28 May 1983, which arose from the conditions of confinement of two remand prisoners in police cells under the provisions of the Imprisonment (Temporary Provisions) Act 1980. *Nahar* was an application for habeas corpus. Stephen Brown J., refused the applications but expressed the view that 'there must be some minimum standard to render detention lawful'. In *Middleweek* the Court of Appeal built on this approach and Ackner LJ,. said that 'it must be possible to conceive of hypothetical cases in which the conditions of detention are so intolerable as to render the detention unlawful and thereby provide a remedy to the prisoner in damages for false imprisonment. A person lawfully detained in a prison cell would, in our judgment, cease to be so lawfully detained if the conditions in that cell were such as to be so seriously prejudicial to his health if he continued to occupy it e.g. because it became and remained seriously flooded or contained a fractured gas pipe allowing gas to escape into the cell. We do not therefore accept as an absolute proposition that if detention is initially lawful it can never become unlawful by reason of changes in the conditions of confinement.' In *Ex parte Hague/Weldon* Lord Ackner described his dictum as 'erroneous'.

this wrongful interference with his residual liberty constituted the tort of false imprisonment. It was never argued that Hague was entitled to leave prison altogether, merely that his detention was tortious and gave rise to a claim for damages.[46]

Weldon's claim involved an allegation both that his detention in the strip cell infringed his residual liberty and that his conditions of confinement in the strip cell were 'intolerable'. The idea that prisoners retained a 'residual liberty', the wrongful infringement of which might sound in damages, had received the support of Ralph Gibson, LJ, in his judgment in *Weldon* in the Court of Appeal.[47] Recognizing the need to define the nature of a prisoner's residual liberty, he held that the Prison Rules indicated the extent to which a prisoner may enjoy some freedom of movement within the prison together with access to his permitted possessions and association with other prisoners. In his view, the Rules may be regarded as the detailed provisions designed to achieve the purpose stated in Rule 1 ('encourage and assist to lead a good and useful life') and Rule 2 (1) ('treatment designed to encourage self-respect and sense of personal responsibility'). Against this background he concluded that:

There is no reason apparent to me why the nature of the tort [of false imprisonment], evolved by the common law for the protection of personal liberty, should be held to be such as to deny its availability to a convicted prisoner, whose residual liberty should, in my view, be protected so far as the law can properly achieve unless statute requires otherwise[48]

Both the residual liberty ground and the intolerable conditions ground were rejected by the House of Lords. Lord Bridge expressed the view that to talk of 'residual liberty' as a species of freedom within prison enjoyed as a legal right which cannot lawfully be restrained by the prison authorities was 'quite illusory' since:

The prisoner is at all times lawfully restrained within closely defined bounds and if he is kept in a segregated cell at a time when, if the rules had not been misapplied, he would be in the company of other prisoners . . . this is not the deprivation of his liberty of movement, which is the essence of the tort of false imprisonment, it is the substitution of one form of restraint for another.[49]

So far as the intolerable conditions ground was concerned, the House of Lords held that an otherwise lawful detention, such as committal to prison, could not be rendered unlawful by the fact that the conditions of confinement

---

[46] It was argued that just as in the Canadian case of *Cardinal and Oswald* v. *Director of Kent Institution* 24 DLR (4th) 44 the prisoner successfully sought habeas corpus with *certiorari* in aid to secure his release from the 'dissociation unit' into the general population of the prison, it was not necessary to hold that the availability of a remedy in damages for false imprisonment necessarily meant that a prisoner could achieve complete freedom from any kind of restraint!

[47] [1992] 1 AC 130.   [48] Ibid., at p.139 D.   [49] Ibid., at 163 C.

were intolerable partly because such a concept was too imprecise and partly also because, in the words of Lord Jauncey:

a prisoner at any time has no liberty to be in any place other than where the regime permits [and so] he has no liberty capable of deprivation by the regime so as to constitute the tort of false imprisonment. An alteration of his conditions therefore deprives him of no liberty because he has none already.[50]

Accordingly the case of *Middleweek*, in so far as it suggested otherwise, was wrongly decided. The issue which concerned the Court of Appeal in *Weldon*, namely a concern to ensure that a prisoner held captive by a fellow prisoner (as happened for example during the Strangeways disturbance) should be entitled to sue for false imprisonment, also concerned the House of Lords. They dealt with it by distinguishing the Home Office's authority totally to restrain a prisoner's freedom of movement, derived from s. 12(1) of the Prison Act, from that of a prisoner 'imprisoning' a fellow prisoner who had no such authority. Lacking the statutory defence provided by s.12 (1), a prisoner or any other person would be liable in the ordinary way for false imprisonment. The quality of a prisoner's already attenuated liberty thus depends on the identity of the person who unlawfully attenuates or abuses it further!

This led Lord Bridge to a somewhat startling conclusion, namely that a prison officer who acts in bad faith by deliberately subjecting a prisoner to a restraint which he knows he has no authority to impose may render himself personally liable to an action for false imprisonment as well as committing the tort of misfeasance in public office. According to Lord Bridge, such an officer, lacking the authority of the governor, also lacks the protection of s. 12 (1) of the Prison Act, and by deliberately acting outside the scope of his authority 'he cannot render the governor or the Home Office vicariously liable for his tortious conduct'. This analysis therefore leaves open the possibility of actions for false imprisonment being pursued by prisoners but only against fellow prisoners or individual prison officers in respect of whose actions no vicarious liability may attach to the Home Office.[51]

Negligence

Negligence is a tort (a civil wrong) and has been described as 'the failure to exercise that care which the circumstances demand'.[52] What amounts to negligence will vary according to the facts of any given case. It can consist in failing to do something which ought to be done or in doing something which ought to be done either in a different manner or not at all. Where there is no duty to exercise care, the fact that a person may be said to have acted carelessly has no legal consequence. Where, however, there is a duty

---

[50] [1992] 1 AC 130, at 177 E.
[51] Ibid., at 164 D. See also *Racz* v. *The Home Office*, The Times 17 Dec. 1992.
[52] *Halsbury's Laws*, vol. xxxiv, para. 1.

to exercise care, reasonable care must be taken to avoid acts or omissions which can be reasonably foreseen to be likely to cause injury to persons or property. Negligence, in order to give a cause of action, must be the neglect of some duty owed to the person who makes the claim—'negligence in the air will not do'.[53] The degree of care required in a particular case will vary according to the amount of risk to be encountered and to the magnitude of the prospective injury.

There is no general duty of care to prevent a third party from causing injury to another by the third party's own deliberate wrongdoing. There are, however, special circumstances where the law does impose an affirmative duty of care to prevent a third party from injuring others. One such example is where a special relationship exists between a person and a third party by virtue of which that person is responsible for controlling the third party.[54] This has some significance in the context of prison life, where the Home Office has complete control over and responsibility for the actions of prisoners in its custody.

The prison authorities owe a common law duty of care to prisoners to take reasonable care for their safety.[55] A breach of that duty may arise in a wide range of circumstances. The most common instances will be where a prisoner suffers injury either:

1. as the direct result of the negligence of prison staff (for example prison medical staff);[56] or
2. as a result of being required to work with dangerous equipment or in conditions injurious to health;[57] or
3. as a result of the actions of fellow prisoners in circumstances where there was a failure to provide reasonable supervision or protection against such injury;[58] or

---

[53] *Haynes* v. *Harwood* [1935] 1 KB 146, per Greer, LJ.

[54] See *Home Office* v. *Dorset Yacht Co Ltd* [1970] AC 1004 and *Smith* v. *Littlewoods* [1987] AC 241.

[55] *Ellis* v. *Home Office* [1953] 2 All ER 149, at 154, CA and more recently *Palmer* v. *Home Office* The Guardian, 31 Mar. 1988 (CA).

[56] See, however, *Knight* v. *Home Office* [1990] 3 All ER 237, in which Pill, J., held that the standard of care provided for a mentally ill prisoner detained in a prison hospital was not required to be as high as the standard of care provided in a psychiatric hospital outside prison since psychiatric and prison hospitals performed different functions and the duty of care had to be tailored to the act and function to be performed.

[57] *Pullen* v. *The Prison Commissioners* [1957] 1 WLR 1186 and see also *Ferguson* v. *Home Office* The Times 8 Oct. 1977, in which a prisoner recovered £15,000 when his hand was badly mutilated as a result of inadequate instruction by prison staff in the use of a circular saw.

[58] There are several examples of this kind of action including *D'Arcy* v. *Prison Commissioners* The Times 17 Nov. 1955, where damages of £190 were awarded to a prisoner in respect of an assault by fellow prisoners; *Egerton* v. *Home Office* [1978] Crim LR 494; *Porterfield* v. *Home Office* Independent 9 Mar. 1988; *Palmer* v. *Home Office* The Guardian, 31 Mar. 1988; *Steele* v. *Northern Ireland Office* (1988) 12 NIJB 1; *H* v. *Home Office* Independent 6 May 1992 (CA).

4. as a result of defective or dangerous premises;[59] or

5. as a result of being detained in 'intolerable conditions'.[60]

The examples given in (1)–(4) above are merely factual variations of the application of the general duty of care to prison life. In most negligence claims the problem is not in establishing the existence of the duty of care but rather in demonstrating that the action (or inaction) of prison staff fell below the standard of care which the law imposes upon them and that the injury in question was the reasonably foreseeable consequence of the breach of duty. Thus in negligence claims arising from an assault by a fellow prisoner, the courts have demonstrated a marked reluctance to impose too rigid a constraint on a governor's discretion to permit prisoners known or believed to be dangerous to mix freely with other prisoners and even to have access to potentially lethal weapons. In *Palmer* v. *Home Office*,[61] for example, the plaintiff was stabbed in the stomach by a fellow prisoner who had been convicted of three murders as well as other serious offences of violence and who had been allowed to work in the tailor's workshop, where he had access to scissors. It was conceded in evidence by a governor grade that the assailant was 'very dangerous to anybody in the prison, staff and inmates'. None the less Lord Justice Neill refused to characterise the decision to permit him to work in the tailor's shop as negligent, stating that:

Those in charge of prisoners have a difficult task. Clearly, except in extreme cases, of which obviously there are some, those responsible for prisons cannot keep prisoners permanently locked up or segregated from other prisoners. In addition it is necessary, or certainly desirable wherever possible, to provide suitable employment for individual prisoners.

This approach (the 'Hannibal Lecter' test, perhaps) reveals an obvious and difficult conflict between a liberal approach towards all prisoners, even violent ones, and the desire to ensure that prisoners who are injured while in prison through no fault of their own are not left without a remedy. This latter point is particularly important when it is remembered that prisoners with convictions for violence who are injured by fellow prisoners will

---

[59] *Christofi* v. *Home Office* The Times 31 July 1975, where a prisoner recovered damages for injuries caused by a fall on a broken step.

[60] Despite the dictum of Lord Bridge in *Ex parte Hague/Weldon* [1992] 1 AC 58, at 165 H, which suggests that the basic principles of the tort of negligence might be stretched to accommodate a negligence action by a prisoner who suffers 'intolerable discomfort' as a result of detention in 'intolerable conditions', the better view is that unless a prisoner can prove some physical or mental injury ('nervous shock') he will not have a cause of action in negligence—see also *Calveley* v. *Chief Constable of Merseyside* [1989] 1 AC 1228 and *Hicks* v. *Chief Constable of South Yorkshire* [1992] 2 All ER 65.

[61] See n. 55 above.

rarely, if ever, be able to recover compensation from the Criminal Injuries Compensation Board.[62]

Lost property claims

Claims for lost property allegedly mislaid due to the negligence of the prison authorities are a regular source of litigation in the County Court. Many of them concern items of property which are not valuable in money terms but which may be of great sentimental or symbolic importance to the prisoner concerned. The majority of these claims will have to be conducted by prisoners themselves without the benefit of legal aid under the small claims procedure. Whether or not a prisoner can successfully sue the Home Office for negligence in respect of lost or stolen property will depend on the status of the property in question and its location at the time of its disappearance.

Each prisoner's property is recorded on Property Record cards or sheets upon being received into prison custody. Each time a prisoner is transferred, the Property Record should record that fact and the prisoner's confirmation that his or her property has been transferred correctly. The basic distinction is between 'stored' and 'in-possession' property.[63] At the time when property is allocated to one or other category, prisoners are routinely required to sign a reception certificate and disclaimer. The purpose of this is, first, to certify that all the stored property has been correctly recorded and securely sealed for safe keeping by the prison and, secondly, to acknowledge that the Prison Service will not accept liability for the loss of or damage to property which is kept in the prisoner's possession.

Where property which has been listed as 'stored' goes missing there is little difficulty in fixing the Home Office with liability in negligence—in practice an *ex gratia* payment will usually be made. More difficult is the question of liability for property listed as being 'in-possession' but which goes missing at a time when the prisoner had no control over it. For example, a prisoner who is subject to a sudden transfer under CI 37/90 will have no opportunity to collect together his personal possessions from his cell. Indeed, for the whole period of the transfer, he will normally be kept in Rule 43 conditions with no access to his property. If, when he is eventually reunited with his possessions, he discovers that items are missing it would hardly be reasonable for the Home Office to   rely   on   the   original   signed

---

[62] R. v. *CICB ex parte Thompstone* [1984] 1 WLR 1234. The Court of Appeal held that the proper question for the Board to ask itself is 'is the applicant an appropriate recipient of an ex gratia compensatory payment made at the public expense?' and that there was nothing unlawful in the Board's decision that an applicant's previous convictions for either dishonesty or violence disentitled him to an award. It was reasonable to withhold compensation or reduce it even where the applicant's conduct, character, or way of life had no ascertainable bearing on the occurrence of the injury or its aftermath.

[63] See Rules 41–2 for the provisions governing disposal of prisoners' property.

disclaimer in order to evade liability. And what of the Rolex watch which goes missing from the supposedly secure wing office while the prisoner has gone to the gym? Will a notice in the office to the effect that 'any valuables left in the staff office are left at the owner's own risk' always relieve the Home Office of liability even where an officer has agreed to look after the items in question? Surely not, at least not in circumstances where it can be established that the loss was directly attributable to some act of carelessness on the part of a member of the prison staff rather than the criminal act of a third party.

Assault and battery

There are three kinds of trespass to the person—assault, battery, and false imprisonment. Each is concerned with a different form of interference with a person's liberty. As we have seen, the tort of false imprisonment is not available to a prisoner *vis-à-vis* his legal custodians, the prison authorities, because he is deemed to have lost all his freedom of movement to the state by the fact of imprisonment. By an action for assault and battery, however, a prisoner is able to define the limitations on the State's right to interfere directly with his person. It is an assertion by the prisoner of his freedom from physical molestation save in accordance with the law.

A battery is the actual infliction of unlawful force on another person whereas an assault is an act which causes another person to apprehend the infliction of immediate, unlawful force on his person.[64] Thus, if force is actually applied, directly or indirectly, unlawfully or without the consent of the person assaulted, the assault becomes a battery however slight the force. In popular language the term 'assault' includes a 'battery' though a person may be guilty of an assault without being guilty of a battery.

An intention to injure is not a necessary ingredient of a battery. The deliberate use of force against another will not always amount to a battery. Apart from special justifications, such as self-defence, there are numerous examples in ordinary life where an intended contact or touching is not actionable as a trespass. These are not limited to those occasions where consent is actual or reasonably to be implied—such as shaking hands. They include the ordinary 'collisions' of everyday life, such as the knocking and pushing which occur in, for example, a crowded underground train. In this context it has been suggested that a hostile intent is not the crucial determinant of the existence of a battery but, rather, 'whether an absence of consent on the part of the plaintiff can be inferred'.[64a] An unintentional trespass cannot, it would seem, be pursued by an action for assault and battery but instead as a claim for negligence with proof of a want of reasonable care by the person who has caused injury.[65]

---

[64] *Collins* v. *Wilcock* [1984] 1 WLR 1172 and see also *Wilson* v. *Pringle* [1987] QB 237.
[64a] See for example *F.* v. *West Berkshire Health Authority* [1990] 2 AC 1.
[65] *Fowler* v. *Lanning* [1959] 1 QB 426 and *Letang* v. *Cooper* [1965] 1 QB 232.

The circumstances in which force may be used by prison officers against prisoners are dealt with elsewhere. Suffice it to say that a claim for assault and battery may arise in the following circumstances:

1. where a prisoner is deliberately attacked either by fellow prisoners or by staff;
2. where excessive force is used to carry out an otherwise lawful order, such as forcibly removing a prisoner to the segregation unit following an order made pursuant to Rule 43;[66]
3. where force is used to execute an unlawful order, for example the unauthorized, forcible removal of a prisoner to a strip cell by officers below the rank of senior officer or the imposition of a restraint without proper authority;[67]
4. where a restraint is imposed for too long, i.e. after a prisoner has ceased to be a danger to himself or others.[68]

Aggravated and exemplary damages

A plaintiff who succeeds in a claim for an assault carried out by prison staff may be entitled to an award of aggravated and exemplary damages. Aggravated damages are part of the award of compensatory damages. Their effect is to increase the level of damages above what would otherwise have been recovered and they arise where the manner of commission of the tort 'was such as to injure the plaintiff's proper feelings of dignity and pride'.[69] As with claims against the police for assault, false imprisonment, and malicious prosecution, in which aggravated damages are frequently awarded, the circumstances in which a prisoner finds himself under attack from prison officers are likely to heighten his sense of injury or humiliation. From the defendant's perspective it appears that aggravated damages are designed to punish but strictly they are made to compensate the plaintiff's injured feelings. Where a plaintiff has himself been guilty of some form of reprehensible conduct, this may be reflected in a reduced or nil award of aggravated damages.[70]

It is by an award of exemplary or punitive damages that the court is able to teach the defendant that 'tort does not pay'. In effect the award of

[66] The use of control and restraint methods to force a prisoner to a segregation unit will only be justifiable if the prisoner is resisting such removal.

[67] Where there is no lawful justification for a prisoner's removal to a segregation unit or strip cell then all force used will be actionable as an assault. In this situation the prisoner who resists and is subjected to greater force in order to enforce an illegal segregation order will be in a better position than the prisoner who 'goes quietly', as a result of the decision in *Ex parte Hague/Weldon* which holds that unlawful segregation does not, of itself, gives rise to a cause of action unless malice or conscious abuse of power can be established.

[68] *Rodrigues* v. *Home Office* LAG Bulletin Feb. 1989 14.

[69] *Clerk and Lindsell on Torts* (16th edn.), para. 5–36.

[70] *O'Connor* v. *Hewitson* [1979] Crim. LR 46.

exemplary damages sends out a message to others and is meant to act as a deterrent against further abuse of power. Exemplary damages may not be awarded in every example of tortious conduct but only in those categories of cases identified in Lord Devlin's speech in *Rookes* v. *Barnard*.[71] One such category includes 'oppressive, arbitrary or unconstitutional conduct by the servants of the government' and this plainly includes members of the Prison Department. In *Barbara* v. *The Home Office,* the plaintiff sued the Home Office for an assault arising from an incident when he was forcibly injected with Largactil following his admission to Brixton Prison for medical reports.[72] He claimed damages for his physical injury and wounded feelings as well as exemplary damages on the ground that an intramuscular injection administered by hospital officers without regard to the right of a prisoner to withhold his consent was an abuse of the officers' coercive powers. Leggatt, J., awarded Mr Barbara £100 for his physical injury and £500 in aggravated damages for injured feelings and loss of dignity. No award of exemplary damages was made however because the judge felt that mere negligence (which had been pleaded) resulting in a trespass should not normally be visited with exemplary damages even though, from the point of view of the plaintiff, the trespass itself was oppressive.[73]

Misfeasance in public office[74]

In *Dunlop* v. *Woollahra Municipal Council*[75] Lord Diplock referred to misfeasance in public office as a 'well-established tort'. There are however very few examples of such claims being pursued through the English courts.[76] None the less its importance as a remedy to prisoners has been reinforced by the decision of the House of Lords in *Hague/Weldon*, where it was expressly identified as the appropriate remedy for certain forms of unlawful action by prison staff.

In neither *Hague* nor *Weldon* was misfeasance in public office directly in issue by the time the case reached the House of Lords. Hague had included a claim for damages for misfeasance in public office (as well as for false

---

[71] [1964] AC 1129. See also *Cassell* v. *Broome* [1972] AC 1027. Actions which may be characterised as violent, cruel, threatening, or abusive will normally be regarded as 'oppressive or arbitrary' and, if there is evidence of such conduct, the issue of exemplary damages will be at large—see e.g. *Holden* v. *Chief Constable of Lancashire* [1987] QB 380.

[72] (1984) 134 New LJ 888.

[73] See now the decision of the Court of Appeal in *Gibbons* v. *South West Water Services Ltd.* [1993] 2 WLR 507 and *Racz* v. *The Home Office*, The Times 17 Dec. 1992.

[74] The *OED* defines misfeasance as 'a transgression, trespass . . . the wrongful exercise of lawful authority or improper performance of a lawful act'.

[75] [1982] AC 158. The tort has been traced back to the reign of Henry V and includes malicious abuse of power, deliberate maladministration, and, possibly, other unlawful acts causing injury—see e.g. *Ashby* v. *White* (1703) 2 Ld. Raym. 938.

[76] In Canada the tort seems better developed—see e.g. *Roncarelli* v. *Duplessis* (1959) 16 DLR (2d) 689 and *McGillivray* v. *Kimber* (1915) 26 DLR 164.

imprisonment and breach of statutory duty) in his application for judicial review of the compendious decisions to transfer and segregate him. The misfeasance claim was based on the allegation that he had been segregated deliberately for an improper motive, namely a desire to avoid the embarrassment of charging him with a disciplinary offence where he would have had an opportunity to defend himself. During a disciplinary hearing it would have become clear that prison officers had, at the very least, condoned Hague's alleged protest at the removal of exercise periods for category A prisoners at Parkhurst. The Divisional Court held that avoiding embarrassment to staff could be a proper reason not to lay a disciplinary charge and the misfeasance claim was abandoned before the Court of Appeal.

In Weldon's case it was alleged that his segregation in the strip cell had been done maliciously (i.e. in bad faith) and with knowledge of the absence of authority for it by prison staff. But misfeasance had not been pleaded, only a claim for assault and false imprisonment.

It is now clear that misfeasance in public office may be committed in one of two ways. Proof of the tort requires either:

1. that the officer or authority knew that it did not possess the power to take the action in question; or
2. that the officer or authority was actuated by malice, for example by personal spite or a desire to injure for improper reasons.

Malice and the conscious abuse of power are alternative not cumulative requirements for the existence of the tort. In an important judgment in *Bourgoin SA* v. *Ministry of Agriculture*, (which was upheld on appeal),[77] Mann, J., said:

I do not read any of the decisions to which I have been referred as precluding the commission of the tort of misfeasance in public office where the officer actually knew that he had no power to do that which he did and that his act would injure the plaintiff as subsequently it does. I read the judgment in *Dunlop* v. *Woollahra* in the sense that malice and knowledge are alternatives . . . . There is no sensible distinction between the case where an officer performs an act which he has no power to perform with the object of injuring A (which the defendant accepts is actionable at the instance of A) and the case where an officer performs an act which he knows he has no power to perform with the object of conferring a benefit on B but which has the foreseeable and actual consequence of injury to A (which the defendant denies is actionable at the instance of A). In my judgment each case is actionable at the instance of A.

This analysis was further endorsed by the Court of Appeal in *Jones* v. *Swansea City Council*, one of the very few examples of a claim for

---

[77] [1986] 1 QB 716, at 740 D–G.

misfeasance being litigated to a conclusion. Lord Justice Slade described the essence of the tort as being that:

someone holding public office has misconducted himself by purporting to exercise powers which were conferred on him not for his personal advantage but for the benefit of the public or a section of the public either with intent to injure another or in knowledge that he was acting ultra vires.[78]

The elements of the tort may thus be summarized as follows:

1. the holder of a public office[79]
2. causes damage
3. to a foreseeable plaintiff
4. maliciously (i.e. with intent to injure) OR with knowledge that he had no lawful power for his actions.

For prisoners there will never be any difficulty in establishing that those who exercise immediate power over their lives are holding public office. All persons employed by the Prison Service, whether full time or part time, are discharging the powers of a public office while working in a prison and exercising authority over prisoners. The need to prove 'damage' arises because misfeasance is an action on the case.[80] It does not mean that personal injury or financial loss must be established. Damages can be awarded to vindicate the plaintiff's rights and substantial damages may be awarded for an injury to a person's dignity or for discomfort or inconvenience. A prisoner who is placed in adverse, punitive conditions of confinement maliciously or with knowledge that there is no authority for it under the Prison Rules would be able to recover damages for his experience. There is, as yet, no authority which analyses what limitation applies to the field of 'foreseeable plaintiffs' in actions for misfeasance. Plainly, 'malicious' misfeasance will be limited to those persons whom the holder of the public office intended to injure. Where however the allegation is one of consciously acting without power, the field of potential plaintiffs may be enormous. The judgment of Mann, J., in *Bourgoin* suggests that 'reasonable foreseeability' of harm/injury is the limiting principle.[81]

---

[78] [1990] 1 WLR 54. See also *Micosta SA* v. *Shetlands Islands Council* [1984] 2 Llo. Rep. 525.

[79] Prison officers are appointed by the Home Secretary pursuant to s. 7 (1) of the Prison Act 1952. They have the status of Crown servants but are not engaged under a contract of employment—*R.* v. *Home Secretary ex parte Benwell* [1984] 3 WLR 843. By s. 8 of the 1952 Act, prison officers while acting as such have 'all the powers, authority, protection and privileges of a constable'.   [80] *Beaurain* v. *Scott* (1811) 3 Camp. 387.

[81] It is surely sensible that the nature of the injury/damage in respect of which damages may be awarded will depend on the nature of the power which has been abused. Where the power is the power to segregate prisoners (i.e. remove from normal association and access to privileges) or to place them in a special cell then damages will be designed to compensate for the deprivation and distress occasioned by the impoverished regime of segregation under Rule 43 or Rule 45.

The most difficult element of the tort to prove will, of course, be the mental element of the alleged tortfeasor. Malice rarely exists on paper. It will almost always have to be inferred from other evidence. Precisely because it is so serious, the courts require cogent evidence before making a finding that someone exercising a public office has deliberately inflicted injury on another. Furthermore, because the existing legislative framework vests prison officers and governors with such broad discretionary powers (for example 'to maintain good order and discipline') it will frequently be difficult to establish that an officer/governor has done something he knows is outwith his powers. Examples of such behaviour would be: a prison officer who deliberately places a prisoner in a special cell instead of an ordinary segregation cell because the ordinary cells are full up and a governor who leaves a prisoner in segregation for more than three days without obtaining authorization from the Board of Visitors or the Secretary of State.

The effectiveness of misfeasance in public office as a remedy to prisoners in respect of consciously unlawful or malicious actions by prison staff has been seriously undermined by the decision of the Court of Appeal in *Racz v. The Home Office*.[82] The facts of the case were very similar to those in *Weldon*. Mr Racz alleged that while he was a prisoner at HM Prison Armley in Leeds he was assaulted by hospital officers in the hospital wing and then dumped without authority inside a strip cell for two days, dressed only in canvas clothing. He claimed that he had to sleep on the floor with only a single blanket for warmth and that officers deliberately tipped his food on the floor and ordered him to clear it up. He sued the Home Office in assault, negligence, and misfeasance in public office. It was conceded that in respect of the two days spent in the strip cell and the associated mistreatment he could not succeed in a negligence claim since he could not prove the necessary physical or mental injury resulting from the experience.[83] Accordingly, his only remedy was in misfeasance in public office. The Home Office sought to strike out the misfeasance claim, arguing that it could not be held to be vicariously liable for a deliberate or malicious abuse of power by prison officers.

Basing itself on some remarks of Lord Bridge in *Hague/Weldon*, the Court of Appeal accepted the Home Office's argument and struck out the misfeasance claim. A prisoner seeking to pursue a misfeasance claim must therefore sue the individual prison officer(s) involved (always assuming that it is possible to identify them). In reaching their conclusion on this issue, none of the judges in the Court of Appeal attempted to reconcile their decision with the general principles of vicarious liability. Salmond's classic definition holds that an employer's liability for his employee's acts extends

---

[82] The Times 17 Dec. 1992.   [83] See n. 60 above.

'even for acts which he has not authorised, provided they are so connected with acts which he has authorised that they might rightly be regarded as modes—although improper modes—of doing them'.[84] Even criminal behaviour on the part of an employee does not in itself take his conduct outside the course of his employment.[85]

Instead the Court of Appeal in *Racz* held that the following remarks by Lord Bridge in *Hague/Weldon* were determinative of the issue:-

a prison officer who acts in bad faith by deliberately subjecting a prisoner to a restraint which he knows that he has no authority to impose may render himself personally liable to an action for false imprisonment as well as committing the tort of misfeasance in public office. Lacking the authority of the governor, he also lacks the protection of section 12 (1). But if the officer deliberately acts outside the scope of his authority, he cannot render the governor or the Home Office vicariously liable for his tortious conduct.[86]

None of the other Law Lords expressly referred to the issue of the vicarious liability of the Home Office for misfeasance in public office, which was hardly surprising since the matter had not been canvassed in argument during the course of the appeal. Nevertheless all three judges in the Court of Appeal in *Racz* found, for slightly different reasons, that Lord Bridge's comments expressed the considered view of the House on the matter and that accordingly no liability could attach to the Home Office. Lord Justice Neill and Lord Justice Beldam sought to confine their remarks to the precise facts of *Racz* but it is hard to see how the issue of vicarious liability for misfeasance can differ in the prison context from that obtaining in other situations where a citizen claims he is the victim of an abuse of power by a public servant. It is striking that at no stage do the judgments in *Racz* explain why vicarious liability for misfeasance should differ from that applicable to other torts. Lord Justice Kennedy came closest when he said that misfeasance is 'an unusual tort' but he did not explain why this was so or why a different approach was needed from that applicable to the tort of malicious prosecution. Lord Justice Beldam's reasoning was particularly obscure. He pointed out that the House of Lords had held in *Weldon* that there could be no claim for breach of statutory duty at the suit of an aggrieved prisoner who complained that the Rules had not been observed. In the light of this, Beldam, LJ, said that it would be wrong to allow a claim for misfeasance in public office since 'this seeks in a different guise to render the Home Office liable to a personal action for breach of the prison rules'. If this reasoning is taken to its logical conclusion, no private law

---

[84] Salmond, *The Law of Torts* (15th edn.). See also *General Engineering* v. *Kingston Corporation* [1989] 1 WLR 64, at 72 A–G.

[85] See e.g. *Lloyd* v. *Grace Smith & Co* [1912] AC 716 and *Armagas Ltd* v. *Mundogas SA* [1986] 1 AC 717.

[86] [1992] 1 AC 58, at 164 D.

claims of any kind should be permitted by prisoners. Actions for negligence and assault will almost always involve an allegation that the Prison Rules have been breached in some way.[87] Should they too be abolished?

Of course the effect of excluding the possibility of a misfeasance claim against the Home Office impacts on the level of exemplary damages recoverable since a relevant factor in fixing such an award is the means of the defendant.[88] An individual prison officer is obviously in an entirely different financial position from a major government department and so a prisoner who succeeds in a misfeasance claim will be bound to receive less in exemplary damages then he would otherwise expect. A secondary, but no less significant, concern is that prison officers are rarely going to be in a position to satisfy any substantial judgment against them, thus leaving the prisoner uncompensated for his treatment. Legal Aid Boards may also be reluctant to extend legal aid to sue a defendant who is likely to be a 'man of straw'. *Racz* is consistent with an unfortunate drift towards restricting the private law remedies available to prisoners in respect of unlawful treatment meted out to them in prison.

Racz was also unfortunate in that the Court of Appeal rejected the application for trial by jury. It had been argued that the discretion to order such mode of trial contained in s. 69 (2) Supreme Court Act 1981 should be exercised in Mr Racz's favour for three reasons. First, his claim involved an allegation of arbitrary and oppressive behaviour by government servants, a matter traditionally appropriate for arbitration by judge and jury. Secondly, the possibility of an award of exemplary damages placed the claim in a category which the courts recognized as being influential in the exercise of discretion under s. 69 (2). Finally, it was argued that a misfeasance claim was so similar to an action for malicious prosecution, where jury trial was a right, that it would be perverse to fail to exercise the court's discretion against jury trial. In effect, if the discretion was not exercised in favour of jury trial in a misfeasance claim it was was hard to envisage any circumstances in which it would be appropriate to do so.

On this issue too the Court of Appeal rejected the plaintiff's arguments. In a particularly unfortunate passage, Lord Justice Beldam relied on the briefly reported decision of Leggatt, J., in *Barbara* v. *The Home Office*[89] in support of the principle that an award of exemplary damages might generally be unnecessary in that the plaintiff's injury could be properly compensated by an award of aggravated damages without the need to punish an individual tortfeasor. In fact *Barbara* was an action for assault and

---

[87] A negligence action arising from an allegedly unlawful confinement in a special cell which has caused personal injury will invariably involve an allegation of a breach of Rule 45. An assault claim will always involve an allegation of excessive force in breach of Rule 44.

[88] See e.g. *Makanjuola* v. *Metropolitan Police Commissioner* [1990] ALR 214.

[89] See n. 72.

negligence. Malice was not alleged and accordingly it is hard to see how this decision could be relevant to a misfeasance claim which inevitably involves far more serious allegations of misconduct by public servants. As with malicious prosecution claims against police officers, it is almost inconceivable that such actions should not involve an award of exemplary damages to mark the seriousness of the misconduct.

Breach of the Bill of Rights

The Bill of Rights 1688 is a statute which prohibits the infliction of 'cruell and unusuall punishment'. As such it lays down a matter of fundamental importance protecting the rights and freedoms of the individual, but the statute does not specify a penalty for breach. Does a prisoner who can show that he is the victim of a 'cruell and unusuall punishment' have a private law cause of action for damages for breach of statutory duty or does his remedy lie exclusively in the public law realm? Existing authority suggests that a private law claim is possible, though the point has never been authoritatively decided.

In *Williams* v. *The Home Office*[90] the plaintiff sued for damages for false imprisonment arising from his detention for 180 days in the notorious control unit at Wakefield Prison in 1974. A claim for breach of the Bill of Rights was not specifically pleaded (indeed counsel for the plaintiff expressly disavowed the existence of such a cause of action) but it was asserted that, because the control unit regime amounted to a breach of the 1688 statute, the Home Office could not seek to justify the plaintiff's detention by reference to Rule 43 of the Prison Rules or the Prison Act generally. It was submitted that the words 'cruell and unusuall' must be read disjunctively with the result that the Act prohibits the infliction of cruel punishment and, separately, the infliction of unusual punishments. Any other reading would mean that it would be lawful to inflict a cruel punishment in present day England provided such a punishment was inflicted commonly, an absurd construction. Tudor Evans, J., dismissed the claim for false imprisonment (on grounds partly upheld by the House of Lords in *Hague/Weldon*) and in the process held that the Bill of Rights was clear in

---

[90] [1981] 1 All ER 1211. The control unit regime was created against a background of widespread disruption in the prison system in 1972, the most prolific incident being a serious disturbance at Gartree Prison. A Home Office working party was set up to consider the question of control in dispersal prisons. It felt that existing segregation units were frequently inadequate to deal with certain 'hardcore, disruptive' prisoners because they were not sufficiently insulated or isolated from the main part of the prison. The working party therefore recommended that special control units be set up (in Wakefield and Wormwood Scrubs) which were physically separate and insulated from the main prison. Prisoners allocated to them should not come from the prison in which the unit was situated. The nature of the regime was set out in CI 35/1974 and was approved by the Secretary of State. After widespread concern about the coercive, punitive nature of the regime and its possible side-effects the units were closed down in 1975.

its meaning and showed an intention to prohibit punishments which are both cruel *and* unusual. On the facts he found the Wakefield regime to be neither cruel nor unusual.[91]

In *R. v. Home Secretary ex parte Herbage*[92] the applicant sought interlocutory relief in respect of his conditions of detention in Pentonville Prison which he claimed amounted to the infliction of 'cruell and unusuall punishment'. He alleged that by reason of his gross obesity and resultant inability to walk more than fifty yards without resting he had been allocated a cell on the ground floor mental observation landing of the hospital wing close to severely mentally disturbed inmates. He had been held in solitary confinement there for seven months with no opportunity for exercise or association with others and the noise prevented him from sleeping. In support of his application for an injunction he claimed that he would have a private law cause of action based on a breach of the 1688 Bill. The Home Secretary submitted that the Bill of Rights did not create actionable claims either in public or private law. Herbage succeeded in obtaining leave to apply for judicial review and when the Home Office appealed to the Court of Appeal against the grant of leave, there was some discussion of whether a claim based on breach of the Bill of Rights was justiciable. Lord Justice Purchas described the prohibition on inflicting 'cruell and unusuall' punishment as :

a fundamental right which, in my judgment, goes far beyond the ambit of the Prison Rules. For my part if it were established that a prison governor was guilty of such conduct it would be an affront to common sense that the court should not be able to afford relief under Order 53 . . . Once the central issue is divested of the encumbrance of considerations of breaches of Prison Rules or duties under the Prison Act 1952 and is viewed as a case involving a breach of the Bill of Rights alone then the matter, in my judgment, becomes easier to comprehend. Two questions arise. The first is: what are the conditions in fact in which the applicant is presently detained at Pentonville? The second is: do these amount to 'cruel and unusual punishment'? By way of example, and not wishing to indicate any view of the actual conditions existing, it is generally held to be unacceptable that persons supposedly of normal mentality, should be detained in psychiatric conditions as is said to occur in certain parts of the world. Coming closer to the alleged facts of this case, if it were to be established that the applicant as a sane person was, for purely administrative purposes, being subjected in the psychiatric wing to the stress of

[91] This was a somewhat surprising conclusion. The control unit regime was divided into two stages, each intended to last ninety days. At stage 1 the prisoner did not associate with other prisoners save when he was allowed his one hour's exercise per day. He was not obliged to work but if he did not he would not progress to stage 2, when he would be allowed a degree of association with other prisoners during work, leisure, and education. If he failed to work during stage 2 or caused trouble, the prisoner reverted to stage 1 and was required to complete a continuous ninety-day period of good behaviour before qualifying once more for stage 2. If the prisoner completed a full 180-day period he would be allowed to return to normal association. Tudor Evans, J., said that 'judged by the standards of the English prison system, I do not think that the regime in the unit was cruel'!
[92] [1987] QB 872. The decision of the Court of Appeal is at [1987] QB 1077.

being exposed to the disturbance caused by the behaviour of mentally ill and disturbed prisoners, this might well be considered as a 'cruel and unusual punishment' and one which was not deserved. This raises issues quite different from compliance or non-compliance with the Prison Rules although they may well involve breaches on the part of the Secretary of State of the Prison Act 1952.

The above passage is plainly not authority for the existence of a private law claim for breach of statutory duty since it was made in the context of an application for judicial review where no claim for damages was included in the relief sought. However, it is supportive of a submission that so fundamental a provision ought on general principle to give rise to a private law sanction where it is breached in the absence of any statutory remedy or penalty to secure its enforcement. The matter is not purely academic either. While an applicant can include a claim for damages in the relief sought on an application for judicial review, this is only when it can be established that the applicant could have recovered damages had he proceeded by writ (or summons),[93] in other words that he had an independent private law cause of action in respect of the matter sought to be challenged by judicial review. If a breach of the Bill of Rights is not justiciable as a private law action then, even though declaratory relief might be available under Order 53, no compensation would be recoverable. In practice it is hard to see how the infliction of a 'cruel and unusual punishment' could fail to give rise to a claim for misfeasance in public office but the point remains moot.

JUDICIAL REVIEW

Jurisdiction to intervene in prison life

Since the late 1970s there has been a steady but persistent change in favour of the view that the normal function of the courts is not ousted by the fact of imprisonment. This reversal of fortune from a prisoner's perspective can be traced through a number of landmark decisions which have finally established that the supervisory jurisdiction of the High Court extends to all decisions taken in pursuance of the Prison Rules.

In *R. v. Board of Visitors of Hull Prison ex parte St Germain*[94] the Court of Appeal held that the High Court had jurisdiction to review the disciplinary adjudications of Boards of Visitors (although the majority of the court

---

[93] Supreme Court Act 1981, s. 31 (4), and RSC Order 54, Rules 7 (1), 9 (5). The effect of these provisions is that although damages (in contrast to an injunction or declaration) is not a specific form of relief by way of judicial review, where a decision gives rise to both a claim for judicial review and a claim for damages the two may be joined. If the application for judicial review fails, the claim for damages may be continued as a writ action. Where however the matter involves no public law element, it will be struck out if commenced by writ—*Guevara* v. *Hounslow London Borough Council* The Times 17 Apr. 1987. Cf. *An Board Bainne Co-Operative Ltd. (Irish Dairy Board)* v. *Milk Marketing Board* [1984] 2 CMLR 584.

[94] [1979] QB 425.

held that a governor's disciplinary powers were not susceptible to review).
In *R. v. Secretary of State for the Home Department ex parte Anderson*[95] the
Divisional Court held that the Prison Department's 'simultaneous ventilation' rule, whereby prisoners who wished to seek legal advice about any
aspect of their treatment in prison were required as a condition of receiving
such advice to lodge an internal complaint, was unlawful in that it placed
an unwarranted fetter on a prisoner's right of access to legal advice which
was an inseparable aspect of the right to unimpeded access to the courts. In
R. v. Home Secretary ex parte McAvoy[96] the Divisional Court held that a
decision to transfer a prisoner from one prison to another pursuant to s. 12
of the Prison Act was susceptible to review. In *R. v. Deputy Governor of
Camphill Prison ex parte King*,[97] the 'forward march' of jurisdiction was
temporarily halted when the Court of Appeal ruled that disciplinary adjudications conducted by governors were immune from judicial review.
However in *R. v. Deputy Governor of Parkhurst Prison ex parte Leech*[98] the
House of Lords overruled *King*, holding that jurisdiction did extend to governors' disciplinary functions as well as those exercised by Boards of
Visitors. Finally, the Court of Appeal in *R. v. Deputy Governor of
Parkhurst Prison ex parte Hague*[99] held that operational or managerial decisions affecting the transfer and segregation of prisoners were amenable to
judicial review, thereby opening up all decisions affecting prisoners to challenge.

The new-found willingness to extend the jurisdiction of the High Court
in order potentially to grant relief against the abuse of power in prison was
not, however, taken lying down. At each stage the Home Office sought to
hold back the incoming tide, arguing that the intervention of the courts was
inconsistent with running an efficient, secure prison system. It was entitled
to believe that this argument had some validity. Until Lord Justice Shaw
declared in *St Germain* that 'the rights of a citizen, however circumscribed
by a penal sentence or otherwise, must always be the concern of the Courts
unless their jurisdiction is clearly excluded by some statutory provision',
English judges found it impossible to accept that prisoners could assert any
legitimate claim to be protected from arbitrary and unlawful action once
the prison gate had closed behind them. Prisoners, so the reasoning went,
had only 'privileges' which could be denied them without the need for justification rather than rights or legitimate expectations which entitled them to
legal remedies. And in the judgments of Lord Goddard and, more recently,
Lord Denning and Lord Justice Lawton, the harassed prison administrator

---

[95] [1984] 2 WLR 725.   [96] [1984] 1 WLR 1408.
[97] [1985] QB 735. In *R. v. Governor of the Maze Prison ex parte McKiernan* (1985) 6 NIJB 6
the Northern Ireland Court of Appeal refused to follow *King* and held that governors' adjudications were susceptible to judicial review.
[98] [1988] 1 AC 533.
[99] [1992] 1 AC 58. See also *R. v. Home Secretary ex parte Leech*, The Times 20 May 1993.

found full support for the 'catastrophe' scenario—that security, good order, or discipline would irretrievably break down if governors and prison officers had to perform their duties with a judge looking over their shoulders.[100]

This consequentialist argument against extending the judicial review jurisdiction into prison life was maintained by the Home Office throughout the 1980s in the face of overwhelming evidence that the existence of legal recourse not only did not injure the good and efficient conduct of prison life but enhanced it. In *Leech* the Home Office sought to resist the extension of jurisdiction to include a governor's disciplinary powers by arguing that this would eventually mean that administrative or managerial decisions of the governor would be potential candidates for judicial review. It was submitted that this would entail the court directly interfering in the administration of the prison regime, a consequence so undesirable that the House of Lords must reject it.[101] Experience has shown that this fear was entirely without foundation. Each advance of the High Court's supervisory jurisdiction did not result in the Crown Office list groaning under the strain of unmeritorious applications from 'disgruntled' prisoners with nothing better to do than spend their every waking hour drafting a stream of Form 86As.

The Home Office (or perhaps more precisely the Prison Service) finally accepted this in the course of the Hague litigation. During argument before the Court of Appeal in March 1990 it abandoned its previous claim that the court lacked jurisdiction to review a Rule 43 decision and, six months later in September 1990, issued CI 37/1990. This provided for the future all the procedural safeguards for which Mr Hague had been arguing in his application notwithstanding the fact that he had failed to establish his entitlement to them as a matter of law before the Court of Appeal. Lord Bridge commented in his speech in *Hague* in the House of Lords that the course the litigation had taken had confirmed that the availability of judicial review as a means of questioning the legality of action taken pursuant to the Prison Rules was a 'beneficial and necessary jurisdiction'.

---

[100] Lord Goddard's judgment in *Arbon* v. *Anderson* [1943] KB 252, Lord Denning's judgment in *Becker* v. *Home Office* [1972] 2 QB 407, and Lord Justice Lawton's judgment in *Ex parte King* make up a triumvirate of highly conservative judges who despaired at the idea of prisoners being allowed to litigate their grievances in the courts. Lord Justice Lawton's father had himself been a prison governor!

[101] In his speech in *Leech* Lord Bridge, in rejecting the 'floodgates' arguments of Treasury counsel, said that 'Mr Laws held out the prospect as one which should make our judicial blood run cold, that opening the door to judicial review of governor's awards would make it impossible to resist an invasion by what he called the "tentacles of the law" of many other department of prison administration.' In argument before their Lordships in *Ex parte Hague*, leading counsel for Mr Hague commented that the extension of jurisdiction to include decisions taken under Rule 43 showed that 'the octopus had breached the floodgates'.

The public/private law distinction

In his seminal speech in the *GCHQ* case,[102] Lord Diplock commented that:

for a decision to be susceptible to judicial review the decision maker must be empowered by public law (and not merely, as in arbitration, by agreement between private parties) to make decisions that, if validly made, will lead to administrative action or abstention from action by an authority endowed by law with executive powers.

The necessity for there to be a 'public law' element to a decision before it can attract the judicial review jurisdiction does not however give rise to much (if any) difficulty in the prison context. All decisions taken in pursuance of the Prison Act or Rules and which affect the rights, privileges, interests, or legitimate expectations of prisoners are therefore prima facie amenable to judicial review. Accordingly, the legal adviser concerned to ensure that procedural correctness is preserved need not be concerned with the more esoteric arguments which have arisen in other areas where the public/private law divide is altogether less clear.[103] Disciplinary decisions, segregation, transfer, security categorization, parole, the regulation of access to the outside world whether by visits or correspondence, indeed almost every conceivable aspect of a prisoner's life in prison, all lie within the public law net.

More difficult is the question of when it will be an abuse of process for a prisoner to proceed by way of a private law action, thereby evading the procedural rules which govern applications for judicial review. In a trilogy of cases in 1983–4, the House of Lords examined the implications of s. 31 of the Supreme Court Act 1981 and the amended RSC Order 53 which together define the scope of the High Court's supervisory jurisdiction.[104] One of these, *O'Reilly* v. *Mackman*, was a prison case and arose from the serious disturbance at Hull Prison in September 1976. After it ended a number of prisoners were charged with disciplinary offences under the Prison Rules and appeared before the Board of Visitors in December 1976. All those convicted were awarded severe punishments of cellular confinement and loss of remission. In 1980 three prisoners issued statements of claim in the Queen's Bench Division of the High Court and a fourth issued an originating summons in the Chancery Division joining as defendants the members of the Board of Visitors who had adjudicated upon them four years before. In each case the relief sought was confined to a declaration that the

---

[102] *Council of Civil Service Unions* v. *Minister for the Civil Service* [1985] AC 374.

[103] See e.g. *R.* v. *Panel on Take-Overs and Mergers ex parte Datafin PLC* [1987] QB 815, *Roy* v. *Kensington and Chelsea and Westminster Family Practitioner Committee* [1992] 2 WLR 239, and *R.* v. *Football Association Ltd ex parte Football League Ltd.* [1992] COD 52.

[104] *O'Reilly* v. *Mackman* [1983] 2 AC 237; *Cocks* v. *Thanet District Council* [1983] 2 AC 286; *Davy* v. *Spelthorne Borough Council* [1984] AC 262.

proceedings were void and of no effect on the grounds that the Board had acted in breach of the Prison Rules and the rules of natural justice. The Board members applied to strike out the proceedings on the grounds that they were an abuse of the process of the court. It was never contended on the prisoners' behalf that they could not have proceeded under RSC Order 53 with applications for judicial review. Rather, it was asserted that they were entitled to choose which form of proceeding to initiate since nothing in either s. 31 of the Supreme Court Act or the amended version of Order 53 indicated that judicial review was an exclusive remedy in cases such as theirs.

The House of Lords upheld the Board's argument that, in proceeding by writ and originating summons rather than pursuing applications for judicial review, the prisoners had been guilty of an abuse of process, and all four proceedings were struck out. Lord Diplock's important speech is of general application in the field of administrative law but it also involved a particular analysis of the nature of the remedies available to each prisoner arising from their treatment at the hands of the Hull Board. He pointed out that, under the Prison Rules, remission of a prisoner's sentence was a matter not of right but of indulgence. All a prisoner could assert was a legitimate expectation that he would be granted remission of one-third of his sentence in accordance with Rule 5 (2) so long as no disciplinary award of forfeiture of remission was made against him. In the circumstances none of the prisoners had any independent remedy in private law. In public law, however, their legitimate expectation gave each of them a sufficient interest to challenge the legality of the Board's exercise of its statutory powers on the grounds that it had acted in breach of the rules of natural justice and the Prison Rules.

Having described the procedural reforms to Order 53 brought about by the 1977 amendments (which permitted applications for a declaration or injunction as well as a claim for damages to be included in an application for judicial review), Lord Diplock concluded that:

it would in my view as a general rule be contrary to public policy, and as such an abuse of the process of the court, to permit a person seeking to establish that a decision of a public authority infringed rights to which he was entitled to protection under public law to proceed by way of an ordinary action and by this means to evade the provisions of Order 53 for the protection of such authorities.[105]

Lord Diplock went on, however, to make the point that this 'general rule' was subject to exceptions, and he gave as one example a case: 'where the invalidity of the decision arises as a collateral issue in a claim for infringement of a right of the plaintiff arising under private law.'[106]

---

[105] [1983] 2 AC at 285 D.
[106] Ibid., at 285 F.

Where does this leave a prisoner who wishes to pursue a private law action for assault, negligence, or misfeasance in the course of which it may be necessary to analyse the legality of a decision purportedly taken in pursuance of the Prison Rules, a 'public law' decision? Take, for example, the case of a prisoner who is segregated pursuant to an initially valid order by a governor under Rule 43 (1). After three days no authority is obtained from either the Board of Visitors or the Secretary of State for his further segregation in breach of Rule 43 (2). On the fourth day the prisoner refuses to return to his segregation cell after an exercise period and force is used by officers to return him to the cell. Is it an abuse of process for him to commence an ordinary action in the County Court for assault rather than proceed first by way of judicial review to quash the decision to continue his segregation?

The answer must surely be 'no'. It is not a condition precedent to the establishment of the prisoner's private law cause of action for the tort of assault that an appropriate public law decision be set aside. It is therefore quite different from the situation which arose in *Cocks* v. *Thanet District Council*,[107] where the House of Lords struck out an action commenced in the County Court for declaratory relief and damages arising from an alleged breach by a local authority of its statutory duty to secure permanent accommodation for the plaintiff and his family imposed by the Housing (Homeless Persons) Act 1977. Lord Bridge explained that it was inherent in the scheme of the 1977 Act that an appropriate public law decision be taken by the local authority as a condition precedent to the establishment of any private law duty to house a homeless family which might then be enforced by an injunction or give rise to a liability in damages. For example, the authority would have to decide if the family was in priority need or intentionally homeless and only if the answers were favourable would a duty to house arise. Precisely because tortious liability depends on a successful application for judicial review, it is an abuse to proceed immediately with a private law claim.

In the example we are concerned with, however, the invalidity of the prisoner's continued segregation arises as a collateral issue. The Particulars of Claim would plead the use of unlawful force by prison staff to enforce his segregation. By way of Defence the Home Office would plead (if it could) that all force used was lawful either to enforce a valid Rule 43 order or, more generally, to maintain good order or discipline in the face of defiance of a lawful order from prison staff. In Reply, the plaintiff would plead that there was no lawful justification for his continued segregation because no valid authority had been obtained from the Board of Visitors or Secretary of State. This latter issue would necessarily have to be decided in

---

[107] See n. 104.

order to determine liability. But the plaintiff's cause of action does not depend on such determination for its very existence. The essence of the claim is for breach of a private law right, that is, a claim for assault.[108]

Exhausting alternative remedies

In both *King* and, later, in *Leech* the Home Office had argued that the High Court had no jurisdiction to grant judicial review of a governor's disciplinary powers primarily because of the existence of alternative remedies within the statutory framework which regulates the prison system. Specifically, s. 4 (2) of the Prison Act 1952 had expressly imposed on the Home Secretary the duty to ensure compliance by his officers (including the governor) with the Act and the rules made thereunder. Accordingly, it was argued, there was no need for the High Court to intervene at any point short of a failure by the Secretary of State to perform that duty. And it was that failure which alone was susceptible to judicial review.

This argument was (as we have described) emphatically rejected by the House of Lords with Lord Bridge stating that:

The existence of an alternative remedy has never been sufficient to oust jurisdiction in judicial review and, in the last analysis, the reliance on section 4 (2) of the Act of 1952, however ingeniously the argument is presented, amounts to no more than saying that the Secretary of State's statutory duty to ensure compliance with the law by the governor affords an alternative remedy for any public law wrong which a prisoner charged with an offence against discipline may suffer at the governor's hands.[109]

In *Hague* the Home Office renewed its objection to an extension of the High Court's supervisory jurisdiction to cover 'managerial' decisions such as segregation under Rule 43. Before the Divisional Court it was argued once again that s. 4 (2) and the possibility of a prisoner petitioning the Home Secretary or complaining to the Board of Visitors meant that, for policy reasons, as well as on general principles concerning the availability of judicial review as a remedy, the court should not entertain applications for judicial review of transfer and segregation decisions. In effect it was sought to distinguish the principles set out in *Leech* on the grounds that they were confined to disciplinary proceedings where there was a duty to observe the rules of natural justice. This argument was again rejected and by the time the case reached the Court of Appeal a new approach was suggested by the Home Office. It was conceded that jurisdiction existed but that the court should exercise the utmost circumspection in exercising it.

---

[108] See *Lonrho PLC* v. *Tebbit and the Dept of Trade and Industry* [1991] 4 All ER 973, in which the Vice-Chancellor refused to strike out a private law action in respect of governmental delay in releasing Lonrho from its undertaking, restricting its ability to bid for shares in the House of Fraser which owned Harrods.

[109] [1988] 1 AC 533, at 562 D.

The court should distance itself from operational/managerial decisions by only entertaining applications made *after* a prisoner had unsuccessfully complained to the Board of Visitors or had petitioned the Home Secretary in respect of the decision under challenge. In other words it was sought to establish a principle that unless a prisoner had exhausted such internal 'remedies' as exist under the Act and the Rules the court would be bound to exercise its discretion to refuse relief. The availability of judicial review would be policed not in terms of basic jurisdiction but rather by the court's power to withhold what is in any event always a discretionary remedy.

Fortunately this 'second stage' approach was also rejected by the Court of Appeal, though Lord Justice Taylor was willing to agree that when reviewing managerial decisions in the prison context the court should approach the exercise of discretion with great caution.

In the light of *Leech* and *Hague*, is there any principle of law which demands or suggests that before proceeding to challenge a decision, whether disciplinary or managerial, a prisoner should exhaust such internal remedies as exist such as complaint to the Board of Visitors or the Home Secretary? The answer is that there is no rule which requires such a step to be taken but in many cases it may well be wise to use the internal mechanisms before proceeding by way of judicial review. Whether or not internal recourse should be attempted will depend on the nature of the decision under challenge and the kind of relief or remedy available via the various complaints mechanisms. The reasons are as follows.

Judicial review is a remedy of last resort.[110] The Divisional Court will not generally grant judicial review if an applicant has not exhausted established appeal procedures, save in exceptional circumstances. But the right to petition or complain to the Home Secretary or a Board of Visitors is not an established appeals procedure nor is it properly to be described as a 'remedy', although it may in some cases afford appropriate redress. In *Leech* Lord Bridge was scathing about the idea that the right to petition the Home Secretary in respect of a disciplinary adjudication could possibly be regarded as an adequate remedy, pointing out that:-

save perhaps in a case taken up by a Member of Parliament, the matter will come before a civil servant in the Home Office who will consider on the one hand the prisoner's petition, on the other hand the relevant records and reports supplied by the governor. If those disclose an issue of fact, I hope that it is not unduly cynical to suppose that in the majority of cases the civil servant is likely simply to believe the governor's account. But even if he wishes to resolve any issue of fact in a

---

[110] See *Re Preston* [1985] AC 835, R. v. *Chief Constable of Merseyside ex parte Calveley* [1986] QB 424, and R. v. *Secretary of State for the Home Dept. ex parte Swati* [1986] 1 WLR 477. The cases (of which there are many) on this subject reveal an inconsistent approach. Though there are dicta in the House of Lords to the effect that alternative remedies must always be exhausted, in practice the courts are usually more generous in granting judicial review than these suggest.

judicial way, he probably lacks the experience and certainly lacks the procedural machinery, including the power to require evidence on oath, enabling him to do so . . . If a prisoner has a genuine grievance arising from disciplinary proceedings unfairly conducted, his right to petition a faceless authority in Whitehall for a remedy will not be of much comfort to him.[111]

These are strong words, but it is important to recall that they were made in the context of a dismissal of the Home Office's argument that there was no jurisdiction to review a governor's disciplinary decision. Earlier Lord Bridge had whole-heartedly endorsed the judgment of Lord Justice Shaw in *St Germain* which included the following passage on the relevance of internal remedies to the availability of judicial review:

The opportunity for a prisoner to seek from the Secretary of State redress for a grievance (rule 7 and rule 56) does not amount to a right of appeal for review of an unwarranted decision by a board of visitors or a prison governor. The fact that such means of possible redress has not been pursued before application is made to the court may in some cases be regarded as a discretionary obstacle to the grant of relief by the courts; but it cannot be an absolute bar.[112]

The effect of these statements would seem to be that where it can be shown that a prisoner's complaint could be fully and properly dealt with by some internal avenue of complaint, both in terms of the machinery available to investigate it and the power to reverse, correct, vary, or quash the decision under challenge, then there is a real danger that a failure to pursue such 'alternative remedy' may lead to relief being refused by the Court in the exercise of its discretion.[113] It is safer not to go beyond such a broad statement of principle save to say that Lord Bridge's speech in *Leech* makes clear that where a complaint involves disputed issues of fact, then the existing internal remedies available under the Act and the Rules will never be regarded as adequate or truly alternative to the court's function.

DISCOVERY: DOCUMENTS TO WATCH FOR

In a private law action for damages, discovery and inspection of documents is automatic.[114] In an application for judicial review, discovery is not automatic, but otherwise almost identical principles apply. In the words of Lord Diplock in *O'Reilly* v. *Mackman*, 'discovery is obtainable on application whenever, and to the extent that, the justice of the case requires'. Whether

---

[111] [1988] 1 AC 533, at 568 B.   [112] Ibid, at 555 E.

[113] As a result of the 1990 amendment to the Prison Rules, Rule 56 now gives the Secretary of State power to 'quash any finding of guilt and [he] may remit any punishment or mitigate it either by reducing it or by substituting another award which is in his opinion less severe'. Under the previous rule he had no power to quash, only to remit or mitigate it.

[114] See RSC Order 24.

in a private law claim or an application for judicial review, the court will only order discovery and inspection of documents where it is satisfied that the documents sought relate to any matter in question in the cause or matter and that the order is 'necessary either for disposing fairly of the cause or matter or for saving costs'.[115]

In prison cases more than most, discovery and inspection is frequently a vital stage in the process of establishing a winning claim. This is mainly because until a prisoner obtains access to the documents in the hands of the prison authorities he will have few, if any, contemporaneous records relating to the matter in dispute. In a claim for damages for personal injuries he may have been able to obtain an order for pre-action discovery of his medical records but, in other claims, no other documents will normally be available.[116]

In any civil litigation, it is essential to have a detailed understanding of the kind of documentation which a defendant is likely to have compiled and which will be relevant to the issues in dispute. Whether a plaintiff is considering an application for pre-action discovery or scrutinizing a List of Documents at the stage of discovery and inspection, an awareness of your opponent's record-keeping methods is a distinct disadvantage. It will enable a request to be made for specific documents and thus avoid the allegation that what is being pursued is a generalized fishing expedition in the hope that something may emerge.

This is particularly true in prison litigation. The Prison Service produces a bewildering array of forms, records, and documents, many of them overlapping. Below is a list of those documents which are likely to be most frequently relevant in claims for negligence, assault and misfeasance in public office as well as in applications for judicial review:

1. *Prisoner's personal record (F2050)*: a loose-leaf file kept in a yellow cardboard wallet which is supposed to be a comprehensive record of a prisoner's time in prison. It should include details of his conviction, sentence, and earliest date of release (EDR). A History Sheet should record each transfer as well as any applications/petitions and governors' observations from prison to prison. A Medical Record should record the medical officer's view of the prisoner's state of health and work classification on each reception and a Medical History sheet should record the results of any reception medical examination. The prisoner's disciplinary record will be recorded and, on what is called page 24, 'Information of Special

---

[115] RSC Order 24, Rule 13.

[116] The Access to Health Records Act 1990, which by s. 3 creates a statutory right of access to a health record on the part of the patient, does apply to a health professional in the public service of the Crown (see s. 2 (4)). Accordingly prisoners who wish to pursue a claim where such records are going to be relevant should make an application to the holder of the record, i.e. the Director of the Prison Healthcare Service, whose address is Cleland House, Page Street, London SW1P 4LN.

Importance' should be listed.[117] This includes security information (escapes, attempted escapes, suspicious circumstances connected with visits/correspondence) as well as suicide attempts/concerns or special medical problems. Any periods of time in segregation or good order and discipline transfers should also be logged on page 24.

2. *Continuous Inmate Medical Record*: the IMR should record all entries by prison doctors relevant to a prisoner's medical history in prison. It is a loose-leaf folder inside an orange cardboard wallet. The IMR is different from the Medical Record and Medical History sheets which are kept in the F2050.

3. *Hospital case papers/Kardex*: if a prisoner is located in a prison hospital, a separate set of records will normally be kept. Any treatment/drugs prescribed will be recorded on special Treatment Cards.

4. *Suicide Prevention*: each prisoner should be screened for suicide risk on reception and at each transfer by a hospital officer and a doctor using Form F1996, the Reception Screening Form. If any member of staff is concerned at any time that a prisoner is or may be suicidal, Form F1997 (Referral Form) should be completed to ensure an assessment of risk by the medical officer.[118]

5. *Disciplinary proceedings*: these give rise to a number of different records. They include the original memoranda compiled by staff (usually done on an F1384), the Report to the Governor of Alleged Offence by Inmate (F254), the Notice of Report (F1127), and the Record of Adjudication (F256).

6. *Use of Force Reports:* whenever a member of staff uses force against a prisoner he or she should give a brief factual report explaining why force was necessary.

7. *Report of Injury to an Inmate (F213)*: this is self-explanatory. It should be filled in by the medical officer whenever it is believed that a prisoner has been injured. It includes a diagram of the human body for the sites of any injuries to be recorded.

8. *Rule 43 segregation*: whenever a prisoner is segregated the authority for it and the reasons should be recorded on Form F1299A. Any authority for its continuance obtained from either the Board of Visitors or the Secretary of State should also be logged.

9. *Register of Non-Medical Restraints (F1980)*: this is the form used whenever a prisoner is placed in any form of mechanical restraint or is ordered to be located in a special cell under Rule 45. The reason for it

---

[117] The selection of p. 24 for this information remains a mystery since it bears no relation to the logical sequence of page numbering within the record itself.
[118] For a full understanding of the detailed procedures governing suicide prevention in prisons see CI 20/1989.

must be recorded and approved by the medical officer as well as the time the restraint or location commenced and the time it ended.

10. *Property cards*: these should record the property which is retained by the prisoner in his personal possession and that which is kept by the prison as 'stored property'. Cell clearance sheets should be made up whenever a prisoner is 'ghosted' from one prison to another pursuant to CI 37/1990 for reasons of good order or discipline and he has no opportunity to collect his property from his cell.

11. *Miscellaneous wing records*: every wing should keep some form of diary/record of significant events. This may include staff rosters, though in many cases these are kept separately. The segregation unit and hospital wing will always have a more detailed form of record-keeping than a normal wing. Most prisons have a Gatekeeper's Daily Occurrence book to record movements in and out of the main gate. In some prisons there seems to be a practice of compiling Incident Reports (sometimes designated as Forms LP 180) to record significant incidents such as escape attempts or mass disturbances.

PREPARING EVIDENCE IN PRISON ACTIONS

As the preceding section has shown, the prison authorities begin with a distinct advantage in any kind of civil litigation in that they have access to a wide range of contemporaneous documents compiled in respect of almost every conceivable event within a prison. A prisoner-plaintiff who has been assaulted or been the victim of negligent medical treatment has, by contrast, many disadvantages. He has no legal right of access to an independent doctor unless and until litigation has been commenced. Given the delays in obtaining legal aid, this may be many months or even years after the incident which has given rise to the complaint. By then a medical examination may be useless. Obtaining statements from witnesses may cause special difficulties. Fellow prisoners may be transferred to a different prison or released. Alternatively, the prisoner-plaintiff may be placed in solitary confinement after the incident which has given rise to the claim and thus find it impossible to contact potential witnesses.

In these circumstances it is essential that a prisoner who believes that he is the victim of any form of injustice should take as many steps as possible to strengthen any subsequent legal action by following a number of basic rules:

1. Take steps at the earliest moment to write a clear, chronological account of the relevant incident or sequence of events. If a prisoner has difficulty in reading or writing he should seek assistance from a fellow prisoner or sympathetic member of staff. Sign and date the statement and ask someone (preferably a member of the prison staff such as a probation officer or the chaplain) to countersign and date it. If the statement is made sufficiently

soon after the incident in dispute, it can be used as a memory-refreshing document in court. It will be bound to be more reliable than a statement given to a solicitor many months later. Make sure that the statement is not vague and generalized. It is no good in an assault claim merely to say 'Prison officers X and Y viciously attacked me.' The statement must record, as much as is possible, who did what, giving details of what blows were landed where, whether they were kicks or punches, and how many were delivered. Where the identity of any assailant is unknown, give as detailed a description as possible ('Officer 1 was six feet tall, blond haired, bearded with a Welsh accent').

2. Establish at the earliest possible moment the identities of any witnesses and try to obtain written statements from them while memories are fresh. These too should be dated, signed, and preferably countersigned. Where the prison authorities make it difficult to obtain such statements, make a formal complaint and keep a record of it. Follow up the complaint with a letter to an MP if this does not produce results.

3. Where any injury whatsoever is sustained, demand access to any medical report compiled by the prison medical officer. If it does not fully record all the complaints made and the injuries sustained ask for the report to be amended. If this is refused, ask that the request is formally recorded. Failing this, make an immediate complaint about the quality of the medical report and follow it up with a letter to an MP. In actions for assault, the prison medical officer's report will normally be the only one available to support (or destroy) the prisoner's claim and now that a medical report which supports all injuries pleaded must be served at the time that any claim for personal injuries is commenced, it is obviously vital that it is as full as possible. A complaint two years after the incident that the medical report is inaccurate will carry far less weight than an immediate, formal objection to its content.

4. Where the complaint relates to a lengthy course of conduct such as persistently poor medical treatment or long-term segregation in intolerable conditions, keep a contemporaneous diary of events (preferably countersigned each day). Such a document can be a vital weapon in establishing credibility, accuracy, and reliability as well as being an important counter to the records maintained by the prison authorities.

5. Contact a solicitor as soon as possible whenever legal action is contemplated. A solicitor will be able to provide advice on how to preserve and collect evidence relevant to the claim but he or she cannot remedy any failures to follow steps (1)–(4) above, which all rest on immediate action by the potential prisoner-plaintiff/applicant.

LEGAL AID

Legal aid is available in respect of both civil actions and applications for judicial review. A prisoner who wishes to obtain legal aid should first contact a solicitor, preferably someone with experience in dealing with prisoners' rights cases. On receiving a request for legal advice or assistance from a prisoner, the solicitor should normally seek authority from the Legal Aid Board to have a Green Form application signed by post. The Green Form scheme enables two hours' worth of advice to be given but to obtain it requires the applicant's signature. Once authority for this has been obtained, the solicitor should send both the Green Form and legal aid application forms to the prisoner for signature with instructions about how to fill them out. It is important to remember that prisoners must record their prison earnings (however paltry), since forms filled in with the words 'no earnings' will be 'bounced' by the Legal Aid Board. The Green Form scheme will enable the solicitor to cover the work necessary to submit the legal aid forms in order to obtain a full legal aid certificate. If it is thought necessary at this stage to have a conference or gather other relevant evidence, specific authority must be sought from the Legal Aid Board for a Green Form extension to cover the cost of this. Most area offices are sympathetic to such extensions where the need for them can be demonstrated by well-reasoned correspondence.

It is normal for the initial legal aid certificate to be limited to obtaining an Opinion from a barrister on the merits of the proposed claim/application and for proceedings to be drafted only if that Opinion is favourable. The Legal Aid Board will have to be persuaded:

1. that the applicant has reasonable grounds for taking, defending, or being a party to proceedings ('the legal merits test'). This means that, on the facts put forward and the law which relates to them, there is a case or defence which should be put before a court for a decision; and

2. that it is reasonable in the particular circumstances of the case for the applicant to be granted civil legal aid ('the reasonableness test').

3. to comply with this test, a prisoner will have to show that he is likely to derive some benefit from the proceedings, but this is not restricted to financial benefit. Most applications for judicial review by prisoners involve no question of financial gain.[119]

It must be borne in mind that financial assistance can also be obtained from bodies other than the Legal Aid Board, such as the Commission for Racial Equality and the Equal Opportunities Commission. Indeed one reason for refusing legal aid is the availability of financial assistance from

---

[119] See the Civil Legal Aid (General) Regulations 1989, paras. 28–34.

other sources, and so a prisoner who has a claim which raises issues of sex/race discrimination by the Prison Service ought to consider an approach to the relevant statutory bodies who deal with these issues.

Eligibility for legal aid is also subject to a financial test. The Board looks at an applicant's income and capital before deciding on eligibility. While prisoners will have no difficulty in complying with the minimum income test, any capital sums above the minimum[120] may lead to the requirement to make a one-off contribution towards the cost of legal aid. A solicitor will advise on the appropriate financial tests.

---

[120] In Jan. 1993 the minimum capital level for legal aid with a nil contribution was £3,000.

4

International Law Avenues of Redress

INTRODUCTION

In the previous chapter we surveyed the ways in which prisoners can pursue claims relating to their conditions and treatment through the domestic legal system. We saw that although the number and effectiveness of available avenues have increased in recent years they still fall some way short of a comprehensive mechanism for the definition and protection of prisoners' rights. Until relatively recently that would have been the end of the matter as regards available forms of redress. However, since the end of World War II prisoners have also been able to turn to a developing body of institutions and legal doctrine for the protection of human rights at an international level. The impact of these international human rights mechanisms in the United Kingdom has been considerable. Indeed it is arguable that without the prodding of the European Convention on Human Rights neither the administration nor the judiciary in the United Kingdom would have done very much to recognize any legal rights for prisoners and that a book such as this would not have been possible.

That prisoners should be an object of concern of human rights treaties and institutions is hardly surprising since it was the treatment of prisoners in the period immediately before and during World War II that was one of the main inspirations for the human rights law-making that came after it. Revelations of what had happened in the concentration camps of Nazi Germany and its occupied territories gave rise to a widespread feeling that it was no longer enough to leave each government to define the civil rights of its citizens and that some kind of international declaration of rights and supervision mechanism was necessary. The concrete expression of these views was the (non-binding) United Nations Universal Declaration of Human Rights in 1948. It was to be followed by a series of treaties which have subsequently become more content and region specific.[1] Although initially it was assumed in most of these agreements that government-to-government negotiation would be the mechanism by which the rights guar-

---

[1] Notably the International Covenant on Civil and Political Rights (1966), European Convention on Human Rights (1950), American Convention on Human Rights (1969), and African Charter of Human and Peoples' Rights (1981). For an overview of these see T. Buergentahl, *International Human Rights* (St Paul, 1988) or P. Sieghart, *The International Law of Human Rights* (Oxford 1983).

anteed would be ensured, increasingly human rights institutions have developed a degree of autonomy from the governments that created them and individuals have sought to assert their rights directly against sovereign states.

Just as the fate of prisoners was a strong motivating factor in the creation of international human rights treaties so prisoners have featured strongly in the activities of the institutions set up to ensure the enforcement of those treaties. Applications from people in detention make up the majority of cases considered by both the European Commission of Human Rights and the United Nations Human Rights Committee.[2] Although in some early applications to the European Commission it was argued that the rights guaranteed by the Convention should not apply fully to prisoners, that there were 'inherent limitations' on such rights imposed by the fact and significance of detention after a criminal conviction,[3] the Commission swiftly rejected this. While regularly acknowledging that the context of imprisonment may have an effect on the extent of the right guaranteed, especially that it must be borne in mind in the application of the limitation or 'clawback' clauses,[4] the European Commission and Court have gone on to define and protect a wide range of rights in the prison context. These decisions have had a significant impact on the prison system of the United Kingdom in areas such as disciplinary procedures,[5] access to the courts and the outside world,[6] release procedures for life sentence prisoners,[7] the treatment of detainees,[8] rights to marry,[9] and secrecy regarding prisons.[10]

The focus in this chapter though is on matters of procedure rather than content. More detailed consideration is given to how the content of international human rights norms comes to bear on prisoners' rights in the United Kingdom in the context of particular issues of discipline, transfer, segrega-

---

[2] See J. Fawcett, 'Applications to the European Convention on Human Rights', in M. Maguire, J. Vagg, and R. Morgan, *Accountability and Prisons: Opening up a Closed World* (London, 1986), 63.

[3] See *Golder* v. *United Kingdom* (1975) 1 EHRR 524, Series A No. 18.

[4] The phrase 'clawback' clause is that of Rosalyn Higgins in R. Higgins, 'Derogation Under Human Rights Treaties' (1976–7) 48 *BYIL* 281. It refers to provisions such as that in Art. 8 (2) of the European Convention on Human Rights, which after guaranteeing the right of privacy goes on to add, 'There shall be no interference by a public authority with the exercise of this right except such as is in accordance with the law and is necessary in a democratic society in the interests of national security, public safety or the economic well being of the country, for the prevention of disorder or crime, for the protection of health or morals, or for the protection of the rights and freedoms of others.'

[5] *Campbell and Fell* v. *United Kingdom* (1984) 7 EHRR 165, Series A No. 80.

[6] *Golder* (1975) 1 EHRR 524, Series A No. 18; *Silver* v. *United Kingdom* (1983) 5 EHRR 347, Series A No. 61.

[7] *Weeks* v. *United Kingdom* (1987) 10 EHRR 293, Series A No. 114; *Thynne, Wilson and Gunnell* v. *United Kingdom* (1990) 13 EHRR 666.

[8] *Ireland* v. *United Kingdom* (1978) 2 EHRR 1, Series A No. 25.

[9] *Hamer* v. *United Kingdom* (1979) 4 EHRR 139.

[10] *Harman* v. *United Kingdom* (friendly settlement reported in Stocktaking (1986)).

tion, etc., which are discussed in subsequent chapters. Most of this chapter is devoted to examination of the procedural mechanisms of perhaps the best known and most influential of the international human rights conventions, the European Convention on Human Rights. We go into some detail on this in the light of its importance and the fact that many lawyers in the United Kingdom are still relatively unfamiliar with it. However, we also look at the procedures of the European Convention for the Prevention of Torture and Inhuman or Degrading Treatment or Punishment, at the European Prison Rules, at the United Nations International Covenant on Civil and Political Rights and at the United Nations Convention against Torture and Other Cruel, Inhuman or Degrading Treatment or Punishment.

Before examining these various international law avenues it is worth making a brief comment on their status in domestic law. Despite the impact of the European Convention on Human Rights on many areas of law and practice, prison law among them, in the United Kingdom not even this well-known treaty gives rise to rights that are enforceable through the domestic courts. Like any other treaty ratified by the United Kingdom it binds the United Kingdom at international level only and is not enforceable in the domestic courts unless it has been 'incorporated' as a result of a statute being passed in the United Kingdom to give effect to it.[11] Courts in the United Kingdom have indicated that where Parliament has legislated and the words of the statute are clear then it must be given effect to even if the statutory provision in question is in contravention of a human rights treaty which the UK has ratified.[12] Since Parliament is presumed to act in accordance with the United Kingdom's international obligations, however, a human rights treaty may be invoked to assist interpretation where the statutory provision in question is ambiguous, even if the statute was not passed to give effect to the treaty.[13] In the case of *R. v Secretary of State for the Home Department ex parte Brind*[14] the applicants argued that Ministers and officials exercising discretionary powers granted to them by statute were bound to exercise these in conformity with the European Convention on Human Rights, that no reasonable Minister or official could do otherwise. This proposition is of particular interest in the prison context, given the fact that the Prison Act 1952 does little more than confer power on the Secretary of State to make Prison Rules and these in turn confer considerable discretion on governors and prison officers. It was not a proposition the House of Lords in *Brind* agreed with, however. Lord Ackner, speaking

---

[11] In the prisons context a good example of this is the Repatriation of Prisoners Act 1984, passed to give effect to the Convention on the Transfer of Sentenced Persons. This is discussed in greater detail in Ch. 6.

[12] See *Salomon v. Customs and Excise Commissioners* [1966] 3 All ER 871.

[13] See *R. v. Chief Immigration Officer, Heathrow Airport, ex parte Salamat Bibi* [1976] 3 All ER 843.

[14] [1991] 2 WLR 588.

for a unanimous House of Lords, indicated that in his view this would be incorporating the Convention 'by the back door' and that it would go beyond the judicial function.[15] All the Lords seemed to acknowledge that the Secretary of State had to take the Convention into account in formulating rules where this was relevant, though it is not clear that this applies to lower officials exercising discretionary power conferred on them.[16] Hence the position remains that neither the European Convention nor the International Covenant on Civil and Political Rights can be invoked as a source of enforceable rights in domestic courts, although there is a strong movement to change this situation politically, and some of the senior judiciary in the United Kingdom have already indicated support for this.[17] There are also indications that the failure of the United Kingdom to incorporate the Convention into its domestic law (it is one of only five member states which has yet to do so) is meeting with increasing disapproval in Strasbourg. In the 'Spycatcher' case a number of European Court judges suggested that the United Kingdom was in danger of failing to meet its obligations under Article 13 of the Convention through not incorporating or clearly enshrining a guarantee of Convention rights in its law.[18]

While the rights guaranteed by international human rights law cannot be asserted in the domestic courts, a range of international forums exist whereby concerns that prisoners' rights in the United Kingdom are being violated can be aired. Successful assertion of rights in such forums in turn creates pressure for change in domestic law.

THE EUROPEAN CONVENTION ON HUMAN RIGHTS

OVERVIEW

The European Convention on Human Rights was signed on 4 November 1950 and came into force on 3 September 1953. The United Kingdom was the first party to ratify it, in 1951. The Convention was the outcome of a resolution of the Consultative Assembly of the Council of Europe in August 1949 to produce a collective guarantee of human rights. Though the Convention's preamble states that it was framed 'to take the first steps for

---

[15] That the *Brind* decision did not mark an end to judicial invocation of the Convention is shown however by the decision in *Derbyshire CC* v. *Times Newspapers* [1992] 3 WLR 28. There the Court of Appeal explicitly invoked Article 10 of the Convention, which it found to be 'stated on high authority' to be the same as the common law, to determine the scope of common law libel and rule that a local council could not sue for defamation. The court regarded the state of the common law as uncertain and the Convention as a relevant guide to its content.

[16] See on this *Fernandez* v. *Secretary of State for the Home Dept.* [1981] Im. AR 1

[17] See the views of Lord Bridge in *Attorney-General* v. *Guardian Newspapers (No 2)* [1988] 3 All ER 545

[18] *Observer and Guardian* v. *United Kingdom* (1992) 14 EHRR 153, Series A No. 216. See esp. the views of Judges De Meyer, Pettiti, and Valticos.

collective enforcement of certain rights stated in the Universal Declaration'
it seems clear that one of the reasons for the drafting of a specifically
European treaty was a perception that the 1948 Universal Declaration on
Human Rights was too imprecise to be legally enforceable. Thus far the
Convention has been ratified by twenty-five member states, the Czech and
Slovak Republics being the most recent to do so, and a number of other
states have signed or applied to join. The Convention is a text of sixty-six
articles, of which Articles 2 to 14 set out the rights that the parties under-
take to guarantee to all within their jurisdiction.[19] In addition the
Convention is now supplemented by nine protocols, though not all of these
have been ratified by all the contracting states.[20]

The responsibility for ensuring that the rights guaranteed under the
Convention are protected lies first with the member states, but claims that
they are not doing so can be examined by two bodies established by the
Convention. The first is the Commission for Human Rights, whose primary
task is to examine whether the criteria for admissibility of applications have
been met. It may also express a preliminary view on the merits. Applications
that are judged admissible and do not result in a friendly settlement may
then be referred to the European Court of Human Rights for a decision on
the merits. Decisions of the Court, unlike those of the Commission, are binding on member states. The Convention also makes use of an institution
which pre-dates it, the Council of Europe, to give a binding ruling on some
cases. When and how cases end up before the Court or the Committee of
Ministers will be discussed in further detail below.

As has been stated earlier, when the Convention was first drafted it was
expected that international human rights guarantees would be enforced
through state-to-state relations. Article 24 of the European Convention creates a procedure whereby one party to the treaty can allege breaches of the
articles contained therein by another. This procedure has been invoked by
state parties in a number of cases, normally concerned with the treatment
of those detained under suspicion of criminal involvement.[21] However, the
real development of the Convention has been in the area of complaints by
individuals of violations of the Convention under Article 25. These out-
number applications by states by over 19,000 to just eleven. The United
Kingdom has accepted compulsory jurisdiction of the Court in respect of
such applications since 1965. Without the large volume of individual applications and the generally robust approach that the Strasbourg institutions
have taken to the protection of human rights the European Convention on
Human Rights might have had a very limited impact indeed.

---

[19] See the text of the European Convention on Human Rights in App. 5.
[20] The United Kingdom has ratified Protocols 2, 3, 5, and 8.
[21] See *Ireland* v. *United Kingdom* (1978) 2 EHRR 1; *Denmark, Sweden and Norway* v.
Greece Yearbook 12 (1969); *France* v. *Turkey* Yearbook 28 (1985) 150.

MAKING AN APPLICATION

Initiating a Convention application is remarkably simple. The applicant need only send a letter to the Secretary of the Commission in Strasbourg setting out the facts of the case and a statement of the articles the applicant alleges have been breached. He should also indicate why he considers the admissibility criteria (which will be discussed in greater detail below) have been satisfied. Although an applicant can conduct his case entirely on his own, applicants represented by lawyers have a much higher success rate. Initially any application will be assigned to a lawyer in the Commission's secretariat, who may advise an applicant that the case does not disclose even an arguable breach of the Convention or request further information before registering it. If an applicant insists, however, his application must be registered. When this happens it will be given a case number (hence references in the reports to case 650/89, case 650 which was registered in 1989, for example) and a member of the Commission will be given the task of being rapporteur in respect of this application.[22]

The Commission is composed of one member from each of the twenty-five states parties. The members of the Commission sit on it in their individual capacity and are independent of the government of the state from which they come. They are required to be people of high moral character who either possess the qualifications for high judicial office or are persons 'of recognised competence in national or international law'. In practice academic lawyers have tended to predominate. Commission members are elected for six-year terms and can be re-elected. The Commission is a part-time body and currently meets for fourteen to sixteen weeks a year. To support it in its work the Commission is assisted by a full-time secretariat staffed by over thirty lawyers from among the member states.

The main task of the Commission is to decide on the admissibility of applications under the Convention. The Rapporteur in each case will direct their efforts primarily to doing this and may seek further information from the applicant and/or the respondent state. Ultimately the rapporteur makes his report to the Commission, either to the full Commission or to a committee of three members of it.[23] Where a committee of three is established it may, if unanimous, declare an application inadmissible or strike it off its list. The Commission may do likewise by a majority vote.[24] No appeal is available to the applicant from such a decision.

If the Commission does not decide to strike out an application or declare

---

[22] See Rule 47 (1) of the Commission's Rules of Procedure.

[23] The possibility of referring the case to a committee of three is provided for in Art. 20 of the Convention and Rules 27–9 of the Rules of Procedure.

[24] The Commission may be composed of chambers of seven under Rule 49 of its Rules of Procedure.

it inadmissible at this preliminary stage then it will, in accordance with Rule 42 (2) of its Rules of Procedure, either seek further information from the state against which the application is sought or more formally communicate the application to the state and 'invite that Party to present written observations to the Commission on the application'. Boyle observes that 'Such communication ought to be the minimum goal of any application under the Convention prepared by a lawyer'.[25] The Commission will communicate such observations, which are usually substantial and will relate to both admissibility and merits, to the applicant or his lawyer for them to make any observations in reply. Time-limits are placed on both the government's reply and that of the applicant but extensions of time may be requested and are generally granted, provided a good reason is given for such delay.

After receiving the government's observations and the applicant's reply the Commission will decide whether it requires further information either in writing or by way of a hearing. Where it decides to have an oral hearing the Commission will not limit the parties to the admissibility issues but will also hear them on the merits to save time. Since 1990 the Commission has had the power to set up chambers of seven members to hear cases that can be dealt with on the basis of established case-law or raise no serious questions of interpretation. Otherwise the hearing will be before the full Commission, which has a quorum of at least ten members.[26] The applicant will be invited to make a submission either in person or by a representative. The state will also be invited to make representations. The Commission has a power to call witnesses itself and may also call witnesses at the request of the parties. It has no power however to compel witnesses to attend.

In ascertaining the facts the Commission has power to 'take any action which it considers expedient or necessary for the proper performance of its duties under the Convention'.[27] This may include delegating the task of making an on-the-spot-visit to one or more of its members, or hearing witnesses elsewhere than in Strasbourg, as occurred during the *Ireland* v *United Kingdom* case when, for security reasons, a number of military witnesses were heard at an air force base in Norway.[28] Though the Commission may request documents and other information from states it appears to have no power to compel them. States failing to supply information may however be in breach of their obligation under Article 28 of the Convention to 'furnish all necessary facilities' to the Commission.

Once the hearing stage is complete the Commission may conclude that

---

[25] See K. Boyle, 'Practice and Procedure on Individual Applications in the European Convention on Human Rights', in H. Hannum, (ed.), *Guide to International Human Rights Practice* (London, 1986), 146.

[26] See Rule 23 of the Commission's Rules of Procedure.

[27] See Rule 34 (1) of the Commission's Rules of Procedure.

[28] See *Ireland* v. *United Kingdom* (1978) 2 EHRR 25.

the application is inadmissible, in which case it goes no further. No appeal is available against this decision. If it concludes that the application is admissible it will indicate this to the parties and place itself at their disposal with a view to securing a friendly settlement. At the same time the Commission will draw up a draft report indicating whether it feels that any articles of the Convention have been violated. Decisions on these questions are by majority vote. This draft report will be submitted to the Committee of Ministers and to any state involved in the case, but it does not appear to be sent to the individual applicant, though Article 2 of Protocol 9 provides for this position to be altered. The report will not however be published until after the Commission has decided that no friendly settlement can be achieved and has referred the case to the Court or the Committee of Ministers, and then it is for the Court or Committee to decide whether it is published. Even when it is published the report of the Commission is not legally binding on the parties as to its decisions on the merits. Under the Convention system the Commission's role is limited to deciding admissibility. A finding of inadmissibility, especially on the ground of an application being 'manifestly ill founded' will however act to discourage applications on a similar point.

If the Commission has declared a case admissible it is under an obligation contained in Article 28 (2) of the Convention to place itself at the disposal of the parties with a view to securing a friendly settlement 'on the basis of regard for Human Rights as defined in this Convention'. The Commission will usually invite the parties to a meeting to discuss a settlement but tends to leave the question of whether there will be a settlement largely to the parties themselves. Where the parties indicate that they are willing to settle the Commission may well then become involved in assisting them to establish the terms of the settlement, sometimes shuttling between the parties, who may never meet. On other occasions the defendant state and the applicant may reach an agreement without the involvement of the Commission.

Settlements are normally reached on the basis of the payment of money to the applicant and/or some change in the applicant's situation, for example reduction of a prison sentence, or allowing the applicant to remain in the country. On some occasions such settlements have been accompanied by government commitments to administrative, or more rarely legislative, reforms. In the *McComb* case for example the United Kingdom government agreed to institute a procedure whereby prisoners' legal mail would only be examined in their presence.[29] As the Commission's task is to ensure that settlements are reached on the basis of respect for human rights it theoretically has the power to veto settlements which allow the violation that gave

[29] 50 D. & R. 81 (1988).

rise to the application to remain in existence. However the Commission has never refused a settlement reached by the parties, a stance which has led to some criticism. In the Northern Irish case of *Farrell* v. *United Kingdom*,[30] for example, the Commission accepted an agreement by the government to pay the applicant, the widow of one of three men shot dead by soldiers, £37,500 without any admission of liability or commitment to review the law on lethal force.[31] Also, in the interstate case of *France et al.* v. *Turkey*[32] the settlement accepted largely consisted of an obligation on Turkey to make a number of confidential reports regarding its progress towards eradicating some serious breaches of human rights, including torture and detention without trial, without any means of taking further action should these prove inadequate.

Where a friendly settlement is reached the Commission is under an obligation to make a report on this to the Committee of Ministers. The Committee will normally authorize the publication of a summary of this. The Commission's provisional finding on the merits in a case such as this will not be published.

If a friendly settlement is not achieved then the Commission may refer the case either to the Court or to the Committee of Ministers. Cases may also be referred to the Court by a member state, usually the defendant state.[33] Currently individual applicants have no power to refer cases to the Court but this will change if and when Protocol 9 becomes operative. Nothing in the Convention indicates to which body the Commission should refer a case which has not proved susceptible of settlement but in practice the Commission has tended to refer cases which raise new and difficult points of law to the Court while sending those which concern already established types of breach to the Committee. This again marks a reversal of what was originally expected under the Convention, as it was anticipated that the Committee of Ministers would play the greater role. In recent years, as the Court has decided an increasing number of cases, its prestige and role have been considerably enhanced.[34]

The Court also sits at Strasbourg on a part time basis. The Convention indicates that its members must be 'of high moral character and must either

[30] (1983) 5 EHRR 466.
[31] T. Opsahl, 'Settlement based on respect for Human Rights under the European Convention on Human Rights', in *Proceedings of the Sixth International Colloquy about the European Convention on Human Rights* (Dordrecht, 1988) 970, comments about the *Farrell* case however that 'from the Commission's point of view, this was not a case of general interest, but one concerning a unique incident'
[32] Yearbook 28 (1985) 150.
[33] However, any state party can refer a case and in *Soering* v. *United Kingdom* (1989) 11 EHRR 439, Series A No. 161 the case was referred by Germany, of which the applicant was a national.
[34] The Court was in existence for eighteen years before it gave its 100th decision in 1987; its 200th was delivered four years later.

possess the qualifications required for appointment to high judicial office or be jurisconsults of recognised competence'.[35] Judges of the Court are appointed for nine years in the same way as members of the Commission. The Court normally hears cases in chambers of nine judges.[36] Where it feels it appropriate, however, the chamber may relinquish jurisdiction in favour of the plenary Court. The Commission's report normally forms the basis of the Court's deliberations and the Commission will delegate one of its members to assist the Court. The defendant state will also appear before the Court and since 1983 the applicant in an Article 25 application also has a right to be represented. The changes in the Court's Rules of Procedure which gave the individual this right of representation also created the opportunity for third parties to seek the leave of the President of the Court to submit an *amicus curiae* petition, providing that they are seen as being a 'person concerned' with the case in a particular sense.

The Court will normally enquire of the parties whether they wish an oral hearing or for the procedure to be entirely written. Where an oral hearing takes place the Court can call witnesses itself or at the request of the parties. Like the Commission it can issue summonses for witnesses but has to rely on the co-operation of the national authorities as to whether these will be complied with or not. Though the Court will normally accept the findings of fact as made by the Commission it is not bound by these and is of course capable of putting a different legal characterization on those facts. The Court also took the view in the *Vagrancy* cases[37] that it was competent to review admissibility findings, but has very rarely used this power.

The Court reaches its decisions by majority and has provision for the publication of dissenting views and separate concurrences. Supervision of the execution of decisions is reserved to the Committee of Ministers. The Convention is silent about the Court's power to issue recommendations but thus far it has refused invitations to do so, restricting itself to declaring that the state's law or practice in question violates the Convention. Under Article 50, however, the Court can make an order for the award of compensation where it finds a breach and concludes that the law of the state in question allows partial or no reparation for the violation in question. Applicants who seek compensation should ask for it in their memorials to the Court; but awards of compensation are discretionary, and in a number of cases the Court has held that the decision itself is adequate compensation for the violation. In general cases claiming compensation for provable financial loss or personal injury have proved more successful than claims

---

[35] See Article 39 (3).   [36] Art. 43, as amended by Protocol 8.
[37] *De Wilde, Ooms and Versijp* v. *Belgium* (1971) 1 EHRR 373, Series A No. 12. However, this approach has recently led to disputes within the Court, with some of its members arguing that admissibility decisions should be wholly left to the Commission; see the dissents of judges Martens and Morenilla in *Cardot* v. *France* (1991) 13 EHRR 853, Series A No. 200.

for more intangible items like excessive periods in detention. The applicants' costs have often been recovered under Article 50. However, the Court's approach to this issue, as with its general approach under Article 50, has been criticized by some commentators as lacking in principle or consistency.[38]

Alternatively the Commission may refer the case to the Committee of Ministers, which is made up of the Foreign Ministers of the member states or their representatives. The individual applicant's position, weak as regards references to the Court, is almost non-existent with regard to references to the Committee. There is not even any obligation to inform him of the progress of the application and the Committee will normally only inform the applicant of the decision reached and will not give a full statement of the reasons on which it was based. The Committee, which meets in camera, also currently decides to what extent the applicant will be informed of the Commission's report. Usually however the Commission's report is published after the Committee has reached its decision. The Committee has adopted a set of rules for the conduct of its examination of cases. These allow it to call experts and witnesses. They do not allow the individual applicant to appear and indeed the Committee has normally ruled inadmissible attempts by the applicant to send written comments to the Committee.

Committee decisions must be reached by a two thirds majority and the state against which a case has been brought is entitled to take part in the examination and vote. No provision exists for the situation where a state is found in breach but not by the two-thirds majority, but as the Committee normally affirms Commission decisions referred to it few problems of this nature have arisen.

The Committee has the task of supervising the enforcement of both its decisions and those of the Court. Generally it performs this function by seeking reports from defendant states on the steps they have taken to comply with the Convention. Its performance of this role has not been free from criticism; for example its enforcement of its own decision in *Cyprus* v. *Turkey*[39] consisted of urging the two parties to resume talks and thereafter declaring the issue closed without this resumption of talks occurring. The Committee can make recommendations but these are not binding on the parties. Ultimately the main sanction available for non-compliance lies in the statute of the Council of Europe, which like the Committee pre-dated the Convention. This allows for the expulsion of a member from the Council. So far this has not occurred, the Committee coming closest to it over the *Greek* case.[40] In that case Greece's withdrawal from the Council rendered the issue moot.

---

[38] See C. Gray, *Remedies in International Law* (Oxford, 1987), 158.
[39] See Yearbook 22 (1979) 440.   [40] Yearbook 12 (1969).

THE ADMISSIBILITY CRITERIA

The admissibility criteria have proved to be a substantial hurdle for many applications, less than 5 per cent of those applications registered have ultimately been found to be admissible by the Commission.[41] Hence lawyers making applications need to ensure that they meet these criteria. These are largely contained in Articles 25–7 of the Convention. The most significant are arguably the following.

1. That the applicant be a victim

States invoking the Article 24 procedure against other member states do not have to show that they, or any of their nationals, have been affected by an alleged breach of the Convention. They are granted a general right to complain of alleged breaches in order to enforce the Convention. As regards individual applications, however, Article 25 states that applications may be made by 'any person, non governmental organisation and groups of individuals' who claim to be the victims of a breach of the Convention. Individuals or groups cannot bring an action as an *actio popularis*, they must show that they particularly are affected in some way by the breach of the Convention. The lack of an *actio popularis* is perhaps a defect in a Convention dedicated to ensuring the most effective protection of the human rights of all in the member states, since obvious human rights violations may go unchallenged for lack of a victim prepared to pursue a case through the lengthy procedures that Strasbourg requires. Non-governmental organisations (NGOs) in particular could play a useful role in highlighting and challenging such violations. In the prison context this might be especially valuable in respect of conditions in remand prisons, where prisoners may not see the value of launching cases that may take five years to reach a decision. Under the Convention NGOs may only bring a claim if they themselves have been a victim of an alleged violation or if they represent someone who has been.

The notion of victim in Article 25 is not limited to direct victims. In a series of decisions the Commission has indicated that it also encompasses 'indirect victims', someone who is a near relative of the victim or a third party who is so closely connected with the victim that the violation concerned is also prejudicial to her, or in so far as she has a valid personal interest in the termination of the violation. The Commission has also recognized a right to petition for those who have not yet suffered from a breach of the Convention but run a significant risk of doing so in the future. This view has been developed in a number of cases, mainly involving the right to

[41] The Commission's own 'Survey of Activities' in 1991 found that only 1,038 of 19,216 registered applications between 1955 and 1990 had been declared admissible. The rate however was rising from less than 2% in 1955–75 to 7% in 1986–90.

privacy and family life guaranteed by Article 8, where the significant factors appear to have been that the victim be a member of an identified class (such as homosexuals in *Norris* v. *Ireland*[42] or schoolchildren in a school where corporal punishment was practised, as in *Campbell* v. *Cosans*[43]) and that the potential violation already has an effect on their lives (e.g. in *Norris* a homosexual's vulnerability to harassment and blackmail or in *Bruggemann and Scheuten* v. *Federal Republic of Germany*[44] the limitation on a woman's sexual freedom raised by the prohibition of abortion). In addition those who suspect but cannot prove themselves to be victims may be entitled to bring claims where it is government secrecy which prevents them from ascertaining whether they are victims, as in the telephone-tapping case of *Klass* v. *Federal Republic of Germany*.[45]

2. Exhaustion of Domestic Remedies

Article 26 of the Convention indicates that the general rule of international law, that international tribunals may not be seized of a case until all domestic remedies have been exhausted, applies to the Convention. Early in its jurisprudence the Commission made it clear that the applicant's obligation is to exhaust all domestic remedies that are 'effective and sufficient' to provide a remedy for the breach of the Convention alleged.[46] This reference to remedies that are effective and sufficient is especially important in the prison context given the existence of internal remedies and the limited effectiveness of judicial review. It gives rise to a number of points.

First it means that the applicant need not exhaust avenues of remedy that are discretionary in character. Only remedies that are mandatory are deemed effective. Thus applicants need not pursue applications for pardons or petitions of mercy or appeals to bodies that are merely advisory in character. While effective remedies are not confined to judicial remedies it would appear that to be effective administrative remedies must be capable of being binding on the authorities. Internal Prison Service grievance procedures would not appear to amount to an effective remedy in these terms. The position is less clear as regards the Prisons Ombudsman, as this institution is designed to be independent of the Prison Service. As the Ombudsman is only empowered to make recommendations which can be overruled by the Secretary of State there must remain doubts as to whether he or she will be an effective remedy and must be exhausted before a prisoner goes to Strasbourg.

The second implication is that even those remedies which might be thought effective can be rendered otherwise by circumstances. In particular

---

[42] (1988) 13 EHRR 186, Series A No. 142.
[43] (1982) 4 EHRR 293, Series A No. 48. [44] Yearbook 19 (1976) 382.
[45] (1978) 2 EHRR 214, Series A No. 21.
[46] *De Becker* (1962) 1 EHRR 43, Series A No. 4.

judicial remedies may be rendered ineffective where it is clear by established jurisprudence of the national courts in question that no remedy is available on the facts as alleged by the applicant. Thus in the case of *Campbell and Fell* v. *United Kingdom*,[47] where prisoners challenged the conduct of disciplinary proceedings in a prison, the Commission held they did not have to seek judicial review in the English courts before bringing the case to Strasbourg as the settled jurisprudence of the English courts at the time the application was brought was that judicial review of decisions made in the prison was unavailable (this was a pre-*St Germain* application, though in the recent *Kavanagh* case concerned with refusal to transfer an Irish prisoner back to Northern Ireland the Commission again found judicial review not to be in need of exhaustion[48]). In this situation, as with the issue of what forums are effective, the test the Commission applies is whether the application has a 'reasonable likelihood of success', where there is a doubt the applicant ought to invoke the doubtful remedy. The Commission has made it clear that it will judge this itself, and that acting on advice given in good faith by a lawyer that a remedy will not be sufficient and effective will not be enough to exhaust domestic remedies where the Commission decides that the remedy did have a reasonable chance of success.[49] However, it is generally safe to assume that a lawyer's opinion of the chances of success in domestic law will be accepted where it is not clearly unreasonable.

The Commission has indicated that applicants must have raised all the issues that they wish to raise in the Convention proceedings before the domestic tribunal in order to show that they have exhausted domestic remedies. Ignorance of a remedy or negligence in failing to raise it, for example through missing a time-limit, will not be an excuse for failing to exhaust it.

However, applicants will be excused from exhausting a domestic remedy where they claim that they are victims of an administrative practice, a claim most usually made in cases concerning allegations of ill treatment. The Commission has indicated that an administrative practice will often render domestic remedies ineffective 'by the difficulty of securing probative evidence and administrative enquiries would either not be instituted or, if they were, would be likely to be half hearted and incomplete'.[50] The Commission sees an administrative practice as consisting in the repetition of acts which form a pattern and which are officially tolerated at 'the level

---

[47] (1984) 7 EHRR 16, Series A No. 80.

[48] Appl. 19085/91, decision of the Commission 9 Dec. 1992. The case is discussed further in Ch. 6.

[49] For a view that unequivocal expression by legal counsel that all domestic remedies are exhausted will be sufficient to comply with the rule see D. Gomien, *Short Guide to the European Convention on Human Rights* (Strasbourg, 1991), 135.

[50] *Greek* case Yearbook 12 (1969) 194.

of the direct superiors of those immediately responsible or that of a higher authority'. In the *Donnelly*[51] case, however, the Commission held that the authorities paying compensation to those injured by ill treatment, coupled with a willingness to investigate the incidents, was sufficient to displace the claim that there had been an administrative practice; paying of compensation coupled with official indifference would not have been sufficient.

3. The Six Months Rule

Under Article 26 of the Convention, in addition to exhausting all domestic remedies the applicant must submit his application to the Commission within six months of the 'final decision' in his case. The two would appear to be linked in that the final decision would normally be that which exhausts all domestic remedies. The link has caused some problems in that an applicant may exhaust a remedy only to be told by the Commission that this remedy was not an effective remedy and that the final decision was taken earlier. That earlier time may be in excess of six months and the applicant may then find himself out of time. On the other hand an applicant, fearing that he may fall foul of the six months rule, may submit an application earlier, only to be told that the application is inadmissible for failure to exhaust domestic remedies. However, since applications inadmissible by reason of failure to exhaust domestic remedies may always be resubmitted once these have been exhausted, an applicant who is doubtful about the effectiveness of the domestic remedy he is pursuing might at the same time lodge an application with Strasbourg. It is only required that domestic remedies be exhausted by the time the Commission comes to rule on admissibility, not when the application is submitted. For the purposes of the six month rule the initial letter to the Commission counts as notification rather than the date of registration.

Where no effective domestic remedy exists the six months are taken to run from the time of the action allegedly causing the violation. However in the *De Becker* case the Commission indicated that this did not apply to cases of a continuing violation. What the Commission appeared to have in mind here is a violation brought about by operation of legislation, where no effective remedy existed, rather than a judicial or executive decision.[52] Yet what exactly was contemplated by this was not entirely clear and the issue of what constitutes a continuing violation remains underdeveloped. It does however suggest that prisoners complaining of violations of human rights arising out of their conditions or regime need not be too concerned about the effect of the six months rule.

---

[51] Yearbook 16 (1973) 212.

[52] De Becker had been sentenced to permanent deprivation of civil rights by a Belgian court for collaboration in World War II.

4. Manifestly ill founded

This provision, contained in Article 27 (2), is the catch-all admissibility criterion. A number of Commission decisions indicate that the standard the Commission applies is whether the application discloses a prima-facie breach of the Convention.[53] Applications ruled inadmissible on this ground may be so as a result of a finding that the facts are not as alleged by the applicant or that even if as alleged they do not disclose an arguable breach of the Convention. Since no appeal is available against this decision and since a finding of inadmissibility on this ground is likely to discourage further applications on the point many commentators are unhappy with this ground of inadmissibility, or at least with the fact that the decision to dismiss an application on this ground can be reached by simple majority vote.[54] This unhappiness is compounded by the suggestion that the Commission uses this ground to dispose of cases that appear close on other admissibility criteria.

A number of other criteria of admissibility can be dealt with relatively quickly. Article 27 (1) (*a*) indicates that applications cannot be accepted if they are anonymous. However Commission officials have proved reasonably flexible regarding this and have admitted applications where from the documents they have been able to ascertain the identity of the applicant. Also applicants' wishes for anonymity as regards the public are respected— hence the number of *X* v. *United Kingdom* and similar cases. Applications may also be declared inadmissible on the grounds that they raise 'substantially the same matter as has already been examined by the Commission' (Article 27 (1) (*b*)); but it seems clear that this is designed only to discourage identical applications rather than applications from different people which are based on similar facts. Even a second application by a person regarding the same issues as an application already dismissed may escape the sanction of this provision if it discloses new facts. Applications that are an 'abuse of petition' are also subject to being ruled inadmissible. This appears to refer to applications begun purely for propagandistic purposes or to escape the consequences of a legitimate decision.[55] This provision would not appear to endanger 'test case' litigation. However, it seems that breach of the Convention's rules about confidentiality during the consideration of an application may be seen as an abuse of the right of petition.

---

[53] See *Pataki* case Yearbook 3 (1960) 356.

[54] See P. Van Dijk, and G. Van Hoof, *Theory and Practice of the European Convention on Human Rights* (2nd edn., Dewenter, 1990), 107.

[55] See e.g. the *Ilse Koch* case Yearbook 5 (1962) 126, where the wife of the former commandant of Buchenwald concentration camp, who had been convicted of elementary human rights offences, submitted a petition claiming she was innocent without making any reference to any aspect of the Convention.

INTERIM MEASURES

When making an application to the Commission an applicant may also wish to consider whether to request that the Commission take any interim measures in his case. Such measures would be especially important where someone is facing deportation or expulsion from a member state or where they are under the threat of torture or death, since the application may take several years to be resolved. The Convention does not make any provision for interim measures, unlike, for example, the American Convention on Human Rights, Article 63 (2) of which empowers the Court (acting alone or at the request of the Commission where a case has not yet been referred to it) to adopt such provisional measures as it deems pertinent 'in cases of extreme gravity and urgency'.[56] However, in Rule 36 of its Rules of Procedure the Commission has stated that it, or the President if it is not in session, may 'indicate to the parties any interim measure the adoption of which seems desirable in the interest of the parties or the proper conduct of the proceedings before it'. The reference to the Commission 'indicating' such measures to the parties shows that such a measure is not intended to be binding and in a recent case the Court indicated that failing to comply with such a recommendation would not be a violation of Article 28.[57] Prior to this decision in most cases parties have been willing to comply with the Commission's indication, the impact of *Cruz Varas* on states' attitudes remains to be seen.[58]

LEGAL AID

In the early days of the Convention most applications were made by non lawyers.[59] However, by the late 1980s over 50 per cent of registered applications were made by lawyers. As cases involved increasingly complex legal points the need for legal aid, which was not available in domestic systems for Convention applications, was recognized. The legal aid scheme is contained in an addendum to the Rules of the Commission. This indicates that legal aid will be available where the Commission is satisfied that it is necessary for the proper discharge of the Commission's duties and that the applicant has insufficient means to meet all or part of the costs involved.

Legal aid will be available on application to the Commission to cover

[56] See e.g. *Provisional Measures in the Case of Bustios and Rojas regarding Peru* (1990) 11 Human Rights Law Journal 257, where the Inter-American Court ordered Peru to safeguard the lives of two journalists about whom an application was pending

[57] See *Cruz Varas* v. *Sweden* (1992) 14 EHRR 1, Series A No. 201.

[58] In *Cruz Varas* it was observed that 182 interim indications had been given and complied with in expulsion cases, though the same was not always true in extradition cases.

[59] In 1955–70 applications introduced by lawyers were always less than 20% of total applications.

expenses from the time that the applicant is asked for written comments on the government's observations at the admissibility stage. The applicant can receive legal aid if he employs a barrister, solicitor, law professor, or 'professionally qualified person of similar status'. The fees available are intended to be set at the rate of civil legal aid in the state from which the applicant comes and aid will be granted also to cover the lawyer's and applicant's travelling and subsistence expenses.

THE FUTURE OF THE CONVENTION SYSTEM

In many ways the Convention's success is now putting it under great strain. Because it is seen by many people as an effective measure of redress for human rights violations in their own countries the number of applications is rising sharply. As more lawyers become aware of the procedural and substantive jurisprudence of the Convention applications are increasingly well prepared and as a result more applications are declared admissible and come before the Court for decision. The full ratification of Protocol 9, allowing individuals to petition the Court directly, is only likely to increase that number. In addition more countries are applying to join the Convention system, with a long queue forming in central and eastern Europe as a result of the end of the Cold War. The Convention began with eight countries; the new Human Rights building currently under construction in Strasbourg is designed to accommodate judges and staff from around forty member states.

The outcome of all this is greater delay in dealing with applications and a potential pressure, which some human rights lawyers feel they can already identify in reality, on the Commission to declare more cases inadmissible rather than refer them to the Court. It is now generally accepted that if these delays are not to become altogether unacceptable, with a consequential weakening effect on the prestige of the Convention, reform of the procedural structure is a matter of urgency. This is likely to focus on eliminating the duplication of functions between the Commission and Court. Two options seem to be emerging. One is to give the Commission power to issue binding judgments and make the Court effectively a court of appeal from it on limited grounds. The second is to abolish the Commission. Whichever approach is chosen it seems likely that the next few years will see a considerable change in the Convention's procedural mechanisms.[60]

---

[60] For a summary of these issues see the symposium on merging the Commission and the Court in (1988) 9 *Human Rights Law Journal* and A. Drzemczewski, 'The Need for a Radical Overhaul' (1993) 143 *New Law Journal* 126.

THE EUROPEAN CONVENTION FOR THE PREVENTION OF TORTURE AND INHUMAN OR DEGRADING TREATMENT OR PUNISHMENT

Whereas the European Convention on Human Rights invokes a wide range of concerns, the European Convention for the Prevention of Torture or Inhuman or Degrading Treatment or Punishment focuses on one of the most serious human rights violations, one that cannot be derogated from even in time of war or emergency. It is also a human right that is always of particular concern to people in detention. The Convention was signed in 1987 and entered into force on 1 January 1989. It has twenty three members, including the United Kingdom.[61]

The Convention introduces a new enforcement mechanism into international human rights law by providing for regular visits by an independent Committee to 'places where persons are deprived of their liberty by a public authority' (Article 2) in any of the member states. Unlike the European Convention on Human Rights there is no provision for individuals to bring their claims of torture or inhuman or degrading treatment before the Committee. As will be seen below the United Nations Torture Convention provides for this to occur by way of an optional protocol. This does not mean however that individual prisoners or lawyers representing them have no input into the Committee's deliberations and reports. Again unlike the UN Torture Convention the European equivalent provides no definition of torture or inhuman or degrading treatment or punishment. The jurisprudence of the Strasbourg institutions on Article 3 of the European Human Rights Convention (which is discussed in greater detail in Chapter 5) obviously plays a part in their working definition but in reports made by the Committee so far it has not seemed to confine itself to this.

The Convention provides for the creation of a Committee with a number of members equal to the states parties. The members are required by Article 4 (2) of the Convention to be 'persons of high moral character, known for their competence in the field of human rights or having professional expertise in the areas covered by the Convention'.[62] Although nominated by the member states each of these experts serves in an independent capacity. To protect that independence the Committee has taken the decision that members shall not take part in a visit to their own country. The remit of the Committee, set out in Article 1 of the Convention, is broad: 'The Committee shall, by means of visits, examine the treatment of persons

---

[61] See, generally, A. Cassese, 'The New Approach to Human Rights: The European Convention for the Prevention of Torture' (1989) 83 *AJIL* 128.

[62] For a discussion of the composition and operation of the Committee see M. Evans, and R. Morgan, 'The European Convention for the Prevention of Torture: Operational Practice' (1992) 41 *ICLQ* 590.

deprived of their liberty with a view to strengthening, if necessary, the protection of such persons from torture and from inhuman or degrading treatment or punishment.' Reports of the Committee thus far indicate that this broad mandate will be fully utilized. The first report on the United Kingdom, for example, looked not just at prisoners' physical conditions and the disciplinary regime but also at such matters as medical care, reception procedures, staff–inmate relations, and the training of prison officers.[63]

The Committee's main mechanism for fulfilling its remit is the programme of visits. Under Article 7 it is required to carry out visits to places of detention (which has generally meant prisons and police stations though mental hospitals may also be examined under this provision) in Convention countries. Under Article 17 (3) it is indicated that the Committee shall not visit places which representatives of the International Committee of the Red Cross (ICRC) 'effectively visit on a regular basis by virtue of the Geneva Conventions of 12 August 1949 and the Additional Protocols of 18 June 1977 thereto'. This raises an interesting question in respect of prisons in Northern Ireland which the ICRC regularly visit. However since such visits take place on an *ad hoc* basis, the conflict in Northern Ireland not being seen by the ICRC as either an international or non-international armed conflict within the meaning of the Geneva Conventions, Northern Irish prisons which contain a high number of people convicted of terrorist offences would appear to be within the Committee's remit. In its practice the Committee has divided Convention countries into 'large' and 'small' and has usually sought to visit two large and one small country in each year. It may also conduct *ad hoc* visits to particular countries or places of detention within a country. Under Article 9 a state may make representations to the Committee against a visit, or against a visit to a particular place of detention, at that time. Such representations may only be based on specific grounds set out in Article 9 (1) which include national defence, public safety, serious disorder at the place of detention, the medical condition of a person, or that an urgent interrogation relating to a serious crime is in progress. Where such representations are made the state is then required to enter into consultation with the Committee 'in order to clarify the situation and seek agreement on arrangements to enable the Committee to exercise its functions expeditiously'. The Committee is required to notify a country that a visit is taking place and thereafter Article 8 (2) places a number of specific obligations on the member state visited. These include giving the members of the Committee (at least two members of which make each visit) unrestricted access both to its territory and to places where people are deprived of their liberty and the right of unrestricted movement

---

[63] See *Report to the United Kingdom Government on the Visit to the United Kingdom Carried out by the European Committee for the Prevention of Torture and Inhuman or Degrading Treatment or Punishment*, CPT/Inf (91) 15.

within both. Under Article 14 the Committee may ask certain experts (often medical or prison administration personnel) to accompany its visit. Again these will not be experts from the state visited. Although states cannot deny access to Committee members (subject to Article 9) they may declare that a particular expert will not be allowed to participate in a visit to their territory. In addition to access the Committee must also be furnished with full information on places where people are detained and any other information it needs to carry out its task. Other provisions of Article 8 guarantee the Committee the right to communicate freely with people it feels can provide relevant information and to interview in private people deprived of their liberty. The practice so far has been for the Committee to meet with both government and non-governmental groups (including prison reform and human rights groups) at the beginning of its visit, before it has visited any place of detention. Once the places of detention have been visited the Committee will then have another meeting with government officials, where it will outline its provisional findings and seek further information, before leaving the country.

After the completion of a visit the Committee draws up a report on the facts as it has found them during the visit and any recommendations it has regarding measures which might better ensure the objectives of Article 1. Such a report is sent to the member state concerned and remain confidential unless the state requests that it be published. Such a request may also be accompanied by the member state's comments on the report, which must also be published. At the beginning of 1993 all states which had been visited had requested that the Committee's report be made public. The reports have generally contained a request for governments to report back to the Committee within a year on the steps they have taken to comply with the Committee's recommendations. Where they fail to do so, or prove uncooperative in other ways, then under Article 10 (2) the Committee may make a public statement if two-thirds of its members agree.

Although the European Convention on Torture is in its early days already it has made a useful contribution to the strengthening of measures to protect prisoners' human rights. The detailed reports produced by the Committee have provided a valuable source of public information on conditions in places of detention in member states and its recommendations provide a focus for those in those member states seeking reform of law and practice in prisons. The cumulative impact of its comments on specific countries and the more general report it submits to the Committee of Ministers each year under Article 12 should provide a guide to European 'good practice' in relation to prison conditions and regimes and may in time come to have greater influence on the decisions of the European Court of Human Rights than the European Prison Rules (to be discussed below) appear to have enjoyed thus far. Like many international institutions, however, the Committee on Torture has almost no enforcement powers beyond negotiation with the member state and public disapproval of failure to comply with its recommendations. If those recommendations are complied with this should prove sufficient. If they are not and a state consistently fails to meet the standards set by the Committee it will be interesting to see whether the members of the Committee will be able to develop a more effective strategy for ensuring compliance within the boundaries of the powers they have at present.

EUROPEAN PRISON RULES

Neither the European Convention on Human Rights nor the European Convention on Torture seeks to provide a detailed set of rules applicable to the treatment of prisoners. That gap is filled at the European level by the European Prison Rules, adopted by the Committee of Ministers in 1987. They are not part of a treaty and thus do not constitute a binding international law obligation; nevertheless the government of the United Kingdom has indicated its objective of complying with them. These Rules were a revision of the European Standard Minimum Rules adopted in 1973, themselves a regional adaptation of the 1955 United Nations Standard Minimum Rules for the Treatment of Prisoners. The tradition of drawing up model prison rules on an international level goes back to 1935, when the League of Nations adopted the first set of Standard Minimum Rules.

Unlike some European countries the United Kingdom has not incorporated the European Prison Rules into its domestic law and hence the obligations contained in the Rules are not enforceable in United Kingdom courts. Also the European Commission of Human Rights has indicated that breach of the European Rules will not amount to a violation of Article 3 of the Convention.[64] However, the Rules may be taken into account in interpreting the obligations of states under the Convention. Indeed the only enforcement mechanism that exists with respect to the European Rules is oversight by the European Committee for Cooperation in Prison Affairs, a five-member expert body established by the European Committee on Crime Problems. This considers state reports on the extent to which the Rules are being implemented every five years. The Committee has not however published any of the United Kingdom's reports since 1983.

Nevertheless the Rules are a valuable reference source for those campaigning for changes in the Prison Rules and seeking the development of a code of standards. Their overall philosophy, largely set out in Rules 1–6, is influenced by notions of respect for human dignity, ensuring that punish-

[64] Appl. No 7341/76. Discussed in A. Reynaud, *Human Rights in Prisons* (Strasbourg, 1987), 34.

ment is limited to deprivation of liberty, and a stress on treatment aimed at re-educating and re-socializing a prisoner that are prominent in many European prison systems. As will be pointed out on several occasions in this book, the Prison Rules and prison administration practices fall short of what is required by the European Rules but it must be remembered that this is falling short of what is largely a minimum standard. As one commentator observed, the European Rules 'set standards which provide a threshold that satisfies basic considerations of humanity without imposing unacceptable burdens upon governments which are themselves constrained by resource considerations and political priorities'.[65]

UNITED NATIONS INTERNATIONAL COVENANT ON CIVIL AND POLITICAL RIGHTS

Though the European Convention is, to most people in the United Kingdom, the better-known and more influential human rights convention, arguably the United Nations International Covenant on Civil and Political Rights (ICCPR) is the more comprehensive and certainly is subscribed to by more countries around the world, 101 to the European Convention's twenty-five. The ICCPR was one outcome of the attempts of the United Nations Commission on Human Rights to produce an international Bill of Rights after 1945. Though the (non-binding) Declaration of Human Rights was adopted relatively swiftly by the UN General Assembly in 1948 it took another eighteen years to produce a binding treaty. Indeed the Commission found it impossible, for political and legal reasons, to produce one treaty to incorporate all the rights it sought to give recognition to. In the end two treaties were adopted by the General Assembly in 1966, the ICCPR and the Economic, Social and Cultural Rights Covenant (ESCRC). Both received the requisite number of ratifications to come into force in 1976. Although some of the rights contained in the ESCRC, such as rights to health or education, have relevance to the situation of prisoners the more obviously relevant Convention is the ICCPR.

Most of the rights contained in the ICCPR have equivalents in the European Convention on Human Rights but the United Nations treaty is more expansive on a number of points. Most obviously of relevance to the situation of prisoners is its Article 10, which provides that people deprived of their liberty 'shall be treated with humanity and with respect for the inherent dignity of the human person' and goes on to indicate in Article 10 (3) that the 'penitentiary system shall comprise treatment of prisoners the

---

[65] See K. Neale, 'The European Prison Rules: Contextual, Philosophical and Practical Aspects', in J. Muncie and R. Sparks, *Imprisonment: European Perspectives* (London, 1991, 203.

essential aim of which shall be their reformation and social rehabilitation'. Article 10 also indicates that unconvicted prisoners shall be separated from the convicted and adult prisoners from juveniles. A number of other rights in the ICCPR also find no equivalents in the ECHR, such as the right of citizens to take part in public affairs (Article 25), to equality before the law (Article 26), and the rights of ethnic, religious, or linguistic minorities to enjoy their own culture, profess and practise their own religion, or use their own language (Article 27). Considerations of whether the rights guaranteed under the ICCPR are being upheld in the United Kingdom's prisons will be considered at appropriate points in the course of this book.

As a treaty ratified by the United Kingdom but not incorporated by an Act of Parliament the ICCPR has the same status in domestic law as the ECHR. As regards seeking to enforce its rights through international forums the United Kingdom has not ratified the Optional Protocol to the Convention which entitles the Human Rights Committee established under the Convention to receive complaints by individuals against that state party. Therefore enforcement is primarily by way of the periodic report that states party to the Covenant are required to submit to the Human Rights Committee under Article 40 of the ICCPR. The Committee is composed of eighteen human rights experts, nominated by states, but who serve in an independent capacity for periods of four years. Members of the Committee are required by Article 28 (2) to be 'persons of high moral character and recognized competence in the field of human rights, consideration being given to the usefulness of the participation of some persons having legal experience'.

States are required to report 'periodically' to the Committee; currently its practice is to have them report every three years (the United Kingdom has submitted three reports, the last in 1991). The Committee gives states guidance on what they must report and usually reports contain descriptions both of general legal measures to protect human rights in the reporting party and of measures addressed to particular articles of the Covenant. These reports are then considered by the Committee at one of its hearings, usually devoting two or three days to each country report. Governments send a delegation to these hearings to present the report orally and be questioned on it by the Committee members. In recent years Committee members have made increasing use of material submitted by both domestic and international human rights organizations as the basis for many of their questions, although the organization submitting the information will not be explicitly referred to. While Article 40 (2) requires states to indicate to the Committee 'factors and difficulties, if any, affecting implementation of the Convenant' it is not surprising that state reports tend to present that state in the most favourable light possible. Alternative sources of information are therefore vital if the Committee is effectively to perform its role of thoroughly examining the extent to which the Covenant is being implemented.

The Covenant indicates in Article 40 (4) that the Committee shall 'transmit its reports, and such general comments as it may consider appropriate, to the States Parties'. There has been lively debate within the Committee as to whether this entitles the Committee to send individual reports to each state and to make specific comments thereupon or whether the reference to 'its reports' being made available to the 'states parties' prevents such particularized reports. Those who take the latter position see Article 40 (4) as referring only to the annual report the Committee makes to the General Assembly under Article 45.[66] As a result of a 1980 compromise the Committee has confined itself to making general comments addressed to all states parties and does not send reports to individual states parties or direct comments to particular states. Committee members may however indicate their view that a state is in breach of a particular article of the Covenant during the oral examination of a state's report. Most members of the Committee currently appear happy with going no further than this. The absence of formal reports and evaluations of a state's compliance they see as valuable to the development of an ongoing dialogue with a state regarding the implementation of the Covenant. Currently all states seem willing to participate in this dialogue by sending delegations to the Committee when it considers their report.[67] This method of working does however mean that the impact the Human Rights Committee has on actual law and practice pertaining to human rights in states parties is largely concealed behind closed doors.

THE UNITED NATIONS CONVENTION AGAINST TORTURE AND OTHER CRUEL, INHUMAN OR DEGRADING TREATMENT OR PUNISHMENT

The extent to which torture is viewed as perhaps the most universally abhorred violation of human rights can be seen by the fact that the United Nations also has a Convention specifically devoted to it.[68] The United Nations Convention in this area preceded the European treaty, being adopted by the General Assembly in 1984 and coming into force in 1987. A total of sixty-five states had ratified it by the end of 1992. Though the

[66] See D. McGoldrick, *The United Nations Human Rights Committee: Its Role in the Development of the International Covenant on Civil and Political Rights* (Oxford, 1991), 89–96.

[67] McGoldrick, *The Human Rights Committee*, 82, notes that only one country has failed to send a delegation and compares this favourably with the implementation of reporting requirements under the Convention on the Elimination of Racial Discrimination.

[68] As has the Organization of American States, whose Convention to Prevent and Punish Torture was opened for signature in 1985 and entered into force in 1987.

Convention refers to 'other cruel, inhuman or degrading treatment or punishment' its main focus is on torture and it provides a definition for this in Article 1 as

any act by which severe pain or suffering, whether physical or mental, is intentionally inflicted on a person for such purposes as obtaining from him or a third person information or a confession, punishing him for an act he or a third person has committed, or intimidating or coercing him or a third person, or for any reason based on any discrimination of any kind, when such pain or suffering is inflicted by or at the instigation of or with the consent or acquiescence of a public official or other person acting in an official capacity. It does not include pain or suffering arising only from, inherent in or incidental to lawful sanctions.

Most of the substantive obligations the Convention places on states parties relate purely to the prevention and punishment of torture. These include prohibiting extradition or expulsion where there are substantial grounds for believing the person would be in danger of being tortured if returned (Article 3), making torture a criminal offence in domestic law (Article 4), establishing the principle of universal jurisdiction over those suspected of torture (Articles 6 and 7), and deeming torture-related offences to be extraditable offences in extradition treaties (Article 8). Some obligations overlap the categories of torture and the other forms of treatment prohibited in the Convention (which are more likely to be relevant to those detained in the United Kingdom). These include training of law enforcement personnel and of those medical personnel and public officials involved in the detention of people in the legal obligations under the Convention (Article 10), keeping under systematic review interrogation rules and practices as well as arrangements for custody and detention with a review to preventing torture etc (Article 11), and ensuring that individuals who allege such mistreatment will be able to have their complaint promptly and impartially examined (Articles 12 and 13).

Like the Civil and Political Rights Covenant the Torture Convention provides for a system of periodic reporting to a Committee of experts. The Committee established under the Convention consists of ten members elected to serve in their independent capacity for terms of four years. They are required to be 'experts of high moral standing and recognized competence in the field of human rights' and Article 17 (2) provides that states should bear in mind the usefulness of nominating people who also serve on the Human Rights Committee. This is a legacy from the original Swedish draft for the Torture Convention which provided for the Human Rights Committee to examine the reports. Currently one member from the Human Rights Committee also serves on the Torture Committee. In addition to considering state reports, which are required to be submitted every four years, there is also provision for states to indicate that they are willing for the Committee to receive interstate (Article 21) or individual (Article 22) complaints that they are failing to meet their obligations under the Convention. The United Kingdom has made a declaration only in respect of Article 21. Article 20 also entitles the Committee to undertake an inquiry where it receives 'reliable information which appears to it to contain well founded indications that torture is being systematically practised in the territory of a state party'. Such an inquiry would be confidential, though at the end of such proceedings the Committee could include a summary account of them in its annual report. No such inquiries have been undertaken to date.

The main mode of operation of the Torture Committee is therefore again the non-judgemental consideration of state reports. Like the Human Rights Committee the Torture Committee has shown itself increasingly willing to receive information from non official sources when formulating questions to put to state delegations during consideration of their reports. Unlike the Human Rights Committee, its remit in Article 19 (3) requires it to make general comments on each report and to submit these to the state party concerned. It may also include these comments, together with any observations on these comments by the state party, in its annual report to the General Assembly. This is a public document and reports so far have disclosed some fairly strong criticism of the failure of some states to assure the Committee that they were meeting their obligations under the Convention.[69]

OTHER INTERNATIONAL MECHANISMS

The treaties and procedures outlined above are the most important potential avenues of redress in the international arena for prisoners complaining that their rights have been infringed. As can be seen some of these offer remedies on an individual level but most are designed to consider at a structural level the extent to which governments have complied with their international human rights obligations and to point out matters which need to be addressed. The raising of concerns about the extent to which prisons in the United Kingdom meet such international standards is thus best done in such forums through reports and intervention by national and international non-governmental groups.

Certain other international mechanisms exist and can be mentioned briefly.[70] The United Nations has established a number of non-binding

---

[69] See the comments of Rapporteur Burns pronouncing himself 'not satisfied' by the reasons given by the United Kingdom for failing to introduce further safeguards in respect of treatment of terrorist suspects in police detention in Northern Ireland, CAT/C/SR.92.

[70] The best overall description of other international remedies can be found in N. Rodley, *The Treatment of Prisoners in International Law* (Oxford, 1987), esp. Ch. 5.

codes that affect the area of arrest and detention. These include the Standard Minimum Rules for the Treatment of Prisoners (1955), the Code of Conduct for Law Enforcement Officials (1979), and the Standard Minimum Rules for the Administration of Juvenile Justice (1985). The United Nations Sub-Commission for the Prevention of Discrimination and the Protection of Minorities has since 1973 included on its agenda every year an item on the human rights of people subjected to any form of detention. This report includes recommendations on how to improve the protection of human rights of those in detention. The Sub-Commission's parent body, the Commission on Human Rights, established in 1985 a Special Rapporteur on Torture. The Special Rapporteur has the task primarily of responding to information that comes to him regarding acts of torture anywhere in the world. He carries out this mandate by requesting immediate information from the state in question and by making recommendations to those governments concerned. A summary of the Special Rapporteur's actions each year is published and includes reference to the countries from which information was sought and to which recommendations were made.

5

Living Conditions in Prison

INTRODUCTION

The topic of prison conditions is a large and diverse one. Many of the issues mentioned in other chapters in this book, notably matters of discipline and powers to maintain good order, have a significant impact on the conditions in which prisoners serve their sentences. In this chapter we will however concentrate on prisoners' day-to-day living conditions, on matters which in the outside world people can provide for themselves or through the help of state agencies but on the inside must be provided by the prison authorities. Thus issues of food, work, education, health care, and living space all pose difficult questions of entitlement and resource allocation. As in most prison systems the courts in the United Kingdom have generally seen these issues as entirely within the discretion of the prison authorities and have been very reluctant to interfere with the decisions reached by the prison administration. This is not the position in the United States. There the courts have responded to severe overcrowding and deteriorating living conditions by declaring individual prisons, or on occasions an entire state prison system, to constitute 'cruel and unusual punishment' under the Eighth Amendment to the Constitution.[1] Such findings have then served as the catalyst for further legal or political action designed to change living conditions in the prison.

Prison conditions in the United Kingdom have not yet reached the depths found in many American state systems in the early 1970s. However, in the last decade concern has been growing as local prisons especially become increasingly overcrowded. As a result physical conditions in many of these prisons, which are often of nineteenth-century vintage, deteriorate and pressure on staff resources means a reduction in work and recreation opportunities. Complaints have abounded of prisoners spending most of their time (up to twenty-three hours a day) in their cells, with little to do and not even adequate access to washing or toilet facilities. In the late 1980s Her Majesty's Inspector of Prisons Judge Stephen Tumim published a series of reports castigating the state of Britain's prisons. In his 1988 Annual Report,

---

[1] As of 1 Jan.1992 the prison systems of forty-three American states were wholly or partly subject to court orders to improve prison conditions. *National Prison Project Journal*, 7/1 (1992), 6.

for example, he observed how in many establishments 'cell windows and outside walls are smeared with excrement and . . . parcels of faeces—wrapped in newspaper or items of clothing—litter the ground outside'.[2] In 1991 the European Committee on Torture, making its first visits to the United Kingdom under the European Convention for the Prevention of Torture, declared that the 'cumulative effect of overcrowding, lack of integral sanitation and inadequate regimes' at three British prisons amounted to inhuman and degrading treatment.[3] Such reports suggest that there is a need for some sort of redress for prisoners held in intolerable conditions in British prisons. After examining the state of the law on a number of specific conditions issues this chapter examines possible avenues of legal challenge to conditions of confinement in the jurisprudence of the federal courts in the United States. It then considers whether any similar challenge could be brought in British or European law.

ACCOMMODATION AND RECREATION

The current legal provisions regarding prison accommodation are somewhat thin. Section 14 of the Prison Act 1952 indicates that a cell may only be used if an official acting on behalf of the Secretary of State has issued a certificate in respect of it. This certificate may indicate the period for which a prisoner may be separately confined in the cell and place a limit on the number of prisoners that can be held in the cell.[4] Certificates now however refer to both the Certified Normal Accommodation (CNA), which indicates how many prisoners a prison or particular cell may accommodate without overcrowding, and the Operational Capacity, which includes 'a certain degree of overcrowding', authorized by the Area Manager. Moreover under Prison Rule 23 (2) the number specified in a certificate may be exceeded with the leave of the Secretary of State. In 1992 over 9,000 prisoners were being held two to a cell designed for one and a further 1,200 were detained three to a cell designed for one.[5] Circular Instruction 19/1992, which currently governs certification, notes that some prisons have been able to operate with the same certificates for many years but provides that certificates should now be reviewed each year.

The English position is therefore somewhat weaker, in both law and practice, than what is required by the European Prison Rules. Rule 14 (1)

---

[2] See e.g. *Annual Report of the Chief Inspector of Prisons for England and Wales 1988–9*, HC. 491 (1990).

[3] See *Report to the United Kingdom Government on the Visit to the United Kingdom Carried out by the European Committee for the Prevention of Torture and Inhuman or Degrading Treatment or Punishment from 29 July 1990 to 10 August 1990*, CPT/Inf (91) 15.

[4] See CI 19/1992 on Accreditation.

[5] See *Prison Overcrowding: Some Facts and Figures*, NACRO Briefing No. 28 (1992).

of these indicates that accommodation at night will normally be in individual cells 'except in cases where it is considered that there are advantages in sharing accommodation with other prisoners' (the English Circular Instruction indicates the wisdom of having some double cells for suicide prevention purposes). In The Netherlands since 1947 housing more than one prisoner in a cell has been prohibited by statute.[6]

A similar lack of specificity can be seen in the rules on space, lighting, heating, ventilation, sanitation, and bedding. Only the last is expressly referred to in the Prison Rules and there all that Rule 24 indicates is that prisoners should be given bedding that is 'adequate for warmth and health'. The Standing Orders provide some more detail but not that much more. Standing Order 14 (4) (*a*) indicates that prisoners must be given the opportunity to take one bath or shower a week, SO 14 (25)(*c*) that cells should be scrubbed once a week, SO 14 (27) that bedding should be laundered 'regularly', and SO 24 (54) that prisoners should have 'routine' access to toilets at night. But Standing Orders do not indicate what the minimum cell size should be or what quality of lighting it should have, though they do indicate that the temperature in cells should be kept at 16° degrees centigrade. Nor do they say anything about the level of noise, which is a major problem in prisons. Although Prison Rule 27 (1) indicates that a prisoner is normally entitled to one hour's exercise in the open air each day the Standing Orders do not elaborate on this by specifying how much time a prisoner should normally spend outside his cell. Other jurisdictions have been much more forthcoming. In the American federal prison system minimum standards have been produced that, for example, indicate that prisoners should have at least sixty square feet of cell space and that if they have less than eighty square feet they should not spend more than ten hours a day in their cells. That having standards is not a panacea is however indicated by the fact that around 60 per cent of American prisoners serve their sentences in less than the federal minimum.[7]

Greater interest in having clearer standards regarding accommodation has arisen in the United Kingdom with the increase in prison overcrowding and its attendant problems of reduced access to sanitation and prisoners spending an increasing amount of time in their cells. Although the Home Office argued in its response to the 1991 Woolf Report that overcrowding had fallen over the past thirty years this decrease has been unevenly spread.[8] Thus, while long-term dispersal prisons are largely operating at below capacity, smaller local prisons, housing mostly remand prisoners, are

---

[6] Noted in A. Rutherford, *Prisons and the Process of Justice* (Oxford, 1986), 112.

[7] Figure quoted ibid. 8.

[8] See Home Office, *Custody, Care and Justice: The Way Ahead for the Prison Service in England and Wales*, Cm. 1647 (London, 1991) 59.

often holding over 150 per cent of their capacity.[9] In these prisons prisoners are generally locked up two to a cell for between eighteen and twenty-three hours. Despite Home Office commitments to end the practice 'slopping out' remains prevalent in these prisons, and even despite the current prison building programme it is estimated that by 1999 a quarter of prison places will still be without access to night sanitation.[10] Whether any legal challenge can be mounted to such conditions appears to amount to whether they can be described as 'intolerable' in domestic law, or 'inhuman or degrading treatment or punishment' in violation of Article 3 of the European Convention on Human Rights, issues to be discussed below under the rubric of challenges to prison regimes. It is worth noting that the US Supreme Court in *Rhodes* v. *Chapman*[11] indicated that the practice of putting two prisoners in a cell designed for one did not *per se* constitute a violation of the prohibition on 'cruel and unusual punishment' contained in the Eighth Amendment to the Constitution. Later courts in the United States interpreting this decision have however indicated that, when combined with other factors, overcrowding may amount to a constitutional violation.[12]

SEARCHES

Prison Rule 39 (1) requires that every prisoner should be searched on reception into a prison and 'subsequently as the governor thinks necessary'. This is obviously a very wide discretion and the rules make no distinction between the different types of search involved. Section 46 of the Canadian Corrections and Conditional Release Act 1992, for example, divides searches into 'frisk searches', 'non intrusive searches', 'strip searches', and 'body cavity searches'. Only 'frisk' and 'non intrusive' searches can be carried out without reasonable suspicion. The only limitations imposed on the manner of the search contained in Rule 39 are that it must be conducted in as 'seemly a manner as possible as is consistent with anything being concealed', and that a prisoner should not be subject to a strip search in the sight of another prisoner or an officer of a different sex. Non-strip searches may be carried out by officers of a different sex but Home Office guidance

---

[9] In 1992 while the prison system as a whole was 1% overcrowded eleven prisons were over 50% overcrowded. For example Gloucester Prison was at 182% of operating capacity, Chelmsford at 169%, Birmingham at 163%, and Durham at 162%. See *Prison Overcrowding*.

[10] This was the view taken by S. Casale, and J. Plotnikoff in *Minimum Standards in Prisons: A Programme of Change* (London, 1989), 4, reviewing the existing prison-building programme. The Home Office pledged to end 'slopping out' and ensure full access to sanitation by the end of 1994. See Home Office, *Custody, Care and Justice*, 59.

[11] 452 US 337 (1981).

[12] See e.g. *Tyler* v. *Black* 811 F. 2d 424 (8th Cir. 1987), where the court held that double celling in small cells with solid 'boxcar'-like doors was cruel and unusual punishment.

indicates that while female prison officers may be involved in searches of male prisoners, male officers should not be involved in searching female prisoners.

Although the power is wide, like any discretionary power its exercise is subject to judicial review. Hence prison staff in deciding to search should direct their minds to the purposes for which the power is granted, essentially to look for items smuggled into the prison (notably drugs) or things which might assist an escape. Searches whose purpose was purely to harass a prisoner would undoubtedly be seen as an abuse of discretion. Similarly the power to search is subject to the requirement of reasonableness, as regards both the decision to search and the manner of its execution. Thus while a random cell search might be upheld by the courts as furthering reasonable aims of institutional security a decision to subject all prisoners to a strip search or body cavity search without any advance reason to suspect the need for this might be viewed as a decision no reasonable governor would reach. In a number of decisions the courts have ruled routine police strip searches, carried out under common law powers, to be unlawful as the police officers did not address their minds to the issue of whether such a search was necessary in the instant case.[13] Where an unlawful search is of a prisoner's person rather than his cell the prisoner may be able to obtain damages for assault.

Excessive searches which might be said to degrade a prisoner (for example regular strip searching, especially if physical force is used to carry out such searches) might also constitute a breach of Article 3 of the European Convention on Human Rights. Though the European institutions have usually shown a fair amount of latitude to domestic authorities in respect of Article 3 claims relating to sentenced prisoners, on the grounds that a certain degree of loss of dignity is inevitable in the circumstances of prison and that security concerns may justify many practices that otherwise would have a degrading effect, actions which are such that they could 'destroy the personality and cause severe mental and physical suffering' have been regarded by the Commission as actionable.[14]

In the United States the Supreme Court has ruled that prisoners do not have any privacy interest as regards their cells and hence that the protections of the Fourth Amendment do not apply to cell searches.[15] The court has not ruled on whether the Fourth Amendment, which protects against unreasonable searches, applies to prisoners in respect of personal searches, but lower courts have ruled that blanket rules requiring body cavity

---

[13] See e.g. *Lindley v. Rutter* [1981] QB 128 and *Brazil v. Chief Constable of Surrey* [1983] 3 All ER 537.

[14] See X v. *United Kingdom* D. & R. 21 (1981) 99.

[15] See *Hudson v. Palmer* 468 US 517 (1984).

searches of prisoners,[16] or visitors,[17] without any showing of probable cause (the American equivalent of reasonable suspicion) violate the Constitution. Searches of visitors to prisons in England and Wales are permitted by Rule 86 of the Prison Rules, though scrutiny of the exercise of this discretion may be even more exacting that of the power to search prisoners. A County Court indicated in 1992 that a strip search of a visitor to a prison gave rise to an action for damages because it was not conducted in as seemly a manner as possible and involved violation of bodily integrity.[18]

FOOD

The Woolf Report observed that the quality of prison food featured in more complaints from prisoners than anything else.[19] Rule 21 (4) of the Prison Rules indicates that food should be 'wholesome, nutritious, well prepared and served, reasonably varied and sufficient in quantity'. Until 1988 remand prisoners were entitled to receive food from friends or relatives. The rule change which prohibited this was held to be *intra vires* the Prison Act.[20]

Standing Order 7A (25) indicates that prisoners are entitled to a diet which accords with the demands of their religion 'as agreed between a relevant religious body and Prison Service Headquarters'. There is also a requirement that the food be inspected 'regularly' by the medical officer and 'at frequent intervals by the Board of Visitors'. The Standing Orders also provide for the inspection of catering facilities and arrangements for the transportation and serving of food by Home Office Health and Safety Officers according to standards laid down in Food Hygiene (General) Regulations 1970. The Standing Orders also indicate that the catering area of each establishment should have at least one hygiene check a year and that the inspection system is itself subject to unannounced random independent validation checks by environmental health officers. Not until the 1990 Food Safety Act removed Crown Immunity in respect of prison kitchens could court proceedings be taken in respect of standards in prison establishments which fall below those required in the Food Act 1984. Also local environmental health officers do not themselves have a right of access to prisons in their area.

Under Prison Rule 22 prisoners are prohibited from having any alcohol

[16] See *Weber* v. *Dell* 804 F. 2d 796 (2nd Cir. 1986).

[17] See *Blackburn* v. *Snow* 771 F. 2d 556 (1st Cir. 1985).

[18] *Bayliss and Barton* v. *Home Secretary and Governor HM Prison Frankland* Legal Action (Feb. 1993) 16.

[19] See *Prison Disturbances: April 1990* (Woolf Report), Cm. 1456 (London, 1991), 397.

[20] See *R.* v. *Secretary of State for the Home Dept. ex parte Simmons* The Times 25 Oct. 1988.

except with the permission of the medical officer and may have tobacco only as a privilege.

CLOTHING

Under Prison Rule 20 prisoners must be provided with clothing that is 'adequate for warmth and health'. This requirement does not apply to unconvicted prisoners and women prisoners, who may wear their own clothes, as may sentenced prisoners in Northern Ireland. An application by a prisoner serving a sentence in a United Kingdom gaol that being required to wear prison clothing constituted a breach of the right of privacy as protected by Article 8 of the European Convention was declared admissible. The Commission went on to hold though that the restriction was justifiable in the interests of public safety and for the prevention of crime.[21] However the Woolf Report, observing that there is a significant amount of freedom for long term prisoners in dispersal prisons as to what they wear, has recommended that all prisoners should eventually be entitled to wear their own clothes and that to begin this process prisoners should be entitled to wear their own clothes during visits.[22] Though this has not yet been adopted a pilot scheme allowing convicted prisoners to wear their own clothes is under way at a number of prisons, and governors at Category C and D prisons have been given discretion to allow inmates to wear their own shoes and underclothes.

The Rules also require the provision of suitable clothing for work. However, apart from a reference in Standing Order 14 (21) to sufficient underclothing being available for prisoners to have a change at least twice a week there are no guide-lines on how often a prisoner should be able to change his clothes. The American federal standards, to take one example, specify changes three times a week in small establishments and daily elsewhere.

WORK

Prison Rule 28 (1) states that all convicted prisoners are required to do a maximum of ten hours' useful work a day. Unconvicted prisoners are not required to work but should be provided with work if they wish. Although

---

[21] See X v. *United Kingdom* D. & R. 33 (1983) 5. In *McFeeley* v. *United Kingdom* (1980) 3 EHRR 161, the Commission rejected the claim of Northern Irish prisoners that requiring them to wear prison uniform violated their freedom of expression as it denied their claim to be prisoners of war.
[22] See Woolf Report, 397.

Article 4 of the European Convention on Human Rights prohibits forced or compulsory labour it contains an exception in Article 4 (3)(*a*) for ordinary work done in the course of detention. Many complaints in prisons concern the lack of work and its limited character. The average number of hours per week worked by prisoners is approximately twenty-two and in remand prisons it falls to around eleven.[23] Also prisoners continue to be employed on repetitive and low-skilled tasks. In 1983 the Chief Inspector of Prisons, in a report on a riot at Albany Gaol, observed that one of its causes was that long-term prisoners were still involved in sewing mailbags.[24] The European Committee on the Prevention of Torture commented, 'much of the little work that was available in the three prisons [visited] was of a dull, repetitive nature and did not involve the acquiring of any skills that might be of some use to the inmate outside prison'.[25] These conditions may well be in breach of the European Prison Rules. Rule 72 (3) of these indicates that 'sufficient work of a useful nature shall be provided to keep prisoners actively involved for a normal working day', while Rule 72 (4) indicates that 'so far as possible the work provided shall be such as will maintain or increase the prisoners' ability to earn a normal living after release'.[26] Neither the Prison Rules in England and Wales nor the Standing Orders contain any reference to these issues.

As the Woolf Report notes, prisoners' pay has also been an issue of constant dispute. The Report noted that prisoners' average pay was £2.65 a week, well below many other European countries, and recommended an increase to £8. In response to this the government introduced a new prisoners pay scheme in late 1992. This provides for a basic rate of £2.50 per week even if a prisoner is unemployed due to an absence of work available in the prison. Prisoners doing any work must be paid an Employed Rate (ER), which is currently set at £4 per week. Thereafter governors have the discretion, subject to agreement with Area Managers, to set pay rates for different jobs at their prison. These should produce an average wage of £6 per week for each prisoner, a target which limits the governor's budget. The Prison Service must be contacted where a prisoner's standard rate is set at more than £12. The scheme also provides for sick pay and bonuses. Standing Order 7B (7) indicates that prisoners will only be paid if their work achieves an acceptable standard and output and it is still a discipli-

---

[23] See Home Office, *Custody, Care and Justice*, 71.

[24] See *Report of HM Chief Inspector of Prisons for England and Wales 1983*, HC 618 (London, 1984), para. 1.03.

[25] *Report by the European Committee for the Prevention of Torture*, at para. 55.

[26] This rule represents one resolution of the debate that was particularly prominent in the 18th and 19th centuries as to whether prison work should be hard and repetitive, so as to punish better, or varied and useful, better to equip the prisoner for life after prison. See D. Garland, *Punishment and Welfare*, (Aldershot, 1985), 12–16.

nary offence under Rule 47 (17) for a prisoner to be 'idle, careless or negligent at work'.

Prisoners are not however regarded as employees of the prison in the course of their work.[27] This has significant consequences in that it means that the Health and Safety at Work Act 1974 is largely inapplicable in prisons and cannot be invoked by prisoners who suffer through breaches of it, even though Standing Order 6A (15) allows for inspections by the Health and Safety Executive and requires the giving of every assistance to them. The courts have also indicated that the Factories Acts are inapplicable to prison industries.[28] Prisoners injured at work are therefore thrown back on the common law duty to provide a safe system of work and have succeeded in negligence actions arising out of their injuries at work.[29] The Standing Orders also provide that where a prisoner's injury at work results in incapacity upon release the prisoner may be made an *ex gratia* payment at the same rate as disablement benefit under the Industrial Injuries Act following certification by a Department of Social Services Medical Board.

Although prisoners are not regarded as employees they may invoke legislation that protects people against discrimination as regards employment. In *Alexander* v. *Home Office*[30] Alexander, who was of Caribbean origin, successfully sued for damages after he was refused a higher-paying job in a prison kitchen on the basis of two assessment reports which contained racially discriminatory remarks. Alexander challenged his treatment invoking s. 20 (1) of the Race Relations Act 1976, which prohibits discrimination in the provision of goods and services to the public or a section of the public, rather than s. 4, which relates to discrimination in employment. The Home Office originally defended the case on the grounds that prisoners were not a 'section of the public', but abandoned this point at trial. The availability of race discrimination legislation to prisoners is particularly significant given that black people form up to 25 per cent of the prison population at some prisons and the Chief Inspector of Prisons has on several occasions drawn attention in his reports to racial tensions in prison.

EDUCATION

Prison Rule 29 indicates that every prisoner able to profit from education should be encouraged to do so and that programmes of evening educational study should be arranged at every prison. It also indicates that prisoners should be afforded 'reasonable facilities' to engage in private study. Standing Order 7B indicates that the prison education officer should organize education, training, and library facilities in co-operation with the Local

[27] See *Pullen* v. *The Prison Commissioners* [1957] 3 All ER 470.　　[28] Ibid.
[29] See *Ferguson* v. *Home Office* The Times 8 Oct. 1977.　　[30] [1988] 2 All ER 118.

Educational Authority and Public Library Authority. The education officer therefore has the responsibility of arranging classes for both daytime and evening. The Woolf Report observed that educational spending had risen significantly and that the Home Office now estimated that around 17 per cent of prisoners were engaged in full-time study. However, prisoners still complained that there was a lack of access tô classes due to staff shortages. Complaints also relate to significant variance in the curriculum at different prisons, which means that prisoners are not able to continue courses they have started after they are transferred.[31]

Prisoners may also pursue correspondence courses and work for national qualifications, including Open University degrees, at the discretion of the governor. In order to do so they may obtain and retain, at their own expense, books and up to two periodicals. In its response to the Woolf Report the Home Office indicated that it would extend access to National Vocational Qualifications.

Education is voluntary for all adult prisoners, but young offenders of compulsory school age are required to attend education or vocational training courses for fifteen hours a week. Standing Orders indicate that all courses should be devised in accordance with a number of objectives including the improvement of prospects for successful social resettlement, employment prospects on release, and the prisoner's morale, attitudes, and self respect. Unlike some European prison systems, however, achieving educational qualifications does not have any formal effect on improving a prisoner's prospect of release.[32]

At time of writing the prison education system seems likely to undergo a significant change as it is one of those aspects of the prison system that the government has decided to market test. Prison education officers are to be empowered to contract out prison education to the private sector. This will not however alter the obligation on the prison system to provide education.

Education is one area where the United Kingdom appears to be falling well short of its obligations under the European Prison Rules. These provide for a 'comprehensive education programme in every institution' (Rule 77), for education to be regarded as a regime activity that attracts the same status and remuneration as work (Rule 78), and for education programmes to be integrated as far as possible into the educational system of the country (Rule 81). The last requirement was considered during the review that led to market testing but appears to have been rejected in favour of allowing free competition for education provision.

---

[31] See Prison Reform Trust, *Prison Report*, 19 (1992), 4.
[32] In France for example under Art. 721-1 of the Code of Criminal Procedure the *juge de l'application des peines* (the judge who oversees the execution of the sentence) may reduce a sentence by up to a month for each year if a prisoner passes a school, university, or professional examination.

FACILITIES

Facilities is the new term for what have historically been known as privileges, a term that has particular significance in the prison context. As is discussed more extensively in Chapter 2 of this book, there has been an important debate over whether prisoners have any rights or just privileges which are dependent on co-operative behaviour. As a result of litigation and political action it now appears clear that prisoners have certain rights, for example to fair disciplinary hearings or to make applications to the courts, which cannot be removed at the discretion of the prison authorities. However, this section deals with privileges in a more precise sense, with the system of privileges or facilities that every prison is required to establish in accordance with Prison Rule 4.

These have now been set out in Standing Order 4 and include a range of things, of which the most significant are perhaps personal possessions, tobacco, books and newspapers, a radio, and association with other prisoners. Most facilities are available to all prisoners. A governor may however add to the list with the authorization of the Area Manager.

Perhaps the most important facilities relate to items of personal possession that prisoners are entitled to retain in their cells. Standing Order 4 (7) indicates that prisoners are allowed to have 'sufficient property in possession to lead as normal and individual an existence as possible within the constraints of the prison environment and the limitations under this and other standing orders'. Items which come within this category are stated in Standing Order 4 (9). This indicates that, among other things, prisoners will be allowed to retain in their possession at least six newspapers and periodicals, at least three books, a sound system, smoking materials up to eighty cigarettes, a battery shaver, and a manual typewriter (but not an electronic typewriter or computer). Prison governors may add to what is already a very detailed list. Facilities may be withdrawn either as a result of a disciplinary punishment or where the governor considers that the number of items in a cell hinders searching, prevents the cell from being tidy, presents a risk to health, safety, security, or discipline, or that they cannot be moved with a prisoner on transfer. Removal of facilities as a result of a disciplinary hearing is obviously subject to judicial review of the procedures by which the disciplinary verdict was reached. However, removal by way of a governor's exercise of discretion could also be challenged by way of judicial review, notably by way of the *Wednesbury* 'unreasonableness' standard. Facilities may be regarded as a 'legitimate expectation', whose deprivation might attract procedural fairness review.[33]

[33] See, generally, B. Hadfield, B 'Judicial Review and the Concept of Legitimate Expectation' (1988) 39 *NILQ* 103.

Standing Order 4 also governs the acquisition of further possessions. It indicates that prisoners may purchase further books, newspapers, or audio cassettes. Except for prisoners in Category C or D establishments, these must be ordered direct from publishers, booksellers, or registered record clubs. Governors may waive this requirement in respect of books provided by friends or relatives. The Standing Order gives a governor power to withhold a book, newspaper, or audio cassette if he or she considers that the content presents a threat to good order or discipline or 'where the medical officer considers that possession of the material is likely to have an adverse effect on the prisoner's physical or mental condition' (Standing Order 4 (35)). The Standing Order also indicates that the governor may impose restrictions on material which he considers 'is likely to cause offence by reason of its indecent or violent or racist content or is racially or sexually discriminatory'. This may seem a welcome policy declaration against racism and sexism but it also indicates how attenuated a prisoners right to privacy is, even in his own cell. In the outside world racial and sexual discrimination is generally permitted in the 'private' sphere (the household exemption from anti-discrimination statutes for example, or the different treatment of public display of pornography from its private possession). In prison it seems virtually no privacy rights remain.

A decision to prevent a particular book or newspaper coming into the prison could also be challenged on judicial review grounds and might be subject to review for violation of the European Convention on Human Rights' articles on the rights to privacy or to receive information. However the Commission has generally upheld official action in this area after fairly cursory review.[34]

SAFETY

As has been indicated earlier the prison authorities have a duty towards the prisoner to provide a safe system of work, a duty that may be enforced by common law actions for negligence. A duty has also been recognized to exist in respect of attacks on prisoners by other prisoners.[35] However, since the actions are based on negligence the context is all-important, as the key question is whether the authorities have done all that could reasonably be asked of them to prevent the prisoner being injured. We have discussed this

[34] P. Van Dijk and G. Van Hoof, *Theory and Practice of the European Convention on Human Rights* (2nd edn., Dewenter, 1990), 427.

[35] See *Ellis* v. *Home Office* [1953] 2 QB 135. See also *H* v. *Home Office* the Independent 6 May 1992, where a prisoner was awarded damages after negligent disclosure of his past convictions for sex offences led to him being assaulted by other prisoners. He was unable to recover anything in respect of his having to go on Rule 43 segregation subsequent to the assault.

issue in greater detail in Chapter 3 but it is worth briefly summarizing its implications for issues of prisoner safety at this point.

Cases on sex offenders, who are generally accepted to be at a higher risk of attack than most prisoners, illustrate the difficulties that are experienced in arriving at an appropriate standard. In *Egerton* v. *Home Office*[36] May J. refused to find that the authorities had been negligent, despite the fact that the officers supervising Egerton were unaware of his offences or the fact that he had just come off Rule 43 following a similar attack, on the grounds that even an experienced officer could not have foreseen the actual attack that took place. However in *Steele* v. *Northern Ireland Office*[37] a prisoner accused of sex offences recovered damages for an assault by other prisoners even though he had refused advice from the prison staff to be accommodated in a protective wing. In the United States courts have held that failure to protect prisoners from assault can amount to cruel and unusual punishment in violation of the Constitution where it results from deliberate indifference to protect a prisoner from an obvious risk of violent harm which is or should be known.[38] However, generally American courts are reluctant to impose too high a burden on prison administrations for the violence that is a daily reality of many of their prisons.

Threats to prisoner safety may come from prison officers as well as other prisoners. Where prison officers use force which has no legitimate objective or is excessive for the legitimate objective they seek to achieve they may be subject to either civil or criminal actions for assault. They may also face disciplinary proceedings.

HEALTH CARE

1. GENERAL ISSUES

Under s. 7 (1) of the Prison Act 1952 every prison is required to have a medical officer. The Prison Rules indicate that all requests to see the medical officer should be recorded by the officer to whom they are made and promptly passed on to the medical officer. As well as the general duty of having 'the care of the health, mental and physical of prisoners in that prison', as Rule 17 (1) states, the medical officer has a number of other specific duties established by the Rules. These include reporting to the governor on any prisoner whose health is likely to be injuriously affected by continued imprisonment or any conditions of imprisonment, reporting on any prisoners who he thinks might be suicidal, regularly inspecting the food, deciding on a prisoner's fitness for work or exercise, deciding whether

---

[36] [1978] Crim. LR 494.   [37] [1988] 12 Northern Ireland Judgements Bulletin 1.
[38] See e.g. *Morgan* v. *District of Columbia* 824 F. 2d 1049 (DC Cir 1987).

a prisoner should remain on Rule 43 segregation or is fit for a disciplinary punishment of solitary confinement, and deciding whether a prisoner may remain under restraint. This broad range of tasks and in particular the prison medical officer's involvement in the authorization of disciplinary punishments or GOAD powers have led some commentators to question whether medical officers are capable of presenting themselves to prisoners as primarily concerned with the prisoner's health as opposed to the managerial objectives of the governor and staff. As one put it:

He [the prison doctor] is part and parcel of the prison hierarchy, intimately involved in the maintenance of order and discipline. His consent is necessary before a prisoner can be kept under special restraint or before a prisoner can be subject to an award of cellular confinement. No individual, however skilled and compassionate a doctor, can maintain a normal doctor–patient relationship with a man whom the next day he may acquiesce in subjecting to solitary confinement.[39]

Concerns about the capacity of medical officers to respond adequately to prisoners' medical needs is exacerbated by the fact that the Health Care Service for Prisoners (HCSP) is not part of the National Health Service and hence its doctors are not responsible to and cannot be censured by the local Regional Health Authority. Several groups have called for its integration into the NHS, including the Royal College of Psychiatrists in 1979, but neither the Woolf Report nor the House of Commons Social Services Select Committee went as far as this, although the latter made a large number of recommendations for the improvement of the service. After a 1990 scrutiny report the government recommended closer association with the NHS and the adoption of NHS standards but without formal integration. In any case the HCSP already makes significant use of NHS staff, with over two-thirds of medical work in prisons being carried out by such staff, including all dental work.[40] The Rules however only provide for unconvicted prisoners to be seen by medical personnel of their choice providing that their request to do so is reasonable and that they are willing to pay any charges. Convicted prisoners have this right only in respect of legal proceedings to which they are a party.[41]

As regards the standard of medical care that a prisoner is entitled to receive the courts have indicated that it is that

of the ordinary skilled man exercising and professing to have that special skill. A man need not possess the highest expert skill at the risk of being found negligent. It is well established law that it is sufficient if he exercises the ordinary skill of an ordinary competent man exercising that particular art.[42]

[39] See M. Brazier, 'Prison Doctors and their Involuntary Patients', [1982] *Public Law* 282–300, at 285.

[40] See Home Office, *Care, Custody and Justice*, 248.   [41] See Rule 37A (3).

[42] This standard is derived from *Bolam* v. *Friern Hospital Management Committee* [1957] 2 All ER 118, cited in *Knight* v. *Home Office* [1990] 3 All ER 237, 244.

This suggests that prison medical staff will be held to the standard of doctors outside prison and that in any litigation respecting the standard of care evidence of other medical practitioners will be crucial. In *Knight* v. *Home Office*, however, Pill J. indicated that this standard of care will be assessed relative to the resources available to medical staff in the prison, and that while to fail to provide any medical services at a large prison would clearly be a breach of duty it must be acknowledged that 'resources available for the public service are limited and that the allocation of resources is a matter for parliament'.[43] In this particular case he concluded that to require a prison to provide the same level of staff to observe a suicide risk who was subject to a hospital order as would be available in a psychiatric hospital was not a breach of duty.

Pill J's conclusion that while a psychiatric hospital had a role of curing people a prison had a duty only to detain people in custody (out of which a duty to prevent them injuring themselves arose), and his decision not to specify what standard a prison might be judged by leaves the law in an unsatisfactory state. It could be argued that a prison, while it keeps physically or mentally ill people in prison rather than transferring them to a hospital, should be assessed by the same standards that apply to the institutions that such people would have gone to had they not been detained. To do otherwise harks back to the idea that prisoners forfeit more than their liberty when detained. One wonders whether Pill J. would have come to the same conclusion in a case involving a remand prisoner. American courts have observed that 'deliberate indifference' to prisoners medical needs may constitute cruel and unusual punishment. By 'deliberate indifference' they appear to mean a significant deviation from accepted professional standards. It is clear that pleading inadequate resources will not be a defence where such a deviation is proven.[44]

Concerns over prisoners' rights to refuse treatment have been almost as significant as those over their rights to receive treatment. In the 1970s in particular several writers on prisons expressed concern about the use of psychotropic drugs in prison.[45] The issue received a hearing in the courts in *Freeman* v. *Home Office*.[46] There the plaintiff, a life sentence prisoner, alleged that he had been forcibly injected with drugs and that in any case he could not be said to be 'voluntarily' giving his consent as he was a prisoner in a coercive environment, certainly not to be giving it voluntarily where he was not fully informed of the potential consequences of the drug.

---

[43] Ibid., at 243.

[44] The 'deliberate indifference' standard is set out in *Estelle* v. *Gamble* 429 US 97 (1976). That inadequate resources will not be a defence is indicated in *Hamm* v. *DeKalb County* 774 F. 2d 1567 (11th Cir. 1985).

[45] See J. Sim, *Medical Power in Prisons* (Buckingham, 1990), ch. 5, for a critical discussion of the use of drugs in prison.

[46] [1984] 1 All ER 1036.

The Court of Appeal upheld the trial judge's rejection of the plaintiff's factual allegations and concluded that a prisoner was as capable as anyone else of giving his consent to medical treatment. It also rejected the idea that the doctrine of 'informed consent' had any place in English law, so that while a prisoner must be told the general nature of the treatment it will not vitiate his consent, and therefore constitute an assault, if the doctor does not inform him of all the potential side-effects.[47] The court therefore ruled that the only issue in every case was one of fact, whether the prisoner had consented. Such a decision exhibits a lack of sensitivity to the prison context. While it would be erroneous to argue that prisoners can never voluntarily consent to treatment (such a position would render the practice of medicine in prisons almost impossible), prisoners' absence of choice and access to independent advice plus the power that the medical officer exercises over other aspects of their lives suggest a particular vulnerability and a need for greater safeguards to ensure consent is voluntary.

Current Standing Orders indicate that 'in general' prisoners are free to accept or reject any treatment offered to them. However the Standing Order does purport to authorize invasive procedures without the prisoner's consent where 'without such procedures the prisoner's life would be endangered, serious injury to him or her or to others would be likely, or there would be an irreversible deterioration in the prisoners condition'.[48] There must be some doubts over the legality of such action where the prisoner remains competent to decide whether to accept or refuse treatment, especially if the medical treatments employed as a form of restraint have significant side-effects. Article 7 of the the International Covenant on Civil and Political Rights indicates that prisoners cannot be subject to medical or scientific experimentation without their consent. Rule 27 of the European Prison Rules goes further by indicating that prisoners 'may not be submitted to any experiments which may result in physical or moral injury'. Some commentators have suggested these provisions may actually do a disservice to some prisoners, for example by preventing them from participating in HIV drug trials.[49]

2. FORCE FEEDING AND HUNGER STRIKES

The issue of the right to refuse treatment also arises in a particularly sharp way regarding prisoners who go on hunger or thirst strike. A decision

---

[47] In *Sidaway* v. *Governors of Bethlem Royal Hospital and the Maudsley Hospital* [1985] AC 871, the House of Lords held that the doctrine of 'informed consent' had no place in United Kingdom law. There remains in every case the issue of what a doctor as a reasonable professional person ought to have told the patient lest he or she be sued for negligence.

[48] SO 13 (25).

[49] See J. McHale and A. Young, 'Policy, Rights and the HIV Positive Prisoner', in S. McVeigh and S. Wheeler, *Law, Health and Medical Regulation* (London, 1992), 111.

resulting from the pre-World War I force feeding of a suffragette suggested that there was a legal duty on the prison authorities to force feed prisoners who would otherwise starve to death.[50] However this decision has been frequently criticized, in part because it invokes the public policy of preventing a crime, which suicide was until 1961.[51] It does not now appear to be regarded as authoritative by the Home Office as the current Standing Order regarding hunger strikes indicates that force feeding is very much a clinical decision for the medical officer in consultation with an outside consultant and that 'there is no rule of practice which requires any medical officer to resort to artificial feeding'.[52] Only in the case of prisoners whose capacity for rational judgement is impaired by illness do the Standing Orders incline towards force feeding and even then they reiterate the importance of clinical judgement. This appears to have been the approach taken during the 1981 hunger strike at the Maze Prison in Northern Ireland when ten prisoners starved themselves to death and artificial feeding was only resorted to in the case of those prisoners who lost consciousness and whose families consented.

3. MENTALLY DISTURBED PRISONERS

One recurrent issue in prison medicine is the extent of provision for mentally disturbed prisoners. According to a 1991 Home Office commissioned Report nearly 40 per cent of sentenced prisoners have a psychiatric disorder. Partially this appears to be due to the reluctance of courts to remand offenders to hospitals, as they may do under the 1983 Mental Health Act, as opposed to prison. This reluctance appears often to be influenced by the reluctance of mental hospitals to admit those convicted of crimes.[53] Prisons are not designated as hospitals for the purposes of the 1983 Mental Health Act and while provision also exists under this Act to transfer people sent to prison to psychiatric hospitals there appear to be significant delays even when this procedure is set in motion and continuing difficulties with hospitals' reluctance to admit or retain prisoners. Thus many mentally disturbed people remain in prison, often in an environment which has been described as 'inadequate for the provision of any but the most rudimentary psychiatric care'.[54] In this depressing scenario the courts refusal to prescribe a

---

[50] *Leigh* v. *Gladstone* (1909) 26 TLR 139.

[51] See e.g. G. Zellick, 'The Forcible Feeding of Prisoners: An Examination of the Legality of Enforced Therapy' [1976] *Public Law* 153–87.

[52] See SO 13 (40).

[53] For a recent discussion of the lack of places in Secure Units see Prison Reform Trust, *Prison Report*, 21 (1993), 8.

[54] See J. Gunn, T. Maden, and M. Swinton, *Mentally Disordered Offenders* (London, 1992).

standard of psychiatric care which prisons are under a duty to achieve leaves mentally disturbed prisoners especially vulnerable.

4. HIV POSITIVE PRISONERS

A more recent problem of medical care has been the presence of HIV positive prisoners and AIDS in prison. Given the high number of intravenous drug users that are imprisoned, the fact that some prisoners will engage in homosexual sex while in prison, and the fact that prison populations are in any case drawn predominantly from the younger and more sexually active sections of the community, prison has the potential to lead to significantly greater risks of HIV transfer. One study of European prisons puts the overall prevalence of HIV in prisons at about 10 per cent of the prison population.[55] To date the numbers of prisoners diagnosed HIV positive in the United Kingdom has been relatively small, though some commentators argue that this is a result of inadequate research on the issue.[56]

Standing Orders require that the diagnosing of a prisoner as HIV positive should be immediately reported to the headquarters of the medical directorate. However the policy by which prisoners are identified as HIV positive has been criticized as unsatisfactory. Currently there is no mandatory testing, which might give rise to significant privacy problems, unlike the United States, where fifteen states currently require it. Prisoners are asked if they wish to have a test on entering the prison. Even if they decide not to the medical officer may request a test where he or she feels as a result of the prisoner's answers about his life-style that he may have become infected. Where prisoners who are requested to have a test refuse they may be segregated into a special unit on arrival and remain there until they agree to take a blood test. As it seems requests are only being made to male homosexuals and intravenous drug users some AIDS researchers have argued that the policy may fail to identify a significant number of HIV positive prisoners.[57] The European Committee on Torture also expressed concern regarding a lack of privacy and confidentiality during the initial medical examination when a prisoner is admitted to prison. This may inhibit prisoners from volunteering to take tests.

Prisoners identified as HIV positive are dealt with under a modified version of the Viral Infectivity Restriction (VIR) guide-lines developed by the Prison Department in 1985 to deal with prisoners who have Hepatitis B. These indicate that prisoners should be located on normal landings pro-

[55] Study cited in P. Turnbull, K. Dolan, and G. Stimson, *Prisons, HIV and AIDS: Risks and Experiences in Custodial Care* (London, 1991).

[56] The Home Office states that 339 prisoners have been found to be HIV positive 1986–91. For a critique of these figures see P. Thomas, 'AIDS/HIV in Prisons' (1990) 29 *Howard J.* 1.

[57] See Turnbull *et al.*, *Prisons, HIV and AIDS*, 43.

vided that they have their own cell and that while they are allowed to work they are not to have contact with machinery, sharp instruments, or the preparation of food. They are also excluded from physical contact sports. Information regarding the prisoner's medical condition is provided to prison staff on a 'need to know' basis. While the guide-lines do not require that HIV prisoners be denied association with other prisoners in practice this seems to have occurred, with HIV prisoners being located in prison hospitals or special wings. While some prisoners have expressed a preference for this type of regime, given a significant level of abuse from other inmates and officers, others have argued that such segregation is unnecessary and that 'in many prisons [segregation] results in prisoners with HIV leading an extremely deprived existence with little to distract them from morbid contemplation of their situation'.[58] Woolf was especially scathing about the treatment of HIV prisoners and felt that the limited facilities available in HIV units positively discouraged prisoners from disclosure. He also saw no need to depart from normal principles of doctor–patient confidentiality. Segregation of HIV positive prisoners could be challenged by way of judicial review, although the courts may be reluctant to second guess prison authorities' claims that it is a reasonable response to the situation. In the United States prisoners challenging such segregation policies have had mixed results in the courts, but there does appear to have been an overall improvement in the situation in recent years.[59]

5. SUICIDES IN PRISON

Suicides in prison occur at about four times the rate in the general population.[60] Moreover the number has been rising in recent years, from 102 in 1982–6 to 171 in 1988 91.[61] The Home Office has recognized that this is a problem by producing a suicide prevention manual and requiring every prison to have a Suicide Prevention Management Group. The current Circular Instruction relating to suicide prevention, CI 20/1989, stresses the role of hospital and medical officers in identifying suicide risks and has been criticized as failing to sensitize all prison officers to the possibility of suicides and how to deal with this.

We have already seen that failure to reach the same standard of care in dealing with suicide risks as an outside hospital will not necessarily render

---

[58] See U. Padel, *HIV, AIDS and Prisons* (London, 1988), 16.

[59] In *Harris* v. *Thigpen* 941 F. 2d 1495 (11th Cir. 1991) segregation of HIV positive prisoners which resulted in disclosure of their status was upheld as constitutional. Some District Courts have upheld prisoners' claims not to be transferred to separate areas for HIV positive prisoners by invoking privacy rights, e.g. *Doe* v. *Coughlin* 697 F. Supp. 1234 (ND NY 1988). See, generally, J. Kelly, 'AIDS, Prisoners and the Law' (1992) 142 *NLJ* 156–9.

[60] See, generally, A. Liebling, *Suicides in Prison* (London, 1992).

[61] See Prison Reform Trust, *Prison Report*, 20 (1992), 17.

prison authorities liable to the family of a prisoner who commits suicide.[62] However the courts have rejected the notion that because suicide is the action of a prisoner the authorities are absolved of all legal responsibility in respect of it. In *Kirkham* v. *Chief Constable of Greater Manchester*[63] the wife of a prisoner who hanged himself was able to recover damages from the police for negligence. Her husband had been identified by the arresting officers as a suicide risk but in breach of normal practice they failed to pass this information on to the prison to which he was remanded, with the result that he was placed in an ordinary cell rather than in the hospital wing for observation. The case would appear to indicate that where a prisoner commits suicide after the prison authorities fail to follow their own suicide prevention measures then a negligence action may be available. The duty of care was defined by Farquharson, L.J. as 'a duty on the person having custody of another to take all reasonable steps to avoid acts or omissions which he could reasonably foresee would be likely to harm the person for which he is responsible'.[64] The *Knight* case suggests though that the courts will be less willing to examine whether procedures adopted to deal with suicide risk, where such procedures are followed, are adequate to discharge this duty.

WOMEN IN PRISON

Women make up less than 4 per cent of the prison population. As a result although this book is directed towards the law affecting prisoners generally most of the rules and practices it describes impact particularly on male prisoners. It is also probably true to say that some of the general provisions, such as the Rule 43 segregation power, are designed largely with a male prison population in mind.[65] However some differences exist in conditions and facilities available to male and female prisoners which are worth noting at this point.

First some Prison Rules make explicit differences between male and female prisoners. Rule 9 provides for the separation of male and female prisoners; Rule 26 (3) exempts women from having their hair cut without their consent unless authorized by the medical officer. Even where the Rules do not state differences Standing Orders or Circular Instructions may. Convicted women prisoners are allowed to wear their own clothes, and while female prison officers may conduct rub-down searches of male pris-

---

[62] *Knight* v. *Home Office* [1990] 3 All ER 237.   [63] [1990] 3 All ER 246.
[64] Ibid., at 253.
[65] In the *1990 Prison Statistics*, Cm. 1800 (London, 1992), table 9.1, statistics on the use of Rule 43 showed that 302 men and three women were subject to Rule 43.

oners the relevant CI indicates that male prison officers should not be involved in searching female prisoners.

Secondly, as women prisoners are detained separately from men their work and education facilities are different. This has led to complaints that the education and especially work facilities available to women prisoners generally make only 'women's work' available.[66] In the United States women prisoners have successfully challenged differential access to work and education programmes as sex discrimination.[67] As we have seen earlier the courts have ruled that the Race Relations Act applies to prisons and that prisons are regarded as providing goods and services to a 'section of the public' when they allocate jobs or other benefits in prisons. There seems no reason why the similar provisions of the Sex Discrimination Act 1975 should not similarly be available, though the segregated nature of women's prisons makes comparisons with facilities that would be accorded to a man more difficult. Black and white prisoners are at least detained in the same institution. If the Act was held applicable it might also raise questions about allowing convicted female but not male prisoners to wear their own clothes. This would appear to infringe Rule 2 of the European Prison Rules, which provides for impartial application of all prison rules.

A third area in which differences in conditions are significant is medical care. Commentators have observed that on the one hand medical care available for women in prison has neglected the specific medical needs of women (and has rarely been provided by female medical practitioners), while on the other a 'medicalized' approach has been adopted towards regimes for women prisoners. Women who offend have been more readily regarded as being in need of treatment than men.[68]

A fourth area where specific provisions apply to women is that of pregnancy and child care. Standing Orders indicate that where a prisoner is discovered to be pregnant arrangements should be made for her early release, if the birth is likely to coincide with her earliest date of release, or her transfer to a local hospital to give birth. They also suggest that particular care should be taken when reporting the birth to the Registrar of Births that the place of birth is not given as a prison establishment.

For those women who give birth while in prison or have young children at the time of being remanded to prison separation is usually traumatic, especially if the mother is a single parent, as the children will probably then be placed in local authority care or foster homes. Prison Rule 9 (3) provides for the Secretary of State to permit women to keep their baby with them in

[66] See R. Dobash, R. Dobash, and S. Gutteridge, *The Imprisonment of Women* (Oxford, 1986), ch. 7.

[67] See B. Bershad, 'Discriminatory Treatment of the Female Offender in the Criminal Justice System' (1985) 26 *Boston College Law Review* 389.

[68] See Sim, *Medical Power in Prisons*, ch. 6.

prison, and three mother and baby units exist within the prison system at Holloway and Styal closed prisons and Askham Grange open prison. These currently provide places for thirty-seven children, with those at Holloway and Styal taking babies up to 9 months while Askham Grange accommodates babies of 9–18 months. Circular Instructions governing admission to the unit leave the decision of what mothers to admit and whether they should subsequently be removed and separated from their child to the governor of the prison where the unit is located. In R. v. *Secretary of State for the Home Department ex parte Hicking*[69] a challenge to the governor's decision to remove a mother from the Askham Grange unit because he felt her behaviour was unsettling to the child and to other babies in the unit proved unsuccessful. The Court of Appeal rejected a claim that the Home Secretary could not delegate his power under Rule 9 (3) to decide on who should be in a unit to the governor. The trial court also rejected arguments that the governor's decision was in breach of natural justice in that he took the decision without giving the prisoner a hearing. While there was evidence in the case that the governor only reached his decision after careful thought following a pattern of disruptive behaviour by the prisoner and after consultation with the visiting paediatrician, the court's failure to spell out any required procedure to be followed in this situation is especially disappointing given the potential effect of separation on both mother and child. Separation in the outside world would require a court order, the rights of both parent and child to family life protected by Article 8 of the European Convention would seem good reason to require at least a hearing before such a decision is reached in the prison.

CHALLENGING CONDITIONS

THE AMERICAN EXPERIENCE

Like most jurisdictions courts in the United States traditionally adopted a 'hands off' approach to prisoners' complaints of unlawful treatment. In the 1970s, influenced no doubt largely by developments in civil rights law outside the prison arena, this attitude changed and a significant jurisprudence of prisoners' cases developed.[70] The most spectacular aspect of this was the willingness of federal courts to declare that conditions at individual prisons, or indeed entire state prison systems, amounted to 'cruel and unusual punishment' in violation of the Eighth Amendment to the Constitution, though there has been something of a retreat from this interventionist

---

[69] The Times 7 Nov. 1985.
[70] See J. Jacobs, 'The Prisoners Rights Movement and its Impacts 1960–80', in N. Morris and M. Tonry, *Crime and Justice: An Annual Review of Research*, 2 (1980), 429.

approach during the 1980s. At the beginning of 1992 the prison systems of forty-three states were operating wholly or partly under a court order or consent decree arising from a challenge to conditions as being in violation of the Eighth Amendment.[71]

Judicial interpretation of the Eighth Amendment has never settled on a single test for determining when punishment is cruel and unusual. Rather a variety of standards have been employed of which the most prominent are that the punishment is disproportionate to the offence, that it is in excess of a legitimate penal aim, or that it 'shocks the conscience' when viewed in relation to 'evolving standards of decency'.[72] Though the third of these definitions was most favourable to prison litigants the conditions cases brought in the 1970s spent less time on seeking to convince judges that prisoners' treatment violated existing legal standards than on laying before the court a frequently horrific pattern of facts through a combination of inmate and expert witnesses. Prison conditions suits brought in federal courts via 42 USC s. 1983 (which gives a cause of action to anyone deprived of their constitutional rights by anyone acting 'under color of law') were frequently class action suits filed on behalf of all the prisoners in a particular prison or state system, a device which allowed the prisoners advocates to bring the whole of the prison or prison system before the court's view.[73] Prisoners' applications regularly focused on a number of factors in order to convince courts that, taken as a whole, the 'totality of conditions' violated the Constitution.[74] These included overcrowding, inadequate physical facilities, lack of medical care, poor food, inadequate sanitation, lack of protection from violence, and a lack of rehabilitation systems.

Many of these cases disclosed appalling conditions. In Arkansas discipline was maintained by armed 'trusty' prisoners and an open barracks sleeping arrangement was found to encourage stabbings and sexual assault.[75] In Alabama one unit housing 200 inmates had a single toilet, food service was unhygienic, there was gross overcrowding, and prisoners in punitive isolation were packed as many as six to a four foot by eight foot cell with no beds, lights, or running water.[76] In Mississippi inmates with contagious

[71] *National Prison Project Journal*, 7/1 (1992).

[72] See for a general discussion I. Robbins and M. Buser, 'Punitive Conditions of Prison Confinement: An Analysis of *Pugh* v. *Locke* and Federal Court Supervision of State Penal Administration Under the Eighth Amendment' (1977) 29 *Stanford LR* 892–930, at 901–2.

[73] Class action suits are permitted under Rule 23 of the Federal Rules of Civil Procedure providing that the following requirements are satisfied: (1) that the class is so numerous joinder of all the parties is impracticable; (2) that there are questions of law or fact common to the class; (3) that the claims or defences of the representative parties are typical of the claims or defences of the class; (4) that the representative parties fairly and adequately protect the interests of the class; (5) that the party opposing the class has acted or refused to act on grounds generally applicable to the class.

[74] See Robbins and Buser, 'Punitive Conditions', 906–15.

[75] See *Holt* v. *Sarver* 309 F. Supp. 362 (DC Ark. 1970).

[76] See *Pugh* v. *Locke* 406 F. Supp. 318 (DC Ala. 1976).

diseases were found untreated in the general population, eighty prisoners were required to wash at three oil drums cut in half, and the sewage system had been condemned by state health and pollution agencies.[77] Initial judicial responses were to declare these conditions unconstitutional and order the states to change them. When however it became clear that state authorities were either not making any reforms or were making them very slowly judges used the equitable remedy of injunction to order more concrete and sweeping changes to the system. Remedies ordered have included the closing of institutions or the release of prisoners but more usually have taken the form of detailed plans to remedy overcrowding, inadequate staffing, lack of medical care, or poor sanitation. Frequently judges have appointed special committees or masters to oversee the implementation of such plans. Judges have also awarded damages to prisoner plaintiffs and contempt fines against recalcitrant prison authorities.[78]

These prison cases, along with similar constitutional challenges to the conditions in state mental hospitals and to segregation in public schools, gave rise to a new phenomenon often termed 'institutional reform litigation' in United States law.[79] While offering hope to prison reformers the innovative and broad-ranging remedies given by federal judges also posed new questions about the judicial role and led to claims that judges were transgressing into areas of prison management they did not fully understand and were crossing the lines between judicial, legislative, and administrative functions.[80] Though it seems that most judges were unwilling to involve themselves too deeply in prison management the Supreme Court in the 1980s was to send a number of signals discouraging such intervention.

Few of the 1970s cases had reached the Supreme Court but in those that did the court tended to affirm lower court decisions without indicating what the appropriate standard for a constitutional violation was. In *Hutto* v. *Finney*[81] it affirmed part of the order given by the District Court in the Arkansas litigation as within judicial competence with a dissent only by Justice Rehnquist. In *Estelle* v. *Gamble*,[82] a challenge to medical conditions in Texas prisons, another eight to one majority indicated that 'wanton indifference' to prisoners' medical needs would amount to cruel and unusual punishment but left it unclear as to whether this standard applied to all conditions litigation. The retreat began in the 1979 decision of *Bell* v.

[77] See *Gates* v. *Collier* 501 F. 2d 1291 (5th Cir. 1974).
[78] See, generally, Note '"Mastering" Intervention in Prisons' (1979) 88 *Yale LJ* 1062–91. Also S. Sturm, 'Resolving the Remedial Dilemma: Strategies of Judicial Intervention in Prisons' (1990) 138 *University of Pennsylvania LR* 805.
[79] A good starting-place for a study of such litigation is the Alabama mental hospital case *Wyatt* v. *Stickney*, 344 F. Supp. 373 (DC Ala 1972) and see Note, 'The Wyatt Case: Implementation of a Judicial Decree Ordering Institutional Change' (1975) 84 *Yale LJ* 1338.
[80] For a general discussion of these issues see A. Chayes, 'The Role of the Judge in Public Law Litigation' (1976) 89 *Harvard LR* 1281–316.
[81] 437 US 678 (1978).   [82] 429 US 97 (1976).

Wolfish.[83] There Rehnquist wrote for a six to three majority reversing an Appeals Court order finding a number of prison conditions, including double celling in a pre-trial detention facility in New York, to be unconstitutional. The Appeals Court had however based its decision not on the Eighth Amendment but on the Fourteenth Amendment's prohibition on deprivation of liberty without due process. The Appeals Court reasoned that, as the state's only interest as regards unconvicted prisoners was to keep them secure until trial, any further restrictions were in violation of due process unless justified by 'compelling necessity'. The Supreme Court majority rejected this approach, arguing that the state also had an interest in maintaining institutional order. Providing that restrictions were not imposed on detainees as punishment the authorities need only show a 'reasonable relation' between the restriction and the aims of security or institutional order to defeat a due process claim.

Although *Bell* v. *Wolfish* was a set-back for prisoners' advocates it did not offer a ruling on the appropriate standard to be employed in the more frequent Eighth Amendment conditions claims.[84] However the greater deference to prison authorities that the decision displayed was to be confirmed two years later in *Rhodes* v. *Chapman*.[85] The *Rhodes* case arose from a challenge to overcrowding in an Ohio prison that was 38 per cent in excess of capacity and required prisoners to share cells of sixty-three square feet. The Supreme Court by an eight to one majority overturned orders of the lower courts that these conditions violated the Eighth Amendment. Justice Powell indicated that conditions would only amount to a violation if they involved 'the wanton and unnecessary infliction of pain [or were] grossly disproportionate to the severity of the crime warranting imprisonment'.[86] He also indicated that in reaching such decisions deference should be paid to the views of prison administrators as to what was required to achieve legitimate penal goals. Justice Brennan's concurrence stressed that the court's decision affirmed the 'conditions of confinement' approach and that the courts had power to declare conditions unconstitutional. However he felt that the test of whether conditions violated the Eighth Amendment was not one of whether certain abstract goals, such as the elimination of overcrowding, were achieved but whether such conditions actually threatened inmates' physical or emotional well-being. He also felt it was legitimate to take account of the likely remedy if conditions were found unconstitutional. Here it would probably have been the moving of prisoners to an older and even more overcrowded prison that was already operating under a court

[83] 441 US 520 (1979).

[84] As regards conditions litigation *Bell* was also seen as an unusual case in that the gaol was a 1975 facility which, in the words of the lower court, 'represented the architectural embodiment of the best and most progressive penological planning' and enabled prisoners to have considerable time out of their cells.

[85] 452 US 337 (1981).

[86] Ibid., at 347.

order. Only Justice Marshall dissented, indicating a belief that deference should be shown to the views of the trial judge who had visited the prison and concluded that the conditions were cruel and unusual.

The *Rhodes* decision was a more severe set-back for prisoners' advocates. It indicated that the fact of overcrowding in itself would not amount to cruel and unusual punishment. Rather, as with any other aspect of imprisonment, the court would consider the overall context and the impact that such overcrowding had on a prisoner's physical and mental well-being. However its language allowed support both for the views of lower court judges who felt deference should generally be shown to prison administrators and for those who felt that, providing more severe impact on prisoners' lives than was evident in *Rhodes* was shown, scope for intervention remained. Changes in the federal judiciary after the election of President Reagan meant that the former view surfaced more often but in the 1980s a significant number of conditions suits continued to be successfully tried or more often settled by way of consent decree.[87] In 1986 however the Supreme Court gave judgement in the case of *Whitley* v. *Albers*,[88] a case arising out of the use of force by prison guards to quell a riot, which seemed to threaten a further limitation in the scope of the Eighth Amendment. In rejecting the prisoner's claim the court reiterated the wanton infliction of pain standard and added that this standard must be applied with 'due regard to differences in the kind of conduct against which an Eighth Amendment objection is lodged'.[89] It also indicated that as regards steps taken to quell a riot the key issue was whether the force was used in good faith or whether it was used maliciously or sadistically to cause harm. Prisoners' advocates feared that this 'malicious cruelty' standard might be applied to all conditions challenges but in *Wilson* v. *Seiter*[90] in 1991 the Supreme Court rejected this claim. The court did however indicate that in ordinary conditions cases it was not enough to prove that the conditions posed threats to the prisoners' health and safety, one had also to prove a 'culpable state of mind' on behalf of prison officials with regard to these conditions.

The *Wilson* case leaves conditions litigation again in a state of some uncertainty. On one reading the 'culpable state of mind' requirement will allow prison officials to defend conditions suits by pointing to their lack of

[87] Consent decrees provide for a judicially endorsed settlement of the case. This settlement may provide for a judge to make further orders. Parties may apply to the court for subsequent modification of the decree but only when significant changes in circumstances can be displayed.

[88] 475 US 312 (1986).

[89] Ibid., at 320. The Supreme Court recently indicated in *Hudson* v. *McMillan* (1992) 112 S. Ct. 995 that the 'malicious and sadistic' use of force by prison officials to inflict injury violates the Eighth Amendment where the injury is more than minimal.

[90] (1991) 111 S. Ct. 2321.

resources and claiming that they have no desire for these conditions to continue, a defence that has not generally been available in the past. On
another, since states of mind are normally established by objective facts, it
will not prove a significant limitation as in most prisons the officials will
quickly become aware of conditions that may be unconstitutional and their
failure to do anything about them will then amount to deliberate indifference. However the attitude of the federal judiciary seems likely to be less
inclined to judicial intervention in prison conditions in the next few years
and the *Wilson* language may be seized upon by judges appointed by
Presidents Reagan and Bush to limit intervention.

If this does occur and American judges move back to a more 'hands off'
approach, or at least one which encourages claims targeted at particular
policies or rules rather than broad conditions suits, then it will witness the
end of one of the most remarkable attempts to bring the rule of law to
bear on prison life. Views remain divided as to whether this intervention
has been beneficial. States with prisons under court orders still spend less
per prisoner than those with no such prisons,[91] and there is evidence that
the adversarial character of prison litigation has produced significant resistance to change.[92] Other observers have suggested that intervention by federal judges with a limited understanding of prison dynamics has produced a
vacuum of power in prisons, which has been filled by prison gangs which a
demoralized staff do little to restrain.[93] However there have clearly been
some success stories, where prison litigation has brought about a marked
improvement in prisoners' living conditions,[94] and evidence that the pressure of litigation has assisted reformers within prison administrations to
make the changes they want.[95] Yet at the beginning of the 1990s there were
more people in American prisons than ever before and the pressure on
space and resources continued to produce cramped, unhealthy, and fright-
ening conditions for many prisoners.[96] One of the main lessons of the
American experience is therefore that the improvement of conditions within
prisons cannot be separated from issues about who should be sent to prison
in the first place.

[91] See C. Smith, 'Federal Judges' Role in Prisoner Litigation: What's Necessary? What's
Proper?' (1986) 70 *Judicature* 144–50, at 148–9.

[92] See S. Ekland-Olson, and S. Martin, 'Organizational Compliance with Court-Ordered
Reform' (1988) 22 *Law and Society Review* 359—83.

[93] See e.g. J. DiIulio, 'The Old Regime and the Ruiz Revolution: The Impact of Judicial
Intervention on Texas Prisons', in J. DiIulio, *Courts, Corrections and the Constitution* (New
York, 1990), 51.

[94] See L. Yackle, *Reform and Regret: The Story of Federal Judicial Involvement in the
Alabama Prison System* (New York, 1989).

[95] See Jacobs, 'The Prisoners' Rights Movement', for discussion of the use made by reforming prison administrators of conditions suits.

[96] The prison population in the United States has risen from just over 500,000 in 1980 to
just in excess of 1,000,000 in 1991 and at a rate of 426 per 100,000 of the population is by
some way the world's highest. See *National Prison Project Journal*, 6/1 (1991).

CHALLENGING CONDITIONS IN THE UNITED KINGDOM

There are a number of reasons why an American-style challenge to prison conditions is unlikely to succeed in United Kingdom courts. One is the lack of a constitutional provision such as the Eighth Amendment which all state authorities must conform to. As the Arkansas court observed in *Holt* v. *Sarver*,[97] 'If Arkansas is going to operate a Penitentiary System, it is going to have to be a system that is countenanced by the Constitution of the United States.' Although, as will be discussed in more detail later, the United Kingdom has a prohibition on 'cruell and unusuall Punishments' in the Bill of Rights 1688, this has only the status of an ordinary statute and sits alongside the Prison Act 1952, which vests responsibility for operation of the prison system in the Secretary of State. Courts have often stated therefore that responsibility for what happens in prisons belongs to the Secretary of State and that apart from ensuring that he exercises his powers in accordance with the limits laid down by Parliament they have no authority to interfere. Such a position would be more difficult to take where a constitution grants all citizens certain basic rights and charges the courts with the duty of protecting them.

As we have already indicated in Chapter 3 procedural difficulties also inhibit any challenge to conditions in United Kingdom prisons. There is no equivalent to the American class action and hence it is more difficult to bring the facts of the entire prison environment, as opposed to the conditions experienced by particular prisoners, before the court. This can however be overcome to a certain extent by joining the actions of a number of similarly situated prisoners. More seriously perhaps there is no equivalent of the s. 1983 action and so prisoners have to frame their conditions challenges more precisely. They then run up against the difficulties, discussed in the chapter 3, of whether to base their claim in public law (for example breaches of the Prison Rules, improper procedures, or unreasonable action) or private law (such as negligence or assault claims). The choice, assuming that the court allows a choice to be made, involves different rules on discovery, standing, and remedies.

Nevertheless some routes of challenge are available. One is via the Bill of Rights provision. Some support was given to it by the Court of Appeal in *R.* v. *Secretary of State for the Home Department ex parte Herbage (No 2)*.[98] There the prisoner, who was severely overweight, challenged his detention on the ground floor of the medical wing of Pentonville Prison. The ground floor contained cells reserved for more severely disturbed prisoners. Herbage, who was not disturbed but was placed on the floor as his physical condition made it difficult for him to walk up stairs, claimed he was

---

[97] 309 F. Supp. 362 (DC Ark. 1970).   [98] [1987] QB 1077.

subjected to constant screaming, banging, and shouting from the other prisoners and as a result found it impossible to sleep. Although his claims never actually came to trial a majority of the Court of Appeal on an application against discovery indicated that they felt breaches of the Bill of Rights could not be justified by reference to the Prison Rules, as these were delegated legislation and the Bill of Rights took precedence over them. Purchas LJ also added, without ruling on the facts in the instant case, that holding a sane person in the psychiatric wing of a prison for purely administrative reasons might well be 'cruell and unusuall' punishment. When the issue of whether conditions could amount to 'cruell and unusuall' punishment was posed more directly in an earlier case the court had not shown any enthusiasm for it.

This occurred in *Williams* v. *Home Office (No. 2)*, a challenge to the 'control unit' system introduced to deal with difficult prisoners in the 1970s. The control units operated under the Rule 43 segregation power and so prisoners could be removed to them without any formal hearing. A prisoner's time in the control unit was divided into two stages. In the first he spent ninety days without any association and although he was not required to work the ninety days did not run until he did. In a second ninety-day period a limited amount of association was permitted but if a prisoner failed to work or caused disruption he was returned to start the first stage again. Only after spending 180 days satisfactorily in the unit would a prisoner be returned to the normal population.

Williams spent 180 days in the unit in either total or near total isolation from other prisoners. After his release from the unit he brought an action claiming that the establishment's nature and operation were unlawful on a number of grounds, including that it amounted to 'cruell and unusuall punishment'. Tudor Evans J. concluded that the words had to be read disjunctively, so that the punishment to be unlawful had to be both cruel and unusual. Examining whether the conditions were cruel he adopted a test which looked at whether they fell below an 'irreducible minimum, judged by contemporary standards of public morality'.[99] He also accepted that treatment which was disproportionate to the conduct or character of the prisoner in question could be cruel. Applying these tests he found that the measures were not disproportionate given that Williams was considered to be a dedicated troublemaker. Nor did he find them cruel; the unit was similar in design to that recommended by a report of the Advisory Council on the Penal System, the conditions in it were not that different from those experienced on Rule 43 segregation, and prisoners spent much less time in the unit than in similar units in the United States and Canada which the courts there had held to be lawful. Despite the finding the Home Office had

[99] [1981] 1 All ER 1211, 1245.

already closed the control units in 1975.[100] The disjunctive reading of the Bill of Rights language suggests that any subsequent challenge has a high threshold to meet as even if condition within the prison system are cruel they may not be unusual; but Tudor Evans J.'s reference to an 'irreducible minimum' suggests that if significant evidence of unsanitary, violent, unhealthy, or grossly overcrowded conditions were brought before a judge disposed to find a remedy for them the Bill of Rights could yet provide a vehicle for a successful challenge to prison conditions.

Another vehicle of challenge has been through actions for false imprisonment on the grounds that the prisoners' conditions of confinement have become so intolerable that the court must order release from those conditions. In the case of sentenced prisoners this would not lead to release from prison but might lead to transfer from segregation to the general prison population or from one intolerable prison to a better one. This approach gained some support in a number of challenges to conditions in police detention, notably in *Middleweek* v. *Chief Constable of Merseyside*, where Ackner LJ indicated that

it must be possible to conceive of hypothetical cases in which the conditions of detention are so intolerable as to render the detention unlawful and thereby provide a remedy to the prisoner in damages for false imprisonment. A person lawfully detained in a prison cell would, in our judgement cease to be lawfully detained if the conditions in that cell were such as to be seriously prejudicial to his health if he continued to occupy it, eg because it became and remained seriously flooded or contained a fractured gas pipe allowing gas to escape into the cell.[101]

However, when the House of Lords came to consider this issue in *Hague* v. *Deputy Governor Parkhurst Prison*[102] Ackner, now Lord Ackner, joined the rest of his brethren in rejecting this statement and in holding that s. 12 (1) of the Prison Act 1952 provided a complete defence to any claims that a prisoner's conditions rendered his or her detention unlawful. All their Lordships accepted though that a prisoner might still have an action in negligence in respect of his conditions. Lord Bridge indicated that such an action might lie without proof of physical injury or prejudice to health 'if a person lawfully detained is kept in conditions which cause him for the time being physical pain or a degree of discomfort which can properly be described as intolerable'.[103] The rest of the Lords appeared somewhat more

---

[100] In 1989, in response to the Control Review Committee, a number of Special Units were established at four prisons. They can currently accommodate up to forty-five prisoners in total. These were aimed at providing a 'more structured' regime for difficult prisoners. Detention in these units limits access to work and education facilities in the general prison but a greater degree of association is permitted among prisoners in the Unit than was true of control units and the authorities aim to provide a more generous regime. See, generally, Woolf Report, paras. 12.272–12.292.

[101] [1992] 1 AC 179 (the case was decided in 1985).   [102] [1992] 1 AC 58.

[103] Ibid., 165.

cautious and anxious to confine such actions, as Lord Goff put it, 'only in respect of the type or types of damages which, on accepted legal principles will give rise to an action'.[104] This approach would appear likely to limit claims to circumstances where prisoners have suffered some actual injury and would be unlikely to provide a remedy for conditions which are dirty, overcrowded, or oppressive without causing tangible injury.

The negligence approach may therefore offer limited opportunities for prisoners to challenge conditions. The majority in the House of Lords appeared to argue for a narrow duty of care and while Lord Bridge seemed to envisage a broader scope for challenge he failed to provide much in the way of an indication of what sort of conditions he regarded as intolerable. Even if he were to adopt the examples given by Ackner LJ in *Middleweek* this would suggest that conditions would have to become almost life threatening before courts would intervene. Moreover, since in none of the litigated cases have the courts actually decided that the conditions were intolerable, there has been no judicial consideration of the prison authorities' most likely defence, namely that the conditions were brought about through lack of resources and that they had acted reasonably in such circumstances. As we point out in Chapter 3 aggravated or exemplary damages may be difficult to obtain in prison cases. The courts also take the view that injunctions are not available against the Crown or its agents in civil actions.[105]

While courts may evolve standards that allow prisoners to challenge conditions through civil actions, perhaps from consideration of whether the conditions fall below the standard required by Article 3 of the European Convention on Human Rights, at the moment the options are limited. Another route, which at least offers greater flexibility in remedies, is to challenge specific rules or practices as unreasonable by way of action for judicial review and to invoke the impact on conditions as evidence of their unreasonableness. Putting three or more prisoners into a cell designed for two might be challenged on such grounds. Without some indication, either from national or international standards, that such actions are regarded as independently unreasonable, however, courts seem likely to defer to the prison authorities for an account of what is reasonable in the circumstances.

---

[104] Ibid., 167.
[105] See Crown Proceedings Act 1947, s. 21 (2). It was recently affirmed that this position had not been altered by s. 31 of the Supreme Court Act 1981. See *Factortame* v. *Secretary of State for Transport* [1989] 2 All ER 692 and *M* v. *Home Office* [1992] 4 All ER 97.

CHALLENGING PRISON CONDITIONS IN INTERNATIONAL LAW

Although the United Kingdom has no domestic constitution to which prisoners may appeal it is a signatory to several international human rights treaties, notably the European Convention on Human Rights and the United Nations International Covenant on Civil and Political Rights which contain prohibitions on torture and inhuman and degrading treatment or punishment.[106] The obligations contained in these treaties are not directly applicable in United Kingdom law but courts have indicated that they should be taken into account in interpreting legislation as Parliament is presumed to be acting in accordance with its international legal obligations.[107] Prisoners may also make individual applications to the Strasbourg institutions. In addition the United Kingdom is also a signatory to both the United Nations and European Conventions against torture.[108] In the case of the European Convention this carries the obligation to allow inspections of Britain's prisons by the Committee established under the Convention.

The broad prohibition against torture and inhuman or degrading treatment and punishment contained in Article 3 of the European Convention on Human Rights has already been the object of significant consideration by the European Commission and Court.[109] These decisions have not settled on any very precise definition of the article, but as regards the general approach to interpretation the Commission has indicated that the Convention is a 'living instrument which must be interpreted in the light of present day conditions'.[110] A number of decisions have also indicated that the Article 3 elements are cumulative: what is torture will also be inhuman and degrading treatment. Perhaps the fullest description of the latter was that of the Commission in the case brought by Denmark and Sweden against Greece. There the Commission indicated that treatment would be degrading to a person if it 'grossly humiliates him before others or drives him to act against his will or conscience'.[111] In the same decision the Commission described inhuman treatment as 'at least such treatment as deliberately causes severe suffering, mental or physical, which, in the particular situation is unjustifiable'.[112] In the case brought by Ireland against the

---

[106] The European Provision is in Art. 3 of the Convention, the Civil and Political Rights Covenant Provision is Art. 7. Both prohibit torture and cruel, inhuman, or degrading treatment or punishment. The UN provision also specifically prohibits the subjecting of anyone without their consent to medical or scientific experimentation.

[107] See Ch. 4 for further discussion of this issue.

[108] These are Convention against Torture and Other Cruel, Inhuman or Degrading Treatment or Punishment UN Doc. A/RES/39/46 (1984), and European Convention for the Prevention of Torture and Inhuman or Degrading Treatment or Punishment (1987).

[109] For a discussion of this jurisprudence see, generally, P. Duffy, 'Article 3 of the European Convention on Human Rights' (1983) 32 *ICLQ* 316–46.

[110] See *Tyrer* v. *United Kingdom* (1978) 2 EHRR 1, Series A No 26.

[111] (1969) 12 Yearbook of European Convention on Human Rights 499.   [112] Ibid., 186.

United Kingdom relating to the treatment of detainees during internment in Northern Ireland the Court endorsed these definitions but added that for the treatment or punishment to attain the level of torture there must be 'deliberate inhuman treatment causing very serious and cruel suffering'.[113]

These definitions are broad in scope and much will therefore depend in each case on the evidence offered. The authors of a leading textbook on the Convention argue that medical evidence of conditions causing physical or mental suffering to a prisoner will be especially influential in encouraging the Commission or Court to come to a finding that there has been a breach of Article 3.[114] They also indicate that despite the references in the jurisprudence to the need for the treatment or punishment to be deliberately inflicted the emphasis has usually been on the effect the treatment has had on the prisoner rather than mental state or motivations of those responsible for the acts or omissions which give rise to ill treatment.

In prisons conditions cases however the Commission has required quite a severe level of deprivation to amount to a breach of Article 3. In a number of cases challenging solitary confinement the Commission has been very reluctant to find a breach where there has not been absolute isolation from other prisoners coupled with sensory isolation. Even then if the prisoners are judged particularly dangerous the conditions may be considered justified.[115] Generally in considering claims that conditions in solitary confinement breach Article 3 the Commission has indicated that relevant factors will be the length of time spent in solitary, the prisoners' state of health, and the facilities available to him while in solitary.[116] Challenges to more general conditions have also been met sceptically. In the *Greek* case the Commission indicated that overcrowding could amount to inhuman treatment but this was part of a general finding relating to a number of conditions and occurred in the context of extreme overcrowding.[117] In the McFeeley case brought by a number of prisoners in the Maze Prison in Northern Ireland who were staging a protest against the failure to grant them prisoner of war status the Commission indicated that dirty and unsanitary conditions could amount to a violation of Article 3 but concluded that they had been brought about by the prisoners themselves.[118]

---

[113] *Ireland v. United Kingdom* (1978) 2 EHRR 25, Series A., No 25, para. 167.

[114] See P. Van Dijk and P. Van Hoof, *Theory and Practice of the European Convention on Human Rights*, 233.

[115] See e.g. *Krocher and Moller v. Switzerland* D. & R. 34 (1983), where applicants were under permanent surveillance by closed circuit television and were denied access to newspapers and radio in addition to being kept isolated. Commission, noting that these were dangerous terrorists who might try to escape, found no breach of Art. 3.

[116] See *Hilton v. United Kingdom* (1978) 3 EHRR 104.

[117] The Commission found that at least two people were regularly held in very small basement cells designed for solitary confinement for periods of between thirty days and nine months. There was no recreation and virtually no light for the prisoners held in these cells.

[118] See *McFeeley v. United Kingdom* (1980) 3 EHRR 161.

That conditions giving rise to mental stress and anguish can amount to a breach of Article 3 is shown by the Court's decision in the *Soering* case. There a prisoner's likely exposure to the 'death row phenomenon' of spending several years on death row if convicted of murder in the United States was held to amount to inhuman treatment and was grounds for refusing his extradition.[119]

The United Nations Human Rights Committee (which oversees enforcement of the Civil and Political Rights Covenant) has also considered a number of cases arising out of prison conditions. In addition to Article 7, which prohibits torture or cruel, inhuman and degrading treatment, the Committee can also draw on Article 10 of the Convention, which provides that 'All persons deprived of their liberty shall be treated with humanity and with respect for the inherent dignity of the human person.' The Committee does not appear thus far to have developed a separate jurisprudence for each article, generally finding that there has been a violation of both Articles 7 and 10. Though the Committee has found violations, relating to solitary confinement, sensory deprivation and unhealthy conditions, this has generally been in cases of particularly brutal treatment.[120] The limited number of cases and the fact that the United Kingdom has not yet ratified the Optional Protocol which allows individuals to petition the Committee means that at the moment the Covenant is of limited significance to prison conditions in the United Kingdom. Currently the main enforcement mechanism is that every three years all signatory states are required to present a report on their compliance with the Covenant to the Committee and may face sharp questioning on areas where they may be in breach.[121]

International concern with the problem of torture, protection from which is one of the few rights which cannot be derogated from in most international human rights treaties, has led to the creation of Conventions at both United Nations and Council of Europe level. The United Nations Convention works on the same enforcement model as the Civil and Political Rights Covenant. Signatory states are required to make regular reports to the Committee established under the Convention as to the extent to which they are complying with the obligations under it. These include

[119] (1989) 11 EHRR 439, Series A No. 161.

[120] See e.g. *Grille Motta* v. *Uruguay* Report of the Human Rights Committee, GAOR, 35th Session, Supplement No. 40 (1980,) annex XI para. 16 (prisoner subjected to electric shocks and *Submarino*—putting his hooded head repeatedly into foul water); *Estrella* v. *Uruguay* Report of the Human Rights Committee, GAOR, 38th Session, Supplement No. 40 (1983), annex XXII para. 1.6 (pianist subjected to frequent beatings, made to stand with legs apart for up to twenty hours and subjected to mock amputation of his hands with an electric saw). For further information see N. Rodley, *The Treatment of Prisoners Under International Law* (Oxford, 1987), ch. 9.

[121] See, generally, D. McGoldrick, *The United Nations Human Rights Committee: Its Role in the Development of the International Covenant on Civil and Political Rights* (Oxford, 1991).

providing independent investigation of all claims of torture or inhuman or degrading treatment, ensuring that torture is a criminal offence, and providing education for all criminal justice personnel in the obligations of the Convention. If a state accepts the competence of the Committee, which the United Kingdom has not, then individuals may bring complaints before it. So far the Committee has not developed any further guide-lines on the application of the Convention to prison conditions.[122] The European Convention goes further on the enforcement scale by creating a Committee that has a right of inspection in prisons and places of detention in signatory states. The Committee also publishes an annual report on its activities and may, with the consent of the government involved, publish a report on its particular inspection. The European Torture Convention does not provide a definition of torture but in its work thus far the Committee has taken a fairly broad approach to the concept, for example in its report on Brixton and Leeds Prisons, where it indicated that overcrowded and unsanitary conditions amounted to degrading treatment.[123]

FUTURE DEVELOPMENTS IN CONDITIONS LAW

Judicial reluctance to uphold challenges to conditions has a number of sources but one is clearly a sense that prison officials rather than judges are in the best place to assess adequate cell space, sanitation arrangements, education programmes, or the use of solitary confinement. Judges lack a set of standards by which to assess independently the habitability of gaols. Yet it is not impossible to provide such standards and they would also be of use to prison officials, prisoners, politicians, and prison reform groups in assessing the state of the prisons. The American Correctional Association and the American Bar Association have both developed and regularly updated standards and these have often been invoked by judges, lawyers arguing prison cases, and prison officials. In the United Kingdom NACRO has developed a Code of Standards and a timetable for their implementation.[124] This provides guidance on things like minimum cell size, time spent in cells, sanitation arrangements, food, exercise, and safety.[125] Such a code could provide benchmarks against which prisoners, prison officials, and the public could measure the performance of prisons. It could also be a resource for judges deciding whether prison conditions had become

---

[122] See N. Lerner, 'The United Nations Convention on Torture' (1986) 16 *Israel Yearbook on Human Rights* 126.

[123] *Report . . . by the European Committee for the Prevention of Torture*; see also M. Evans and R. Morgan, 'The European Convention for the Prevention of Torture: Operational Practice' (1992) 41 *ICLQ* 590.

[124] See S. Casale and J. Plotnikoff, *Minimum Standards in Prisons*.

[125] For example cell space should be sixty square feet each, should have a minimum of twelve hours out of cells, and should have access to sanitation at all times.

'intolerable' or whether practices or rules adopted within a prison were 'unreasonable'. In 1982 the government endorsed the idea and undertook to produce a draft code, but this failed to appear, and in 1987 the Home Office indicated that it was no longer of the view that a code of standards would be helpful. However House of Commons Select Committees regularly commented throughout the 1980s on the desirability of such a development and in 1990 the Woolf Report again recommended a code of standards, though adding that it would not be wise to make them legally enforceable at this stage. Woolf added that they could nevertheless be used in an accreditation system, with a prison only attaining accredited status when it met all of the standards.

In its White Paper response to Woolf the government made a fresh commitment to the code of standards principle. This resulted in a consultative document in early 1992. The consultative document indicates that the primary aim of the code should be to ensure that 'prisons are decent places for all who live and work in them and all who visit them'. It envisages standards to cover a wide range of matters including accommodation, categorization, clothing, grievances, health care, race relations, and security, control, and safety. It is not envisaged that these will be legally enforceable and the issue of accreditation is still under discussion. Indeed, while a role is envisaged for governors, Boards of Visitors, and the Chief Inspector of Prisons in monitoring the implementation of standards, exactly how this is to come about remains very much to be decided. The consultative document displays significant anxiety about setting standards too high and hence recommends that at least two levels of standards be specified, basic and standard, and that these be set relative to the type of institution to which they are applicable (for example training prisons should have better education facilities than remand centres). There is a timetable for bringing the code into operation by late 1993.

Codes of standards will also need to be related to another of Woolf's ideas, that of prisoner contracts, or compacts as the Home Office now refers to them.[126] Another Home Office consultation paper has been produced on this. Unlike the code this would be a two-way process, setting out obligations that the institution has to the prisoner but also those that the prisoner has to the institution. Woolf appeared to see contracts as ways of increasing the range of facilities available to a prisoner, beyond what would be set out in the minimum standards, in return for a prisoner agreeing to extra obligations beyond those set out in the discipline rules.[127] The consultative paper does not indicate if this is to be the case. It merely indicates that compacts will establish certain basic obligations that the prisoner owes

---

[126] See Woolf Report, paras. 12.120–12.130.
[127] Woolf gives an example from the Dutch system where a prisoner gets access to greater facilities in return for agreeing to urine tests for drugs after returning from home leave.

the authorities and the authorities owe the prisoner. If compacts are introduced at the same time as codes of standards then the relationship between the two will remain to be worked out. Will compacts merely be another way of stating the basic obligations contained in the code or will they, as Woolf suggested, be a way of expanding a prisoner's rights and duties? Another matter in need of development is what redress either prisoner or prison authorities will have for breaches of compacts. Currently it seems the former will only be able to look to the existing grievance procedure, the latter to the existing disciplinary system.

In explaining the shift from a notion of contracts to a notion of compacts the consultative document makes the interesting observation that 'This change of title acknowledged a prisoner's inability to make a meaningful legal contract. Prisoners are not equal partners and are not voluntary recipients of prison services—obligations would exist regardless of whether a "contract" had been accepted and signed.'[128] This statement on one level reflects the realities of prison life. By the fact of their detention prisoners lack the possibilities of choice that legitimize contractual obligations.[129] Yet on another level it ignores the idea that legally (at least in international law) prisoners have certain rights as well as the prison authorities having obligations. The absence of rights thinking is also evident in the code of standards consultative paper, which includes the European Prison Rules as one of the potential influences on the standards to be drafted but not the European Convention on Human Rights or the International Covenant on Civil and Political Rights. Such an absence means that the consultative paper can contemplate, but ultimately seems to reject, setting the basic standards at whatever currently prevails, even though in some prisons such conditions probably violate Article 3 of the European Convention. In the final chapter we shall again consider what is the most appropriate basis for the content of legal regulation of prisons; an approach which starts from a foundation of prisoners' basic rights or one whose point of departure is the obligations of the authorities.

A second major development could however change the whole focus of who is responsible for conditions in prisons. In April 1992 the UK's first privately run gaol opened at The Wolds near Hull. A second private prison is to be built at Blakenhurst and at time of writing tenders are being sought for Manchester Prison. The current government seems strongly committed to putting particular services within prisons, such as food, education, and

---

[128] Consultation paper on Prisoners Contracts (Home Office, 1992), para. 3.

[129] The extent to which a notion of contract is realistic in the coercive circumstances of prison and the issue of whether it may be employed oppressively against prisoners is considered by Patricia Williams in respect of a South Carolina judge's decision in 1985 to let three convicted rapists choose castration over prison, a decision subsequently overruled by the North Carolina Supreme Court in *State* v. *Brown* (1985) 326 SE 2d 410. See P. Williams, *The Alchemy of Race and Rights* (Cambridge, Mass., 1991), 33.

court escorts, out to tender as well as contracting out entire prisons. The legal framework for such privatizations was created by the 1991 Criminal Justice Act, which gave the Home Secretary power to enter into contracts for the running of remand prisons (with a power to extend it beyond the confinement of remand prisoners by statutory instrument, which was done in 1993). However the Act does not provide for the entire delegation of all the functions of the prison service to the private contractor. Instead contracted-out prisons are to be run by a director, who is appointed by the contractor subject to the Home Secretary's approval, and a controller, who is directly appointed by the Secretary of State.[130] While the director is responsible for most of the day-to-day running of the prison he or she does not take on governor's functions in relation to discipline or, except in cases of urgency, in relation to removal from association for good order and discipline reasons. These reside with the controller, who also has the function of keeping the running of the prison under review and of reporting to the Secretary of State any complaints against prison custody officers. The latter are the contractor's employees, whom the Home Secretary has certified as authorized to carry out custody duties.[131] The regime in the prison is expressly made subject to the Prison Rules but in addition the contract between the Home Office and the contractor sets out specific standards to be complied with on a wide range of issues. For reasons of commercial confidentiality the Home Office has refused to make public the first contract it concluded with a private contractor but an indication of its contents can be gleaned from the published tender documents.[132]

The legal nature and consequences of contracted-out prisons are discussed in Chapter 2. At this point it is worth however examining the impact that contracting out a prison is likely to have on the conditions in which prisoners are detained. One initial impression is that the very act of setting out contractual obligations provides detail on what can be expected in a variety of areas of prison life. This contrasts with the patchwork quilt of Prison Rules, Standing Orders, and Circular Instructions which still leaves significant areas of prison experience lacking any set of entitlements or objectives. Perusal of the published tender documents also suggests that such prisons should provide a significantly improved regime. Each of the published tenders so far indicates that prisoners should spend at least twelve hours a day out of cell. In announcing the Blakenhurst contract the Prisons Minister indicated that the contract would exceed tender standards and allow fifteen hours out of cell.[133] Prisoners in the contracted prisons will be entitled to a daily change of underwear (as opposed to twice weekly

---

[130] Criminal Justice Act 1991, s. 85 (1). [131] Ibid., s. 89 and Schedule.
[132] See the tender documents for The Wolds Prison (1991), Blakenhurst Prison, and Manchester Prison (1992), available from the Home Office.
[133] Quoted in *NACRO Criminal Justice Digest*, 75 (1993).

in current Standing Orders), to a presumption that they can wear their own clothing, and to better visiting arrangements, up to an hour a week. However, some of the standards fall short of what might have been hoped. Only the Manchester tender provides for a toilet in each cell, none requires hospital officers to have nursing qualifications, and none provides for any specific provisions relating to the treatment of HIV positive prisoners.

From the perspective of prisoners' rights an issue of particular interest is the sanctions available to the Home Office if the contractor fails to comply with these contractual standards. Under s. 88 of the Criminal Justice Act 1991 the Home Secretary may intervene and appoint a controller to take over management of a prison where he or she feels that the director has or is likely to 'lose effective control' of the prison or where it is necessary to prevent injury or serious damage to property. This is a drastic step and seems unlikely to be the first response taken to any breach of the contract. Neither the statutory provisions nor the tender documents provide any indication of how the Home Office intends to monitor the operation of the contract and deal with failure to comply with aspects of it. Nor is it clear whether information on the extent to which the contract is complied with will be published, something essential for ensuring the accountability of the Home Secretary on decisions to renew contracts. Indeed thus far Home Office Ministers have refused to answer parliamentary questions on issues such as costs or staffing levels. They have claimed such issues are subject to commercial confidentiality and indicated that such questions might be better directed to the private contractor.

Greater efficiency (through better use of staff resources) and improved conditions for prisoners have been the main rationale for the contracting out initiative.[134] The American experience, where some states have turned to the private sector to build and run prisons in response to court decisions outlawing overcrowding and poor conditions, has often been cited. However even the government's Green Paper conceded that the American experience is inconclusive regarding improved conditions, and other commentators have questioned the gains in economic efficiency.[135] As regards the legal context the Home Office retains overall responsibility for the contracting-out prisons complying with the Prison Act 1952 and rules made under it.[136] The Home Secretary is also responsible to Parliament with respect to such prisons. However in respect of private law injuries, such as

---

[134] See *Private Sector Involvement in the Remand System*, Cm. 434, (London, 1988), 9.

[135] See M. Ryan and T. Ward, *Privatization and the Penal System* (London, 1989), 34.

[136] s. 84 of the Criminal Justice Act 1991 indicates that any contracted out prison is to be run in accordance with the Prison Rules and the Prison Act 1952, s. 4 (2) of which makes the Home Secretary responsible for ensuring that the Prison Rules are complied with. The tender documents indicate that the extent to which Prison Service Standing Orders and Circular Instructions are available to prisoners is a matter for negotiation, though ultimately the Home Office has power to give direction on this.

assaults or injuries at work, prisoners should be able to take legal action against the contractor, who could not benefit from Crown Immunity. Due to the concept of privity of contract (that only parties to a contract gain rights under it) prisoners would not be able to take action against the contractor for failure to comply with the standards in the contract, but they might be able to seek judicial review of the Home Office if it failed to do so, although the failure of the Home Office to publish contracts could prevent any such action from getting off the ground. Such claims could be based on the idea that prisoners have legitimate expectations under the contract or at least the tender documents, expectations which the state has an obligation to ensure the contractor respects. What remains significantly unclear is the responsibility for exercise of discretionary powers such as those to search, deny visits, or censor mail. Since the Prison Rules are not directly enforceable these are probably only susceptible of judicial review by way of a reasonableness standard. Yet judicial review is not normally available against private entities, while the Home Office might argue that it is not responsible, since such decisions are the responsibility of the contractor.[137] The Home Office's residual responsibility for the provision of adequate medical care and safe living and working conditions if a contractor fails to pay any damages awarded is also a matter that could give rise to legal dispute. For all these reasons the development of the contracted-out sector might be accompanied by a significant amount of litigation. At the start of 1993, however, aspects of the contracting-out policy in respect of services such as education and escort duties had already run into difficulties. The escape of several prisoners from private escort custody and claims that drug use and violence were more prevalent in The Wolds private remand institution raised again questions as to the wisdom of this approach.[138]

[137] The recent decision in R. v. *Jockey Club ex parte Aga Khan* (1993) 143 NLJ 163 shows the Court of Appeal retreating from the application of judicial review to non-state bodies. The court appears to suggest that to be amenable to judicial review a body must be exercising power given to it by the government. Since the authority to run contracted out prisons derives from statute, even though the companies operating the prisons are the creation of private enterprise, the operators of a contracted-out prison should still be susceptible to judicial review.

[138] The government's programme to contract out the provision of services in prisons has recently run into difficulties; for example the contract for the operation of the supply and transport department was withdrawn after legal advice that application of the European Communities Acquired Rights Directive and Transfer of Undertakings Regulations meant that staff would have to be guaranteed the same rates of pay and benefits in the private sector as they had in the public sector, see *Guardian*, 8 Mar. 1993. A report published by the Prison Reform Trust on the first year of The Wolds prison suggested that there were some positive features of the regime but also that the level of violence was higher and that availability of drugs was greater than in most prisons. Both problems were linked to staff shortages. See, generally, Prison Reform Trust, *Wolds Remand Prison: Contracting Out: A First Year Report* (London, 1993).

6

Access to the Outside World and Maintenance of Family Contacts

INTRODUCTION

Maintaining links between prisoners and the outside world, especially with their families, is seen as a vital aspect of their rehabilitation and preparation for release. Prison Rule 31 indicates that 'special attention' shall be paid to maintaining contacts between prisoners and their families, and that a prisoner should be 'encouraged' to develop contacts with the outside world which 'best promote the interests of his family and his own social rehabilitation'. In its response to the Woolf Report the Home Office observed that visits and home leave were 'key elements in meeting the Prison Service's obligations to those in its care'.[1] The European Prison Rules also place great stress on the importance of maintaining contact with the outside world. Rule 65 (c) for example, which is part of the Rule setting out the main objectives of the prison regime, indicates that the regime should be designed and managed so as 'to sustain and strengthen those links with relatives and the outside community that will promote the best interests of prisoners and their families'. Indeed, with the decline of faith in the rehabilitative capacity of prison itself, contact with the outside world as a means of reducing the dehabilitating effects of institutionalization has come to be seen as perhaps the most important rehabilitative strategy in the prison context. It is also of course important for the prisoner's family, who, while unconvicted of anything, also effectively serve a sentence in terms of restrictions on their lives for the time that a close relative is in prison. Location of prisons plays a large part in determining how substantial this contact will be, as the further away a prisoner is detained from his family the more difficult it will be to maintain contact. In this respect it is interesting to note Lord Woolf's recommendation that where possible prisoners should be detained in 'community prisons', close to the community with which they have the greatest links.[2] Even with close location, though, issues of regime ultimately determine the extent of contact.

[1] See Home Office, *Custody, Care and Justice*, Cm. 1647 (London, 1991) para. 7.35.
[2] See *Prison Disturbances: April 1990* (Woolf Report), Cm. 1456 (London, 1991), para. 11.69. For further discussion of the idea of community prisons see Prison Reform Trust, *Prison Report*, 19 (London, 1991), 2.

Prison policy in this area has in fact almost come full circle. In the seventeenth and eighteenth centuries the dividing line between prison and the outside world was a flexible one. Traders and prostitutes came in to work within the prison walls, prisoners with sufficient money could entertain visitors in their cells. This changed in the late eighteenth and early nineteenth centuries, when reformers urged isolation from the pleasures and corruptions of the outside world as the best means of producing changes in the disposition of criminals.[3] It was this era that witnessed the development of a rule of silence in prisons and which greatly limited contact between prisoners and those on the outside. Although the reform justification for limitations on contact has largely gone, in the late twentieth century prison security has been given as the reason for maintaining restrictions, though institutional habits and conservatism may play a greater role. Nevertheless the last two decades have witnessed a relaxation in the rules governing such matters as letters, visits, and home leave. The European Commission and Court of Human Rights have played an especially prominent role in prompting these changes. Domestic courts have been less to the fore in this area, though they have given several important decisions on prisoners' access to the courts. Yet although the United Kingdom has made progress in this field it still lags behind many European systems in its approach to matters like home leave and conjugal visits. It is on such matters that a significant amount of energy is likely to be expended in the next few years.

PRISONERS' LETTERS AND VISITS: CONTACTS WITH LAWYERS

In the early 1970s prisoners' letters were still routinely read and stopped. The Prison Rules indicated that letters could be read at any time and stopped when deemed to be 'objectionable', a broad category on which no further clarification was available.

The breakthrough in the move away from this restrictive regime came in the area of prisoners' letters to lawyers. In the case of *Golder* v. *United Kingdom*[4] the European Court of Human Rights ruled that the prison authorities could not refuse to forward a letter to a prisoner's lawyer on the grounds that it referred to a matter which had not been raised or decided upon within the prison. Golder had actually written to the Home Secretary seeking permission to contact a solicitor regarding a potential libel claim against a prison officer and had been refused. In Strasbourg the European Court indicated that this refusal violated his right under Article 6

---

[3] See, generally, C. Harding, B. Hines, R. Ireland, R. and P. Rawlings, *Imprisonment in England and Wales: A Concise History* (London, 1985), ch. 4.
[4] (1975) 1 EHRR 524, Series A No. 18.

of the Convention to a fair hearing in a case involving a potential civil right. Although Article 6 did not expressly guarantee a right of access to the courts the majority of the European Court were of the opinion that the rights to a fair hearing established in that article would be meaningless if a member state could deny access to the courts in the first place. The Court also indicated that stopping Sidney Golder from corresponding with his solicitor violated his right to respect for his correspondence guaranteed by Article 8. This aspect of the decision arguably turned out to be even more significant as in doing so the Court rejected the government's arguments that there were inherent limitations on the exercise of rights established in the Convention when it came to prisoners. Instead the Court took the view that any lawful restrictions on prisoners' exercise of their Article 8 rights had to be found in the limitation clauses of Article 8 (2). In this case they did not see how any of these could justify preventing a prisoner from corresponding with a solicitor.

The *Golder* case was a major development, in the context both of correspondence and of the potential of the European Convention on Human Rights to respond to prisoners' claims.[5] As regards the issue of legal correspondence it was to be endorsed by the Commission and Court in the case of *Silver* v. *United Kingdom*.[6] Though this case had a greater impact on non-legal correspondence and will be discussed in greater detail below it did indicate that stopping a letter to a solicitor on the grounds that the complaints it contained had not previously been raised in the internal complaints mechanism—the 'prior ventilation rule'—did violate Article 8 and was not necessary in a democratic society to satisfy any of the legitimate aims set out in Article 8 (2).

In response to these decisions the Home Office, by Standing Orders and Circular Instructions rather than amendments to the Prison Rules, replaced the prior ventilation rule with a 'simultaneous ventilation' rule, whereby complaints could be sent to solicitors providing that they were at the same time being raised in the internal complaints mechanism. This was to fare no better than the prior ventilation rule. Though the European Court in *Silver* appeared prepared to wear this solution,[7] this time it was the domestic courts which found it unacceptable.

The case was R. v. *Secretary of State for the Home Department ex parte Anderson*.[8] Therein a prison governor had prevented a prisoner having an interview with his lawyer to discuss a potential civil claim relating to his

---

[5] See G. Zellick, 'The Rights of Prisoners and the European Convention on Human Rights' (1975) 38 *MLR*, 683–9.

[6] (1983) 5 EHRR 347, Series A No. 61; the Commission's decision is available at (1980) 3 EHRR 475.

[7] As the complaints in *Silver* related to correspondence between 1972 and 1977 the Home Office had amended and published its Standing Orders by the time the case reached the Court.

[8] [1984] QB 778.

treatment in prison because the prisoner had not simultaneously submitted this claim to the internal disciplinary procedure. The governor invoked Standing Orders issued by the Secretary of State under the then Rule 33 (1), which required all complaints to be simultaneously examined through the internal procedure. In the Divisional Court Robert Goff LJ concluded that this provision was an impediment to prisoners' access to the courts. As the Prison Rules could not confer power to impede a prisoner's access to the courts he concluded that the Standing Orders must be *ultra vires* as not authorized by the relevant section of the Prison Act. He noted that prisoners, especially those who lacked knowledge or confidence, would be unlikely to pursue their cases through the courts without the assistance of a lawyer. In addition he observed that requiring prisoners first to submit their claim to an internal procedure might discourage justified claims being brought through the fear of being subject to a Rule 47 (12) disciplinary charge of making false and malicious allegations against an officer (an offence since dispensed with). With this decision the simultaneous ventilation rule also bit the dust.

In reaching his decision Goff LJ drew on the authority of *Raymond* v. *Honey*,[9] which had been decided a year earlier. In that case the House of Lords had concluded that a governor could not prevent a prisoner sending documents to the High Court to pursue a claim for contempt. The claim arose out of the governor stopping his original letter to his solicitor which contained an allegation of theft against an assistant governor. Again the authorities had relied on Standing Orders interpreting Rules 33 and 37. However the House of Lords indicated that such an interpretation could not interfere with a prisoner's right of access to the courts as this was an important right which could only be taken away by legislation which expressly provided for it to be taken away.[10] The primary legislation in question, s. 47 of the Prison Act 1952, did not authorize the removal of a prisoner's access to the courts. Lord Wilberforce indicated that 'a prisoner retains all civil rights and obligations which are not taken away expressly or by necessary implication', words which have been regularly quoted by every prisoners' rights advocate ever since.

As a result of these decisions there have been a number of changes in the rules governing prisoners' correspondence and visits from lawyers. Prisoners' written, but not physical, access to the courts has also been altered substantially. The rules are now contained in Prison Rules 37 and 37A and in Standing Order 5, which was published in order to comply with the ruling of

---

[9] [1983] 1 AC 1.
[10] Lord Bridge invoked the decision of *Chester* v. *Bateson* [1920] 1 KB 829, to say that the right of access to the courts could not be removed by delegated legislation. Only primary legislation which indicated that this was the express intention of the legislature would suffice.

the European Court of Human Rights in the *Silver* case.[11] The Rules and Standing Orders divide correspondence with legal advisers (these are defined by Standing Order 5A 34 as 'the inmate's counsel, solicitor or a clerk acting on behalf of the solicitor') into two regimes. One is where the prisoner has become a 'party to legal proceedings', the other is where he has not. Under Rule 37A (1), for example, correspondence between a prisoner who is a party to legal proceedings and his lawyer may not be read unless the governor has reason to believe that it contains matter which does not relate to those legal proceedings. Where the prisoner has not become a party to legal proceedings, however, for example where a prisoner is writing to his lawyer for advice as to whether he might pursue legal proceedings in relation to matters that have occurred in the prison, then this may be read under Rule 37A (4). Standing Order 5B 35 indicates also that it may be stopped if it contains matter which falls within the general restrictions on correspondence set out in Standing Order 5B 34 (which provides guidance on the meaning of 'objectionable' material referred to in Rule 33 (3)).

The definition of 'being a party to a legal proceeding' was explored by the Court of Appeal in *Guilfoyle* v. *Home Office*.[12] Therein the court appeared to indicate that a prisoner would only attain this status where a writ had been issued. The case itself concerned an application to the European Commission of Human Rights and Lord Denning, MR, expressed the view that a prisoner making such an application could never become a party to a legal proceeding as the Commission was not in a position to give a judgment that was binding, only the Court could do this. Even if he was wrong in this, Denning, MR, felt a prisoner could not be regarded as a party to a legal proceeding until at least the Commission had declared the application admissible (something which might only occur several years after the initial application was made). This view does not however appear to be taken by the Home Office, as Standing Order 5B 35 (*b*) and 5F 7 exempt correspondence with a legal adviser relating to an 'application to the European Commission or proceedings resulting from it' from the general rule on reading correspondence.

Once a prisoner is regarded as being a party to a legal proceeding or is facing a discipline charge, then his correspondence with a legal adviser may not be read unless the governor has reason to suppose that it contains matter which is not related to these legal proceedings. The level of information a governor would need on which legitimately to base such a suspicion is not specified, but such an exercise of discretion should be susceptible to judicial review. This class of legal correspondence should be marked 'SO

---

[11] In *Silver* the Court indicated that to comply with the Art. 8 (2) requirement that all restrictions be 'prescribed by law' the content of such restrictions should be available in the public domain if prisoners are to know them and adapt their conduct to them.

[12] [1981] 1 QB 309.

5B 32 (3)' and may be handed in sealed by the prisoner. The governor may still require that it be examined without being read to ensure that the letter does not contain any unauthorized article. In the case of *McComb* v. *Home Office*[13] a prisoner claimed that his legal correspondence was likely to be read when it was opened to be examined and that the only way to prevent this was to allow him to be present when the prison officers examined his legal correspondence. The Divisional Court refused to grant such an order. However, Standing Order 5B 32 (3) (c), amended after a European friendly settlement, does now provide that such letters may only be examined in the presence of the prisoner unless he has waived this opportunity.

Correspondence with legal advisers regarding applications to the European Commission of Human Rights are treated in the same way as legal correspondence where the prisoner is a party to domestic legal proceedings. The wording of SO 5F 7 'in regard to an application' leaves it unclear as to whether this provision only applies where an application has actually been lodged or whether it also covers seeking advice on whether to make an application. Standing Order 5F 4 indicates that a prisoner may correspond with his legal adviser and other persons 'in connection with the preparation of the application' without first obtaining the permission of the Secretary of State, but this does not really indicate whether such correspondence is defined as being correspondence in regard to such an application. Direct correspondence between the prisoner and the European Commission, including applications, is treated in essentially the same way as legal correspondence where the prisoner is not a party to legal proceedings.

As indicated earlier a prisoner's correspondence with his legal advisers, where the prisoner has not yet become a party to legal proceedings, may be read in the normal way. It may also be stopped if it contains material prohibited by the general provision on the content of correspondence, Standing Order 5B 34, whose content will be discussed in greater depth below. The same provisions apply to applications directly to the European Commission on Human Rights or correspondence resulting from such applications. Where prisoners are contemplating taking legal action the probability (in the case of category A prisoners) that such mail will be read would appear to have a restrictive effect and clearly violates legal professional privilege. However, the Divisional Court has indicated that it does not regard this as an unlawful restriction on the prisoner's access to the courts.

This was the outcome of *Ex parte Leech*.[14] Leech invoked *Raymond* v. *Honey* and *Ex parte Anderson* to claim that the likelihood that his letters to lawyers would be read discouraged him from being entirely honest about

---

[13] R. v. *Governor HM Prison Brixton ex parte McComb* unreported 29 Mar. 1983.
[14] R. v. *Secretary of State for the Home Dept. ex parte Leech* unreported 22 Oct. 1991. But see now the decision of the Court of Appeal which reversed the Divisional Court, The Times 20 May 1993.

potential actions against the prison authorities and hence hindered his access to the courts. He argued that nothing in the Prison Act or Rules authorized such a hindrance. Webster J. concluded that his right of access to the courts was not being denied as there was no bar on confidential visits from his lawyer, and that although his right to confidential correspondence with his lawyer was impaired this was not a 'fundamental or constitutional right' (terms of very uncertain character in English law). It was, he acknowledged, a restriction on an aspect of the 'natural right' of association, but he considered that s. 47 of the Prison Act 1952 gave the Secretary of State power to make Rules infringing on such rights so long as they did so no more than was 'reasonably necessary'. Examining this question under *Wednesbury* principles (which Webster J. acknowledged counsel for Leech had not asked him to do), he concluded that it was not, largely on the strength of an affidavit from a Prison Service official which warned that not all lawyers could be trusted to behave professionally, and gave a single example of a lawyer sending in money in a Rule 37A (1) letter 'which there was reason to believe was intended for use in connection with drug offences'.

The decision is indeed a sad one. Webster J. invokes (and prioritizes) concepts of 'fundamental' and 'natural' rights which have no real foundation in English law (there is not even any reference to the international human rights conventions that might give them some foundation). He then upholds the reasonableness of reading *any* lawyer's correspondence on the basis of *one* reported instance of misconduct, and even then one which is less than clear on the facts and related to an abuse which could be discovered by examining rather than reading correspondence. Little faith is shown in the many members of the legal profession who honourably and professionally advise prisoners.

The decision is also one which is now at variance with the ruling of the European Court of Human Rights in the case of *Campbell* v. *United Kingdom*[15] There the Strasbourg Court upheld the complaint of a prisoner in Scotland that the prison authorities' routine reading of his correspondence with his lawyers violated Article 8 of the Convention. Campbell was a convicted prisoner who had consulted with his lawyer about a variety of possible civil claims against the prison authorities. However in the case of remand prisoners corresponding with their lawyers about their defence in a criminal case the Strasbourg authorities seem more willing to permit official scrutiny of the correspondence. This seems to be the implication of some of the reasoning in *Schoenberger and Durmaz* v. *Switzerland*.[16] In that case the

[15] *Campbell* v. *United Kingdom* (1992) 15 EHRR 137, Series A No. 233-A.

[16] (1989) 11 EHRR 202, Series A No. 137, at para. 24. It is worth noting that Durmaz had already been charged and thus in United Kingdom terminology was a party to a legal proceeding. The Court held, however, that the stopping of a letter from his lawyer Schoenberger because it advised Durmaz to remain silent was a breach of Art. 8 on proportionality grounds. In the decision of *S* v. *Switzerland* (1992) 14 EHRR 670, Series A No. 227 the Court held that

Court again found a violation of Article 8 where a lawyer's letters to his client on remand were intercepted but suggested that greater restrictions on the correspondence of remand prisoners might be justified 'since in such a case there is often a risk of collusion'. There seems little reason in principle why pre-trial contact between a prisoner and his lawyer should be treated differently from post-conviction. Indeed interference with the former raises Article 6 as well as Article 8 issues. In time the European Court may have to re-examine the relationship of these two decisions and perhaps indicate that the common factor is the need for reasonable suspicion of unlawful action in the particular case before privileged legal communications can be subject to examination. The Strasbourg institutions may also have reason to examine the somewhat illogical position that prisoners' direct correspondence with the Commission is subject to being read and stopped if it contains something objectionable,[17] but is exempt from such scrutiny if it is submitted by a lawyer.

Though Prison Rule 37 allows legal visits in sight but out of hearing of a prison officer only where the prisoner is a party to legal proceedings, Standing Order 5A 34 widens this entitlement to cover nearly all legal visits. Apart from visits to discuss legal proceedings to which the prisoner is a party, pending adjudications and applications to Strasbourg, prisoners may also have in sight, but out of hearing, visits from legal advisers in two other circumstances. The first is where the prisoner wishes to consult about possible legal proceedings, the second where he wishes to consult about other legal business (for example about selling a house or making a will) that does not involve possible proceedings. For the former Standing Order 5A 34 (*b*) requires that the prisoner in advance indicate that the visit is to discuss possible legal proceedings but not what such contemplated proceedings are about (however, as the letter requesting the solicitor's visit may be read, subject to any change resulting from the *Campbell* case, the authorities may already know this if the prisoner has provided any details in his request). For the latter the prisoner must disclose in advance the purpose of the visit. Where such legal visits take place the legal adviser will be entitled to take notes and use a cassette recorder (providing the legal adviser gives an undertaking that the recording will be used purely in connection with the

constant monitoring of the applicant's conversations with his defence lawyer was a violation of his Art. 6 (3) right to a defence, perhaps suggesting that different approaches may be taken to oral as opposed to written confidential communications.

[17] The grounds for stopping a letter to the Commission are that it contains obscene material that it would be an offence to send through the post, that it contains material that might assist in the commission of a criminal offence, that it contains escape plans or material which might jeopardize prison security, that it contains obscure or coded messages, or that it contains material which might threaten national security. It is difficult to see what the Commission might do with such material if it received it. However, the existence of such grounds of restriction does have the function of allowing the authorities to read such communications, unlike correspondence with a lawyer about an application.

legal business that the visit related to) as an exception to the normal prohibition contained in Standing Order 5A 26 and 29. The rules governing visits therefore provide for greater protection of legal confidentiality than those governing correspondence, though this situation should alter with the implementation of the *Campbell* judgment.

As legal aid is available for most of the claims that prisoners are likely to want to make, including civil actions and judicial review (on terms which are discussed in the Chapter 3 above), there has been little discussion whether prisoners are denied access to law. This is a matter which has much exercised courts in the United States following the Supreme Court's 1977 decision in *Bounds* v. *Smith*.[18] Therein the court indicated that the Fourteenth Amendment to the Constitution guaranteed prisoners a right of 'meaningful access' to the courts. The state was under an obligation to ensure this and the court indicated that there was a number of ways that it could do this, either by providing legal assistance free to indigent prisoners or by ensuring that there was an adequate law library in each prison which gaolhouse lawyers and individual inmates could then avail themselves of. Little attention has thus far been paid to the state of legal literature available to prisoners in the United Kingdom's prisons.

The issue which gave rise to *Raymond* v. *Honey*, the stopping of a prisoner's letter to the courts, has also been addressed in changes to the Standing Orders. Standing Order 5B 33 (4) provides that applications to court which constitute the issuance of proceedings are not subject to the restrictions on correspondence and should be posted without delay.

Where prisoners wish to appear themselves before courts the judiciary have been less willing to facilitate them. In *Becker* v. *Home Office*[19] it was held that the governor was entitled to recover the costs of producing a prisoner at court to pursue a civil claim from the prisoner. In *Ex parte Greenwood*[20] this was extended to allow the Home Secretary to refuse to direct the prison governor to produce a prisoner to the court to pursue a private prosecution unless the prisoner gave an undertaking in advance that he would pay the costs of production. The situation was more fully considered in *Ex parte Wynne*.[21] There the applicant had obtained leave for judicial review actions against the governor and Home Secretary but had subsequently been refused legal aid. He sought to be allowed to appear in person but when given an application to do so tore it up, possibly because appearance would be conditional on his agreeing to pay the costs of production. His subsequent judicial review action failed on the grounds that without filling in the application he could not claim that the governor had unlawfully refused his application. However in the Court of Appeal Lord

[18] (1977) 430 US 817.   [19] [1972] 2 QB 407.
[20] R. v. *Secretary of State for the Home Dept. ex parte Greenwood* The Times 2 Aug. 1986.
[21] R. v. *Secretary of State for the Home Dept. ex parte Wynne* [1993] 1 All ER 574 (HL).

Donaldson, MR, suggested that in some circumstances requiring an applicant to agree to pay would constitute an unlawful interference with his right of access to the courts.

Lord Donaldson MR observed that the relevant statute, s. 29 (1) of the Criminal Justice Act 1961, empowered the Home Secretary to order production where he felt it was 'desirable in the interests of justice'. Even where it was desirable, and Lord Donaldson felt it nearly always would be, that still left the Home Secretary a discretion. In exercising it he could look at a range of factors, in particular whether the nature of the case and its procedures required the applicant's presence. Where the procedure was entirely written, for example, it probably would not. Where the Home Secretary found such presence to be necessary, if the procedure was oral and the applicant had no counsel for example, a decision not to allow production would be subject to review and the prisoner's inability to pay production costs would not be a good reason for refusal. To do so would be to put prisoners who could afford to pay in a better position than those who could not and in any case Lord Donaldson MR felt that as a prisoner should not have to pay for his own imprisonment and had no control over where he was detained such costs should be limited to those of transport from the prison closest to the court. Where an action was against the Prison Service itself there was a danger that the Home Secretary, in exercising this discretion, might appear a judge in his own cause. Therefore Lord Donaldson suggested he might seek the guidance of the court on this matter.

The other Court of Appeal judges were less willing to endorse the view that conditioning production on payment of costs was unreasonable, Staughton LJ indicating that prisoners were not entitled to equal access to the courts as other people and McCowan LJ suggesting that the right to go to court was one of those forfeited by necessary implication of imprisonment. Nevertheless they also felt it would be wise for the Home Secretary to seek the view of the court, a view supported by a unanimous House of Lords which did not reach the payment issue. All the judges involved in the case appeared to agree that R. v. *Governor of Brixton Prison ex parte Walsh*[22] remained good law. There the House of Lords upheld a governor's refusal to produce a prisoner for a remand hearing because of a lack of adequate staff resources. Despite what Lord Fraser referred to as a 'scandalous' situation of delay of the applicant appearing on charges the court unanimously concluded that the governor had not acted unreasonably in refusing production. Unlike Wynne the prisoner in *Walsh* was not being prevented from appearing at the trial of the matter in which he was concerned and so the ultimate outcome was not being prejudiced. However the

[22] [1984] 2 All ER 609.

court could perhaps have sent a clearer signal that justice delayed is justice denied.

NON LEGAL CONTACTS: CORRESPONDENCE

The present rules on general prisoners' correspondence are a direct outcome of the *Silver* case, decided by the European Commission of Human Rights in 1980 and affirmed by the Court in 1983. The case arose out of numerous complaints from prisoners of censorship of their correspondence after the *Golder* case in 1975. The Commission eventually selected the complaints of six prisoners and one relative, relating to sixty-two letters which had been stopped between 1972 and 1976. These included letters to MPs, lawyers, journalists, and relatives among others. The letters had been stopped on a number of grounds. Some because they raised complaints about treatment that had not been dealt with internally (the 'prior ventilation rule' referred to earlier), some because the prisoner had not asked to correspond with people who were not friends or relatives, and some because they contained matter which was seen as 'objectionable' and would therefore be prohibited under Prison Rule 33 (3).

The Commission held for the applicants that stopping the letters constituted a breach of Article 8 as regards nearly all of the complaints. In particular the Commission took the view that, since any stopping of a letter was prima facie a violation of Article 8's protection of the right of correspondence, to be valid such interference had to come within the justified limitations of Article 8 (2). The first requirement to do so was that the restrictions be 'prescribed by law', which, the Court had indicated in the *Sunday Times* case,[23] meant that the norm must be 'formulated with sufficient precision to enable the citizen to regulate his conduct: he must be able—if need be with appropriate advice—to foresee, to a degree that is reasonable in the circumstances, the consequences which a given action may entail.'[24] Most of the restrictions invoked to justify stopping the letters fell at this stage. The Commission held that they could not be reasonably foreseen from the simple declaration that correspondence would be interfered with if found to be objectionable. Though Standing Orders and Circular Instructions did spell out more precisely the criteria for deciding what sort of correspondence was regarded as objectionable these were not available to prisoners or their relatives. Only restrictions based on material which attempted to stimulate public agitation, circumvent prison regulations, or which contained threats of violence or discussed the crimes of others were held to be foreseeable from the language of Rule 33 (3).

[23] *Sunday Times* v. *United Kingdom* (1979) 2 EHRR 245, Series A No. 30.
[24] Ibid., at para. 49.

In addition to failing to meet the criteria of being prescribed by law the Commission took the view that many of the restrictions failed to justify a violation of Article 8 because they were over-broad and were not necessary in a democratic society to meet the legitimate grounds for restriction set out in Article 8 (2). This finding applied notably to the prior ventilation rule, the restriction on corresponding with those other than friends and relatives without obtaining permission, and a number of the content-based restrictions on correspondence. These included prohibiting letters containing material intended for publication, representations concerning the prisoner's trial conviction, and sentence, complaints about prison treatment, the stimulation of public petitions, and material which contained grossly improper language. Those grounds which had been held to be foreseeable and hence 'in accordance with law' were also regarded as being justified as necessary to prevent disorder or ensure the protection of the freedoms of others.

In response to the *Silver* decision the Home Office undertook to publish Standing Order 5, relating to communications, so that prisoners would have a much better idea of inclusion of what sort of material might lead to a letter being stopped.[25] They also reviewed the criteria on which censorship can take place and Standing Order 5B 2 now expressly states that 'The provisions affecting correspondence set out in this Order are necessary to meet the operational needs summarised above, while complying with Articles 8 and 10 of the European Convention on Human Rights.'

The current Standing Order 5B 34 sets out restrictions on 'general correspondence', which SO 5B 33 (1) defines as correspondence with family, friends, and other organizations. Correspondence which does not come within this categorization is that with legal advisers, the Parliamentary Commissioner for Administration, and the courts. The restrictions remain fairly extensive (see Appendix 3) but those ruled in violation of Article 8 by the *Silver* court have been removed or modified. In the case of indecent or offensive language, for example, this can only be grounds for stopping a letter if it appears that the intention of the writer is to cause distress or anxiety to the recipient or any other person. Material intended for publication can be stopped only if it is intended in return for payment, if it is for an organization which the governor has decided that the prisoner should not correspond with on security and good order grounds under Standing Order 5B 26,[26] if it concerns the prisoner's own crime or the crime of oth-

---

[25] Other states took similar steps to deal with *Silver*. If France for example the Décrets in the Code de Procédure Pénale Art. D 413–20 were altered by decree of 26 Jan. 1983 to bring them into line with the *Silver* requirements.

[26] To take this action the governor must have reason to believe that the person or organization concerned 'is planning or is engaged in activities which present a genuine and serious threat to the security or good order of the establishment or any other Prison Department establishment'. Governors are required to consult with headquarters before coming to this decision.

ers, except where this consists of serious representations or forms part of a serious comment about the criminal justice system, or where it refers to individual members of staff in such a way that they might be identified. Most of the other restrictions concern material which might threaten prison or national security (including coded messages), which take the form of threats to others outside the prison, or which involves business transactions outside a number of business exceptions outlined in SO 5B 34 (10).

Most of these restrictions would now appear to satisfy the European Court's application of Article 8 in *Silver*. The Commission and Court there indicated that the main problem lay in the blanket or over-broad nature of the restrictions. The Court did however acknowledge the legitimacy of the type of interests the government claimed it was seeking to protect by the restrictions. These included the security of the prison, the safety of people inside and outside it, and the maintenance of good order in the prison. This suggested that more narrowly drawn restrictions would be upheld. However, the Court would continue to apply a balancing test to these restrictions, examining them to discover whether they 'correspond to a pressing social need' and are 'proportionate to the legitimate end pursued'. Those on indecent language and material intended for publication might still fall on the wrong side of the line.[27] Letters to MPs, restrictions on which were found to be unlawful in *Silver*,[28] are now the subject of their own Standing Order—SO 5D. This indicates that letters to UK MPs may be read but may be stopped only if they contain material which it would be an offence to send (i.e. obscene material). Letters to MEPs, however, are essentially treated in the same way as general correspondence and hence may be stopped on a much broader range of grounds. Correspondence with the Parliamentary Commissioner for Administration is treated in the same way as letters to UK MPs.

Prisoners are no longer limited to a particular list of correspondents, with the need to get approval to write to anyone not on that list. The presumption now in Standing Order 5B 19 is that a prisoner can correspond with whomever he wishes. However Standing Orders still place some restrictions on whom a prisoner may correspond with. These include minors (where the person with parental responsibility for the minor requests that the prisoner not be allowed to write to the minor), convicted

---

[27] In the Commission's judgment in *Silver* (1980) 3 EHRR 475 it stressed with regard to restrictions on material intended for publication the importance of access to the media in a democratic society. With regard to restrictions based on indecent language it stressed the democratic right of everyone to freedom of expression, even if this involves the use of terms that others might find vulgar or shocking, and added, at para. 406, 'This freedom may be particularly important for persons, such as prisoners, subject to the daily frustrations of a closed community life.'

[28] See also *McCallum* v. *United Kingdom* (1991) 13 EHRR 597, Series A No. 183, where the letters were stopped under the requirements of Scottish Standing Orders that any allegations of ill treatment should first have been submitted to internal complaints review.

prisoners in other institutions or ex offenders (where the governor feels this is undesirable on rehabilitation or security and good order grounds), victims of the offence, and organizations which the governor has reason to believe pose a threat to security or good order at that or any other prison. Prisoners are generally not allowed to write to box numbers or receive anonymous letters.

Where a letter is stopped the prisoner will be informed and, in the case of letters containing prohibited material, invited to rewrite the letter. All letters are subject to being read (with the exception of some types of legal correspondence discussed above) and may be examined to see if they contain any unauthorized articles. The European Court has regularly upheld the practice of reading prisoners' mail,[29] although reform groups have consistently called for the end of censorship.[30] The Woolf Report accepted the general principle that prisoners' mail should not be censored unless there is a reasonable suspicion that such mail contains objectionable material. Noting that 'we have found in informal discussions, both with staff in this country, and in some other countries, an attachment, which we consider unreasonable, to continuing some form of censorship',[31] Woolf recommended that letters should not be read unless the prisoner was in category A or there was reasonable suspicion that the letter contained unauthorized material. Such letters could continue to be examined without being read to ensure that they did not contain any unauthorised material. In responding to the report the Home Office, which had ended censorship at open prisons in 1986 and reduced it to a sample of 5 per cent in category C prisons since 1988, did not go quite as far but undertook to end the routine reading of all prisoners' mail at all but dispersal prisons.[32]

Ending routine reading also has an effect on the number and length of letters that prisoners may send and receive, as staff resources for censorship purposes are the main factor in placing limits on this. In those prisons where routine censorship has been ended no limits are placed on the length of prisoners' letters or how many they may send or receive. Such absence of limitations is considerably more generous than the Prison Rules, as Rule 34 (2) (*a*) entitles a convicted prisoner only to send and receive one letter a week, although under Rule 34 (1) unconvicted prisoners are not subject to any limits in the number of letters they send or receive.

The Rule 34 (2) (*a*) letter is referred to by Standing Orders as the 'statu-

---

[29] See e.g. *Boyle and Rice* v. *United Kingdom* (1988) 10 EHRR 425, Series A No. 131. The United States Supreme Court has also held that prison authorities can read prisoners' correspondence in *Procunier* v. *Martinez* (1974) 416 US 396, In *Thornburgh* v. *Abbott* (1989) 490 US 401 the Supreme Court indicated that it would subject restrictions on incoming mail to less scrutiny than those on outgoing mail.

[30] See JUSTICE, *Justice in Prisons* (London, 1983)

[31] See *Prison Disturbances: April 1990*, para. 14.270.

[32] Home Office, *Custody, Care and Justice*, para. 7.36.

tory letter', postage costs of which are to be met out of public expense and which may not be stopped as or during punishment. In addition Standing Order 5B 6 (2) indicates that convicted prisoners should normally be allowed to send one 'privilege' letter per week and as many more as are compatible with available staff resources for censorship purposes. Such letters may be paid for out of prison earnings or private cash. As privileges the right to send such letters can be removed as part of a disciplinary award. In prisons where routine censorship prevails letters are normally limited to four sides of A4. A third class of letters also exists, known as 'special letters'. These are permitted to deal with a number of specific circumstances, such as to deal with family or personal business affairs, transfer to another prison, or to contact a probation officer regarding employment on release.[33] Letters on transfer should be sent at the public expense and the governor has a discretion to charge other special letters at the public expense. Despite Rule 34 (1), Standing Order 5B 8 limits unconvicted prisoners to two statutory letters a week and as many further privilege letters as they wish.

NON-LEGAL CONTACTS: VISITS

Under Prison Rule 34 (2) (*b*) a convicted prisoner is entitled to one visit every four weeks. Like the letter allowed every week under Rule 34 (2) (*a*) this is referred to by Standing Orders as the 'statutory visit' and it should normally last at least thirty minutes. Like the statutory letter the statutory visit cannot be withdrawn as part of punishment, unlike the 'privilege visits' which may be made available at an institution in addition to the statutory visit. However, under Standing Order 5A 9 a governor may defer a visit while a prisoner is in cellular confinement where the governor believes that because of the prisoner's behaviour or attitudes removal from cellular confinement for the purposes of a visit would be clearly impracticable or undesirable. It follows therefore that a prisoner in cellular confinement should not have a visit deferred as a form of punishment or pressure to modify behaviour. If a visit is deferred the prisoner can be given the opportunity of writing an extra statutory letter instead of accumulating visits.

Unlike letters Standing Orders do not indicate a minimum number of privilege visits that may be allowed. This appears to be left up to the governor of each prison depending on available resources. As a result disparities are likely to arise between prisons which may adversely affect prisoners on transfer from one prison to another.[34] Standing Orders also make

---

[33] The full list is given in SO 5B 7. Prisoners who are non-United Kingdom nationals may also write to their consulate or High Commission on the same basis.

[34] It is worth noting that in the *Boyle and Rice* case (1988) 10 EHRR 425 the Court took the view that a reduction in the number and quality of visits as a result of a transfer did not amount to a breach of Arts. 8 or 14.

provision for 'special visits', to deal with family or personal business, or where the prisoner is seriously ill. Such 'special visits', like legal visits or those by probation officers, MPs, priests or ministers, consular officers, the Parliamentary Commissioner for Administration, or police officers, do not count against a prisoner's allocation of visits. Prisoners may choose to accumulate visits over a period of time and then seek temporary transfer to a prison closer to home in order to take these visits (normally over a period of a month). Category A prisoners, who are often detained a considerable distance from their home and family, are especially encouraged to make use of this facility. Standing Order 5A 11 indicates that a minimum of three and a maximum of twelve visits may be accumulated. The same Standing Order indicates that temporary transfer for accumulated visits should take place no more than once a year.

Unconvicted prisoners are entitled by Rule 34 (1) to as many visits as they wish, which should normally be for up to fifteen minutes a day, Monday to Saturday. However, Standing Order 5A 4 indicates that where available staff resources make compliance with this Rule impracticable the Area Manager can authorize a reduction in the frequency of visits as long as the aggregate of ninety minutes of visiting a week is maintained. Despite the apparently mandatory language of Rule 34 (1) one judicial review challenge is denial of visits to unconvicted prisoners failed where the court took the view that in the circumstances the governor was not acting unreasonably.[35]

As regards who can visit a prisoner Standing Orders 5A 30 and 31 give particular prominence to the visits of close relatives. These are defined as spouses or persons with who the prisoner was living as man or wife before imprisonment,[36] parents, or those *in loco parentis*, children or children to whom the inmate is *in loco parentis*, brothers, sisters, and fiancés or fiancées (providing the governor is satisfied that a genuine intention to marry exists). While the governor may refuse visits from such people or security and good order grounds SO 5A 31 indicates that this should only be done in exceptional cases. Where a prisoner's close relative is in another prison establishment then, subject to requirements of security and the availability of transport and accommodation, visits may take place at three

[35] See e.g. the Northern Irish case of *Mulvenna* v. *Governor HM Prison Belfast* unreported 23 Dec. 1985; the court upheld reduction of visits for remand prisoners to one and a half visits a week on the grounds that this was not a clearly unreasonable response to staffing limitations.

[36] It is not clear that this extends to gay or lesbian partners. In one American case visiting restrictions on gay and lesbian partners were held to be unconstitutional under the due process clause as the institution could not demonstrate any legitimate aims to be served by such restrictions, *Doe* v. *Sparks* (1990) 733 F. Supp. 227. Canadian prisons allow family visits by same-sex partners, after settlement of a case involving a challenge to denying access visits to the family visits programme to same-sex partners under the equality provisions of the Canadian Charter.

monthly intervals. MPs may also visit prisoners as of right on production of a special visiting order issued by the governor.

Other visitors are also allowed but the governor may refuse these on security or good order grounds. He may also refuse them, according to Standing Order 5A 32, where he has good reason to believe that the visit would seriously impede the rehabilitation of the prisoner. Decisions by governors to refuse visits are subject to judicial review as they potentially infringe the rights of both the prisoner and the visitor. In a Northern Irish case, *McCartney* v. *Governor HM Prison, Maze*,[37] the decision of the Secretary of State to deny a convicted IRA prisoner visits from a Sinn Fein local councillor (Sinn Fein being the political party which supports the IRA) was upheld as reasonable despite the absence of any claims that this specific councillor's visit posed a threat to the security of the prison. Limitations on who may visit could also be reviewed under Articles 8 and 14 of the European Convention on Human Rights. Decisions of the Commission in this area show a greater reluctance to scrutinize official restrictions than is apparent with restrictions on correspondence, at least where they do not involve close family members or amount substantially to prohibiting the prisoner contact with the outside world.[38] While visits from legal advisers and priests or ministers will generally be allowed (and may take place out of the hearing of an officer and without restrictions on the use of notes or tape recording) those from journalists in a professional capacity will generally be prohibited, although the governor has a discretion to allow them. Visits by police officers in connection with an offence are required to be conducted in accordance with PACE and Code C of its Codes of Practice. Such visits should be in the sight and may be in the hearing of a prison officer and Standing Order 5B 40 (4) indicates that a prisoner is not required to remain in an interview beyond the time necessary for the police to explain the purpose of the interview.

Standing Orders indicate that visits should take place 'under the most humane conditions possible'.[39] In practice the quality of visiting conditions varies widely.[40] Unless special security considerations are thought to dictate otherwise they should normally take place in an open visiting area where the prisoner and visitors may sit at a table and may embrace. Up to three people will normally be permitted on a visit, with children under 10 years old not counting towards this total. Except where the visit is a legal visit or

[37] (1987) 11 NIJB 94.

[38] Visits, unlike correspondence, are not expressly referred to in Art. 8 and therefore are raised under the more general respect for family life provision. Courts in the United States have generally subjected restrictions on who may visit a prisoner to a reasonableness test which is deferential to the institution's judgment, see *Kentucky Dept. of Corrections* v. *Thompson* (1989) 109 S. Ct. 1904.

[39] See SO 5A 24 (1).

[40] See V. Stern, *Bricks of Shame: Britain's Prisons* (London, 1987), 104–6.

one by an MP, priest, or minister, visits will normally be in the sight and hearing of prison officers. This usually means that prison officers will be in the room where visits take place but the governor may direct closer oversight of a visit where he feels security or good order considerations require this. Notes may not be taken or made in an interview without the permission of the governor although Standing Order 5A 26 indicates that 'in general the degree of control should be commensurate with the level of censorship in force at the establishment'. Nor are tape recordings permitted in visits, though this is subject to exceptions for visits by legal advisers, MPs, consular officials, priests and ministers, or members of the European Committee on Torture.

As has been indicated elsewhere in this book both prisoners and their visitors are liable to be searched, though a search of a visitor probably requires a greater level of suspicion that they are taking an unauthorized article into the prison than the search of a prisoner. If an unauthorized article is found during a search or an officer observes a visitor passing or receiving an unauthorized article then the visit may be stopped, similarly if a visitor is seen to be tape recording the visit. Violation of the prohibition on passing unauthorized articles may be a ground for refusing future visits by the person involved. Where however that person is a close relative a total ban, or a ban for a substantial period of time, as a result of an incident of passing an unauthorised article would appear to go beyond the scope of exceptional circumstances referred to in Standing Order 5A 31 and might violate Article 8 of the European Convention on Human Rights. The validity of any such ban would be especially questionable if the article in question was trivial and was unrelated to escape or to bringing drugs into the prison.

The issue of whether prisoners are entitled to conjugal visits has been raised before the European Commission on Human Rights, which found that a ban on such visits did not violate Article 8 of the Convention.[41] Several European jurisdictions do, however, provide for such visits, or for more extensive family visits, on the prison premises. The Dutch prison system, for example, allows for one such 'private' visit a month.[42] The Home Office announced in 1992 that it would review its prohibition on such visits in the United Kingdom system. In any such review issues arise as to how extensive such a change will be, whether it will be limited to spouses or will extend to non-married partners and to partners of the same sex. There is also an issue as to whether the visits will be limited to partners or will extend to allowing children to accompany partners.

Interestingly, in view of the refusal of conjugal visits, the Home Office at

---

[41] See Appls. 6564/74, 2 D. & R. 105 and 8166/78, 13 D. & R. 241.

[42] See D. Downes, *Contrasts in Tolerance: Post War Penal Policy in the Netherlands and in England and Wales* (Oxford, 1988),176.

one time refused to allow prisoners to marry (or leave to marry outside the prison), partly on the grounds that a prisoner would be unable to cohabit with his or her wife or husband. In *Hamer* v. *United Kingdom*[43] the European Commission on Human Rights rejected this argument, on the ground that marriage was primarily a legal status which did not require cohabitation, as well as others based on security and public interest. They found that refusal to allow a prisoner to marry while detained was a breach of Article 12. The position has been changed by the Marriage Act 1983 and marriages may now take place in places of detention.

In its evidence to the Woolf Inquiry the Prison Service observed that prisoners valued visits more than any other privilege. This was a view Woolf agreed with while noting that visits were a right rather than a privilege. However, arrangements in the United Kingdom appear more restrictive than in many other European countries and contrary to the spirit at least of Rule 43 of the European Prison Rules, which indicate that visits should be allowed from families and members of outside organizations 'as often as possible'. Woolf endorsed the Prison Service's target of allowing convicted prisoners two statutory visits a month (and recommended that relatives' costs of these should be met by the DSS's Assisted Prison Visits Scheme). He also recommended that the aim of the Prison Service should be to ensure convicted prisoners one visit a week.[44] In response to this the Home Office announced that convicted prisoners would be able to enjoy two visits every twenty-eight days and that it would work towards entitling unconvicted prisoners to enjoy three one-hour visits a week. On family visits Woolf endorsed the idea but did not see it as a priority, for example, as opposed to extending the use of telephones or better home leave arrangements. The idea is however now receiving greater Home Office attention and a extension is also being considered of an experimental scheme introduced at Holloway Women's Prison whereby children are allowed to spend all day at the prison in their mother's company. As discussed in Chapter 5 above mothers can keep very young children with them in prison if they are detained in a prison with a mother and baby unit and places are available in that unit.

TELEPHONES

Prisoners have always been entitled to make a request to the prison governor to use a telephone. Such calls have generally only been allowed to legal

---

[43] See *Hamer* v. *United Kingdom* (1979) 4 EHRR 139. In the United States the Supreme Court indicated that restrictions on prisoner marriage could only be justified if imposed for reasons of 'compelling necessity' in *Turner* v. *Safely* (1987) 482 US 78.
[44] See Woolf Report, para. 14.229.

advisers or to close relatives to deal with a family problem. However, since 1988 the Home Office has begun the introduction of cardphones into prisons, a development which Woolf observed could lead to less emphasis on letters and the censorship thereof. Woolf encouraged further extension of the cardphone scheme as a good way of enabling prisoners to maintain more regular and meaningful contact with their families and to provide a better outlet for emotions that build up in the conditions of prison life. Young prisoners in particular have found telephone contact with their families to be an important way of lessening the isolation of prison life. In response the Home Office announced in late 1991 that they would extend the provision of cardphones to all parts of the prison estate, with the exception of special security units.

The use of cardphones is governed currently by Circular Instruction 21/1992, which indicates that prisoners may obtain as many as they wish of the special cards—which do not operate on the public cardphone system—but should not normally retain more than two at one time. Category A prisoners, prisoners on the escape list, or prisoners who are viewed as vulnerable may not retain possession of the cards but once purchased must turn them over to the wing officer. They must then pre book all calls they wish to make using the cardphone system. Clearly concerned by the fear that phonecards will become a new form of currency within a prison, with attendant risks of extortion, the authorities have provided that prisoners should be issued with personal cards which must be returned when they have expired.

All calls using the cardphones can be monitored and may be stopped if they contain any material which would be grounds for stopping a letter under Standing Order 5A 34. They may also be stopped if made to chatline services or Directory Enquiries. Prisoners are given access to STD code information but not to the telephone directory. Category A prisoners are restricted to making calls to people on a pre-approved list. CI 21/1992 indicates that monitoring of most calls should be done randomly. With category A prisoners, or those on the vulnerable or escape list, all calls must be continuously monitored and tape-recorded and such tape recordings may be admissible in legal or disciplinary proceedings.

As access to telephones is such a recent phenomenon the legal principles governing it remain unclear, though they are generally analogized to those relating to correspondence. In cases on telephone tapping the European Court of Human Rights has treated telephone conversations as correspondence and so it seems likely restrictions on the use of telephones by prisoners would be liable to scrutiny for potential breaches of Article 8 of the Convention in the same way that mail censorship is.

PUBLIC SCRUTINY OF PRISONS

Most of the discussion of letters and visits has focused on the issue of getting people and material into a prison. However, as prisons are closed institutions it is also important to ensure that information gets out so that the public have a chance to evaluate what is going on in prisons. Abuses of rights are much more likely to continue if knowledge of them exists only within the prison.

The prison system in the United Kingdom has been notorious for secrecy. As Rutherford notes, even inquiries into the prison system have been denied access to Standing Orders.[45] The Prison Service Annual Reports provide many valuable statistics but lack detail on individual prisons.

Until 1989 the broad scope of s. 2 of the Official Secrets Act 1911 also subjected the disclosure of almost any information by prison staff to criminal sanction. The Official Secrets Act 1989 repealed s. 2 and narrowed the categories of information which are subject to the Act's provisions. The section most relevant to prisons in the 1989 Act is s. 4 (2) (a) (ii). This makes it an offence for any Crown servant or government contractor to disclose without authorization any information, document, or article 'the disclosure of which facilitates an escape from legal custody or the doing of any other act prejudicial to the safekeeping of persons in legal custody'. The offence is also committed if disclosure of such information is likely to have this effect. Defences exist in ss. 4 (4) and 4 (5) of having no knowledge or reasonable cause to believe that disclosure would have the effects mentioned or that the information is within the scope of Article 4 (2). This is obviously a much more restricted provision and would appear only to extend to the disclosure of information likely to facilitate escape from prisons.

In recent years, though, there has been a change, at least at the institutional level, in the provision of information to the public. More Standing Orders have been published, in part because of European Court pressure described earlier in this chapter, some Boards of Visitors have begun to publish reports, and the regular reports of the Chief Inspector of Prisons have provided a large supply of detailed information on conditions in individual prisons. The Woolf Report provides another rich source of information and the reports of the European Committee on Torture have already provided a further detailed and independent evaluation of prison conditions in the United Kingdom. In addition the Home Office has shown a greater

---

[45] *Prisons and the Process of Justice* (Oxford, 1986), 104. The committee involved was the Prison System Enquiry Committee. Rutherford also notes the Webbs writing in the 1920s of 'a silent world, shrouded so far as the public is concerned in complete darkness'.

willingness to allow academic researchers into prisons[46] and to co-operate with international human rights organizations looking at prisons in the United Kingdom.[47]

The picture is less encouraging when it comes to investigations by the media. In the early 1980s the Home Office took contempt proceedings against then NCCL Legal Officer Harriet Harman for giving to a journalist documents discovered from the Home Office in a case relating to control units, even though these had already been read out in open court. Although the Home Office was successful in the domestic courts, the government eventually had to concede a breach of Article 10 of the European Convention on Human Rights and to alter the law on contempt.[48] As referred to earlier the normal prohibition on journalists visiting in their professional capacity, the prohibition on prisoners writing material for payment and the restrictions of the Official Secrets Act (though these have been considerably reduced by the 1989 Act) all work against media investigation of prisons. However, a claim by the Home Office of copyright in a television interview with a prisoner, which would give it a veto on any broadcast, has been rejected by the High Court.[49]

HOME LEAVE

While frequent contact by letter, telephone, and visit can reduce the isolating and dehabilitating effect of prison on both prisoners and their families it remains contact within the prison environment. It is therefore limited, takes place under prison authority scrutiny, and is removed from the pleasures and strains of outside living to which prisoners will have to return. More meaningful contact with families and the outside world in general can be provided by home leave during sentence. Home leave provides greater time for family problems to be worked through (though it can also increase them) and for long-term prisoners in particular can reduce that sense of shock at changes in the world when a prisoner is released after a

[46] For examples of recent research studies see E. Genders and E. Player, *Race Relations in Prison* (Oxford, 1989); A. Liebling, *Suicides in Prison* (London, 1992).

[47] For example the Helsinki Watch Report on *Prison Conditions in the United Kingdom* (New York, 1992).

[48] *Home Office* v. *Harman* [1983] AC 260. The European action produced a friendly settlement, reported in *Stocktaking on the European Convention on Human Rights* (Strasbourg, 1986).

[49] The case involved the potential broadcast of an interview with Dennis Nielsen, see *Guardian* 28 Jan. 1993. In the United States the Supreme Court has held that the media have no right of access to prisons in *Houchins* v. *KQED* 438 US 1 (1978). Art. 19 (2) of the International Covenant on Civil and Political Rights guarantees the right to 'seek' information and this might provide the basis for a claim where the media were excluded from a prison story of public interest.

long time inside. Most prison systems in Continental Europe have recognized this and provided for regular home leave from an early stage in the prisoner's sentence.[50] In the United Kingdom Woolf was critical of the limited home leave arrangements. Though changes in 1992 have made improvements, the schemes available remain less than generous.

Temporary release is provided for in Rule 6 of the Prison Rules. This does little more than confer a power to make temporary releases and indicate, in Rule 6 (3), that a prisoner released under the Rule 'may be recalled to prison at any time whether the conditions of his release have been broken or not'. Detail on release is provided in Home Office Circular Instructions. In its current statement of policy (contained in Circular Instruction 43/1992) the Home Office sees the purpose of home leave as being twofold. These purposes are (1) to help restore a prisoner's self-confidence by placing trust in him in conditions of freedom and (2) to allow a prisoner to maintain family links, make contact with employers, and thus help readjustment to life outside prison. The interests of prisoners' families do not therefore appear to be formally considered. In determining when a prisoner can get home leave a two stage process is followed. First the prisoner must qualify as eligible. Security classification, length of sentence, and length of time already spent in prison are the key factors in this. To a certain extent the coming into force of the 1991 Criminal Justice Act results in different arrangements applying for those sentenced after 1 October 1992. Secondly, since the Home Office regards home leave as a privilege under Rule 6 of the Prison Rules rather than what it describes as an entitlement, even eligible prisoners must apply for leave and a number of risk factors will be assessed in deciding whether to grant their application. Most decisions can be taken by governors but some require Home Office approval. Under the home leave scheme two forms of home leave are available. *Short home leave* is of two days' duration, exclusive of travelling time. *Long home leave* amounts to five days, exclusive of travelling time. Opportunities also exist for compassionate home leave and pre release home leave and will be discussed after consideration of the home leave scheme.

Some prisoners are ineligible for any home leave. These include category A prisoners, prisoners on the escape list, unconvicted or unsentenced prisoners, those regarded as mentally ill, and those serving less than eighteen months. The exclusion of category A prisoners, presumably on the grounds that they are very likely not to return, in England and Wales is interesting given that in Northern Ireland Christmas and summer home leave schemes have been operated in recent years for life sentence prisoners convicted

---

[50] In Germany for example home leave is available for up to twenty-one days a year after six months of the sentence have been served, see R. Vogler, *Germany: A Guide to the Criminal Justice System* (London 1989), 70.

of terrorist offences, without, it seems, resulting in a high rate of non-returning.[51]

Prior to April 1992 the relevant security classification for the purposes of home leave was the security classification of the prison in which the prisoner was detained. This was especially unfair for low-security category prisoners temporarily moved to higher-category prisons purely for operational reasons. Now the prisoner's personal categorization is what matters, something which again shows the significance of classification and how unsatisfactory the lack of review of it is. As regards *short home leave* category B prisoners sentenced before 1 October 1992 to less than three years are ineligible for home leave. Those sentenced to more than three years may apply for home leave after one-third of their sentence. This is their Home Leave Eligibility Date (HLED), which for such prisoners corresponds with the earliest date at which they can apply for parole, their Parole Eligibility Date (PED). The HLED can be extended as a result of a punishment of extra days resulting from a disciplinary conviction. If home leave is granted the prisoner is entitled to one home leave during his sentence which must be taken in the nine months before his Automatic Release Date (ARD), that is, two-thirds of the sentence if he is not given parole, though this may be extended by extra days awarded for disciplinary offences. If the prisoner is sentenced after 1 October 1992 the minimum qualifying condition remains a sentence of three years or more. The HLED remains at one-third of sentence and the maximum entitlement remains at one short home leave. However, there is some change in the time within which it must be taken. For those sentenced to less than four years this is within nine months of the Conditional Release Date (CRD), which is half of the sentence imposed. For those sentenced to more than four years it is within nine months of the Non Parole Release Date (NPD), which occurs at two-thirds of the sentence imposed.

Category C prisoners have more generous short home leave arrangements. At a minimum they must have been sentenced to two years' imprisonment. Their HLED again arises at one-third of the sentence. Thereafter the prisoner may apply again for short home leave at six-monthly intervals, but all such home leave must be taken within two months of the ARD. Arrangements for those sentenced after 1 October 1992 are essentially unchanged, though the time within which home leave is taken is now set by the CRD (for those serving less than four years) or the NPD (for those serving more than four years). The most generous arrangements apply to category D prisoners who have a qualifying period of a sentence of eighteen

---

[51] In 1990 for example Christmas home leave was granted to 434 prisoners, including many life sentence prisoners, only five failed to return. Summer home leave of five days was granted to 111 life sentence prisoners who had served thirteen years or more. See *Report on the Work of the Northern Ireland Prison Service 1990–1*, HC 46 (London ,1991), 25.

months or more. The HLED arises after one-third of sentence and prisoners may apply thereafter every two months until two months before the ARD (if sentenced before 1 October 1992) or the CRD or NPD (if sentenced after 1 October 1992).

Life sentence prisoners are excluded from short home leave if classified as category A or B. Category C lifers may apply and can be granted home leave if they have a provisional release date and their application is approved at Home Office level.[52] Category D lifers are eligible once they have a provisional release date and have served four months in open conditions or have served nine months in open conditions and Home Office approval is forthcoming. Thereafter they may apply on the same basis as a similarly categorized determinate sentence prisoner.

Turning to *long home leave* all otherwise eligible adult prisoners with determinate sentences of eighteen months or more may apply, as may young offenders sentenced to twelve months or more and life sentence prisoners given a provisional release date. Once admitted to the Pre Release Employment Scheme (PRES) prisoners are not entitled to long home leave. Only one long home leave is permitted and this must normally be taken in the two months before ARD, CRD, or NPD, whichever is appropriate. Normally a prisoner must have taken a short home leave before long home leave will be granted. The only change made in these arrangements by the Criminal Justice Act 1991 is that in the case of prisoners sentenced after 1 October 1991 the Home Leave Board (an institution whose composition and function will be discussed below) may consider granting a long home leave up to four months before the CRD/NPD if it considers this in the interests of the prisoner's rehabilitation plans.

A scheme also exists of *pre-parole home leave*. This is available to prisoners sentenced to determinate sentences in excess of eighteen months who have been granted release on licence under the parole scheme (described in Chapter 9 below). It is not available for life sentence prisoners or to those sentenced after 1 October 1992 to more than eighteen months but less than four years as they are ineligible for parole. Once prisoners eligible for the pre-parole scheme have been given a release date they are ineligible for short and long home leave, although they may apply for this if it better suits their rehabilitation plans. Pre-parole leave consists of two consecutive days at home, exclusive of travelling time, and is designed to allow prisoners to contact a potential employer or supervising probation officer.

As stated above the decision whether to grant home leave to prisoners who fulfil the eligibility criteria is a discretionary one. In most cases the discretion is that of the prison governor. The relevant Circular Instruction, CI 43/1992, indicates the criteria on which he should make his decision but

---

[52] The application goes to D1P1 or DSP2.

not the information on which he should bring those criteria to bear. In cases involving life sentence prisoners, however, the governor must refer to the Home Office, which has the final decision.

Special arrangements apply to first home leave applications or applications after a prisoner has committed a serious breach of conditions on a previous home leave. In such cases the governor must first refer the case to a Home Leave Board (HLB). This is composed of a governor, a prison officer who knows the prisoner well, a seconded probation officer who is well acquainted with the prisoner, the lifer liaison officer if the prisoner is a life sentence prisoner, and one other person who knows the prisoner well (such as the chaplain or education officer). The Board are required to consider reports by the seconded probation officer (compiled with the home probation officer) in relation to the prisoner's home circumstances and any relation with the victim, from the wing in relation to the prisoner's conduct and attitudes, and from the medical officer if the prisoner is thought to present a risk (such as of suicide or alcoholism). The HLB should also have before them the prisoner's criminal history and, if the prisoner is a lifer, an up to date sentence plan. CI 43/1992 indicates that the Board may invite the prisoner to attend for some or all of its deliberation. However, since the same Circular Instruction notes that some of the information considered, especially from the wing report, may be of a confidential nature it would seem inconsistent with practice on parole or life sentence release decisions (where similar material is considered) to give the prisoner access to such material or allow him to be present during discussion of it. However, just as we have argued in relation to release decisions that fairness requires that a prisoner be entitled to respond to negative comments in such reports, so we would argue that prisoners should be entitled to the same level of participation in decisions on home leave. The HLB make their recommendations to the governor, who must sanction any decision. In respect of first home leaves to any particular address by life sentence prisoners the Home Office must also give its approval. Where home leave is refused the Circular Instruction indicates that the governor should give written reasons to the prisoner.

Circular Instruction 43/1992 also spells out the factors that a governor or the HLB must take into account in deciding on whether to grant home leave. These include (1) the risk that the prisoner would present to public safety, (2) the risk of further offending, (3) the likelihood of the prisoner failing to comply with any condition in the licence, (4) the propensity to abscond, and (5) the availability of suitable accommodation. The guidelines indicate that the element of risk can never be eradicated but that the governor should assess it to rule out applications where there are reasonable grounds to believe that the licence conditions might be broken. They go on to list factors to take into account in assessing the risk. These include the prisoner's criminal history, home circumstances, the position of the victim(s) and the community, conduct on previous home leaves, and behaviour in prison. Governors are warned that good conduct in prison may bear no relationship to the risk that the prisoner may pose to his family or the general public but there is no corresponding caution that those who show difficulties in adapting to prison may exhibit fewer problems on the outside.

Where home leave is granted it will normally be made subject to conditions. These should address the issue of returning on time and any specific risks in relation to this prisoner, for example in relation to alcohol or being in the vicinity of the victim while on leave. Failure to comply with these conditions, most significantly failure to return at the appointed time, is a disciplinary offence under Rule 47 (8) (*b*). Governors are directed that it is undesirable to impose a condition of reporting to a police station when a telephone call to the prison might suffice but are directed to impose a condition of reporting to the home probation officer in respect of lifers, first time releases, violent or sex offenders who are returning to the vicinity of the victim(s), or where the probation service recommend the imposition of such a condition.

In addition to the home leave scheme Standing Order 7E makes provision for temporary release for compassionate reasons, education or training, or to attend an outside hospital. Compassionate reasons include to be married, for a prisoner to see his own children in a children's home, or to attend a funeral or a terminally ill relative. In respect of the last reason the Order limits relatives to parents, children, siblings, or spouses (including someone the prisoner was living with as a husband or wife immediately before reception). Governors are required to contact Headquarters when considering (non-home leave) temporary release for lifers who have not got a provisional release date, those serving over ten years who are not within twelve months of release, and prisoners convicted at any time of a serious sexual offence.

A final type of temporary release scheme is the Pre Release Employment Scheme (PRES). This allows prisoners to spend the last six months of their sentence in a PRES hostel while working for a local employer. Male prisoners serving life or determinate sentences in excess of four years or female prisoners serving eighteen months or more who are not excluded from the general home leave criteria may apply to participate in PRES twelve months before their expected release date. Their application will be considered by a selection committee composed of prison staff 'best able to judge the applicant's character and abilities'. Again key criteria in reaching the decision are whether the applicant presents a serious risk to the safety of the public and whether he can be trusted to respect the confidence placed in them.

As argued earlier in this section home leave allowances in the United

Kingdom are very limited, especially in comparison to many other European countries. This hardly seems in line with the Home Office's stated purpose of using home leave to maintain ties to family and the outside world and even the 1992 changes, which increased the frequency available for category D prisoners, do not seem to amount to the 'substantial' increase that Woolf called for. The procedure by which decisions on home leave are reached is also very traditional to the prison system. It is discretionary, prioritizes security considerations, is reached purely by Prison Service staff, and allows very little input by the prisoner. It clearly contravenes the intent of the Council of Europe Committee of Ministers Recommendation on Prison Leave that states should 'provide the means by which refusal can be reviewed'.[53] That recommendation has no legal force but decisions to refuse home leave are susceptible of judicial review. Despite the Home Office's view that home leave is a privilege rather than an entitlement it would appear that prisoners who satisfy the eligibility criteria would have a 'legitimate expectation' of being released unless there were reasons relating to their personal circumstances which justified denial of release. Although guidance to governors indicates that eligibility is always to be considered in relation to risk it is arguable that since the eligibility criteria are already structured to exclude high-risk prisoners (by excluding category A prisoners from any release scheme for example) those who meet them are entitled to regard themselves as having a presumption in favour of release. Hence their claims to a greater degree of procedural safeguards are strengthened and they should certainly be entitled to review of refusals to release which are clearly unreasonable, are based on irrelevant factors, or which fail to take into account relevant factors.

TRANSFER

The issue of inter-prison transfer for reasons of good order and discipline will be discussed in Chapter 8 below and will not be raised here. However, transfer may also take place for purely operational reasons, without any formal or informal disciplinary overtones. When it does it may have an adverse effect on the prisoner's ability to maintain family and outside contact as family and friends may find the distances involved in travelling to visit prohibitive. It may also lead to a prisoner being moved to a prison with considerably inferior facilities from the one he currently occupies. The problem comes into particularly sharp relief when prisoners from one jurisdiction in the United Kingdom find themselves imprisoned in another jurisdiction (Northern Irish prisoners in England for example) or imprisoned in

---

[53] Recommendation R (82) 16, para. 10.

a jurisdiction outside the United Kingdom. Legal mechanisms exist in both cases to facilitate the transfer of a prisoner to his 'home' jurisdiction but their exercise is again largely dependent on the discretion of the prison authorities.

Once again the basic statutory provision on transfer is an enabling one. Section 12 (2) of the Prison Act 1952 entitles the Secretary of State to determine the prison to which a prisoner is committed and to direct the removal of a prisoner from the prison in which he is held to any other prison. More detail is provided in Standing Order 1H. This indicates that prisoners may be transferred to relieve overcrowding, among other reasons, but that in selecting prisoners to transfer for this reason appellants and those whose domestic circumstances would be gravely prejudiced should not be chosen. Before a transfer takes place the prison medical officer should certify that the prisoner is fit to travel and should give any special information about the prisoner's health to the medical officer of the receiving institution twenty-four hours in advance of the transfer.

Relatives of prisoners often claim that they are given insufficient notice of transfers and may even arrive for a visit only to find that the prisoner has been moved. Unfortunately this appears more likely to happen to those with furthest to travel, the relatives of category A prisoners held in dispersal prisons. The Standing Order indicates that advance notification of moves should not be given to category A prisoners or their potential visitors. The governor also has the discretion to refuse advance notification to the relatives of any other prisoners where he feels this might pose a threat to security or good order. Should such visitors arrive without being notified of a move the prison is directed to consider their claim for reimbursement of travelling expenses. In respect of other prisoners where the governor does not perceive such a threat prisoners are normally allowed to notify their intending visitors by a first-class letter (not counted against their usual allowance) or, if there is insufficient time, a telephone call.

The reference to the undesirability of moving prisoners where this may severely prejudice their domestic circumstances may have been influenced by the one significant challenge in the courts to transfer decisions. This occurred in R. v. *Secretary of State for the Home Dept. ex parte McAvoy.*[54] McAvoy, a category A remand prisoner on a robbery charge, was transferred from Brixton to Winchester—over sixty miles away—without warning. He claimed that this prejudiced his right to a fair trial as Winchester had poorer consultation facilities and it would be difficult for his chosen counsel in London to visit him there. He also asserted that it would be difficult for his parents to visit him in Winchester because of their ill health. Webster J. accepted that these were relevant considerations and that the

---

[54] [1984] 1 WLR 1408.

Home Secretary would have erred in law if he failed to take account of them. Indeed, he acknowledged that the move might lead to the applicant requiring a new counsel. However, in the end he refused to interfere with the Home Secretary's decision as it was taken for 'operational and security reasons' (primarily it appeared that Brixton was not seen as secure enough for the prisoner and London's other high-security prison at Wormwood Scrubs was felt to be inappropriate). As Webster J. observed,

In my view it is undesirable, if not impossible, for this court to examine operational reasons for a decision made under that section [s. 12]; and to examine security reasons for a decision made under that section could, in my view be dangerous and contrary to the policy of that statutory provision, which is to confer an absolute discretion, within the law, on the Secretary of State to make such executive decisions as he thinks fit for operational and security reasons.[55]

This decision is in marked contrast to Webster J.'s more interventionist rhetoric in *Tarrant*, delivered the same year, but again shows judicial reluctance to interfere with discretionary powers of the governor or Secretary of State. Although in theory it does leave some scope for challenges to transfer decisions where the administration has stated an irrelevant reason, only a very naïve governor or Secretary of State would fail to invoke 'operational and security' reasons after this decision and it seems unlikely that the courts would press them further as to what exactly these were.[56] Indeed in the *Kavanagh* case, which will be discussed in greater detail below, the European Commission on Human Rights indicated that after *McAvoy* prisoners seeking to challenge transfer decisions under Article 8 of the Convention did not have to take judicial review to exhaust their domestic remedies.

If moving a prisoner from one part of England to another may impair his contact with family and friends how much more is this likely to be impaired when the prisoner is detained in another jurisdiction? Given the existence of three separate jurisdictions within the United Kingdom people from one may find themselves detained in another as a result of being convicted of a crime committed in that other jurisdiction. A mechanism for transfer between jurisdictions in the United Kingdom exists in the Criminal

---

[55] [1984] 1 WLR 1408.

[56] Courts in the United States have reached similar conclusions. In *Meachum* v. *Fano* 427 US 215 (1976) the Supreme Court held that there were no constitutionally required due process safeguards where a prisoner was transferred within the Massachusetts prison system, even where the move involved going to more onerous conditions, as a prisoner had no right or expectation created by state or federal law that he would serve his sentence in one prison. The court affirmed this even when dealing with inter-state transfers in *Olim* v. *Wakinekona* 461 US 238 (1983), a case involving a Hawaii–California transfer of over 2,500 miles. This was despite an extensive Federal Bureau of Prisons procedure whereby the prisoner could put his view to a hearing and be represented by counsel. However, in *Vitek* v. *Jones* 445 US 480 (1980) transfer from an ordinary prison to a state mental hospital was held to require due process safeguards.

Justice Act 1961. Under s. 26 of this Act the Secretary of State for the respective jurisdiction may, on the prisoner's request, make an order to transfer that prisoner to serve his sentence in another jurisdiction in the United Kingdom. Section 27 provides for an order to be made in respect of temporary transfer, normally to take accumulated visits.

The law and politics of the transfer issue have been thrown into sharp relief by requests by prisoners from Northern Ireland, convicted of offences in England, to be transferred back to Northern Ireland to serve their sentences. A significant number, probably around forty, of these prisoners have been convicted of IRA-related shootings or bombings and have received very lengthy gaol terms. Requests for transfer from many of these prisoners, whose families live in Northern Ireland, have frequently been turned down by the Home Secretary despite the availability of space in high-security prisons in Northern Ireland. Relatives' groups and prison reform groups have been frequent critics of such refusals on the grounds that they impose unnecessary burdens on relatives who have to travel long distances to visit prisoners, that they make rehabilitation more difficult, and that transfer appears to be denied not on administrative or security grounds but out of a desire to punish the prisoners for their involvement in terrorism.

Legal challenges to the operation of the discretion to transfer have not proved particularly successful. In *McComb*[57] Taylor LJ observed that s. 26 (1) of the Criminal Justice Act 1961 gave the Home Secretary an unfettered discretion in respect of decisions whether to transfer or not. McComb had been sentenced to seventeen years for conspiracy to cause explosions. His request for transfer was refused on the basis of a policy adopted by the Home Secretary in 1989 that transfer might be refused where a prisoner's crimes were thought to be so serious as to make it inappropriate that he should benefit from a substantial reduction in the time served if that were the outcome of transfer.[58] Since at the time McComb was sentenced remission was available for half of one's sentence in Northern Ireland as opposed to one-third in England it was argued that transfer would mean he would serve nearly three years less than if he had stayed in England, as under s. 26 (4) of the 1961 Act transferred prisoners are treated as having

---

[57] R. v. *Secretary of State for the Home Dept. ex parte McComb* The Times 15 Apr. 1991.

[58] The full criteria were (1) that the prisoner had six months of his sentence to serve, (2) that the prisoner was ordinarily resident in the receiving jurisdiction prior to imprisonment or that his close family currently resided there and that there were reasonable grounds for believing that it was his firm intention to take up residence on release, (3) that both sentencing and receiving authorities were reasonably satisfied that the inmate would not be disruptive or pose an unacceptable risk to security. Even if these criteria were satisfied transfer could still be refused if there were grounds that the inmate was seeking transfer primarily to get a reduction in sentence or where 'the inmate's crimes were so serious as to render him or her undeserving of any degree of public sympathy or to make it inappropriate that the inmate should benefit from a substantial reduction in time left to serve, if that would be a consequence of transfer'.

been sentenced in the jurisdiction to which they were transferred.[59] Taylor LJ concluded that this was a legitimate factor for the Home Secretary to take into account in reaching his decision whether to transfer or not and that while family unity was important it was not the overriding criterion. He noted that if the prisoner had a co-accused who had also received seventeen years but whose family were in England he would end up serving significantly less than that co-accused. Nor did Taylor LJ find the decision unreasonable on *Wednesbury* principles.[60]

Another ground of challenge in *McComb* was that refusal of transfer violated the guarantee of a right to family life protected by Article 8 of the European Convention on Human Rights. This was disposed of on the grounds that the Convention was not enforceable in English law and that, in any case, the limitations on the right to family life contained in Article 8 (2) were applicable here. The European Commission appears to agree with this view. In *Kavanagh* v. *United Kingdom* it ruled inadmissible an application by a Northern Irish prisoner sentenced to life with a recommended term of thirty-five years for murder and causing explosions.[61] Observing that nothing in Article 8 gave a prisoner a right to choose where he was detained and that separation of a detained person and his family is one of the inevitable consequences of imprisonment the Commission indicated that only in 'exceptional circumstances' would detention of a prisoner a long way from his family violate the Convention. Noting the security risk presented in transferring the applicant, who was a category A prisoner, the fact that his mother was entitled to thirteen state-assisted visits a year, and that he had been permitted to marry another prisoner and have a number of visits from her the Commission concluded that there were no exceptional circumstances in this case.

These decisions would appear to foreclose legal challenge to the operation of inter-jurisdiction transfer decisions, at least in the absence of very extreme facts. However, politically there has been movement on the issue with the publication of an interdepartmental review on transfer in 1992.[62] This recommended amendment of s. 26 (4) of the 1961 Act so that problems relating to differentials in time to serve between jurisdictions should no longer be an obstacle to transfer (either through transferred prisoners'

[59] As a result of the 1989 Prevention of Terrorism Act prisoners in Northern Ireland convicted of 'scheduled' (essentially terrorist) offences after 1 Mar. 1989 are entitled to only one-third remission on their sentence. There is no parole scheme in Northern Ireland.

[60] This avenue might, however, be open where the potential reduction in sentence was very little and the family circumstances especially pressing. A lifer case might pose slightly different questions. Lifers are not entitled to remission or fixed conditional release dates. How long they serve is decided by review boards in each jurisdiction who make recommendations to the Secretary of State.

[61] Appl. 19085/91, decision of the Commission 9 Dec. 1992. Kavanagh's mother and sister were also applicants.

[62] See *Guardian* 24 Nov. 1992.

release issues being governed by the law applying in the sentencing jurisdiction or by giving the receiving jurisdiction the discretion whether to apply its own rules or not). In the mean time it recommended more extended temporary transfers under s. 27 of the Criminal Justice Act 1961 and removal of the 'undeserving of public sympathy' criterion as an obstacle to such transfers. Release arrangements for prisoners on temporary transfer will continue to be governed by the law applying in the sentencing jurisdiction. How these new arrangements work out remains to be seen.

If imprisonment in a different jurisdiction within the United Kingdom can be a distressing and isolating experience for prisoners and relatives alike, how much more distressing is it to be detained outside the United Kingdom? Prisoners detained abroad may be unfamiliar with the language and culture of the country in which they are imprisoned and visits from relatives will probably be very infrequent. Similar observations apply in respect of foreign prisoners imprisoned in the United Kingdom. Since 1983 however an international treaty, the Convention on the Transfer of Sentenced Persons,[63] has existed to facilitate the transfer of prisoners between countries which are parties to this Convention. In the United Kingdom this Convention was incorporated into domestic law by the Repatriation of Prisoners Act 1984. Between Convention members transfer can take place providing a number of criteria are met. These are (1) that the prisoner be a national of the receiving or 'administering' state, (2) that the judgment is final, (3) that at the time of transfer the prisoner still has at least six months of his sentence to serve, (4) that the prisoner agrees to the transfer (the sentencing state has the obligation to ensure that this is done voluntarily and in compliance with the law governing consent in that state), (5) that both of the states parties agree, and (6) that the acts or omissions which form the basis of the prisoner's criminal conviction in the sentencing state would constitute a criminal offence if done in the administering state. While either state can amnesty or pardon a transferred prisoner decisions on any review of the original conviction are purely for the courts of the sentencing state.

Where a prisoner is transferred the administering state has the option of either converting, by judicial action, the sentence given in the sentencing state to one given by its own courts or of continuing the sentence given by the sentencing state. The United Kingdom has chosen this second option, set out in Article 11 of the Convention. It adds that if the sentence is, by its nature or duration, incompatible with the law of the administering state then that state

may, by a court or administrative order, adapt the sanction to the punishment or measure prescribed by its own law for a similar offence. As to its nature, the

[63] (1983) 22 ILM 530.

punishment or measure shall, as far as possible, correspond with that imposed by the sentence to be enforced. It shall not aggravate, by its nature or duration, the sanction imposed in the sentencing state, nor exceed the maximum prescribed by the law of the administering state.

To give effect to this provision Parliament chose the somewhat inelegant language of s. 3 (3) of the 1984 Act which directs the Secretary of State, in continuing the sentence, to have regard to the 'inappropriateness' of the transfer warrant containing provisions which

(a) are equivalent to more than the maximum penalties (if any) that may be imposed on a person who, in the part of the United Kingdom in which the prisoner is to be detained, commits an offence corresponding to that in respect of which the prisoner is required to be detained in the country or territory from which he is transferred or

(b) are framed without reference to the length—

(i) of the period during which the prisoner is, but for the transfer, required to be detained in that country or territory; and

(ii) of so much of that period as will have been, or treated as having been, served by the prisoner when the said provisions take effect

The impact of this section was considered by the House of Lords in the case of *Read* v. *Secretary of State for the Home Department*.[64] Read had been convicted of importing counterfeit currency into Spain. In view of the fact that the amount involved was not regarded as large he was given the minimum sentence available for this offence, which was twelve years and one day. The sentencing judges regarded even this as excessive and under an available procedure petitioned the government to reduce this to six years. In the mean time Read sought and obtained transfer. The Home Secretary took the view that, as the maximum sentence under English law for an equivalent offence was ten years, he could not reduce Read's sentence to any less than ten years if he was to 'adapt' the sentence under Article 10 rather than 'convert' it under Article 11. It was accepted, however, that ten years was reserved for the most serious offences and Read's did not fall into that category. If he had committed the same offence in the United Kingdom he would have received no more than four years. His lawyers sought a declaration that the Home Secretary was entitled to reduce the sentence he would serve in England to this.

This view proved successful in the Divisional Court.[65] There Parker LJ took the view that Parliament had authorized the Secretary of State to do no more than implement the United Kingdom's international obligations and that to do so he had to find a sentence that was equivalent to that 'prescribed by law' for a similar offence. In the opinion of Parker, LJ, to discover what was 'prescribed by law' in respect to sentencing in the United

---

[64] [1988] 3 All ER 993.   [65] [1988] 1 All ER 759.

Kingdom one looked to the decisions of the courts and not merely to the maximum terms indicated in statutes. Hence to comply with international obligations the Home Secretary could consider a term lower than the English maximum.

Lord Bridge, whose speech was agreed with by the other members of the House of Lords, allowed the Secretary of State's appeal. He noted that the statute was framed to operate any international agreement for the transfer of prisoners that the United Kingdom might enter into. It would even allow the transfer to and detention of prisoners in the United Kingdom for acts done abroad which did not constitute offences in the United Kingdom. It seems unlikely, however, that this would extend to crimes (such as imprisonment for anti-state activities) whose content involves the violation of international human rights treaties to which the United Kingdom is a party. Acknowledging that in this case the Home Secretary had to act in accordance with the Transfer Convention Lord Bridge concluded that 'similar offence' referred to the category of the offence rather than its particular circumstances. This simply meant importing counterfeit money. In deciding what term to award for the offence he felt the Home Secretary was correct in drawing a distinction between continuing the sentence of the sentencing state and converting it. The former involved coming as close as possible to the duration awarded in the sentencing state without exceeding the administering state's maximum. The latter amounted to a resentencing process which would more closely have looked at what was appropriate in the administering state. In reaching his decision he was especially influenced by the explanatory memorandum to Article 10 which indicated that 'the administering state may adapt the sanction to the nearest equivalent available under its own law' and that the 'duration . . . corresponds to the amount of the original sentence'. Hence in a situation, such as here, where the sentencing state imposes a sentence in excess of the English maximum for an offence the Home Secretary can do no more than reduce it to the English maximum.

Disappointing as it is for prisoners such as Mr Read, and although the Convention provision is not free from ambiguity, there is much to be said for Lord Bridge's position. Article 10 indicates that the administering state is bound by the duration of sentence but goes on to add that this is not so where that exceeds the receiving state's maximum. Article 11, however, does not indicate that the administering state is bound by the duration of sentence imposed by the sentencing state but only that in converting the sentence it is not bound by the minimum for the offence provided by the law of the administering state. This would appear to give the administering state greater freedom to reduce the duration of a sentence while the reference to the minimum for that offence would appear to indicate that the term 'offence' refers to the broad category rather than the precise circumstances referred to by Parker LJ.

A second issue in the case was also decided against the prisoner. This was whether time spent in prison in Spain could be counted for remission purposes as time spent in England or whether English remission would only be based on time spent in England, with any remission on Spanish time determined by the provisions of Spanish law. The problem arose primarily because the Spanish authorities failed to provide information on their remission arrangements. This information was forthcoming by the time of the Lords hearing and for Lord Bridge rendered the point moot.[66] However, he went on to suggest that remission on custody served abroad could only be determined in accordance with the law prevailing in that state. This was even where, as here, no information on this had been forthcoming at the time the transfer took place.

---

[66] The Spanish remission arrangements in the end turned out to be much the same as those prevailing in England.

7

Prison Discipline

INTRODUCTION

In this chapter we examine the prison system's internal legal order. We look at the nature of disciplinary offences, the procedures followed in adjudicating on them, and the punishments imposed on those who fall foul of it. This is an area where prisoners and prison authorities have often had very different perceptions. As JUSTICE commented in 1983, 'It is in the context of discipline that the sharpest conflicts between prisoner and prison staff are apt to occur'.[1] There can be little doubt as to the need for the maintenance of order in prisons; without it prisoners and staff may face the climate of fear and violence that has been at the heart of many prison conditions cases in the United States. A code of disciplinary offences and punishments is clearly one way of maintaining that order though, as other chapters in this book make clear, it is not the only one.

However, the operation of the discipline system in prisons has its own potential for disorder, that which grows out of inmates' perception that the rules they are charged with breaking are highly arbitrary and the procedures employed to convict them significantly unfair. The Woolf Report observed that how discipline is administered can have a very significant impact on how prisoners view the fairness of the prison system as a whole. A strong perception of unfairness can be an important factor in the creation of disturbances.[2]

In the United Kingdom prison system two factors in particular raise concerns about the appearance of fairness of the system. First, while breaches of prison rules may have adverse disciplinary consequences for prisoners we have already seen that breaches by prison officers or administrators do not have any clear legal consequences for them. Secondly, the enforcement of prison discipline is primarily in the hands of prison officials themselves. Over 95 per cent of all disciplinary charges are adjudicated on by prison governors, whose ability to manage the prison is heavily dependent on retaining the support of the prison officers who bring the charges. Even though the 5 per cent of most serious charges were until recently heard by Boards of Visitors significant questions were posed regarding the Boards'

---

[1] See JUSTICE, *Justice in Prisons* (London, 1983).
[2] *Prison Disturbances: April 1990* (Woolf Report), Cm. 1456 (London, 1991), para 14.437.

theoretical and in some cases actual independence from the prison authorities, questions which have ultimately proved fatal to their retaining this role (a matter to be discussed in greater detail below).

To preserve the appearance and actuality of fairness such a situation cries out for independent scrutiny of the exercise of disciplinary powers; the type of scrutiny the judiciary, with its training and expertise in matters of procedure and fact-finding, is ideally placed to provide. Yet for most of this century judges in the United Kingdom refused to examine the fairness of prison discipline proceedings. When called upon to do so they invoked analogies with military discipline, arguing that 'It is of the first importance that the cases be decided quickly.'[3] Judicial review, and the procedural standards it might mandate, were seen as only likely to cause delay and thus undercut the main purpose of disciplinary proceedings, the swift restoration of order. This military analogy was not without foundation. As Edward Fitzgerald has observed the model for prison discipline devised at the end of the nineteenth century was taken from the army and the understanding of those who devised this model was that it would be a purely internal affair, with the administration deciding on its own procedure and no interference from outside bodies.[4] What however clearly distinguished prison discipline from that in the army was that prisoners and prison staff often had radically opposed interests and perceptions, unlike soldiers and their commanders. Even more significantly the scale of punishments handed out by prison disciplinary tribunals, notably the loss of remission for periods that ran into years, was far more severe than the analogy of parade ground discipline.

What seems to have changed judicial attitudes was the scale of the disciplinary punishments resulting from the Hull Prison Riots of 1976. These came before the courts in the *St Germain* litigation. The punishments in those cases, some amounting to up to two years' loss of remission, were clearly much more severe than those of the parade ground variety. The judges in *St Germain* appear to have been motivated by this to investigate the quality of procedures followed before such punishments were imposed. Even then the military metaphor was not entirely given up. Rather disciplinary proceedings for more serious offences were analogized to court martial proceedings, where more formal procedures were envisaged, while those on less serious charges continued to be referred to in parade ground terms. Hence for a number of years hearings by Boards of Visitors were subject to judicial review as to the quality of their procedures but governors' hearings, the vast majority, remained immune, a view confirmed in 1985 by the English Court of Appeal decision of *R. v. Deputy Governor Camp Hill*

[3] The view is that of Lord Denning in *Fraser* v. *Mudge* [1975] 3 All ER 78.

[4] E. Fitzgerald, 'Prison Discipline and the Courts', in M. Maguire, J. Vagg and R. Morgan (eds.), *Accountability and Prisons: Opening up a Closed World* (London, 1986), 29–46.

Prison ex parte King.[5] A year after the *King* decision the Northern Irish Court of Appeal reached a different conclusion in *In re McKiernan's Application,*[6] thus creating a period of uncertainty which was only resolved by the House of Lords decision in *Leech* v. *Deputy Governor Parkhurst Prison,*[7] which upheld the Northern Irish view.

The outcome of *St Germain* was that the courts began to conduct an examination of the procedures followed in Board of Visitors hearings. Often they were not impressed by what they found there and a string of decisions followed on the calling of witnesses, cross-examination, the interpretation of offences, the burden of proof, and the circumstances in which a prisoner should have legal representation, which will be examined in more detail below. By 1988 the House of Lords felt able to comment, 'there seems no reason to doubt that . . . the courts' infrequent interventions have improved the quality of justice administered by boards of visitors'.[8] As regards governors' hearings such review has taken longer to get going and judges have always indicated that the standards required will be less strict, reflecting the less severe punishments available to governors and the closer relationship such proceedings have to general prison order. However in Northern Ireland in particular a significant jurisprudence has evolved regarding the procedural requirements of governors' disciplinary offences.

Despite the procedural changes influenced by judicial review and effected in changes to Standing Orders, Home Office Circulars, and training for adjudicators, many critics remained unsatisfied. A particular target of their criticisms was the fact that Boards of Visitors continued to perform the dual role of being a watchdog to ensure standards for the treatment of prisoners were met and the adjudicating tribunal for the most serious disciplinary offences. Although the government rejected a proposal from a committee of inquiry which it established in 1986 that Boards lose their disciplinary functions,[9] in the wake of the Strangeways riot it accepted the same proposal when made in the Woolf Report. At the start of 1992 therefore the prison disciplinary system entered a very new and uncertain phase.

PRISON DISCIPLINE HEARINGS: THE RETREAT FROM A DUAL SYSTEM

Section 47 of the Prison Act 1952 gives the power to make rules relating to discipline to the Home Secretary. The disciplinary offences and basic

---

[5] [1985] QB 735.
[6] See *McKiernan* v. *Governor HM Prison Maze* (1985) 6 NIJB 6.
[7] [1988] AC 533.   [8] Ibid.
[9] *Committee on the Prison Disciplinary System* (The Prior Report), Cmnd. 9641 (London, 1985).

structure for adjudicating them are set out in the frequently amended Rules 47–56. Until 1989 these provided for a distinction between 'disciplinary offences', adjudicated upon by governors, and 'grave' (escape attempts or gross personal violence against other inmates) or 'especially grave' offences (gross personal violence against prison officers and mutiny), which were referred to Boards of Visitors.[10] The main distinction lay in the punishments these two tribunals could order. While governors were limited to the forfeiture of twenty-eight days' remission and three days' solitary confinement, Boards could order forfeiture of up to 180 days' remission and fifty-six days' solitary confinement for conviction on a 'graver offence', and on conviction of an 'especially grave' offence, unlimited loss of remission. Boards could also order more substantial periods of exclusion from associated work, stoppage of earnings, and exclusion from privilege than could governors.

The power to order loss of substantial periods of remission proved especially significant when it came to judicial review challenges. Although the Prison Rules only indicated that prisoners 'may' be granted remission it was clear from after World War II that they would always be granted remission in accordance with the prevailing scheme. As a result the courts came to reject prison authority arguments that remission was only a privilege and that prisoners who lost remission as a result of disciplinary hearings did not lose anything they were entitled to.[10a] Even if remission could not be said, under the prevailing statutory framework, to be a right it was clearly at least a 'legitimate expectation', in that prisoners could expect their release date to be automatically brought forward by the remission to which they were entitled. The effect of removing remission for a disciplinary conviction was to make a prisoner spend more days in prison than he would otherwise have done. Once courts accepted this they were open to the argument that had prisoners been facing another sentence in a criminal court they would at least have had a certain level of procedural protections.[11] Though the courts have been reluctant fully to analogize prison disciplinary hearings to criminal proceedings, especially reluctant in the case of governors' hearings, they have focused on cases involving substantial losses of remission, such as the 720 days in the *St Germain* litigation, as particularly in need of procedural safeguards. Similarly the European Court of Human Rights, in its decision in *Campbell and Fell* v. *United Kingdom*,[12] observed that a loss of

---

[10] Technically the Prison Rules until 1989 required the governor first to refer the case to the Secretary of State who could then delegate the power to hear the case to the board. In practice this is always what occurred.

[10a] See R. v. *Board of Visitors of Highpoint Prison Ex parte McConkey*, The Times 23 Sept. 1982.

[11] It is worth noting that the United States Supreme Court has refused to see loss of good time credits as in any way analogous to a criminal conviction. See *Superintendent, Massachusetts Correctional Institution, Walpole* v. *Hill* 472 US 445 (1985).

[12] (1984) 7 EHRR 165.

remission of 570 days was a particularly important factor in its conclusion that the applicants were facing 'criminal charges' and hence were entitled to be legally represented under Article 6 (3) of the Convention.

Partly in response to these decisions the Rules were amended in 1983 to remove the power to order unlimited loss of remission. However, concerns remained over whether it was a good idea to have Boards perform the dual functions of being a watchdog as regards the prison administration but also the adjudicator of more serious charges, a position in which they would often have to uphold the administration against the prisoner. Board members often defended their adjudicatory function on the grounds that their knowledge of the circumstances at the prison put them in a better position to understand the events and personalities involved in discipline hearings and that their disciplinary function gave them greater respect among prison staff and hence influence when it came to their supervisory function. Critics though argued that prisoners are unlikely to put their trust in Board members who may the next day be ordering stiff disciplinary punishments against them and that in order to maintain their influence with the prison administration Board members may feel under pressure to provide disciplinary convictions. Moreover they argued that in view of the significant punishments which Boards could hand out some legal training or judicial experience should be required.[13] The European Commission in the *Campbell and Fell* case expressed itself of the view that the Board could not be seen as an 'independent tribunal' as required by Article 6 (1) of the Convention in view of the fact that its members were only appointed for short periods by the Home Secretary and could be removed by him. This view was not upheld by the Court.

Concerned by these criticisms, and perhaps even more by the possibility that many prisoners would be legally represented at Board hearings in the wake of the *Campbell and Fell* and *Tarrant* decisions, the government established a committee under the chairmanship of Mr Peter Prior to examine the prison disciplinary system. Among the Prior recommendations was that the Boards lose their adjudicatory function. This would go to specially created Prison Disciplinary Tribunals, each of which would have a legally qualified chair. Initially this view found some favour with the Home Office but the government finally decided against it, strongly influenced it seems by a view that legal representation was not being granted as often as expected before Boards and proved less of a problem than feared when granted.[14] Hence although further revisions to the Prison Rules in 1989 abolished the distinctions between offences, graver offences, and especially grave offences and reduced the maximum loss of remission on any one

[13] For a discussion of these criteria see JUSTICE, *Justice in Prisons*, at 55–6.

[14] For a discussion of post Prior developments see, R. Light, and K. Mattfield, 'Prison Disciplinary Hearings: The Failure of Reform' (1988) 27 *Howard J.* 266–82.

charge to 120 days, they still permitted governors to refer charges to boards where they felt the punishments they could give on conviction would, 'having regard to the nature and circumstances of the offence, be inadequate'.[15]

The issue would not go away and after the 1990 riots the Woolf Committee was charged with looking at the structure of discipline hearings again. It endorsed the view of Prior, which it referred to as 'the most authoritative source on the prison discipline system', that Board members lose their disciplinary function, drawing on materials which questioned not just whether Boards could preserve the appearance of fairness but whether their practice had been satisfactory. In particular Woolf referred to a study of attitudes among prisoners which showed a clear feeling that Boards were the least satisfactory tribunal to appear before,[16] and the Committee's own conclusions after talking with some Board members that the latter felt they would be letting the governor down if they did not return a conviction. It also referred to the conclusions of justices' clerks involved in a pilot scheme to give legal advice to Board members: 'Those who were ready to state a point of view indicated by a considerable majority that the Board of Visitors did not use the same care in approaching LOR [loss of remission] as justices would have approached imprisonment.'[17] However, Woolf differed from Prior as regards what should replace Boards. Concluding that the Prior proposals had been 'on the table for some time without being taken up', it decided a return to first principles was appropriate and that a clearer distinction should be drawn between disciplinary and criminal penalties. The former could continue to be adjudicated upon by governors, the latter should be adjudicated upon by the criminal courts, where the penalties could be greater but so would be the procedural safeguards.

DECISION-MAKING ON DISCIPLINARY CHARGES: THE PRESENT SYSTEM

These Woolf proposals have now largely been adopted and are enshrined in the Prison (Amendment) Rules 1992.[18] These rules abolish the Boards' disciplinary functions. Instead governors are now instructed to lay a disciplinary charge within forty-eight hours, in compliance with Rule 48 (1), but to refer the facts of any suspected criminal offence to the police (more detailed guidelines refer especially to assault, escape, possession of weapons or class A drugs, criminal damage of an amount exceeding £2,000, arson, robbery with violence, mutiny, and any mass disturbance). Guidance in the Standing

---

[15] See Prison Rule 51(1), as amended by Prison (Amendment) Rules 1989, SI 1989/30.
[16] See J. Ditchfield and D. Duncan, 'The Prison Disciplinary System: Perceptions of its Fairness and Adequacy by Inmates, Staff and Members of Boards of Visitors' (1987) 26 *Howard J.* 122–52.
[17] Woolf Report, para 14.394.   [18] SI 1992/514.

Orders then indicates that the governor should open the disciplinary hearing and satisfy him- or herself that there is a case to answer before adjourning it pending the police and Crown Prosecution Service Inquiry. If a decision to prosecute is taken then the case will go to an outside court and the disciplinary charge will not be proceeded with.[19] If it is not then the governor may reactivate the disciplinary hearing; however, Standing Orders state that

If it is clear that the police or CPS have decided that a prosecution cannot be brought because the available evidence is insufficient, and the disciplinary charge is similar to and relies on the same evidence as the potential charge, the governor should dismiss the disciplinary charge. In addition, where the resumption of internal proceedings is likely to create an appearance of unfairness out of proportion to the seriousness of the alleged disciplinary offence, the governor should consider whether the charge should be dismissed. In other cases it is open to the governor to proceed with the charge.[20]

Thus if the CPS has decided not to prosecute as not being in the public interest the way remains open for the governor to prefer disciplinary charges. The range of criminal offences to which prisoners can be subject has been expanded by the creation of the offence of mutiny in the Prison Security Act 1992. Mutiny was one disciplinary charge which because of its complexity and potentially severe penalties gave rise to a right to legal representation for prisoners charged with it. Dropped as a disciplinary offence in 1989 its conversion into a criminal offence was recommended by the Prior Committee. Section 1 (2) of the Act defines mutiny as 'where two or more prisoners, while on the premises of any prison, engage in conduct which is intended to further a common purpose of overthrowing lawful authority in that prison'. Where a mutiny is in progress prisoners who refuse to submit to lawful authority, having been given a reasonable opportunity to do so, without lawful excuse, are defined as participating in it. Mutiny carries a maximum ten-year sentence and prosecutions can only be brought with the consent of the DPP.

Where governors decide not to refer the events to the police they can then deal with the case themselves. In larger prisons governors will often delegate the hearing of charges to deputy governors. As will be discussed in greater detail below another change brought about by s. 42 of the Criminal Justice Act 1991 and the Prison (Amendment) (No. 2) Rules 1992[21] is that instead of ordering loss of remission governors can order the award of extra days. As prisoners no longer stand to 'lose' something on disciplinary

---

[19] Doing so could be in violation of Rule 36 (1) of the European Prison Rules, which state that a prisoner should not be punished twice for the same act. However, in *R. v. Hogan* [1960] 3 All ER 149 the court found that a disciplinary conviction on an escape offence was no bar to subsequent criminal proceedings.

[20] See SO 3D (7).   [21] SI 1992/2080.

convictions it could be argued that natural justice protections are no longer as apposite (courts being generally more willing to require such protections in the case of benefits people lost than those they hoped to get). However, since the effect of a punishment of extra days is still that a prisoner remains detained beyond the date he legitimately expected to be released there should not be a change in judicial attitudes to the need for fair hearings.

In the discipline sphere as elsewhere the advent of private prisons as a result of the 1991 Criminal Justice Act poses new questions for the administration of disciplinary proceedings. The Act indicates that responsibility for discipline is to rest with the Home Office-appointed controller rather than the privately appointed director and Prison Rule 98A indicates that in contracted-out prisons references in the prison rules to the disciplinary powers of governors are to be read as references to the controller.

PRISON DISCIPLINE OFFENCES

The disciplinary offences a prisoner may be charged with are set out in Rule 47. There are twenty-two of these. Some of these are general offences which might be found in any discipline code, such as assault[22] or destroying property belonging to another,[23] while others more clearly relate to the prison environment, such as denying an officer access to any part of the prison[24] or intentionally obstructing an officer in the execution of his duty.[25] When the Prior Committee examined the code of discipline offences in 1985 it recommended revisions to produce a code which

(1) is in clear, simple terms for staff to operate (bearing in mind that it is prison officers, not lawyers, who are responsible for identifying and laying the correct charge)
(2) makes clear to prisoners what constitutes an offence, and states clearly the degree of fault, knowledge etc which needs to be proved
(3) identifies and sets out separately important or common offences for which no separate charge exists at present; and thereby
(4) reduces the need and scope for the controversial catch-all 'good order and discipline' offence.[26]

The revised disciplinary code which emerged after this inquiry in 1989 goes a long way towards meeting these requirements. It abolished two of the most notorious offences, that of 'making a false and malicious allegation against an officer' and that of 'repeatedly making groundless complaints'. These had long been criticized by prison reform groups as

---

[22] Prison Rule 47 (1). [23] Prison Rule 47 (14). [24] Prison Rule 47 (3).
[25] Prison Rule 47 (6). [26] Prior Report, para. 7.16.

discouraging legitimate complaints although the Prior Committee itself, mindful of the fact that libel actions are not a practical option for prison officers unfairly accused, had actually recommended their retention in a modified form. The new code also tightens up the mental element requirements of most offences. For example the old offence of being 'idle, careless or negligent at work or, being required to work, refuses to do so'[27] has been replaced by one which finds a prisoner guilty of an offence against discipline if he 'intentionally fails to work properly or, being required to work, refuses to do so'.[28] The old rule against having an unauthorized article in one's cell has been replaced by a more simple rule which prohibits having an unauthorized article in one's possession.[29]

In addition to Prior the judiciary had begun to indicate in a number of cases that they would interpret prison discipline offences strictly and that, as with the criminal law, they would construe ambiguities in favour of the defendant. Thus in R. v. *Board of Visitors Highpoint Prison ex parte McConkey*[30] McCullough J. quashed a disciplinary conviction of 'offending against good order and discipline' in respect of a prisoner against whom nothing had been established but that he had remained in a cell while others were smoking cannabis. It was neither alleged nor proven that he had actively encouraged others to smoke cannabis. McCullough J. indicated that an element of *mens rea* had to be part of the definition of this offence. Similarly in *Ex parte King* the Court of Appeal concluded that a governor had misconstrued a prison rule prohibiting having an unauthorized article (a hypodermic needle) in one's cell where he had found the prisoner guilty after concluding that the prisoner knew the article to be in the cell. The governor had failed to inquire whether the prisoner had control over it, the court noting that prisoners were often required to share cells with people they were afraid of crossing.[31] However, the proposal of Justice that the prison rules include a general defences article[32] (relating for example to self-defence, mistake, or lawful excuse) has not been adopted, although to continue the criminal law analogy of *McConkey* failure of a governor to consider such pleas should be susceptible to judicial review.[33]

One offence which Prior ultimately did not recommend the removal of was the catch-all offence of 'in any way offends against good order and discipline'.[34] Along with the similar offence of 'disobeys any lawful order'[35]

---

[27] Old Prison Rule 47 (17).   [28] Prison Rule 47 (18).   [29] Prison Rule 47 (9).

[30] The Times 23 Sept. 1982.

[31] R. v. *Deputy Governor Camphill Prison ex parte King* [1985] QB 735. For further support of the view that disciplinary offences ought to be construed in line with the criminal law see R. v. *Board of Visitors Thorpe Arch Prison ex parte De Houghton* The Times 22 Oct. 1987.

[32] *Justice in Prisons*, at 48–9.

[33] Such a defence is especially important where the charge is 'fighting with any person', Prison Rule 47 (4), for example.

[34] Prison Rule 47 (21).   [35] Prison Rule 47 (19).

this offence has been heavily criticized as maintaining a highly discretionary discipline regime and one that may be in contravention of the United Kingdom's obligations under Rule 35 (*a*) of the European Prison Rules that conduct constituting a disciplinary offence shall be 'provided for and determined by the law or by the regulation of the competent authority'. Prior concluded that a closed disciplinary institution could not survive without such a flexible power to take action against conduct prejudicial to discipline:

Ultimately all disciplinary systems rely on the concept of punishing conduct which undermines proper authority or threatens orderly community living; and in reality it is not possible to identify in advance and legislate for every action which is appropriately punishable. In closed institutions there is a particular risk that certain offences may become fashionable unless they can be dealt with swiftly.[36]

This conclusion however leaves in existence a situation where prisoners can be punished for doing something today that was lawful yesterday because a governor or perhaps even simply a prison officer deems it now inimical to good order and discipline. Perhaps nothing in the Prison Rules so clearly shows the limited extent to which the rule of law prevails in prisons or the extent to which discretionary power is enshrined. In response to the Prior Committee's arguments it can be pointed out that governors already have powers, under Rule 43, to deal with short-term conduct threatening order. Indeed in the *Hague* case the Court of Appeal had no difficulties with the use of Rule 43 to segregate a prisoner who might have been charged because it was thought charging him might cause embarrassment to prison officers. If a particular practice becomes established and a sufficient threat to prison staff or other prisoners to warrant disciplinary punishment then perhaps a governor could be given powers to create a local disciplinary rule. Indeed, Justice recommended dealing with the uncertainty of the 'disobeying any lawful order' offence in this way, providing, however, that the local rules were published in advance and that a governor could not devise one simply for the purpose of punishing a particular prisoner.[37]

THE CONDUCT OF PRISON DISCIPLINE HEARINGS

Prison Rule 49 (2) indicates that at any inquiry into a charge a prisoner is to be given a full opportunity of hearing what is alleged against him and a

---

[36] Prior Report, para 7.80.

[37] *Justice in Prisons*, 48. In France under Art. D 255 of the Code of Criminal Procedure all disciplinary rules are formulated by the Prison Director, who must submit them to the visiting judge (the *juge d'applications des peines*) for approval.

full opportunity to present his own case. Courts have indicated that this rule is essentially declarative of the principles of natural justice, which apply to the hearing of disciplinary charges. While the principles of natural justice, or its modern equivalent fairness, are applicable to all administrative decisions with the capacity to affect citizens adversely their content varies with the significance of the decision and the character of the issues to be considered.[38] At one end stands the need simply for a decision maker who has no personal or pecuniary interest in the outcome, at the other the full procedural protections of a criminal trial.

Although the courts recognized Board of Visitors hearings, particularly where they could involve unlimited loss of remission if the prisoner were convicted, as coming close to criminal trials they fell short of applying the analogy fully. Hearsay evidence for example has always been accepted as permissible in all forms of prison discipline hearings. Since governors' hearings have only more recently become the object of judicial scrutiny, at least in England and Wales, the procedural standards required in them are less fully developed. Cases in Northern Ireland, however, suggest that a less rigorous approach will be required and that the courts will be more prepared to consider whether on an examination of the proceedings as a whole the prisoner was given a fair hearing as opposed to focusing on particular aspects of the hearing and signalling what is required by way of procedural fairness in respect of each. Courts have spoken of the need for governors' hearings to be conducted in a 'firm, fair and crisp manner' and have said that they will not be reviewed simply for technical breaches which cause no injustice.[39]

Under Article 6 (1) of the European Convention on Human Rights, 'In the determination of his civil rights and obligations or of any criminal charge against him, everyone is entitled to a fair and public hearing within a reasonable time by an independent and impartial tribunal established by law.' Although people in the common law world are used to associating the phrase 'civil rights' with rights against the state the Convention takes its meaning more from the civil law world, where civil rights connotes civil or private law rights. While the boundary between public and private rights is often blurred (especially when it comes to semi contractual rights against the state[40]) it seems clear that prison disciplinary hearings fall outside the first half of Article 6 (1). However, in *Campbell and Fell* v. *United Kingdom*[41] the European Court indicated that prison discipline hearings could amount to 'criminal proceedings' within the meaning of Article 6.

---

[38] See, generally, P. Craig, *Administrative Law* (London, 1983), 253–98.
[39] See *In re McEvoy and Others' Application* (1991) 8 NIJB 89.
[40] The boundaries of the concept of 'civil rights' are explored by the European Court of Human Rights in *Feldbrugge* v. *Netherlands* (1986) 8 EHRR 425, Series A No. 99.
[41] (1984) 7 EHRR 165, Series A No. 80.

Such a determination is particularly significant as criminal charges attract further explicit procedural requirements under Article 6, including the right to legal representation. Rejecting the government's claim that there was a clear division between disciplinary and criminal offences the Court indicated that it would take three factors into account in deciding whether or not a charge would fall into the criminal category. These were the domestic classification as criminal or disciplinary, the nature of the offence itself and whether it would normally appear in a criminal code, and the nature and severity of the penalty faced. In *Campbell and Fell*, where the prisoners faced mutiny charges and risked unlimited loss of remission (one of them actually losing 570 days' remission), the Court concluded these criteria were satisfied.

Whether after this decision all prison discipline hearings come within the scope of Article 6 (3) has yet to be tested. In a previous decision the Court appeared to accept that any punishment involving the deprivation of liberty implicated a criminal charge.[42] In *Campbell and Fell* it appeared to reject the argument that loss of remission, being a privilege, did not involve any loss of liberty and hence might arguably be indicating that any loss of remission would constitute a deprivation of liberty. However, the Court's language suggests a more cautious approach, it indicated that 'by causing detention to continue for substantially longer than would otherwise have been the case, the sanction came close to, even if it did not technically constitute, deprivation of liberty'.[43] This suggests that not all losses of remission will amount to deprivation of liberty and with a sharper division of disciplinary and criminal regimes post-Woolf European supervision in this area may be reduced.[44] Replacing loss of remission with an award of extra days may also lead to a reconsideration of the situation by the European authorities, but since it remains a sanction that extends the length of detention the principles enunciated in *Campbell* should still be applicable.

LAYING A CHARGE AND NOTICE OF THE HEARING

Under Prison Rule 48 (1) a charge must be laid as soon as possible and, except in exceptional circumstances, no longer than forty-eight hours after the incident.[45] Standing Order 3D indicates that the charge should normally

---

[42] See *Engel* v. *Netherlands* (1976) 1 EHRR 647, Series A No. 22.

[43] (1983) 7 EHRR, at 197.

[44] See V. Treacy, 'Prisoners Rights Lost in Semantics' (1989) 28 *Howard J*. 27–36, for the view that Art. 6 is applicable to all losses of remission. Generally see C. Kidd, 'Disciplinary Proceedings and the Right to a Fair Criminal Trial under the European Convention on Human Rights' (1987) 36 *ICLQ* 856–72.

[45] For example in *In re Hunter's Application* (1989) 1 NIJB 86, Carswell, J., upheld delaying adjudication of a prisoner charged on a Friday until Monday as Standing Orders required a

be laid by the officer who witnessed the incident or against whom the alleged offence is committed. Officers are advised to consult a senior member of staff to ensure that the correct charge is laid. Prisoners charged are required to be given a Notice of Report (Form 1127) which sets out the notice of the offence charged and, according to Standing Orders, should be 'sufficiently detailed to leave prisoners in no doubt as to what is alleged against them'. Failure to inform a prisoner of the charge in advance is clearly a ground for judicial review,[46] and the courts have indicated that simply repeating the words of the alleged offence is insufficient to give the prisoner notice of what is alleged against him.[47] Standing Orders indicate that prisoners should be given at least two hours' notice of the adjudication. In some cases this may be too short, especially where complex factual or legal issues are involved. In the *Tarrant*[48] case dicta suggest that on a complex charge like mutiny twenty-four hours' notice might be necessary.

Laying the correct charge is especially important, as the Court of Appeal indicated in R. v. *Board of Visitors Dartmoor Prison ex parte Smith*[49] that the requirements of Rule 48 (1) were mandatory. There the court quashed a Board decision to convict a prisoner of assault after having acquitted him of causing gross personal violence on the grounds that the lesser offence must be contained in the greater. The court took the view that nothing in the Prison Rules entitled the Board to substitute a different charge for the one laid. One effect of proceeding on the assault charge which had not specifically been laid was that the prisoner had no opportunity to prepare a defence to it. In a Northern Irish case the High Court indicated that prisoners could not be convicted of two separate offences arising out of the same incident.[50] Once the charges have been laid the tribunal should normally hear them in the order in which they have been laid.[51]

The problem of actual or apparent bias has been a particularly difficult one for the prison discipline system. Given that the people involved in hearing charges, governors and Boards of Visitors, have a fairly close relationship with the people bringing the charges, prison officers, it must be asked whether the appearance of impartiality can ever fully be achieved. The

prisoner to be examined by a medical officer before adjudication and a doctor would not normally be called out until a Sunday.

[46] See R. v. *Board of Visitors Long Lartin Prison ex parte Cunningham* Unreported 17 May 1988.

[47] See R. v. *Board of Visitors Swansea Prison ex parte Scales* The Times 21 Feb. 1985.

[48] R. v. *Home Secretary ex parte Tarrant* [1985] QB 251.

[49] [1986] 3 WLR 61. See also R. v. *Board of Visitors Gartree Prison ex parte Sears* The Times 20 Mar. 1985.

[50] For example in *In re Murphy's Application* (1988) 8 NIJB 94 the prisoner was convicted of both using abusive language and offending against good order and discipline by the governor arising out of a single incident of swearing at a prison officer.

[51] A recommendation of Hodgson, J., in R. v. *Board of Visitors Gartree Prison ex parte Mealey* The Times 14 Nov. 1981.

courts rejected a fairly direct attempt to raise this question in R. v. *Board of Visitors Frankland Prison ex parte Lewis*,[52] where the applicant argued that a member of a Board hearing a disciplinary charge of possessing drugs in his cell against him could not be impartial as only a few weeks previously the Board member had been on the Local Review Committee that turned down his parole application. As a member of that committee the Board member would be aware of the applicant's previous convictions for drugs offences. Woolf J. concluded that the appropriate test was whether a reasonable and fair-minded bystander, knowing all the facts, would have a reasonable suspicion that a fair trial for the applicant was not possible.[53] Among the facts that such a hypothetical bystander would have to take into account was the dual role of Board members. Woolf J. noted that where a Board member was aware of a possible appearance of unfairness (which the member here apparently was not as he did not remember the parole hearing) they would have to exercise their discretion as to whether or not to sit. Interestingly, in light of the views that Lord Justice Woolf, as he subsequently became, expressed in his Report on the role of Board members in disciplinary proceedings, here he urged Board members not to be too eager to disqualify themselves as

they are offences which are primarily against discipline and in order to deal with those kind of offences, it is important that those who adjudicate upon them should have knowledge of the workings of prisons in general and the particular prison of which the prisoner concerned is an inmate.[54]

However, while the courts have shown themselves unwilling to consider possibilities of structural bias in the discipline system they have been prepared to review more obvious deviations from established practice. In R. v. *Governor Pentonville Prison ex parte Watkins*[55] Judge, J., indicated that if the prison officer preferring charges was in the room where the adjudication took place before the prisoner entered or remained behind after the prisoner left this would offend the appearance of fairness.[56]

SEGREGATION PRIOR TO THE HEARING

The laying of a disciplinary charge may not end the conflict which gave rise to it. Mainly to prevent the continuing of conflict governors are given pow-

---

[52] [1986] 1 WLR 130.
[53] The test is derived from the judgment of Ackner, LJ, in R. v. *Liverpool City Justices ex parte Topping* [1983] 1 All ER 490.
[54] [1986] 2 All ER 272.   [55] Unreported 4 Dec. 1991.
[56] In *In re Grogan's Application* (1988) 8 NIJB 87, Carswell, J., reached a similar conclusion and quashed a Board's award where a governor remained behind with the Board after the prisoner and everyone else left, even though the judge accepted that the governor did not take any part in the Board's deliberations.

ers to segregate prisoners prior to adjudication. However, criticism of the use of these powers is frequent. Critics argue that rather than being a discretionary last resort to prevent a difficult situation becoming worse they have been used automatically to segregate all prisoners pending adjudications. They have noted that under the old Rule 48 (2) segregation pending a hearing did not require the approval of the Board or Secretary of State, unlike Rule 43 segregation. In addition such prisoners have often been segregated in punishment cells, hence depriving them of facilities and heightening the sense that the prisoner is already regarded as guilty.[57] On a practical level segregation also hampers a prisoner's ability to prepare his defence at adjudication, especially as regards discovering and interviewing potential witnesses.

One consequence of granting legal representation for some Board hearings was that prisoners could spend more time in pre-hearing segregation as both sides took greater care over the preparation of their defence. With the abolition of Board hearings this problem has been removed, and since governors' hearings usually take place within a few days lengthy periods of segregation before adjudication on disciplinary charges should be rare. However, as more prisoners may find themselves before outside courts, whose delay in hearing charges may be even greater than that of Boards, problems of pre-hearing segregation may not be a thing of the past.

The manual on adjudications indicates that governors should segregate only when necessary, for example where there is a risk of collusion, or threats to witnesses or to alleged victims. Recent changes to the Rules may ensure that segregation does not become automatic. Rule 48 (5) now permits pre-hearing segregation only before the governor's first inquiry. Thereafter all segregation, including that pending the hearing of criminal offences, must be on the basis of Rule 43. As has been indicated elsewhere in this text such decisions are susceptible to judicial review. Standing Orders indicate that reasons should be given for any decision to segregate prior to a hearing, that prisoners on pre-hearing segregation should not be housed in punishment cells 'unless there is no other suitable cell available elsewhere' (a frequent problem in older prisons), and that they should not be deprived of any privileges that they are normally entitled to 'except those that are incompatible with segregation'.[58]

LEGAL REPRESENTATION AT ADJUDICATIONS

No area of the requirements of natural justice in the context of Board of Visitors hearings was more fiercely fought than that of legal representation.

---

[57] See the criticisms of JUSTICE, *Justice in Prisons* 59.   [58] SO 3D (11) (*b*).

As one of the authors of this text has argued, it was not without good reason that prisoners' advocates saw this as a particularly significant procedural guarantee to enshrine.[59] The presence of lawyers not only offered prisoners the skills of those with greater experience of arguing before tribunals, it also brought a genuinely external element into the closed world of prison discipline and put Board members on notice that the quality of their adjudications was under scrutiny. In addition it changed the character of discipline proceedings. Rather than convicted prisoners, whose veracity and motivation the Board might suspect, arguing against the word of prison officers, one would have professional lawyers defending their clients. Boards were always likely to listen to their arguments with greater respect.

Early attempts to secure legal representation in discipline hearings fared no better than most prisoners' claims. In 1975 Lord Denning, MR, in an unreserved judgment, invoked the military model and concerns about delay in hearings being inimical to discipline to dismiss brusquely an application on this point.[60] An early European application also proved fruitless, the Commission in *Kiss* v. *United Kingdom*[61] becoming stuck on the point that loss of remission could not constitute a deprivation of liberty within the meaning of Article 5 as the prisoner was lawfully sentenced to prison and remission was not a matter of right.

With the demise of the military model in the first *St Germain* case it was always likely that the issue of legal representation would arise again. The first real breakthrough, however, came in the European Commission of Human Rights decision in *Campbell and Fell* v. *United Kingdom*.[62] There the Commission indicated that the classification of offences as disciplinary under domestic law would not render them immune from being regarded as 'criminal' offences under the Convention if certain conditions were met. The Commission indicated it would look at the initial classification under domestic law, whether the content of the disciplinary offence also normally appeared in criminal codes, and the seriousness of the penalties prescribed for the offence. Applying these criteria to prisoners who faced possible unlimited loss of remission and actually received up to 570 days' loss of remission for mutiny and assault offences it concluded they had been charged with 'criminal offences'. Under Article 6 (3) of the Convention all those charged with criminal offences are entitled to be legally represented.

The European Court of Human Rights was to affirm the decision of the Commission on the Article 6 (3) point in *Campbell and Fell* within two years. Before it did so the Divisional Court in England also recognized that at least in limited circumstances natural justice could require a right to legal representation in Board of Visitors discipline hearings. Like *Campbell*

[59] See, generally, S. Livingstone, 'Prisoners and Board of Visitors Hearings: A Right to Legal Representation after All?' (1987) 38 *NILQ* 144–69.
[60] *Fraser* v. *Mudge* [1975] 3 All ER 78.   [61] 7 D. & R. 55.   [62] (1982) 5 EHRR 207.

and Fell the case of *R. v. Secretary of State for the Home Department ex parte Tarrant*[63] arose out of mutiny hearings following disturbances, this time at Wormwood Scrubs and Albany. Webster J. quashed the convictions handed down by the Board on the grounds that they had failed even to consider whether they had a discretion to grant representation. If they had considered it he indicated that they would have had to take a number of factors into account. These were :

1. the seriousness of the charge and the potential penalty;
2. the likelihood that difficult points of law would arise;
3. the capacity of a prisoner to present his own case;
4. procedural difficulties, such as the inability of prisoners to trace and interview witnesses in advance;
5. the need for reasonable speed in deciding cases;
6. the need for fairness between prisoners and between prisoners and prison officers.

Moreover Webster J. indicated that as regards some offences, such as mutiny where the penalties were severe and points of law complex, no reasonable board exercising its discretion could decline to grant representation.[64]

The *Tarrant* case sparked a significant rethink in the Home Office regarding the whole prison disciplinary system and led directly to the establishment of the Prior Committee. That Committee eventually came down against a right of representation in all disciplinary cases. Prior concluded that having a legally qualified chair, as it recommended for its proposed Prison Disciplinary Tribunal, would protect prisoners against procedural carelessness while the costs and potential delays resulting from having representation in all cases were too much to bear.[65] The government did not disagree with this view and so essentially the position on representation remained unchanged.

However, even before Prior reported the pendulum was swinging against representation. In *R. v. Board of Visitors Blunderstone Prison ex parte Norley*[66] Webster, J., returned to the issue and indicated that where the prisoner was articulate, his defence essentially factually based, and the charge not among the most serious a Board would not as a rule be held to have gone wrong in exercising their discretion to refuse to permit representation. Boards generally became more prepared to refuse representation and

---

[63] [1985] QB 251.

[64] Fitzgerald, 'Prison Discipline and the Courts', at 37, praises this decision as 'a tribute to the resources of the common law that the court was able to find a way round the seeming impasse of *Fraser* v. *Mudge*'.

[65] See Prior Report, paras. 10.9–10.23, for the Committee's views on the issue of legal representation.

[66] Unreported 4 July 1984. See also *R. v. Board of Visitors Swansea Prison ex parte McGrath* The Times 21 Nov. 1984.

few were adversely reviewed as a result.[67] There was also a lack of lawyers with knowledge or interest in the area.[68] After the Prior Report the trend against representation moved even more swiftly. In *R. v. Board of Visitors Risley Remand Centre ex parte Draper*[69] the Court of Appeal upheld the view of the Divisional Court that a potential 180 days' loss of remission was not sufficiently serious to require legal representation if none of the other *Tarrant* factors were present. It also indicated that a fairly narrow view would be taken of what constituted a point of law requiring representation.

The death knell for attempts to establish a right of representation, at least in the United Kingdom context, came with the House of Lords decision in *Hone and McCartan* v. *Board of Visitors Maze Prison.*[70] In a case brought by prisoners in Northern Ireland to raise the issue of whether a right to representation existed Lord Goff, for a unanimous House, indicated that he could not find a right to representation in either natural justice or the European Convention and was much concerned by the problem of delays if representation were granted as of right.

With the demise of Boards' disciplinary functions in 1992 attention shifts to whether legal representation might be required before governors. In 1979 the Royal Commission on Legal Services indicated that legal representation should be available in any circumstances where a person faces loss of liberty of over seven days. Governors still have a power to add up to twenty-eight days to a prisoner's sentence on a disciplinary conviction. The Prior Committee and the Woolf Report recommended against allowing representation before governors and in *Hone and McCartan* Lord Goff indicated that he found it difficult to imagine when the rules of natural justice would ever require representation before a governor. However, in Northern Ireland, governors' adjudications have been quashed because they failed even to consider a request for representation.[71] Yet further decisions have indicated that where governors do give their mind to the issue only very rarely would they be vulnerable to judicial review if they decide not to grant representation. In *In re Reynolds Application*[72] Higgins, J., indicated that only where a prisoner was clearly unable to conduct his or her defence, for example where a prisoner was mentally subnormal, would a governor be under an obligation to grant representation.

Current Standing Orders do however indicate that prisoners appearing before governors on disciplinary charges should be asked if they wish to be legally represented, although when they must be asked this is not specified. In one case concerning a Board of Visitors hearing the prisoner sought judi-

---

[67] See R. Morgan, 'More Prisoners Denied Lawyers' Help' (1987) 91 *New Society* 1283.
[68] Light and Mattfield, 'Prison Disciplinary Hearings', 270.
[69] The Times 24 May 1988.   [70] [1988] 1 All ER 381.
[71] *In re Carroll's Application* (1987) 10 NIJB 23.   [72] (1987) 8 NIJB 82.

cial review because his application had only been considered, and rejected, on the day of the hearing. He argued that once the Board had been gathered for a hearing they would be very reluctant to adjourn to allow a prisoner's legal representative to prepare his or her defence. Rejecting the application the High Court indicated that this was a fairly straightforward case involving an articulate prisoner, where the prisoner would have been unlikely to get legal representation anyway.[73] This decision further downgrades the significance of legal representation and is in line with a number of decisions which suggest that the courts will now consider whether legal representation would have made a difference to the overall fairness of the case (for example by allowing prisoners to raise a defence which they would not have raised or presented coherently themselves) as opposed to quashing decisions automatically if legal representation has been wrongly refused.[74]

PROCEDURE AND EVIDENCE AT A DISCIPLINARY HEARING

On the day of an adjudication a prisoner must be examined by the medical officer as to whether he is fit to undergo an adjudication and in particular whether he is in a fit state of health to undergo a punishment of cellular (solitary) confinement. If the medical officer determines that the prisoner is not fit then he must give an indication of how long he feels this condition will last.

Standing Orders indicate that 'eyeballing' (the practice of prisoners having to stand throughout the hearing with two officers facing them) is not permitted and that prisoners must be allowed to sit at a table and be provided with writing material throughout the adjudication. Prisoners should also be given a copy of Form 1145 (Explanation of Procedure at Disciplinary Charge Hearings) and should have access to the Home Office manual on the conduct of adjudications, which is issued to all governors hearing disciplinary charges. However, it has been held that refusal to allow a prisoner a copy of the manual will not invalidate the adjudication if the court feels that the prisoner is well aware of its contents.[75]

In addition to being asked whether they wish to be legally represented prisoners may also consult with their lawyer in advance of the hearing and the governor hearing the charge has a discretion to allow them to have the

---

[73] R. v. *Board of Visitors Parkhurst Prison ex parte Norney* unreported 18 July 1989.
[74] See e.g. R. v. *Board of Visitors Long Lartin Prison ex parte O'Callaghan* unreported 5 Apr. 1989.
[75] See R. v. *Governor Pentonville Prison ex parte Watkins* unreported 4 Dec. 1991.

assistance of a friend or adviser (a 'McKenzie Friend'[76]) during the hearing. Prisoners do not appear to have made much use of this possibility.[77] The governor must record all requests for assistance and the decision reached on them as part of the whole record of the proceedings which will be put on Form F256.

When the hearing begins the prosecution will produce its evidence first. This normally consists of the report of the prison officer laying the charge. Attempts by prisoners to gain access to the statements made before the hearing by prison officers or prosecution witnesses have not proved wholly successful, courts generally ruling that prisoners are well aware of the nature of the allegations.[78] However, some decisions indicate that where the prisoner claims that there is a discrepancy between what the witness claims now and what he claimed at a previous hearing or in his original report it would be unfair not to allow the prisoner access to that previous report.[79] In *R. v. Board of Visitors Wandsworth Prison ex parte Raymond*[80] Webster, J., indicated that in the interests of fairness prisoners should be allowed access even to welfare reports prepared upon them. Even then he held that failure to do so would not make the hearing unfair and that the Board still had a discretion not to make them available, for example if they were likely to distress the prisoner unduly by revealing serious family problems previously unknown to the prisoner.

In the second *St Germain*[81] case Geoffrey Lane LJ indicated that prison tribunals were not to be treated as equivalent to courts of law as regards the rules of evidence. Therefore hearsay evidence was admissible. However, he went on to add that 'it is clear that the entitlement of the Board of Visitors to admit hearsay evidence is subject to the overriding obligation to provide the accused with a fair hearing'.[82] In particular, in circumstances where hearsay evidence formed a substantial part of the case against the prisoner, he or she should be given the opportunity of questioning those who gave such evidence:

Again, depending on the nature of that evidence and the particular circumstances of the case, a sufficient opportunity to deal with the hearsay evidence may well involve the cross examination of the witness whose evidence is initially before the board in the form of hearsay.[83]

---

[76] Based on the decision in *McKenzie* v. *McKenzie* [1970] 3 All ER 1034.

[77] Prior Report, para. 10.58, noted only eighty-seven requests to June 1985 of which seventy-six were granted, half of them at one open prison.

[78] See, e.g. *R. v. Board of Visitors Albany Prison ex parte Mayo* unreported 18 Mar. 1985 and *In re Crockard's Application* (1986) 7 BNIL 65.

[79] See e.g. *R. v. Board of Visitors Gartree Prison ex parte Mealey* The Times 14 Nov. 1981.

[80] The Times 17 June 1985.

[81] *R. v. Hull Board of Visitors ex parte St Germain (No. 2)* [1979] 3 All ER 545.

[82] Ibid., at 552.　　　　[83] Ibid., at 553.

In one decision it has been indicated that the governor may allow cross-examination to be made only through him.[84] Standing Orders suggest that this should be done only where the prisoner is abusing the right of cross examination; even then the governor is under an obligation to ensure the prisoner's right to present his defence is complied with.[85] Another right established by *St Germain (No. 2)* was the right to call witnesses. Indeed it is arguable that no procedural right has become more clearly established. In *St Germain (No. 2)* Geoffrey Lane LJ indicated that tribunals had a discretion to allow prisoners to call witnesses and they should be reluctant to exercise it against the prisoner except in the case where he or she was simply calling witnesses in an attempt to prolong or disrupt the hearing. Mere arguments of administrative convenience would not suffice. He added:

It would clearly be wrong if, as has been alleged in one instance before us, the basis for the refusal to allow a prisoner to call a witness was that the chairman considered there was ample evidence against the accused. It would equally be an improper exercise of the discretion if the refusal was based on an erroneous understanding of the prisoner's defence, for example, that an alibi did not cover the material time or day, whereas in truth and in fact, it did.[86]

In a subsequent decision Glidewell, J., went so far as to say that refusal to allow a prisoner to call a witness would be prima facie unfair.[87] Arguably the high-water mark of this strong line on the calling of witnesses came in *R. v. Board of Visitors Blundeston Prison ex parte Fox-Taylor*,[88] where Phillips, J., indicated that failure of the prison authorities to inform the prisoner of a witness material to the prisoner's defence that they but not the prisoner were aware of could render the adjudication unfair. The decision shows a strong commitment to making available to the prisoner all material which is relevant to his defence. This contrasts with the rather more ambiguous judicial pronouncements concerning prisoners' access to previous statements made by officers, witnesses, or probation officers. However, where the prisoner is aware of but does not call the witness the Board will not act unfairly if they decide not to call the witness.[89] Nor has the tribunal any power to compel a witness to attend who does not want to do so.

Northern Irish decisions on governors' hearings indicate that the right to call witnesses is likely to remain fundamental even given the lesser degree

---

[84] See *Ex parte Mayo* unreported 18 Mar. 1985.

[85] It is worth noting that the United States Supreme Court refused to recognize any right of cross-examination in *Wolff* v. *McDonnell (1974)* 418 US 539, though ABA Standards and many state prison rules provide for cross-examination.

[86] [1979] 3 All ER, at 550.

[87] *R. v. Board of Visitors Nottingham Prison ex parte Mosely* The Times 23 Jan. 1981.

[88] [1982] 1 All ER 646.

[89] See *R. v. Board of Visitors Liverpool Prison ex parte Davies* The Times 16 Oct. 1982.

of procedural formality required in these hearings.[90] Decisions refusing to allow prisoners to call witnesses are likely to be upheld only where the witness could not assist the prisoner's case, for example where the tribunal has already decided for the prisoner on the point he or she wants to call the witness on.[91]

Despite court rulings, in the context of legal representation and admissibility of evidence, that prison discipline hearings cannot be analogized to criminal trials the Divisional Court in *Tarrant* indicated that the criminal standard of proof, that of proof beyond a reasonable doubt, applies in prison discipline hearings. The Woolf Report refers to this as supporting its view that transferring cases to outside courts will not result in prisoners escaping punishment for disciplinary offences because sufficient evidence is difficult to obtain in the prison context. The approach of the United Kingdom courts on this issue compares very favourably with those in the United States, the Supreme Court having declared in *Superintendent, Massachusetts Correctional Institution, Walpole* v. *Hill*,[92] that it was sufficient that the conviction was supported by 'some evidence on the record'. However, while the criminal standard applies courts do seem willing to allow prison tribunals to draw rather wide inferences from the facts.

DISCIPLINARY PUNISHMENTS

Where a prisoner has been found guilty of a disciplinary punishment the governor may impose one or more of a number of punishments. These are set out in Rule 50 of the Prison Rules as:

(1) Caution
(2) Forfeiture of privileges for up to 28 days
(3) Exclusion from associated work for up to 14 days
(4) Stoppage of earnings for up to 56 days
(5) Cellular confinement for up to 3 days
(6) An award of additional days not exceeding 28 days
(7) Loss of property a prisoner is entitled to have under rule 41 for any period
(8) In the case of unconvicted prisoner who escapes or attempts to escape, loss of the right to wear their own clothes

If a prisoner is found guilty on more than one charge arising out of an incident the punishments imposed may be ordered to run consecutively.

The most frequently utilized punishment, at least in male prisons, has been forfeiture of remission.[93] This fact disappointed both the Prior

---

[90] See e.g. *In re Quinn's Application* (1988) 2 NIJB 10; *In re Rowntree's Application* (1991) 11 NIJB 67.

[91] See, e.g. *Ex parte Watkins* unreported 4 Dec. 1991.   [92] 472 US 445 (1985).

[93] *Prison Discipline Statistics 1991*, Cm. 2066 (London, 1992) indicate that loss of remission

Committee and Lord Justice Woolf. Both suggested that this may have had a lot to do with a lack of alternatives, especially as regards loss of privileges, since in many prisons prisoners had few privileges to lose. They suggested that as conditions improved and prisoners had greater access to privileges this might become a more frequently used penalty and one more likely to have a deterrent effect, especially for long-term prisoners. This faith however must reckon with the fact that it is in open prisons, where more privileges might be thought to exist, that loss of remission is most frequently used.[94] The Prior Committee noted that many prison reform groups argued that loss of remission, since it prolonged loss of liberty, should only be imposed by courts or similarly constituted tribunals. Governors, however, argued that retention of the power to remove remission was essential to deal quickly with disciplinary situations which need to be cooled or 'where an offence at risk of becoming fashionable needs to be stamped out quickly'.[95] As these seem very similar to reasons for segregating prisoners under Rule 43 they illustrate the close connection between formal and informal disciplinary powers. Though loss of remission has now been replaced by a power to award additional days it seems unlikely that this will mark a major change in substance as opposed to form.

Standing Orders indicate that additional days should be reserved for serious or repeated offences or where the adjudicator feels that an exemplary punishment is necessary. Where additional days are imposed for more than one offence they will run consecutively unless the adjudicator indicates they are to run concurrently. However, where the offences are related the total additional days should not exceed twenty-eight. Particular difficulties attend unconvicted prisoners and life sentence prisoners. As the additional days provision only applies to short- and long-term determinate sentence prisoners both of these groups are excluded. As regards unconvicted prisoners Rule 54 (1) indicates that they may be given a prospective award of additional days, to take effect only if they are subsequently convicted and imprisoned. As for life sentence prisoners Standing Orders indicate that where at the time of the adjudication the prisoner has been given a prospective date of release a recommendation may be given to the Home Secretary that this date be postponed. The additional days awarded will also be recorded on the prisoner's record, which will be available to the review committee considering whether a life sentence prisoner can be recommended for release. The tariff set for discretionary life sentence prisoners in accordance with s. 34 of the Criminal Justice Act 1991 appears to be unaffected by any award of additional days.

accounted for 38% of all punishments in male prisons and 31% in female. Stoppage of earnings was the next highest in male prisons at 26%, and was the highest equal in female prisons (31%).

[94] Ibid. It amounted to 64% of all punishments in such prisons.
[95] Prior Report, para. 8.50.

Where additional days have been imposed adult prisoners may apply within six months of the disciplinary decision to have this remitted. An application to remove additional days is not an appeal against the decision or the severity of the punishment. The application is made to the governor, who will consider a range of factors, including the prisoner's behaviour since the punishment was imposed and the need to preserve the deterrent effect of the punishment.[96] Circular Instruction 37/1992 indicates that 'The power to remit ADAs should be used to reward prisoners who take a constructive approach to their imprisonment.' Factors such as making the most of work and education opportunities are expressly referred to as being positive.

After award of additional days cellular confinement is perhaps the most severe punishment. Standing Orders indicate that this should be reserved for serious offences of an anti-social kind and that no such punishment shall be imposed if the medical officer does not certify that the prisoner is fit to undergo such punishment. In addition prisoners undergoing such punishments should be observed by an officer at least once an hour and visited daily by the chaplain and by the medical officer when necessary. The maximum period of cellular confinement that can be awarded is three days and consecutive punishments of cellular confinement on different offences which would take the time served beyond three days should not be ordered. Nor should similar punishments, such as spending loss of association in the segregation unit, be imposed which would have the same effect.[97] Prisoners who are sent to cellular confinement should be detained in an ordinary cell with only a table, stool, and chamber pot (the bed being removed except at night) and should be entitled to all privileges not inconsistent with being in solitary confinement unless they have received a separate punishment of loss of privileges.

Loss of privileges refers to those privileges available under Standing Order 4. Even where this punishment is given Standing Order 3D indicates that, unless specifically indicated, prisoners should not be deprived of educational notebooks, attendance at classes, and correspondence courses. They should also be allowed to have radios, general notebooks, and buy postage stamps. Governors are directed that exclusion from associated work may be looked upon as an alternative to solitary confinement, although the two may be combined. In general where more than one punishment is awarded for separate offences it is for the governor to order whether they will run consecutively or concurrently. Whichever is chosen must be clearly indicated to the prisoner.

---

[96] Prior Report, para. 8.99, called for a clearer list of factors to be taken into account, including (1) prisoners' behaviour since the offence, (2) the need to encourage prisoners to behave better, (3) the gravity of the offence involved, (4) the need to uphold discipline.

[97] See SO 3D (47).

Many disciplinary punishments are suspended under Rule 55 (1) for up to six months.[98] They then may be, but do not have to be, activated if the prisoner is convicted of another offence against discipline. Standing Orders indicate that suspension should normally not be given for an offence where violence has been used and injuries caused.

APPEALS AGAINST DISCIPLINARY CONVICTION OR PUNISHMENTS

Where a prisoner feels that a disciplinary conviction is wrongful or a punishment excessive he may petition the Secretary of State under Rule 56 (1) to seek the remission or mitigation of the punishment. The new rule changes in 1992 also gave the Secretary of State power to quash an award, the lack of which power had been commented on unfavourably in *Leech*[99] These complaints will be dealt with by Area Managers. Area Managers do not conduct a rehearing but instead will examine the record of the adjudication as to whether there have been procedural errors, the decision is wrong in light of material that has subsequently come to light, it is clearly unreasonable, or it is indefensibly severe (where the punishment is out of line with penalties imposed in broadly similar circumstances).[100]

In view of the fact that such applications to quash or mitigate a disciplinary decision remain within the prison administration many prisoners have expressed dissatisfaction with this procedure. Such dissatisfaction is hardly assuaged by a success rate of less than 10 per cent of such applications resulting in any change to the decision or punishment. The Prior Committee shared such concerns and argued that 'The availability of an effective appeal process where any substantial issue or right of liberty is at stake is an important element of the perceived fairness of a disciplinary system. Procedures need to be open, accessible, prompt and decisive.'[101] It also stressed the need for this appeal procedure to be independent of the original decision maker and therefore recommended that in respect of governors' decisions involving loss of remission in excess of seven days this should be to the Prison Disciplinary Tribunal. In respect of appeals from that body it recommended the creation of a Prison Disciplinary Appeal Tribunal. In addition to providing an adequate avenue for grievances about

---

[98] *Prison Discipline Statistics 1991* indicates that over 50% of loss of remission awards were not immediately implemented, despite the fact that SO 3D (58) (c) indicates that forfeiture of remission should be suspended only where there are special extenuating circumstances.

[99] [1988] AC 533.

[100] See Home Office, *The Prison Disciplinary System in England and Wales*, Cmnd. 9920 (London, 1986).

[101] Prior Report, para. 12.7.

decisions it felt an appeal structure could assist the development of consistency in decisions about punishments.

The government originally accepted Prior's notion of a right of appeal from governors' decisions (but not the creation of an Appeal Tribunal from the Prison Disciplinary Tribunal) but the idea was lost with the decision not to put most of the Prior recommendations into effect. Instead petitions against adjudications were brought under the jurisdiction of Area Managers. However, the Woolf Report was to revive concerns about the place of appeals in the disciplinary system. Focusing purely on governors' adjudications Woolf argued that there was still a need for some system of independent review of these. He argued that this function could be performed by the proposed Complaints Adjudicator. Whereas Prior argued that appeals against convictions should be by way of rehearing Woolf was content to allow the Complaints Adjudicator to review the record of the proceedings and conduct a rehearing only if he or she deemed it necessary. On the other hand Woolf envisaged this appeal mechanism applying to all decisions of the governor, whereas Prior limited it to decisions resulting in loss of remission exceeding seven days.

In its response to Woolf the government indicated support in principle for an independent appeal mechanism. It subsequently indicated that considering complaints about disciplinary hearings would come within the remit of the new Prisons Ombudsman.[102] The Ombudsman will not have power to rehear disciplinary procedures but will be empowered to review both the procedure and the merits of the hearing via an examination of the documentary record. The Ombudsman can also visit a prison to take evidence from a prisoner or prison officer. The Prisons Ombudsman will be independent in that the post will be publicly advertised rather than being appointed from within the civil service and will be appointed on fixed five-year contracts. However, there will be no new primary legislation in relation to the Ombudsman and so he can only make recommendations to the Home Secretary as regards the exercise of his existing powers. As we observe generally in Chapter 3 above the extent to which this Ombudsman can provide an independent element may depend on the strength of the personality who occupies this post and how he deals with resistance from the Prison Service and Home Office.

As has been seen earlier in this chapter prisoners may always challenge disciplinary hearings by way of judicial review. Yet such actions go only to whether the procedure of the hearing is fair and courts have said they will only interfere with the substance of the decision if it is one 'no reasonable body could have come to'.[103] This is a very difficult test to satisfy.

---

[102] See Home Office, *Custody, Care and Justice*, Cm. 1647 (London, 1991), para. 8.8.
[103] The criteria set out in *Associated Picture Houses Ltd.* v. *Wednesbury Corporation* [1948] 1 KB 223.

However, it seems likely that many judicial review cases have been prompted by concerns about the adequacy of the content of the hearing and not merely its procedure. With a satisfactory appeal procedure resort to judicial review might be considerably reduced.

CONCLUSION

There have clearly been very significant changes in prison discipline procedures over the past decade. Even before the alterations prompted by the Woolf Report a combination of judicial review and modifications in the Prison Rules had reduced maximum penalties appreciably and ensured that prisoners would have the opportunity to put their side of the case. No longer could governors or Boards dismiss them out of hand. However, examination of the prison discipline statistics raises questions about how much has changed in substance as opposed to form. Throughout the 1980s both the total number of offences punished in prisons and the amount per prisoner steadily rose, at least for male prisoners.[104] Although figures for charges dismissed are not so readily available it does not seem that they have wavered much from the 6 per cent that prevailed in 1991. In other words the development of greater procedural guarantees does not seem to have had any effect on the scale of disciplinary punishment occurring in the United Kingdom's prisons. If prisoners thought in 1980 that too many wrongful disciplinary punishments were being handed out there seems to be little reason why they should think otherwise in 1993.

Some will argue that this is the result of the courts' failure fully to apply the criminal law model to prison discipline proceedings. They would claim that if access to statements, the right to cross examination, the right to legal representation, and the provision of an adjudicator seen to be independent had been provided this would have produced a significant change in the outcome of hearings. There was, however, little chance of the courts ever going this far; the very existence of a separate discipline system suggested that different procedures were appropriate. Moreover there are reasons for believing that even if they had the effects might not have been that dramatic. These lie in the role disciplinary offences and punishments play within a prison. Whereas the intervention of the criminal law in outside society is an infrequent and somewhat distant form of social control disciplinary law in prisons reinforces a constant and pervasive regulation of prisoners' lives. Most disciplinary charges relate to offending against good order and discipline or disobeying a lawful order, in effect to refusing to

[104] The total number of offences punished rose from 63,391 in 1980 to 85,500 in 1991, and from an average of 1.5 to 1.9 over the same period. For males the rise was from 1.4 to 1.9, while females experienced a decline from 3.3 to 2.9. *Prison Discipline Statistics 1991.*

comply with the social order designed by the prison administration. The relationship of disciplinary punishments to social control is particularly clear in women's prisons, where prisoners are much more likely to be put on report for swearing, talking back, or even self-mutilation than in male prisons.[105] Such conduct was seen as at variance with the images of reformed 'normal' women the prison regime sought to achieve. When combined with the structural inequalities of prison hearings, where prison staff are always more likely to be believed when it comes to disputed issues of fact, it seems clear that prison justice will continue to be of a fairly rough kind. Since regular disciplinary punishments are part of the normal regime and since ultimately the courts acknowledge the legitimacy of the normal prison regime they are unlikely ever to interfere with this too much. As Michel Foucault has argued attempts to appeal to legal rights (or sovereignty as he refers to it here) against disciplinary power are likely to prove ultimately fruitless 'because sovereignty and disciplinary mechanisms are two absolutely integral constituents of the general mechanism of power in our society'.[106]

Arguably more progress may come with reforms which change the context in which discipline is administered, such as a reduction in maximum penalties, altering the content of charges, or reviewing practice in conferring charges. Such a significant development appears to have occurred in 1992, notably through the ending of the two-tier disciplinary system and the removal of more serious offences to be heard by outside courts. The impact of this remains to be seen. Clearly more serious offences will now be the subject of proceedings which incorporate the procedural guarantees of the criminal law, will guarantee rights of appeal, and will be heard by adjudicators who are not in any way dependent on the prison administration. Conflicts in prisons which give rise to charges are likely to be the subject of more outside scrutiny by the police, CPS, and courts than prevailed with the hearing of serious disciplinary charges by Boards of Visitors. Such scrutiny may play a role in encouraging the development of a management style which discourages conflict.

On the other hand there is little reason to believe that local panels of magistrates will be any more disposed to believe prisoners against the word of prison staff than Boards were and whereas loss of remission could be

---

[105] See the discussion in R. Dobash, R. Dobash, and S. Gutteridge, *The Imprisonment of Women* (Oxford, 1986), 146.

[106] M. Foucault, 'Two Lectures', in *Power/Knowledge: Selected Interviews and Other Writings 1972–77* (New York, 1980), 108. Hugh Collins describes a similar process in labour law whereby the formal equality of the employment contract is undermined by a web of disciplinary rules in the workplace. When these rules become a source of conflict legal intervention, premissed always ultimately on the legitimate authority of management, does little more than curb the greatest excesses of management power but leaves untouched, and therefore legitimizes, 'normal' forms of arbitrary power. See H. Collins, 'Capitalist Discipline and Corporatist Law' (1982) 11 *Industrial Law Journal* 78–93, 170–7.

restored sentences given by outside courts are fixed. There is also significant uncertainty over exactly how many cases will be referred. Woolf argued that it should be less than 700. Even this is a significant rise above the 200 or so that currently exists and could lead to significantly greater delays in having such cases adjudicated on. In addition Woolf's estimate is based on the belief that many cases currently referred to Boards should be dealt with by governors. However these cases have been referred precisely because governors feel that the twenty-eight days' loss of remission (now to become added days) is insufficient to mark the seriousness of the offence given its context or the prisoner's history.[107] Unless there is a serious rethink of the role of disciplinary punishments in prison management there must be a risk that prison culture, traditionally slow to change,[108] will seek to achieve the same ends through different means—perhaps through greater use of the 'informal' disciplinary system to segregate, transfer, or recategorize prisoners who prison staff feel are receiving insufficient disciplinary punishment. Despite the changes in structure it is the extent to which there is a change in substance which will determine whether the disruptive potential of disciplinary proceedings is diffused.

---

[107] This seems especially so for cases of having an unauthorized article and destroying property belonging to another. For these offences 5–7% of punishments in 1991 exceeded twenty-eight days. For assaults the figure was 21%. *Prison Discipline Statistics 1991.*
[108] The relationship of disciplinary offences and management style is discussed in P. Quinn, 'Prison Management and Discipline: A Case Study of Change', in M. Maguire, Vagg, and Morgan, (eds.), *Accountability and Prisons: Opening up a Closed World*, 204–16.

8

Non-disciplinary Powers to Maintain Good Order and Discipline

Good order and discipline (GOAD) is maintained in prisons by a combination of disciplinary and non-disciplinary powers. In Chapter 7 we described how the disciplinary powers of the governor operate. In this chapter we analyse the array of non-disciplinary powers available to the prison authorities which exist to regulate and enforce what is perceived to be in the interests of the good order and discipline of the prison regime.

By non-disciplinary we mean those powers whose exercise does not depend upon proof of the commission of a specific disciplinary or criminal offence by the prisoner affected. They may not therefore be imposed as a punishment. Precisely because they are broader and can be invoked independently of the 'official' disciplinary system, these non-disciplinary powers are a major source of discontent amongst prisoners. This is largely because they can achieve as drastic a change in the quality of a prisoner's life as a disciplinary award yet their use is unaccompanied by the statutory procedural safeguards which prisoners enjoy under the disciplinary system. From the point of view of the prison authorities, there is the obvious attraction that because the very purpose of the powers is to assist in maintaining security, discipline, and control the courts have displayed a great reluctance at the prospect of 'second guessing' managerial decisions in sensitive areas of prison administration. Even though the House of Lords held in *R. v. Deputy Governor of Parkhurst Prison, ex parte Hague*[1] that the supervisory jurisdiction of the High Court does extend to include all decisions taken in pursuance of the prison rules, this remains an area where the existing legislative framework in practice gives governors and Area officials an almost unfettered discretion to regulate prisoners' lives.[2]

In the light of such judicial circumspection it is perhaps hardly surprising

---

[1] [1992] 1 AC 58.

[2] The Home Office argued before the Divisional Court that the High Court had no jurisdiction to quash a Rule 43 segregation decision. It lost and during the course of argument before the Court of Appeal conceded that jurisdiction did exist but only after a rejection of a complaint to the board of visitors or of a petition to the Secretary of State and then only by way of review of such rejection. The Court of Appeal rejected this 'second stage' approach but accepted that the court should employ great caution before exercising its discretion in issues affecting prison management. No appeal was pursued by the Home Office to the House of Lords on this point.

that many prisoners regard the Home Office claim that these powers are never used as a means of bypassing the official disciplinary system with a degree of suspicion. Conscious or malicious abuse of public law powers is always remediable but in prison proof of such abuse is hard to achieve thus making it difficult for the courts effectively to scrutinize illegal resort to non-disciplinary measures.

The use of non-disciplinary powers affects prisoners in five areas— security categorization, Rule 43 segregation, transfer under Circular Instruction 37/1990, allocation to Special Units, and finally in the use of approved methods of restraint and location in special cells. We will examine each area in turn before finally looking at the approach of the European Court of Human Rights in claims under Article 3 of the Convention as well as American prison law.

SECURITY CATEGORIZATION

Imprisonment in the modern era has always been accompanied by a desire to separate and classify prisoners according to both the reason for their imprisonment and their perceived threat to the good order and discipline of the prison regime (the former frequently influencing the latter). The Prison Act 1865 required prisoners to be separated according to their sex, whether or not they were criminal prisoners or debtors, and whether or not they were undergoing punishment for a prison offence. Separate cells were required to be provided sufficient to accommodate prisoners in each category.[3] Previous legislation had been even more complex and exhaustive, reflecting the Victorian obsession with labelling and regulating the process of punishment, including punishment for debt.[4]

The Prison Act 1952 vests the Secretary of State with a broad power to make rules 'for the regulation and management of prisons . . . and for the classification, treatment, employment, discipline and control of persons required to be detained therein'.[5] Rule 3 of the Prison Rules provides further guidance on the basis for classification:

3. (1) Prisoners shall be classified, in accordance with any directions of the Secretary of State, having regard to their age, temperament and record and with a view to

---

[3] s. 17 (5) of the 1865 Act stated that 'In a prison where criminal prisoners are confined such prisoners shall be prevented from holding any communication with each other, either by every prisoner being kept in a separate cell by day and by night, except when he is at chapel or taking exercise, or by every prisoner being confined by night to his cell and being subjected to superintendence during the day as will, consistently with the provisions of this Act, prevent his communicating with any other prisoner.'

[4] See the seven-fold division of prisoners in Prison Act 1842, s. 17, which included a requirement to prevent communication between the seven classes.

[5] Prison Act 1952, s. 47 (1).

maintaining good order and facilitating training and, in the case of convicted prisoners, of furthering the purpose of their training and treatment as provided by Rule 1 of these Rules.

(2) Unconvicted prisoners shall be kept out of contact with convicted prisoners as far as this can reasonably be done.

(3) Nothing in this Rule shall require a prisoner to be deprived unduly of the society of other persons.

Plainly, the division of prisoners according to sex and whether or not they are convicted or on remand deserves no further comment. It is security categorization which gives rise to controversy and in particular the consequences of allocation to category A. It applies to prisoners of both sex regardless of whether they are as yet unconvicted. The only published directions which the Secretary of State has issued to indicate the basis on which categorization decisions are made are the so-called Mountbatten criteria, based as they are on the recommendations of Earl Mountbatten's Inquiry into prison escapes and security, established in the aftermath of the escape of George Blake from Wormwood Scrubs in 1966.[6] The current criteria have remained unchanged since 1966:

* Category A: those prisoners whose escape would be highly dangerous to the public or to the police or to the security of the state.
* Category B: prisoners for whom the very highest conditions of security are not necessary but for whom escape must be made very difficult.
* Category C: prisoners who cannot be trusted in open conditions but who do not have the ability or resources to make a determined escape attempt and for whom simple basic security precautions would be sufficient.
* Category D: prisoners who can be housed in open prisons without danger.

To these may be added a fifth category, category E, which has emerged since the Mountbatten Report and which applies to those prisoners who have escaped (or attempted it) and who are consequently required to wear yellow striped clothing ('patches') at all times.

CATEGORIES B, C, AND D

Administrative guidance on the allocation of prisoners to categories B, C, and D is contained in Circular Instruction 7/1988 (which is not published and therefore not generally available to prisoners). The general approach as set out in paragraph 4 is that

prisoners should be categorised objectively according to the likelihood that they will seek to escape and the risk that they would pose should an escape succeed. *The security category should take account of these considerations alone.* Factors such as likely conduct (other than a propensity to escaping); ability to mix with other pris-

[6] Cmnd. 3175 (London, 1966).

oners; educational, training or medical needs; and the availability of suitable establishments should not be taken into account at this stage. They are for consideration during allocation. . . . Every prisoner should be placed in the lowest security category consistent with the needs of security.

The criteria to be taken into account when making a categorization decision are then set out in paragraphs 6–8. All prisoners should be regarded as suitable for category D on first categorization unless they are:

1. sentenced to over twelve months for any offence of violence; or
2. convicted of any but the most minor sexual offence; or
3. have a previous sentence of over twelve months for any violent or sexual offence and did not successfully serve part of that sentence in an open prison; or
4. have current or previous convictions for arson or any drugs offence involving importation or dealing; or
5. have a recent history of escapes or absconds.

Once these criteria have been applied, those prisoners who are not allocated to category D should be placed in category C unless they:

1. are sentenced to over seven years for any violent or sexual offence; or
2. have a previous sentence of over seven years for any violent or sexual offence and did not successfully serve part of that sentence in a category C prison; or
3. have a current sentence exceeding ten years; or
4. have a recent history of escape from closed conditions or have significant external resources which they might use to assist an escape attempt.

Category B is reserved for prisoners convicted of a serious violent, sexual, or drug related offence. According to CI 7/1988, 'few other prisoners will merit Category B'.

Decisions about categories B–D are made by 'Observation, Classification and Allocation Units' at Area level and depend largely on the content of current and previous custodial records, pre-trial reports and antecedents, the nature of the charge, and any remarks made by the sentencing judge. All remand prisoners or those convicted and awaiting sentence are (unless they have been provisionally placed in category A) placed in category U (Unclassified) and will normally be assumed to require accommodation suitable for category B.[7]

---

[7] CI 7/88 states that 'categorisation should continue to take place in 2 stages—an initial provisional assessment on paper to be followed by an interview with the prisoner on the outcome of which the final categorisation decision should be made. This interview may be dispensed with for prisoners sentenced to less than 6 months, unless the Observation, Categorisation and Allocation Unit considers an interview necessary' (para. 13). The Woolf Report recommended that all remand prisoners should be regarded as equivalent to category C rather than category

CATEGORY A

A decision to place a prisoner in category A is made by the category A committee, a permanent body comprised of Prison Service officials which meets every three weeks. When prison establishments first receive prisoners charged with certain serious offences they are required to report their cases to the category A Committee, which will consider whether they should be placed provisionally in category A. If a prisoner is classified as provisionally category A he is subject to broadly the same security conditions as a prisoner who is confirmed in category A. Upon conviction, the security category is reviewed. Within category A there is a division between 'ordinary Cat. A' and 'High Risk Cat. A'. The latter are subject to even greater restrictions on their freedom within the prison.

Security categorization is a vital determinant of the kind of regime which a prisoner is likely to enjoy. At one extreme, allocation to category D leads to location in an open prison with a relaxed regime, relatively easy contact with the outside world, and home visit privileges. At the other end of the scale, category A status leads to a relatively impoverished regime with great emphasis on observation, regulation, and security. In *Payne* v. *Home Office* Cantley J. listed six disadvantages to a prisoner flowing from allocation to category A:-

1. there are only a relatively small number of prisons suitable for his safe accommodation; this may result in his being detained in a prison which is more distant from persons from whom he wishes to have visits than some less secure prison would be; 2. he can have visits only from a solicitor, a probation officer, a prison visitor or a person who has been passed as suitable by the Home Office; 3. his cell is a specially secure one and it is liable to be searched more frequently than other cells; he is also under closer surveillance, and this may sometimes result in his sleep being disturbed when the officer who is looking into his cell cannot be sure in a dim light that there is more than a carefully arranged heap of bedclothes on his bed; 4. he cannot attend general vocational training classes or concerts, nor can he attend the ordinary church services, although he has regular visits from the chaplain and can take communion in his cell if he wishes; 5. he can attend only educational classes of not more than two students, and so he has less frequent opportunity to attend educational classes; 6. he is not likely, to say the least, to be put on parole while he is a category A prisoner[8]

B prisoners 'unless there is reason to regard them as needing category B or category A conditions of security' (*Prison Disturbances: April 1990*, Cm. 1456 (London, 1991), para. 15.47).

[8] Unreported 2 May 1977. The plaintiff, Roger Payne, appeared in person before the court. Four years later, and with legal representation, he challenged the Parole Board's refusal to give reasons for a rejection of an application for parole (see *Payne* v. *Lord Harris of Greenwich* [1981] 1 WLR 754). Cantley J. complimented Mr Payne on his 'well-sustained argument of admirable lucidity and learning which, both in manner and in matter, would have done credit to a professional lawyer and advocate'.

Notwithstanding the drastic consequences attendant upon allocation to category A (in his case for seven years), Payne failed in his action for a declaration that such allocation was invalid by reason of breaches of the rules of natural justice—specifically the fact that he was not told of the material before the category A committee, nor given an opportunity to make representations, nor informed subsequently of the reasons for the decision.[9] The court accepted the existence of a general duty to act fairly in reaching categorization decisions (and was content to accept the Home Office's assurance that it did always act fairly) but held that 'the full panoply of the rules of natural justice is wholly inappropriate to the classification of prisoners'. In reasoning which exemplifies the 'hands off' approach towards prison affairs, Cantley, J., held that any application of the right to be heard would 'as a matter of general principle . . . seriously hamper, and in some circumstances could frustrate, the efficient and proper government of prisons . . . Very considerable difficulties would arise if the nature of the case to be considered by the classifying authorities were to be communicated to the prisoner in any detail sufficient to enable him to present relevant observations of his own.' He also held that Parliament had by s. 47 (2) of the 1952 Act expressly provided for procedural fairness requirements in the prison disciplinary process and must thereby have intended that such requirements were not applicable to the process of classification.[10]

Despite its antiquity, *Payne* remains the only fully reasoned decision in this area of prison law and indicates that a categorization decision will only be quashed where the decision can be characterized as perverse. In its marked reluctance to apply basic principles of procedural fairness (such as an opportunity to know and respond to the reasons advanced for a proposed allocation to category A) it is however in line with more recent decisions on, for example, parole and Rule 43 segregation, where the courts have steadfastly refused to imply into existing legislation procedural fairness requirements.[11] It is hard to justify this reluctance other

[9] It was by no means obvious why Payne had remained in category A for so long. He had been convicted of murder in 1968 but the majority of convicted murderers avoid category A status throughout their imprisonment. He had two previous convictions for relatively minor offences of assault, it was accepted by the Home Office that he had been a model prisoner, and he had never been the subject of psychiatric treatment. Accordingly the argument that 'although you haven't been told the reason it is plain enough' could not possibly apply to Payne's case, emphasizing the injustice of his apparently arbitrary allocation.

[10] Identical reasoning was employed by the Court of Appeal when rejecting Christopher Hague's argument for a right to be heard in the context of Rule 43 segregation—see per Taylor, LJ., [1992] 1 AC at 110C.

[11] R. v. *Deputy Governor of Parkhurst ex parte Hague* [1992] 1 AC 58. The procedural fairness issues were only argued before the Divisional Court and the Court of Appeal. By the time the case reached the House of Lords in July 1991 the Prison Department had already issued CI 37/1990 revoking CI 10/1974, which afforded all prisoners the procedural fairness protections for which Hague had been arguing.

than by frankly acknowledging that prisoners are not regarded by the courts   as suitable beneficiaries of procedural protections which would ordinarily attach to decisions of such importance. The test of a right to be heard with the necessary concomitant of a right to adequate notice of what is to be alleged is not the source or character of the power but its impact on the person affected by its use.[12] The draconian effect of allocation to category A surely meets this test, involving as it does a grave limitation on a prisoner's already attenuated liberty. Yet security categorization remains, in every sense, a closed world. Parliament's requirement in s. 47 (2) that the Rules provide for due process in disciplinary adjudications does not imply a licence for arbitrary decision-making elsewhere in the Rules. There is no universal test of 'necessity' for the implication of procedural fairness requirements into decision-making. The leading authorities all treat the obligation of fairness as a basic requirement of decision-making except where the statutory words or purpose clearly exclude it.[13] The reluctance to import detailed provisions guaranteeing procedural fairness in this area of non-disciplinary powers echoes the similar reluctance to concede that the supervisory jurisdiction of the High Court extended to the prison disciplinary process. It is surprising that such attitudes remain despite the discovery, following *St Germain* and *Leech*, that the existence of legal recourse not only does not injure the good conduct of prison but enhances it.

It is almost unheard of for a prisoner to see any of the reports submitted to the category A committee but in *Hague* the Home Office chose to exhibit a number of such reports on Hague (who was in category A) from various governors and prison officers. They revealed a disturbing tendency to label his involvement in litigation against the Home Office and his use of the complaints mechanism as examples of his subversive nature! One report from a principal prison officer stated that 'he does not present the Wing staff with many problems but is a thorn in the side of management/Board

[12] See *Kanda* v. *Government of Malaya* [1962] AC 322 and also R. v. *Deputy Governor of Parkhurst ex parte Leech* [1988] 1 AC 533, at 578 E–G. As for what generates a right to be heard see *Wade on Administrative Law* (6th edn.), at 529 ff., esp. *Durayappah* v. *Fernando* [1967] 2 AC 337.

[13] *Cooper* v. *Wandsworth, Board of Education* v. *Rice, Wiseman* v. *Borneman* [1971] AC 297, and esp. *Lloyd* v. *McMahon* [1987] AC 625, in which Lord Bridge said at 702 G: 'My Lords the so called rules of natural justice are not engraved on tablets of stone. To use the phrase which better expresses the underlying concept, what the requirements of fairness demand when any body, domestic, administrative or judicial, has to make a decision which will affect the rights of individuals depends on the character of the decision-making body, the kind of decision it has to make and the statutory or other framework in which it operates. In particular it is well-established that when a statute has conferred on any body the power to make decisions affecting individuals, the Courts will not only require the procedure prescribed by the statute to be followed, but will readily imply so much and no more to be introduced by way of additional procedural safeguards as will ensure the attainment of fairness.'

of Visitors with a continual stream of applications and petitions etc'. The very fact that such comments are believed to be relevant for inclusion in a report to the category A committee is indicative of the dangers attendant upon a decision-making process which takes place immune from effective scrutiny by the courts.

The refusal to give reasons for a categorization decision remains a source of discontent. It is increasingly recognized that a duty to give reasons for a decision which has to be taken in accordance with law is an integral part of the decision-making process especially where the decision affects personal freedoms.[14] There is no general rule *against* the giving of reasons any more than there is a general rule in favour of it. If the circumstances call for it the law will require it.[15] The argument frequently advanced by the Home Office that a prisoner is properly protected from arbitrary decision-making by the availability of recourse to the Board of Visitors or to the Secretary of State via the complaints system simply emphasizes the need for prisoners to be provided with reasons in order to make sensible representations to those bodies. Just as in *Kanda*, where the court reasoned that the right to be heard involved a right to be informed of the charges, so in the context of security categorization a right to petition the Board of Visitors or Secretary of State involves a need to know why a prisoner is considered to be in need of category A treatment. Rejection of the right to be heard (prior reasons) strengthens the case for being given reasons retrospectively.

The Woolf Report recommended that 'as a matter of good and sensible administration and management', the Prison Service should adopt the practice of giving reasons to a prisoner for any decision which materially and adversely affects him. It went on, however, to accept that this approach could not be regarded as inflexible and stated that 'it would not be practicable to require the giving of reasons where this is not desirable in the public interest . . . for example in the case of category A prisoners it may be necessary to defer giving reasons on security grounds'.

It is not entirely clear if this qualification was intended to be confined to deferring the giving of reasons for placement in category A itself or whether, generally speaking, deferral *may* apply in respect of any decision which adversely affects a prisoner already allocated to category A. It must surely be the latter, because to exclude from the general requirement of giving reasons a decision as important as allocation to category A while

---

[14] See in particular *Wade on Administrative Law*, 934–7, and *R. v. Home Secretary ex parte Singh* unreported (DC) 22 May 1987, in which Woolf LJ criticized the Home Office reluctance to provide full reasons for rejecting asylum claims and pointed out that the giving of reasons will frequently avoid the need for recourse to judicial review. See also *R. v. Civil Service Appeal Board ex parte Cunningham* [1991] 4 All ER 310.

[15] See the judgment of Deane J. in *Public Service Board of New South Wales v. Osmond* (1986) 60 ALJ 209 as well as the speech of Lord Mustill in *R. v. Home Secretary ex parte Doody*, HL, 24 June 1993.

accepting its applicability to other 'security' decisions, such as Rule 43 segregation, would be to create an unacceptable anomaly.[16]

RULE 43, TRANSFER IN THE INTERESTS OF GOOD ORDER OR DISCIPLINE, AND SPECIAL UNITS.

When a prisoner is believed to be a threat to good order or discipline, he may be dealt with in one of three ways by the prison authorities. First, he may be segregated under Rule 43 of the Prison Rules in the prison where he is perceived to be a threat. Secondly, he may be transferred under Circular Instruction 37/1990 either short term to a local prison with the expectation of immediate Rule 43 segregation on arrival or medium term (six months) to a local prison on normal location. Finally, there is the possibility of allocation to a Special Unit for prisoners believed to present special problems of control and security.[17] The first two options give rise to a number of overlapping considerations many of which received detailed consideration in the important case of *R. v. Deputy Governor of Parkhurst Prison ex parte Hague.*[18]

RULE 43.

Under Rule 43 (1) of the Prison Rules (Rule 46 of the Young Offender Institution Rules) a prison governor has the power to remove a prisoner from normal association 'where it appears desirable for the maintenance of good order or discipline or in his own interests'. In this chapter we are solely concerned with segregation in the interests of good order or discipline (GOAD), 'own interests' segregation giving rise to different considerations.[19]

---

[16] CI 7/1988 contains no requirement to provide any prisoner with reasons for his allocation to a particular category. This is in contrast to CI 26/1990 and CI 37/1990, both of which reflect a new willingness to encourage the giving of reasons for any segregation or transfer decision along lines expressly recommended by the Woolf Report.

[17] CI 37/1990 refers in paras. 37–51 to a fourth option, namely medium-term transfer from a training prison to a local prison where the prisoner will be placed on normal location. As this option does not anticipate or envisage the imposition of Rule 43 segregation as a result of the transfer it requires no further comment.

[18] The judgments delivered in both the Divisional Court and the Court of Appeal are included in the Appeal Cases report at [1992] 1 AC 58.

[19] The Woolf Report criticized the fact that Rule 43 covers both GOAD and 'own protection' segregation, commenting at para. 12.196 that 'It is unhelpful that the Rule should be applied to two very different categories of prisoners. What the establishment should be seeking to achieve in relation to the two categories is quite different. In the case of the prisoner separated for reasons of good order or discipline, the removal from association needs to be fairly strictly imposed if the object of the exercise is to be achieved. In the case of the vulnerable prisoner, on the other hand, it is accepted that any more separation than is necessary to

The full text of Rule 43 is as follows:

43 (1) Where it appears desirable, for the maintenance of good order or discipline or in his own interests, that a prisoner should not associate with other prisoners, either generally or for particular purposes, the governor may arrange for the prisoner's removal from association accordingly.

(2) A prisoner shall not be removed under this Rule for a period of more than 3 days without the authority of a member of the board of visitors or of the Secretary of State. An authority given under this paragraph shall be for a period not exceeding one month, but may be renewed from month to month except that, in the case of a person aged less than 21 years who is detained in prison, such an authority shall be for a period not exceeding 14 days, but may be renewed from time to time for a like period.

(3) The governor may arrange at his discretion for such a prisoner as aforesaid to resume association with other prisoners, and shall do so if in any case the medical officer so advises on medical grounds.

The governor's power to segregate is limited to a period of three days but this can be extended for up to a month with the authority of either a member of the Board of Visitors or a representative of the Secretary of State (in reality, a senior Prison Service official). In practice prisoners can be segregated for unlimited lengths of time throughout their sentence with the periodically (monthly) renewed agreement of these authorities. The former Chief Inspector of Prisons, Sir James Hennessy, commented in his 1985 special report on the use of segregation that it 'can entail living under an impoverished and monotonous regime which may even be psychologically harmful'.[20] Though the Home Office prefers the term segregation, the description solitary confinement more accurately describes the reality of what is involved.

Prisoners on GOAD segregation are held in virtual isolation, shut in their cells in the punishment block for twenty-three hours per day, and are deprived of most opportunities for work, education, and recreation. The Woolf Report confirmed this unhappy state of affairs, commenting that 'while segregation under Rule 43 is not intended to be a punishment the use of the Rule will invariably adversely affect the inmate who is made subject to it. In most establishments, anyone segregated under Rule 43 will be subjected to regime restrictions very similar to those undergoing punishment.'

Rule 43 has three material limbs. The initial provision in paragraph (1) makes clear that it is the governor of the prison in which the prisoner is for the time being (i.e. in whose custody he is under s. 13 Prison Act 1952) who alone may arrange for his removal from association within that

protect the prisoner is undesirable. The object from the start should be to return the prisoner to association if this is possible.'

[20] *A Review of the Segregation of Prisoners Under Rule 43*, Report by HM Chief Inspector of Prisons (London, 1985) (Hennessy Report); see also para. 2.29 of the Report.

prison.[21] Paragraph (2) requires reasoned consideration to be given by the Board of Visitors member or the Secretary of State to the question whether and for how long the initial three-day period of segregation is to be extended.[22] In other words neither the Board of Visitors member nor the Secretary of State should act as a rubber stamp of the governor's initial decision. Paragraph (3) makes clear that governors must keep the need for segregation under constant review despite the grant of authority for an extension by up to one month and must permit the resumption of association on receipt of medical advice.

The second and third of these provisions have no purpose other than to protect the individual prisoner's interests. The first provision operates both in the interests of sound prison administration by permitting non-disciplinary segregation and in the prisoner's interest by confining that power to the governor in whose custody he is.

The terms of Rule 43 make it clear that association with other prisoners is not a privilege under the Rules but the norm (otherwise there would be no need for Rule 43). Being on Rule 43 is neither a status nor a classification which a prisoner carries from prison to prison. It is the fact of being subjected to a temporary regime by prescribed means and for ascertainable statutory purposes within a particular prison.

Comprehensive guidance both to governors and to Boards of Visitors on the operation of Rule 43 is to be found in Circular Instruction 26/1990.[23] Originally drafted in the wake of the *Report of the Prison Department Working Group on the Management of Vulnerable Prisoners* (May 1989), it was amended in 1991 in the light of recommendations contained in the Woolf Report and represents a significant shift in attitude in terms of both the occasions when segregation may be justified and the safeguards which should accompany its use.

The final amended version of CI 26/1990 accepts the Woolf Report's conclusion that 'segregation should not be used unless it is absolutely necessary . . . It must not be regarded as the instant remedy for problems of a control nature.' It goes on to warn that 'inmates who are under segregation are more likely to be at risk of suicide . . . It is important therefore that feel-

---

[21] *Ex parte Hague* [1992] I AC at 105–108, per Taylor, LJ.

[22] Ibid. CI 6/1993 alerts Boards of Visitors to the fact that before a Board member authorizes continued segregation he must make a personal visit to the prisoner concerned. It states that 'The Co-ordinating Committee of Boards of Visitors have told us that some Board members authorize continued segregation by telephone without seeing the inmate. This is contrary to the training they receive. Also it is in the interests of justice that a Board member should see the inmate in person before authorizing continued segregation. . . . An initial telephone authorization should be given only in exceptional circumstances—such as freak weather conditions—which prevent a Board member attending the establishment within 3 days.' The Circular is a welcome example of the Prison Service's willingness to extend its policy requirements beyond the meagre procedural fairness requirements of *Ex parte Hague*.

[23] See Appendix 4.

ings of isolation are eased as much as possible by regular face-to-face contact and out-of-cell activity.'

Annex C of CI 26/1990 indicates that where the governor wishes to extend segregation beyond three days neither he nor the Board of Visitors member who is required to authorize such extension ought routinely to assume that the maximum permitted period of one month is the appropriate period to authorize. Indeed it points out that in 'good order or discipline cases' authorization for the maximum period should rarely be necessary and that 'the Board member will decide, in the light of the governor's advice, for how long a period to authorise continued segregation and will not automatically give authority for the maximum of one month. The member may decide to give authority for only a few days initially, pending the opportunity to interview the inmate, or for any period less than the maximum which may seem sufficient in the circumstances of the case.'

The Woolf Report was critical of the role of the Board of Visitors in authorizing extensions to segregation and called for this aspect of their functions to be ended, pointing out that 'whatever is the true position, when a Board member gives such authority, prisoners do not see him as acting to safeguard their interests but as the arm of management . . . this is not consistent with or helpful to the Board's watch-dog role' (para 12.270). The Report went on to recommend that in future it should be the responsibility of the Area Manager to authorize any extension to segregation and that any more than one extension beyond twenty-eight days should only be justified in exceptional circumstances. Boards of Visitors should carefully supervise and monitor the exercise of the power to segregate. These recommendations have not, as yet, been accepted by the Prison Service.

The use of Rule 43 segregation as a punishment is an improper and unlawful abuse of power even though, as has been said, the regime to which Rule 43 prisoners are subjected is frequently indistinguishable from that endured by prisoners undergoing the punishment of cellular confinement. Paragraph 6 of CI 26/1990 informs governors of the broad preemptive purpose behind their Rule 43 power:

Segregation under rule 43 is designed to assist governors to prevent trouble and governors may use it for this purpose in respect of known subversive inmates . . . either on reception or at any subsequent stage of their time in custody. It is not necessary to wait until a prisoner has actually jeopardised control. It is sufficient that he should have shown that it is his intention to do so; and it is right to take into account a history of disruptive behaviour, either inside or outside the institutional setting.

Such advice immediately reveals the dangers inherent in a power which requires only past reputation to trigger its use. According to the Prison Service, it is not necessary for a prisoner actually to have done anything which threatens good order and discipline in the prison to which he has recently been transferred. On reception the governor can take account of what it is alleged he did at his previous prison (where he may have been discontented for numerous reasons no longer relevant once he was moved) or even his behaviour outside prison and decide on that basis to order segregation. Of course the very act of segregating before anything has happened may well then trigger behaviour which will be seen retrospectively to justify the decision to use Rule 43 and so the merry-go-round continues.

From the prisoner's point of view the blurred division between supposedly non-punitive Rule 43 segregation and punitive cellular confinement is reinforced by the knowledge that the courts have held that it is not an abuse of power for a governor to segregate a prisoner instead of requiring a disciplinary charge to be laid, even where the reason for not laying a charge is dubious. Christopher Hague's case clearly demonstrated the injustice of this approach. He was placed on Rule 43 and transferred from Parkhurst to Wormwood Scrubs because it was alleged that he was stirring up trouble amongst other category A inmates at Parkhurst over the decision occasionally to withdraw exercise facilities from all category A prisoners for security reasons. There was no dispute that Hague and others had complained about this through legitimate channels but the decision to transfer him pursuant to CI 10/74 (the predecessor to CI 37/1990) was based on specific allegations, which Hague disputed, that he had deliberately gone on exercise at a time when he knew he should not have and that he subsequently told the deputy governor of his intention to continue his defiance. On any view Hague's alleged conduct both in substance and in law amounted to an offence contrary to Rule 47 (21) ('in any way offends against good order and discipline') yet the governor admitted that a decision was deliberately taken not to charge Hague because this might embarrass those prison officers who, as Hague had claimed, had seen him on exercise but chose not to give him a direct order to return to his wing. In the Divisional Court, Ralph Gibson LJ held that avoiding embarrassment to prison staff, whose function it was to maintain discipline, could be a legitimate reason for not preferring a charge. He went on to state that:

There is also . . . no obligation in a prison governor to prefer a disciplinary charge as a precondition of acting upon information which demonstrates commission by a prisoner of an offence against discipline before he may act upon that information as a ground for placing the prisoner on rule 43 for transfer under CI 10/74. Such a precondition would impose on the exercise of the governor's discretion, in his task of maintaining discipline, a fetter which is not expressly imposed by the prison rules and which, in our judgment, is not only not implicitly required by those rules, upon their proper construction, but seems to us to be contrary to the fullness of discretion which in such matters those rules intend a prison governor to have.

This aspect of Hague's attack upon his segregation was not pursued in the Court of Appeal and thus represents the current law. It is submitted that it represents an extremely unfortunate bias in favour of administrative discretion in an area where the courts ought to be alert to protect prisoners from potential abuses of power. The avoidance of embarrassment to officers may be a good enough reason for exercising the discretion not to lay a disciplinary charge but it is surely an impermissible reason for resorting instead to segregation under Rule 43. This objection is not met if Rule 43 required prisoners to be granted a fair hearing because segregation can continue indefinitely whereas disciplinary awards need not involve segregation at all and cellular confinement cannot exceed three days if awarded by a governor.

So far as the regime to which Rule 43 prisoners are subjected is concerned, CI 26/1990 seeks to emphasize that it 'should be as balanced and well integrated as for the rest of the population within the limits of affording them protection and maintaining good order and discipline, and nothing less than that'(para 19). Accordingly prisoners should not deliberately be subjected to a restricted regime by being denied facilities or privileges available to others. A prisoner who is placed on Rule 43 in conditions indistinguishable from those imposed on a prisoner undergoing punishment, possibly in an adjacent cell, could thus legitimately seek to challenge a refusal to grant him ordinary facilities and privileges save in so far as these can be shown to be wholly incompatible with his segregation. The wording of Rule 43 (1) ('should not associate with other prisoners, *either generally or for particular purposes*') makes it clear that not every decision to segregate can justify total isolation from fellow prisoners. Each case requires separate reasoning and justification.

The use of Rule 43 GOAD segregation is markedly different as between male and female prisoners. The 1990 *Prison Statistics* for England and Wales revealed that on 30th June 1990 a total of 305 prisoners were subject to such segregation. This total comprised 233 adult males, sixty-nine males under 21, two adult females, and one female prisoner under 21. Of the adult males, 161 had been in segregation for up to one month, fifty-seven for up to three months and fifteen for over three months.[24]

CIRCULAR INSTRUCTION 37/1990: TRANSFER IN THE INTERESTS OF GOOD ORDER OR DISCIPLINE

If an allegedly disruptive prisoner cannot effectively be segregated in the prison where he is confined, a second option available to prison management is a (short-term) transfer pursuant to Circular Instruction 37/1990. The practice of short-term transfers in the interests of good order and discipline (known to prisoners as 'ghosting') gives rise to considerable

[24] *Prison Statistics: England & Wales*, Cmnd. 4800 (London, 1990).

controversy. Such transfers will normally be from a dispersal prison to certain designated local prisons where a small number of cells are kept permanently available for such use. It is applied to those prisoners who are regarded as seriously disruptive or subversive and for whom location in the dispersal prison's segregation unit is not considered appropriate or practical either because of their capacity to exercise a disruptive influence from the unit or because their very location there might provide a focal point for unrest in the general prison population.[25]

The procedure to be followed in the event of such a short-term GOAD transfer is now set out in Circular Instruction 37/1990 ('The Transfer of Inmates in the Interests of Good Order or Discipline') which wholly supersedes CI 10/74, the Circular which gave rise to the decision in *Hague*.[26] The *Hague* case is central to this area of law, involving as it does a consideration both of the procedural fairness requirements attendant upon segregation/transfer and of the issue of private law remedies arising from its use.

The result of the *Hague* litigation was that the Court of Appeal held that CI 10/74 was unlawful in that it breached two of the protective provisions contained in Rule 43. In the first place, it sanctioned the practice of dispersal prison governors launching a prisoner on to what was frequently a succession of transfers in segregation with no subsequent decision ever taken by governors in the receiving prisons. As the prisoner arrived at the receiving prison he was immediately and automatically placed in segregation without any decision being taken by the receiving governor solely on the basis that he was a '10/74' man. The Circular led to the Rule 43 label attaching to a prisoner as he moved from prison to prison instead of requiring an assessment by each governor as he acquired custody of the individual concerned.

Secondly, it contemplated that authority to segregate for the full twenty-eight-day period would routinely be granted by the Secretary of State whenever a dispersal governor opted to use CI 10/74, wholly bypassing the role of the Board of Visitors and the requirement inherent in Rule 43 (2) that a reasoned decision had to be made not only as to whether authority for continued segregation should be granted but for how long it should be granted.

CI 37/90 remedies the unlawful advice set out in CI 10/74 pointing out:

(i) that while the dispersal prison governor may authorise segregation at the dispersal prison as an interim step before the transfer takes place, a decision as to segregation at the local prison must rest entirely with the local prison governor;

(ii) if, in the judgement of the local prison governor, the prisoner requires segregation beyond the end of that period the board of visitors must be asked to consider authorising continued segregation;

(iii) if at any time while the prisoner is detained at the local prison, the local

---

[25] CI 37/1990, para. 22.
[26] The full text of CI 37/1990 is to be found at Appendix 4.

prison governor concludes that segregation is no longer necessary, he or she must forthwith arrange for the prisoner to resume association[27].

Of course the chances of the local prison governor deciding not to order segregation upon receiving a prisoner transferred under CI 37/90 must in practice be slim, not least because, as paragraph 28 of the Circular emphasizes, if he decides not to authorize segregation it will be his responsibility to arrange other suitable accommodation at his establishment. The state of overcrowding in most local prisons makes it highly likely that Rule 43 segregation will frequently be the easier option for the harassed local prison governor faced with a prisoner already regarded as a control problem by the transferring governor with whom, almost invariably, he will have spoken.

CI 37/1990 makes it clear that the expectation is that at the end of the one-month 'cooling-off' period the prisoner will be returned to his dispersal prison but it is open to the dispersal governor to seek to arrange for a further transfer to a different prison or refer the prisoner for allocation to a Special Unit (see below). It is this possibility of persistent short-term transfers accompanied by Rule 43 segregation which has given rise to much criticism. The Woolf Report accepted the need for some kind of transfer option as a means of dealing with a specially difficult situation which the governor cannot resolve within his own prison but warned that the provisions of CI 37/90 needed to be closely monitored by the Prison Service to ensure that particular prisoners are not placed on a permanent 'carousel' within the prison system (para. 12.253).

PROCEDURAL FAIRNESS AND RULE 43

There is now a significant difference between what the law requires and what administrative policy demands in terms of the procedural fairness requirements attendant upon a decision to segregate a prisoner, whether or not such a decision arises following a transfer initiated for reasons of good order or discipline. As a matter of law, a prisoner has no right to be heard either before he is initially segregated by the governor under Rule 43 (1) or before authority to extend segregation beyond three days is sanctioned by a member of the Board of Visitors or the Secretary of State under Rule 43 (2).

In *Hague* both the Divisional Court and the Court of Appeal rejected the argument that the draconian effect of segregation against a prisoner's will justified a legal right to a hearing in order to respond to the proposed reasons for his segregation. At the time of Hague's appeal to the Court of Appeal, a governor's power to segregate was limited to an initial period of

[27] Ibid., para. 26

twenty-four hours and it was therefore conceded that this limitation, combined with the fact that segregation decisions would usually be urgent, made it impracticable to argue for a full-blown hearing before the governor's initial decision. At the stage of seeking authority for continued segregation however no such 'emergency' rationale was applicable and thus, it was submitted, an informal hearing before the Board of Visitors member was both practicable and necessary.

The Court of Appeal was unimpressed by this argument. Relying on dicta in *Payne* v. *Lord Harris of Greenwich* (a parole case), Lord Justice Taylor said that:

apart from the urgency of decisions under rule 43, there may well be other public policy grounds for not giving reasons in advance to the prisoner so as to enable him to make representations. Giving reasons would often require unwise disclosure of information. Such disclosure could reveal to prisoners the extent of the governor's knowledge about their activities. It could reveal the source of such information, thereby putting informants at risk. It could cause an immediate escalation of trouble.[28]

The court held further that the specific inclusion of natural justice requirements in Rule 49 of the Prison Rules and their absence from Rule 43 gave powerful support to the Home Office argument that the 'full panoply' of the rules of natural justice was not intended to apply to segregation decisions in contrast with disciplinary proceedings which may result in punitive action. Of course the 1990 amendment to Rule 43 (giving a governor the power to segregate a prisoner for up to three days without the need for authority from the Board of Visitors or the Secretary of State) now makes it that much harder to argue that at the stage of seeking authority for a continuation of segregation there could ever be any question of an urgent decision having to be taken which would be obstructed by a legal requirement to provide a hearing (however informally) to the prisoner concerned. Nevertheless it remains the case that prisoners have no right in law to be heard by either the governor or the Board of Visitors before it is decided to initiate or continue Rule 43 segregation.

As with the right to be heard, so too with the requirement to give reasons after a segregation decision has been made. The Divisional Court had held in *Hague* that the general duty upon a governor to act fairly when making a decision to segregate required, by necessary implication, adequate notice to the prisoner concerned within a reasonable time of the reasons for the decision. Without such an obligation it was pointed out that the pris-

---

[28] In R. v. *Parole Board and Home Secretary ex parte Wilson* [1992] QB 740 Lord Justice Taylor referred to *Payne* as a case 'decided in 1981 when established views on prisoners rights were very different from those of today' and ruled that it would be unjust to apply it in 1992 to deny a discretionary lifer access to reports and other material before the Parole Board when considering the question of releasing him on licence.

oner's right to complain to the Board of Visitors against his segregation under Rule 95 (1) would be rendered ineffective.[29]

The Court of Appeal reversed this favourable finding. Pointing out that there was no general rule in public law that reasons be given for administrative decisions, it held that the same policy reasons which made it inappropriate to give reasons before a segregation decision was made applied with equal force after the event.

There was no appeal to the House of Lords on any of the procedural fairness issues because, soon after judgment was delivered by the Court of Appeal, the Prison Department replaced CI 10/1974 with CI 37/1990. The new Instruction provides clear and emphatic guidance to governors about the need to give reasons for both a transfer and a segregation decision. Paragraph 13 states that 'inmates must be told the reasons for their transfer as far as is practicable and as soon as possible. In general this should be done before, or at the time of, transfer.' It goes on to explain that prisoners have no absolute right to be informed of a decision to transfer them before it is taken or even before or at the time the transfer takes place and that control or security factors may properly cause a deferral of the giving of reasons.

Once transfer has taken place, the Instruction extends the obligation to give reasons so that 'an inmate who is segregated at the receiving establishment must be told the reasons as far as is practicable and as soon as possible. In general, this should be done before, or at the time that, the inmate is placed in conditions of segregation' (para. 14). Where reasons for transfer or segregation are not given before or at the time of its implementation, they must be given 'as soon as possible thereafter and at the latest within 48 hours of the transfer or segregation taking place unless, in a most exceptional case, the inmate's demeanour necessitates delay—in which case at the earliest opportune time' (para. 15). Finally the Instruction requires that the reasons must be given by a member of staff not lower than governor V level who is familiar with the facts of the case. Reasons must be given in writing if requested and should be as full as is reasonably possible subject always to the interests of good order, discipline, and security (para. 16).

Similar guidance is contained in CI 26/1990, which applies to all decisions to segregate irrespective of whether they follow a transfer. The 1991 addendum to CI 26/1990 accepts and adopts the general recommendation of the Woolf Report about the importance of giving reasons to prisoners for important decisions, stating that:

in future, any inmate who is segregated must be advised in writing of the reasons as far as is practicable, and as soon as possible . . . An inmate who is not given reasons before or at the time must be given reasons as soon as possible thereafter (and

---

[29] [1992] 1 AC 58 at 94A.

at the latest within 24 hours unless, in a most exceptional case, the inmate's demeanour necessitates delay. (paras. 14–15)

Importantly, paragraph 16 requires a copy of the written reasons to be held in the segregation unit so that it is readily available for prison staff and members of the Board of Visitors when visiting.

There is a discrepancy between CI 37/1990 and CI 26/1990 over the maximum time (forty-eight as opposed to twenty-four hours) which may elapse before reasons are given other than in the most exceptional case, but now that governors have power to segregate for up to three days this makes little difference. The important point so far as a prisoner is concerned is that by the time a decision has to be taken by the Board of Visitors member (or exceptionally the Secretary of State) about continuing segregation, the prisoner concerned must be in possession of written reasons for his segregation. This at least ensures that meaningful representations can be made to whomever is considering an extension of segregation.

CI 26/1990 falls short of requiring the Board of Visitors member to provide a hearing to a prisoner before authorizing continued segregation but comes close to it. In the 1991 addendum the Prison Department states that:

a member of the Board of Visitors whose authority is being sought for continued segregation should be encouraged to see and speak with the inmate, if necessary at some length, before deciding whether to give such authority. In addition Board members should be invited to see and speak with any inmates under segregation in the course of regular rota visits . . . even when cases are not due for formal review.

The language of 'encourage' and 'invite' properly reflects the fact that Board members are not subject to Prison Service instructions but the intention is clear.

The effect of the administrative guidance contained in CI 37/1990 and CI 26/1990 is that, although prisoners may not, post-Hague, assert a legal right to be given reasons for a transfer or segregation decision or a right to be heard before segregation is continued on the authority of a member of the Board of Visitors, they do now have a legitimate expectation of such procedural protections. It follows that a failure to give reasons for a transfer/ segregation decision after forty-eight hours or a failure to provide some kind of hearing to a prisoner before authorizing continued segregation beyond three days could leave the prison authorities vulnerable to a successful application for judicial review.[30]

ALLOCATION TO A SPECIAL UNIT

The third option available to prison management when seeking to deal with an allegedly disruptive prisoner is placement in one of the three

---

[30] See the passage in *Halsbury's Laws* (4th edn.), i(1): *Administrative Law*, at para 81.

Special Units at Hull, Parkhurst, and Milton Keynes prisons. The idea of developing Special Units came from the 1989 Report of the Control Review Committee (*Managing the Long Term Prison System*), which addressed the question of how to deal with the minority of prisoners whose control problems could not be dealt with in normal conditions. In September 1990 there were a total of thirty-four prisoners in such Units out of forty-five available places. Selection for the Units is slow and exhaustive. While in the Unit, prisoners are not subject to Rule 43, although security is intense. The Woolf Report described the Units as attempting 'to provide a generous regime to counterbalance their claustrophobic environment' with a consequent need for intensive staffing, and commented that 'there is more than a philosophical difficulty about apparently providing the best conditions and facilities for those who behave the worst'. While the case for some form of Special Unit was accepted, Woolf recommended that the aim should be to return the prisoners concerned to 'normal' prison life before release.

APPROVED RESTRAINTS AND SPECIAL CELLS

In addition to powers to segregate and transfer prisoners in the interests of good order or discipline, the Prison Rules contain two emergency provisions (Rule 45 and 46) which enable prison management to deal with an allegedly violent, disruptive prisoner by confining him temporarily in special accommodation or subjecting him to an approved method of restraint. The two powers can be—and frequently are—used in tandem.

Their use must be seen in the context of the general principle governing the use of force against prisoners by prison staff to be found in Rule 44:

44 (1) An officer in dealing with a prisoner shall not use force unnecessarily and, when the application of force to a prisoner is necessary, no more force than is necessary shall be used.
(2) No officer shall act deliberately in a manner calculated to provoke a prisoner.

It is to be noted that nothing in the Prison Act or the Rules stipulates *when* force is necessary. Prison officers have the powers of constables[31] (who may use reasonable force when acting in the execution of their duties) and, though prison officers enjoy no special power authorizing the use of force, it must be assumed that they may use force against prisoners in so far as it can be shown to be reasonably necessary to enforce lawful orders or regulations pursuant to the Act and the Rules. It has been said that a lawful order 'is presumably one which can derive support from the Prison Rules

[31] Prison Act 1952, s. 8.

or any rule or regulation of the prison or is necessary to the discharge of the proper functions of the prison authorities in keeping the prisoners in custody in a safe, orderly and efficient manner'.[32]

Accordingly such guidance as is given to staff in Standing Orders and Circular Instructions about dealing with violent or refractory prisoners is merely illustrative of the general principle that only such force as is reasonably necessary to effect a lawful purpose (including, of course, self-defence) may be used at any time. Force may never however be used to punish a prisoner, corporal punishment for disciplinary infractions having been abolished.[33]

SPECIAL CELLS

Section 13 (6) of the Prison Act provides that 'In every prison special cells shall be provided for the temporary confinement of refractory or violent prisoners.' (It was presumably thought that the use of special cells for this purpose required express parliamentary sanction over and above the general power in s. 47 (2) to make rule for the 'discipline and control' of prisoners.) Accordingly Rule 45 states that: 'The governor may order a refractory or violent prisoner to be confined temporarily in a special cell, but a prisoner shall not be so confined as a punishment, or after he has ceased to be refractory or violent.' As with Rule 43 it is immediately clear that a prisoner may be placed in a special cell under this provision without any disciplinary proceedings being instituted against him. Indeed, the accommodation may never be imposed as a punishment following conviction for a disciplinary offence precisely because it is regarded as fit only for temporary use to deal with an emergency situation. However, unlike Rule 43, Rule 45 may not be used pre-emptively. Before a prisoner can lawfully be placed in a special cell he must actually be behaving in a 'refractory or violent' manner. It is not enough that a prison officer anticipates such behaviour. And so if prison staff choose to act under this power before such behaviour has been exhibited, their actions would be unlawful and any force used to move the prisoner concerned to the cell would render them vulnerable to an action for assault.

Guidance on the use of Rule 45 is to be found in Standing Order 3E

---

[32] *Halsbury's Laws* (4th edn.), vol xxxvii, para. 1163.

[33] CI 44/1981 requires prison officers who resort to force when dealing with a prisoner to complete a Use of Force Report which should be a 'brief, factual report to the governor explaining why it was necessary to use force, the kind of force and how much force was used'. The Circular goes on to state that if one officer decides to submit such a report all officers involved should do so. Reports should always be submitted if it is believed that the inmate has been or could have been injured or if any of the officers present has been injured. Use of Force Reports are intended to be in addition to filling out an F213 (Report of Injury to an Inmate).

('The Management of Violent or Refractory Prisoners')[34] as supplemented by Circular Instruction 55/1990. Confusingly the Order refers in general terms to 'special accommodation' rather than the term used in the Rule ('special cell') and then proceeds to describe 'special accommodation' as including two types of cell—a cell which has been designated as a special cell and an unfurnished cell.

In practice a special cell is usually equipped with a double door, has no access to natural light, is virtually sound-proof, and is much smaller than a normal cell. Instead of a bed, there is a raised wooden platform. Prisoners refer to it as 'the strong box' or 'the box', a description which more accurately describes the degree of sensory deprivation involved.

An unfurnished cell is stated in SO 3E to be either:

a cell which is specifically designated as one to be used exclusively as an unfurnished cell for temporary confinement . . . or a cell which is designated and usually used for ordinary accommodation purposes but from which the usual furniture has been removed . . . and which is either totally unfurnished or does not contain basic items of furniture such as a table and chair and is thus to be considered, for all practical purposes, to be unfurnished.

This somewhat verbose definition appears to suggest that any cell can temporarily be used to confine a prisoner under Rule 45 simply by removing a table and chair even if it is the prisoner's normal cell. This is surely not the purpose of the Rule 45 power and in practice most prisoners who are believed to be so 'refractory or violent' as to justify the use of Rule 45 are placed in a special cell on the punishment block.

A special cell has no medical justification and requires no medical authorization in order to justify its use. It is a control measure which, according to SO 3E, may only be used 'if it is necessary in order to prevent the prisoner causing self-injury, injuring another prisoner or staff, or damaging property, or creating a serious disturbance'. These criteria are almost identical to those contained in Rule 46 (1) which governs the use of restraints (see below). Although Rule 45 speaks of the governor ordering confinement in a special cell, the emergency nature of the power necessarily means that the initial decision will almost always be taken by prison officers. SO 3E states that where the governor in charge cannot be contacted the decision can be taken at the level of a senior officer, who must refer his decision at the earliest opportunity to the governor. The Standing Order also sets out detailed requirements for the recording of all decisions under Rule 45, for notifying the medical officer to enable an early examination of the prisoner, and also for informing the Board of Visitors so that a member may visit and enquire.

Rule 45 expressly forbids the use of special cell accommodation after the

---

[34] The full text of SO 3E is to be found at Appendix 3.

prisoner has ceased to be refractory or violent, but SO 3E states that if it is sought to continue its use beyond twenty-four hours, the written authority of the Board of Visitors must be obtained. The temporary, emergency nature of the power makes it hard to see how detention beyond twenty-four hours could ever be justified, particularly in the light of medical evidence suggesting that whatever calming effect the special cell may initially produce it tends to exacerbate aggressive feelings when used for too long.

Whenever a prisoner is placed in a special cell, it seems to be common practice in most prisons for him routinely to be stripped of all his clothing (or at least to his underpants) and given paper 'protective clothing'. It is hard to see that this practice can ever be justified save in circumstances where there is a risk of the prisoner injuring himself or committing suicide by hanging with the clothing, in which case it hardly seems suitable to use the special cell at all. SO 3E states that 'where a prisoner is placed in special accommodation or in a mechanical restraint . . . and, as part of the protective arrangements, it is necessary to deprive the prisoner of normal clothes, the prisoner should be provided with a suit of protective clothing of an approved type so that he or she can be decently dressed'. The important word here is 'necessary'. A prisoner who is forcibly stripped for no reason other than the fact that he is being placed in a special cell could legitimately sue for assault unless it can be shown that removing his ordinary clothing is clearly necessary to prevent him harming himself. Forcing a prisoner to wear paper clothing invariably increases the sense of humiliation and degradation brought about by location in a special cell and, where it is not necessary, invariably gives substance to the charge that the purpose is to punish.

RESTRAINTS

The use of mechanical methods of restraint is expressly sanctioned by Rule 46 though, unlike the issue of special cells, the Prison Act is silent on the subject, leaving it as a matter within the general rule-making power to maintain discipline and control. It is appropriate to set out Rule 46 in full:

46 (1) The governor may order a prisoner to be put under restraint where this is necessary to prevent the prisoner from injuring himself or others, damaging property or creating a disturbance.
(2) Notice of such an order shall be given without delay to a member of the board of visitors, and to the medical officer.
(3) On receipt of the notice the medical officer shall inform the governor whether he concurs in the order. The governor shall give effect to any recommendation which the medical officer may make.
(4) A prisoner shall not be kept under restraint longer than necessary, nor shall he be so kept for longer than 24 hours without a direction in writing given by a

member of the board of visitors or by an officer of the Secretary of State (not being an officer of a prison). Such a direction shall state the grounds for the restraint and the time during which it may continue.

(5) Particulars of every case of restraint under the foregoing provisions of this Rule shall be forthwith recorded.

(6) Except as provided by this Rule no prisoner shall be put under restraint otherwise than for safe custody during removal, or on medical grounds, by direction of the medical officer. No prisoner shall be put under restraint as a punishment.

(7) Any means of restraint shall be of a pattern authorised by the Secretary of State, and shall be used in such manner and under such conditions as the Secretary of State may direct.

As with special cells, administrative policy on the use of approved restraints is contained in Standing Order 3E. It lists four approved methods of mechanical restraint:-

1. a body belt with iron cuffs for a male prisoner; with leather cuffs for a female prisoner;
2. standard handcuffs, for a male prisoner;
3. leather wrist straps, for a female prisoner;
4. ankle straps, for a male or female prisoner.

No other method of restraint may lawfully be used against a prisoner. Lest there be any doubt about the exceptional nature of this power, SO 3E states that

the use of mechanical restraints is particularly undesirable and every effort must be made to avoid recourse to them. The use of special accommodation as an alternative to mechanical restraints should be considered before any decision is taken to use [them] . . . This is not intended to preclude the use of mechanical restraint and special accommodation in conjunction, where the circumstances of the case so demand.

In fact it is by no means unusual for a restraint to be applied to a prisoner inside a special cell and in many cases it is hard to see how this can be justified. For example, a disgruntled prisoner emerges from his cell in the segregation unit and in a fury empties the contents of his chamber pot over a member of staff but then shows no further aggressive intent and moves to return to his cell. He has certainly committed a disciplinary offence and may properly be charged with assaulting the officer concerned. But would staff be justified in grabbing him and then dragging him to a special cell before unceremoniously incarcerating him inside a body belt? The answer must be 'No'. However enraged they may be, they are not entitled to mete out summary justice in advance of the governor's adjudication. In any event neither the special cell nor the body belt can be used to punish. Since the prisoner has ceased to behave in a 'violent or refractory' manner and indicated his desire to return to his cell, any force applied to him would be unlawful and certainly an effort forcibly to drag him to a special cell and place him in a body belt would be incapable of justification. Yet there are many recorded examples of just this kind of behaviour by a prisoner prompting a show of force on the part of staff, who no doubt feel that a single act of violence against them may justify immediate relocation to a special cell and, if their efforts meet with resistance, the application of a body belt to protect them on leaving the cell.

The case of *Rodrigues* v. *Home Office*[35] is a good illustration of the need for strict justification of the use of powers contained in Rules 45 and 46 and indeed is the only example of a prisoner successfully suing the Home Office for damages for assault arising from the use of an approved method of restraint. The plaintiff (R) was a 23-year-old prisoner serving a five-year prison sentence at Wandsworth Prison for a variety of offences of dishonesty and violence. He was regarded by the authorities as a control problem and had spent a number of weeks in segregation when an incident occurred which led to him being taken to a special cell. When he was unlocked for exercise he became annoyed about a request to search him and slammed his cell door on a prison officer, slightly injuring the officer's hand. Notwithstanding that R was not offering any further violence and indeed was now inside his cell, the senior officer on duty ordered that he be restrained and taken to the special cell, where he was stripped to his underpants. Later the same day, the deputy governor and medical officer visited R in the special cell and when the door was opened he lunged towards the senior officer and threw a piece of broken floor tile. A struggle ensued and R was restrained on the floor before he was put in a body belt. He remained in the belt for approximately twenty-four hours. The opinion of psychiatrists on behalf of both R and the Home Office was that the average time for detaining any person in a body belt was two to three hours and that its use was justified only as a last resort when it was immediately necessary to prevent a prisoner from behaving violently. Her Honour Judge Norrie held, giving judgment for R, that although the initial decision to place R in the body belt was reasonable because of his violent behaviour, staff were under a duty to remove it as soon as possible. She stated that use of a body belt was a wholly exceptional measure and that mere threats and abuse by a prisoner were not enough to justify its application. The decision taken by the deputy governor to order restraint overnight for a twelve-hour period without review was unlawful. On the evidence of prison staff, R's behaviour thereafter did not justify the use of the body belt. It was particularly significant that on three separate occasions staff removed a single arm from the belt in order to let R feed himself while still detained in the special cell

---

[35] A decision of a County Court judge (Her Honour Judge Norrie), 16 Dec. 1988, Croydon County Court. There is a brief case report in the LAG Bulletin Feb. 1989, at 14.

and no violence took place but still staff restored the restraint. They claimed that R remained a threat to staff but could only point to abusive language as evidence of this. R was awarded £750 in damages for assault based on a period of approximately twenty hours wrongful detention in the body belt.[36]

The case shows that prison management must provide an effective procedure for the constant monitoring of any prisoner detained in a body belt and must persist in efforts to remove it right from the outset. There is no standard, reasonable period which may be imposed to give 'a taste of restraint' and identical reasoning applies to location in a special cell.

The use of mechanical and non-mechanical restraints (special cells, protective rooms, and body belts, etc.) has grown considerably over the past decade. In 1981 a total of 703 male prisoners were subject to some form of restraint whereas in 1990 this had increased to 1,618. Of these figures the overwhelming majority were on non-medical grounds which indicates either a greater willingness by staff to resort to extreme methods of control or a sharp increase in violent, disruptive incidents involving prisoners. The current Chief Inspector of Prisons has frequently expressed concern that staff do resort to control and restraint techniques, as well as the use of body belts, too often and that 'use of force' reports are not correctly filled out by staff, thus making it difficult properly to assess if staff had acted reasonably.

EUROPEAN AND AMERICAN LAW

Article 3 of the European Convention on Human Rights, which prohibits the infliction of torture or inhuman or degrading treatment or punishment, has not hitherto proved to be an effective protection against the use of non-disciplinary sanctions such as long-term segregation. Since the European Court has defined torture as 'deliberate inhuman treatment causing very serious and cruel suffering', those cases which have reached the Commission or the Court arising from the use of non-punitive solitary confinement have focused on what constitutes inhuman or degrading treatment. In general it is clear that there is no absolute standard for the kinds of treatment and punishment prohibited by Article 3. In its ruling in *Ireland* v. *United Kingdom*, the Court said that:

ill treatment must attain a minimum level of severity if it is to fall within the scope of Article 3. The assessment of this minimum is, in the nature of things, relative; it depends on all the circumstances of the case, such as the duration of the treatment,

---

[36] In the course of the case Judge Norrie accepted the invitation of counsel for the plaintiff to visit Wandsworth Prison and actually see the special cell. On arrival at the prison she asked to be placed inside a body belt and then to be locked inside the cell for a few minutes to gain some appreciation of what the experience entailed.

its physical or mental effects and, in some cases, the sex, age and state of health of the victim.[37]

More specifically the Commission has defined inhuman treatment or punishment as any treatment which 'deliberately causes severe suffering, mental or physical, which in the particular situation is unjustifiable'.[38] Degrading treatment or punishment is caused 'if it grossly humiliates a person before others or drives him to act against his will or his conscience'[39] or if it 'constitutes an insult to the applicant's human dignity'[40] or if it is 'designed to arouse in victims feelings of fear, anguish and inferiority capable of humiliating and debasing them and possibly breaking their physical or moral resistance'.[41]

The Commission has generally taken the view that solitary confinement is in principle undesirable, particularly where the prisoner was detained on remand, and requires exceptional reasons for its imposition. Regard must be had to the surrounding circumstances including the particular conditions, the stringency of the measure, its duration, the objective pursued, and its effects on the person concerned. Also relevant is the extent to which a basic minimum possibility of human contact has been left to the detainee.[42] In *Krocher and Moller* v. *Switzerland* the Commission stated: 'The question that arises is whether the balance between the requirements of security and basic individual rights was not disrupted to the detriment of the latter.'[43] In that case the prison conditions included isolation, constant artificial lighting, permanent surveillance by closed-circuit television, denial of access to newspapers and radio, and the lack of physical exercise. While expressing serious concern about the compatibility of such a combination of measures with the terms of Article 3, the Commission concluded that they did not amount to inhuman or degrading treatment and accepted the argument that such conditions were necessary to ensure security inside and outside the prison. The applicants were considered to be dangerous terrorists who were liable to escape if given the chance. In other cases the Commission has accepted that stringent security measures within a prison may be justified where a prisoner was 'extremely dangerous',[44] 'able to manipulate situations and encourage other prisoners to acts of indiscipline',[45] or where he

---

[37] 2 EHRR 25.   [38] *Greek* case Yearbook 12 (1969) 186.   [39] Ibid.
[40] Appl. 8930/80, *X, Y & C* v. *Belgium* unpublished.
[41] *Ireland* v. *United Kingdom* 2 EHRR 25 at para. 167. See also *Soering* v. *United Kingdom* (1989) 11 EHRR 429 at para. 100.
[42] Appl. 6038/73, *X* v. *Federal Republic of Germany*, Coll. 44 (1973) 115; Appl. 6166/73, *Baader, Meins, Meinhof and Grundmann* v. *Federal Republic of Germany*, Yearbook 18 (1975) 132 (144–6); Appls. 7572, 7586, and 7587/76, *Ensslin, Baader and Raspe* v. *Federal Republic of Germany* Yearbook 21 (1978) 418 (454–60); Report of *Krocher and Moller* v. *Switzerland* D. & R. 34 (1983) 24 (51–5).
[43] D. & R. 34 (1983) 24 (51–5).
[44] Appl. 9907/82, *M* v. *United Kingdom* D. & R. 35 (1984) 13 (34).
[45] Appl. 8324/78, *X* v. *United Kingdom* unpublished.

had used firearms at the time of his arrest. It is also clear that a relevant factor taken into account by the Commission is whether the measures imposed on a prisoner are the result of a deliberately uncooperative attitude on the part of the applicant.

It is clear that any Article 3 claim will require cogent medical evidence to establish not merely a direct causal link between the prison conditions complained of and the injured or deteriorating health of the applicant but also that the conditions were such as to 'destroy the personality and cause severe mental and physical suffering'.[46] Of course it may be difficult for a doctor or psychiatrist who has no knowledge of a prisoner's mental or physical health when he was at liberty to assess the precise impact of solitary confinement upon him as opposed to, say, the general effects of long term imprisonment. In the case of *Ensslin, Baader and Raspe*, all members of the German terrorist group the Red Army Fraction who were held in high security conditions in Stammheim Prison, the Commission said that medical reports did not 'make it possible to establish accurately the specific effect of this isolation in relation to their mental and physical health, as compared with other factors'.[47]

The Commission has ruled that complete absolute sensory isolation combined with complete social isolation does constitute inhuman treatment for which no security considerations can be prayed in aid by way of justification. In making this clear however a series of decisions has distinguished this form of treatment from 'removal from association with other prisoners for security, disciplinary and protective reasons', which does not by itself constitute conduct in breach of Article 3.[48] The opportunity on Rule 43, for example, to meet prison officers, medical officers, lawyers, and relatives and to have contact with the outside world via newspapers and radios prevents it from amounting to complete social or sensory isolation. A depressing feature of the Commission's approach to Article 3 cases has been its reluctance to require strict compliance with modern penological views as set out in the Minimum Rules for the Treatment of Prisoners (1973) and, more recently, the European Prison Rules.[49] In the case of *Eggs* v. *Switzerland*, for example, the Commission took the position that 'the conditions of detention which in certain aspects did not come up to the standard of the "Minimum Rules" did not thereby alone amount to inhuman or degrading treatment'.[50]

---

[46] Appls. 7572/76, 7586/76 7587/76, *Ensslin, Baader and Raspe* v. *Federal Republic of Germany* Yearbook 21 (1978) 418 (454).

[47] Ibid.

[48] Report of 25 Jan. 1976, *Ireland* v. *United Kingdom* B. 23/I, 379; Appl. 8317/78, *Mcfeeley* v. *United Kingdom* D. & R. 20 (1980), 44 (82); report of 16 Dec. 1982, *Krocher and Moller* D. & R. 34 (1983), 24 (53); and Appl. 10263/83, *R* v. *Denmark* D. & R. 41 (1985), 149.

[49] The European Prison Rules were adopted by the Committee of Ministers on 12 Feb. 1987.   [50] Appl. 7341/76, Yearbook 20 (1977) 448 (460).

The American courts have in recent years shown a similar reluctance to intervene in special prison regimes imposed for non-disciplinary reasons. During the early 1970s a number of federal district courts in the USA established minimum procedural fairness ('due process') requirements for non-disciplinary transfers either into administrative segregation within the same prison or from one prison to another ('interstate transfers'). In some cases, it was held that the same safeguards which applied in formal disciplinary proceedings should be complied with. For example in *Gomes* v. *Travinso* the First Circuit District Court held that due process for the interstate transfer of a prisoner required (1) notice that a transfer is contemplated, (2) notice of the reason for the proposed transfer, (3) a personal hearing before a decision maker, and (4) a reasonable opportunity, at the hearing, to controvert actual assertions concerning the prisoner that have been advanced in support of the transfer.[51] The rationale for this approach was that decisions to classify, reclassify, and transfer prisoners could significantly affect their eligibility for parole or their access to the most advantageous job and accordingly it was essential that decision-making was not arbitrary, capricious, or discriminatory.

In two decisions in 1976 the Supreme Court held that the transfer of a prisoner from one prison to another within the same prison system, whether or not for punitive reasons, did not infringe a 'protected liberty interest'.[52] Such decisions were discretionary matters for the prison authorities to decide and prisoners could not assert any right or legitimate expectation that they would not be transferred. In the same year however a Californian District Court held in *Wright* v. *Enomoto* that:

when a prisoner is transferred from the general prison population to the grossly more onerous conditions of maximum security, be it for disciplinary or for administrative reasons, there is severe impairment of the residuum of liberty which he retains as a prisoner.[53]

The court went on to hold that the procedural safeguards applicable to such transfers had to be no less rigorous than those required in transfers for disciplinary reasons.

This progressive analysis was ultimately rejected by the Supreme Court in the 1983 decision of *Hewitt* v. *Helms*, which remains the leading authority.[54] The Court held by a 5 : 4 majority that there was no independent liberty interest at stake for a prisoner in his being subjected to administrative segregation. Hence the due process clause of the Fourteenth Amendment to the Constitution (which provides that no one may be deprived of liberty or

[51] 510 F. 2d 537 (1st Cir 1974) 238.
[52] *Meachum* v. *Fano* 427 US 215, 96 S. Ct. 2532, 49 L. Ed 2d 451 (1976) 239, 261 and *Montayne* v. *Haymes* 427 US 236, 96 S. Ct. 2543, 49 L. Ed. 2d 466 (1976) 239, 261.
[53] 462 F. Supp. 397 (DC Calif. 1976).
[54] 459 US 460, 103 S. Ct. 864, 74 L. Ed. 2d 675 (1983) 122, 240, 314.

property without due process of law) did not apply to a prisoner who had been segregated prior to a disciplinary hearing following a prison riot. Chief Justice Rehnquist stated that such segregation was 'the sort of confinement that inmates could reasonably anticipate receiving at some point in their incarceration'. However, the majority did hold that if the individual state regulations created an expectation in the prisoner that he would not be transferred without some sort of formal process then he would have a liberty interest which required due process protection. The Supreme Court found that the Pennsylvania statute involved did create such an expectation (it stated that an inmate would only be transferred if there was a threat of a serious disturbance or a serious threat to the individual or others). The court then went on to consider what due process required in such a case. The majority held that this was satisfied if there was an informal, non-adversarial review. All the prison authorities had to do was give the prisoner notice of the case against him and allow him to make written representations. Even the requirements of written representations could be waived if the authorities felt oral representations would be sufficiently informative. In a strong dissenting opinion, Justice Stevens felt that the due process clause was implicated at any time a prisoner was singled out for significantly worse treatment and that due process required at least a periodic review of the segregation together with a written statement of reasons to the prisoner for such continuation.

While the Rehnquist view in *Hewitt* remains the law, courts have subsequently shown sympathy for aspects of Justice Stevens's dissent. For example the Supreme Court has not reneged on the view it expressed in *Vitek* v. *Jones* that where prisoners are transferred to a totally different environment (from a prison to a mental hospital) they are entitled to a full range of procedural protections, including the right to counsel.[55] This was because commitment to a mental hospital could 'engender adverse social consequences' and 'stigma'. Other courts have ruled that segregation into 'intolerable conditions' may breach the Eighth Amendment to the Constitution which prohibits cruel and unusual punishment and may furnish grounds for release from such conditions under a habeas corpus petition.[56]

The Supreme Court refused to extend the *Vitek* v. *Jones* principle to interstate transfers of prisoners. In *Olim* v. *Wakinekona* the court was asked to rule on whether the transfer of a Hawaii prisoner to a Californian prison without a hearing by an impartial committee (as required by Hawaii prison regulations) violated the prisoner's due process rights.[57] It held that, even though the transfer involved long distances, it was more akin to a transfer from one prison to another within the same state than a transfer to

---

[55] 445 US 480, 100 S. Ct. 1254, 63 L. Ed 2d. 552 (1980), 241, 261, 268.
[56] *Howard* v. *Wheaton* 668 F. Supp. 1140 (ND Ill. 1987).
[57] 461 US 238, 103 S. Ct. 1741, 75 L. Ed 2d 813 (1983) 241.

a mental hospital. Justice Marshall dissented and pointed out that the 4,000-mile transfer was far more drastic than that which ordinarily accompanies imprisonment and amounted, in effect, to banishment from his home. (A similar feeling is experienced by prisoners from the North who find themselves serving lengthy terms of imprisonment on the Isle of Wight.)

The use of separate prisons to house prisoners officially categorized as violent or disruptive has been a growing feature of the US prison system over the last decade. The regimes in such units are particularly harsh with prisoners locked in their cells for twenty-three hours per day. The one hour out of cell is for specific purposes only such as a family or medical visit, and at all times the prisoner is heavily shackled and closely escorted by two or three prison officers. While in their cells only minimal facilities are afforded. Prison officials openly concede that the only purpose of this level of security is to inflict punishment.

The National Prison Project of the American Civil Liberties Union has described the expansion and widespread acceptance of 'supermax confinement' as a 'bleak, damaging and potentially dangerous prison practice'.[58] And the Human Rights Watch Report on *Prison Conditions in the United States* has referred to the 'Marionization' of American prisons, a reference to the US Penitentiary in Marion, Illinois which was long considered the toughest prison in America. An example of this trend was the construction of the Pelican Bay prison complex in California in the late 1980s which included a special unit to house 1,000 prisoners. The *California Prisoner* reported that a form of high-tech incarceration had been created in the unit in which 'prisoners are watched on screens in a central control room. Their movements are monitored by video cameras. Doors open and close electronically. Prisoners move at verbal commands issued over a loudspeaker. The cells have no windows and a steel door with rows of 2-inch round holes.' No equivalent regime affecting so many prisoners exists at the moment within the prison system of England and Wales but the lesson from existing legal rulings in the domestic courts and in the ECHR is that too frequently the balance between the perceived requirements of security and basic individual rights will be determined in favour of the former.[59]

---

[58] *National Prison Project Journal*, 7/4 (Fall 1992).

[59] A striking contrast to the more conservative approach of the domestic courts, the ECHR, and the American courts can be seen in the decision of the Supreme Court of South Africa in *Whittaker & Morant* v. *Roos* [1912] Appellate Division 92. The plaintiffs were charged with conspiring to cause explosions in Johannesburg during the course of a bitter strike on the tramway system and were placed in solitary confinement cells for some two weeks while awaiting trial. Under prison regulations, the cells were exclusively reserved to punish convicted prisoners and the plaintiffs were awarded damages for their unlawful confinement in actions in delict (the equivalent in Dutch Roman law of a tort). In an extraordinary judgment Lord de Villiers, the Chief Justice, said that 'It surely cannot be the policy of the law that wealth or

position should give . . . advantage to an accused person. It often acts harshly enough that a man without money or friends is kept imprisoned until he can be tried, while the man of position is free until tried with every opportunity of preparing his defence. That harshness would become absolutely intolerable if directors of prisons or governors of gaols had it within their power to add to the misfortune of imprisonment the disgrace and torture of confinement in a small punishment cell of four feet by nine without sufficient light, with no bed to sleep upon, with insufficient means of exercise, without books to read, without the companionship of a single soul and without even the consolation afforded by a smoke. The object of the imprisonment before trial is to obtain the appearance of the accused at the trial, and not to punish him'. It is a sobering thought that a South African court in 1912 was willing to award a remedy in damages to prisoners for an experience which today would almost certainly not result in similar compensation from either the domestic courts or the ECHR.

9

Release from Prison

INTRODUCTION

For all prisoners the fundamental concern is when liberty will be regained. Most prison systems have some mechanism whereby release can be achieved earlier than the sentence of the court would appear to demand, either by a scheme of automatic remission of sentence or by selective release on parole or both in combination.[1] In this chapter we examine the existing schemes for determining the early release of prisoners in England and Wales which have resulted from the provisions of the Criminal Justice Act 1991 ('the 1991 Act'). The 1991 Act, as it affects life sentence prisoners, was itself the product of two important decisions in the European Court of Human Rights ('ECHR'). Before analysing the current schemes, we will examine how remission and eventually parole came to be standard features of a sentence of imprisonment. This exercise is not merely of historical interest. Many of the principles articulated by the courts in relation particularly to parole will remain relevant as the courts grapple in future with the new provisions.

THE NINETEENTH-CENTURY REFORMS[2]

The system of transporting English criminals to America before the Revolutionary War of Independence and later to Australia in the nineteenth century demonstrated the potential benefits of the early release of prisoners from a sentence served entirely in prison. Transportation involved a staged process of detention abroad in the colonies during which discipline was gradually eased. It culminated in the grant of a 'ticket of leave', an 'indulgence' dependent on good behaviour which enabled a convict to attain a large degree of freedom within the colony, though he could not leave until

---

[1] In almost every European jurisdiction there is a single early release mechanism rather than a parole and remission system working in tandem. In the USA the parole system has been under attack for the past fifteen years but no state has abolished parole altogether since 1984. Despite recent reforms, Canada retains a parole/remission scheme which most closely mirrors the pre-1991 system in England and Wales.

[2] Sean McConville's fascinating study of the development of the English prison system from 1750 to 1948, *A History of English Prison Administration* (London, 1981), is the basis for this summary of the gestation period of parole and remission.

his sentence had ended. The success of the system convinced prison legislators that such a release scheme was not only practical but also cheaper than continued imprisonment to the end of the sentence.[3]

The phased abandonment of transportation, which began during the 1840s, and the introduction of the sentence of penal servitude in 1853 meant that new measures had to be designed to deal with the release in the home country of convicts who were previously subject to transportation. The issue of allowing convicts to be released at home was politically highly sensitive.[4] Prior to the 1850s a small number of convicts had been released directly from the Hulks either as a result of pardons or, in the case of those serving short sentences, upon the expiry of their sentences.[5] These convicts were those who were unfit for transportation and a practice developed whereby they were recommended for pardons after serving half their sentences. But the idea of releasing at home large numbers of convicts before their sentences had ended was something which shocked and alarmed wide sections of public opinion who were largely unaware that such a scheme had operated in the colonies under transportation.

During the early 1850s the guiding philosophy of the recently formed Directorate of Convict Prisons under its Director, Sir Joshua Jebb, was to structure the sentence of penal servitude so as to keep alive 'an invigorating hope and a salutary dread at every stage of the progress of the prisoner'.[6]

---

[3] The 1779 Gaols Act had helped to establish imprisonment with hard labour as a standard punishment for felony. The 'discovery' of Australia by Cook led to the use of transportation to Australia, which began in 1787 and continued to be the mainstay of penal treatment for long term prisoners ('convicts') up until 1857.

[4] In 1847 the Home Secretary, Sir James Graham, had warned reformers of the danger to public security if transportation was abandoned and replaced by releasing convicts at home: 'I fear that the British public, which has the expense, will lose all the benefit of annual relief to Society from the Transportation or Exile of Criminals, if yearly . . . the Ranks of the Thieves and Cutthroats are to be recruited by public servants from our Hulks and Gaols. The new System may be less expensive; but I doubt whether the Community will gain, if the value of Life and Property is to be considered' (quoted in McConville, *A History of English Prison Administration*, i. 387).

[5] The American War of Independence brought an end to transportation in 1776 and so the government invented the Hulks as an alternative. They were phased out in 1857. More recently in the 1980s the government resorted to using a former cross-channel ferry to house asylum seekers who were being detained pending a decision on their claims, an unfortunate reminder of Victorian values.

[6] In 1842 Lord Stanley, the Colonial Secretary, had issued policy instructions in respect of the organization of the system of transportation indicating that he did not 'contemplate a state of things in which the convict, suffering under the sentence of the law, should ever be excluded from the hope of amending his condition by blameless or meritorious behaviour, or from the fear of enhancing the hardships of it by misconduct. On the contrary, to keep alive an invigorating hope and a salutary dread at every stage of the progress of the prisoner . . . appears to us to be an indispensable part of the discipline to which he should be subjected. Further, we contemplate the necessity of subjecting every convict to successive stages of punishment, decreasing rigour at each successive step, until he reaches that ultimate stage . . . of a pardon, either absolute or conditional, though not ever entitled to demand the indulgence of right.' (quoted ibid. i. 400)

Jebb favoured a system of remission of sentence as the best means of sustaining the 'invigorating hope' in the heart of every convict but until 1857 his advice went unheeded by the Home Office. The passage of the 1857 Penal Servitude Act finally led to the establishment of a formal scheme whereby sentences of three years were subject to one sixth remission for good behaviour, the proportion gradually increasing so that sentences of fifteen years or more were subject to one-third remission. Eligibility for remission depended on a system of good marks and conduct badges which gave public recognition of a prisoner's progress towards eventual early release on a ticket of leave.

Despite a moral panic in the late 1850s caused by an outbreak of garrotting, which increased public alarm at the idea of unreformed convicts roaming the streets, the principle of early release survived the 1863 Royal Commission on Penal Servitude. But the Commission's recommendation that remission should be dependent not merely on good conduct but rather on the quantity of productive work achieved was accepted.[7] The remission scheme was amended several times during the last quarter of the nineteenth century. In 1891 it was fixed that male convicts would be eligible for one-quarter remission and female convicts for one-third. And in 1898 the anomaly that remission was only available to convicts and not to ordinary prisoners (who had been convicted of less serious offences) was ended.[8] There was, however, a difference between the terms of remission granted to prisoners and convicts. While ordinary prisoners were completely free from the day of their early release, convicts remained vulnerable to recall to prison until the end of their sentence if they were guilty of 'misconduct', even if this fell short of the commission of a criminal offence. In this sense, remission of a convict's sentence bore some resemblance to the later system of parole with its concept of a licence liable to be revoked in the event of any breach of its conditions.

MOVEMENT TOWARDS A PAROLE SYSTEM

For the next forty years the early release of prisoners remained largely unaltered in its structure. In 1940 the pressure on prison places brought about

[7] McConville comments that the marks system was primarily concerned with ensuring obedience and productivity and that reformation was viewed as an incidental benefit. He cites the remarks of the Directorate of Convict Prisons in 1867 to the effect that 'It is difficult, and indeed impossible, to express any decided opinion upon the real amount of good conduct effected upon the minds of the convicts by the efforts made for their reformation; secluded as they are from the world, they are more parts of a machine than free agents' (ibid. i. 403).

[8] The Gladstone Committee recommended that remission and the marks system should be extended to ordinary prisoners partly on the grounds of equity and partly for rehabilitative reasons.

by detention under wartime regulations led to an increase in remission for both prisoners and convicts to one-third. In 1948 penal servitude was abolished and with it went the convict licence.[9] From then on, apart from lifers or recidivists undergoing corrective training or preventive detention, all adult prisoners became eligible for release at their remission dates without being subject to any form of licence. In 1973 new rules provided that time spent on remand would attract remission as well as counting towards the length of sentence. Remission remained at one-third until in 1987 another period of intense pressure on prison places led to an announcement by the Home Secretary that for those serving sentences of twelve months or less, remission would be increased to 50 per cent.

While remission had been a consistent feature of imprisonment since the mid-nineteenth century, a parole system is of relatively recent antiquity.[10] Unlike the USA, where a parole scheme was first introduced at New York's Elmira Reformatory in the 1870s, it was not until the passage of the Criminal Justice Act 1967 that prisoners became entitled to consideration for release on parole.[11] And whereas the American scheme was developed against a background of largely indeterminate sentencing, the scheme which was applied in England and Wales was implanted into a structure of almost exclusively fixed sentences with the 'parolable' portion precisely specified.[12] Prior to the introduction of parole, prisoners were able to be released for short periods for certain express purposes. This included visits to seriously ill relatives, funerals, work in outside jobs as discharge approached, or attending classes not provided by the prison system. In addition some prisoners could take home leave shortly before they were released as a means of re-establishing contact with families. But only a small proportion of the population qualified for these privileges. Parole, by contrast, was potentially available to all-comers.

The idea of a system based on the selection of prisoners suitable for early release on licence was of course not entirely new as we have seen. The novelty was to extend it to all determinate sentenced prisoners. In the 1965

[9] After the passage of the Criminal Justice Act 1948, under-21-year-olds and Borstal trainees remained subject to supervision under licence.

[10] The word 'parole' is derived from the French expression for 'word of honour'.

[11] 'In the initial American parole plan, youthful reformatory inmates were sentenced to indeterminate terms. Elmira's Board of Managers was authorized to parole anyone at any time after 1 year's confinement. An elaborate review process preceded the decision to approve or deny release on parole. Inmates were required to earn a specified number of marks in order to be eligible for parole. If released, individual offenders were expected to obey certain rules and to submit willingly to supervision by selected persons in the outside community. Failure of a parolee to obey the rules could result in return to Elmira,' quoted in Henry Burns, *Correctional Reform: Britain and the USA Compared and Contrasted* (1978).

[12] The American system of indeterminate sentencing caused a substantial decline in the judiciary's sentencing role since parole boards were empowered to release an inmate at any time between the fixed minimum and maximum terms. In the absence of appeals against sentence, parole in the USA has always had an explicitly resentencing purpose.

White Paper *The Adult Offender* (which followed closely the passage of the Murder (Abolition of Death Penalty) Bill 1965) it was proposed that a 'prisoner who has shown promise or determination to reform . . . should be able to earn a further period of freedom on parole [i.e. over and above the one-third remission] of up to one third of his sentence'. It went on to express the rehabilitative ideal that:

prisoners who do not of necessity have to be detained for the protection of the public are in some cases more likely to be made into decent citizens if, before completing the whole of their sentence, they are released under supervision with a liability to recall if they do not behave . . .What is proposed is that a prisoner's date of release should be largely dependent upon his response to training and his likely behaviour on release. A considerable number of long term prisoners reach a recognisable peak in their training at which they may respond to generous treatment but after which, if kept in prison, they may go downhill . . . These arrangements would afford the strongest incentive to reform and greatly assist the task of prison administration.

The White Paper's proposals found expression in the Criminal Justice Act 1967 and the basic features of the scheme remained in place for the next twenty-five years. The Act provided for the appointment by the Home Secretary of a Parole Board and beneath it a network of Local Review Committees (LRCs).[13] Save in one respect, the function of the Board was advisory. The Board's duty was to advise the Home Secretary about the release on licence of determinate and life sentence prisoners, the conditions to be attached to such licences, and the recall to prison of prisoners whose cases are referred to the Board for consideration. It was assisted in its task by the LRCs, whose job it was to commence the process of reviewing the suitability for release on parole of every prisoner who had become eligible for it.[14] The only direct contact which a prisoner had with the reviewing process was a personal interview with a single member of the LRC to whom any representations had to be made. The LRC's report and recommendation would then be forwarded to the Parole Board for it to consider and make a recommendation. Finally the case would be referred to the Home Office Parole Unit where the ultimate decision would be taken, usually by civil servants but sometimes by the appropriate Minister or even, in rare cases, by the Home Secretary personally. No reasons were given for a decision to refuse parole nor did a prisoner have any right of access to material before the various bodies considering his suitability for release on licence.[15]

[13] See Criminal Justice Act 1967, ss. 59–62.

[14] The size and composition of LRCs varied from place to place but every panel considering cases had to include one of the prison governors, a probation officer from outside the prison, a member of the Board of Visitors, and two independent members chosen by the governor.

[15] See *Payne* v. *Lord Harris of Greenwich* [1981] 1 WLR 754 and the Local Review Committee Rules 1967 (SI 1462), which set out a comprehensive code governing the procedures

The tripartite structure of Home Secretary, Parole Board, and Local Review Committee did not disguise the fact that in each case the ultimate responsibility for release remained with the Home Secretary under what was described as a 'dual key' system of responsibility. As the 1987 Carlisle Report observed, 'the Home Secretary could not grant parole without a positive recommendation from the Parole Board but could deny parole even where a positive recommendation had been made'.[16] The one exception to the Board's purely advisory role was in the case of a prisoner recalled to prison following the revocation of his licence. If in such a case the Board recommended that person's immediate re-release, the Home Secretary was obliged to give effect to that recommendation.[17]

The parole system came into operation on 1 April 1968. Every prisoner serving a determinate sentence became eligible for release on licence after he had served one-third of his sentence (including time spent on remand) or at least twelve months after sentence, whichever was the longer. The licence lasted until the day the prisoner would ordinarily have been released, that is, after serving two-thirds of his sentence. It followed that the minimum qualifying period (MQP) which the prisoner had to serve was twelve months before he could be released on parole even if he was eligible for parole after serving one-third (the remand time did not count towards the MQP). Between 1967 and 1973 lifers were initially considered for release, in general, after they had served no more than seven years in custody. This arrangement was varied in 1973, however, when a new scheme was introduced whereby all lifers who had served between three and four years of their sentence had their cases considered by a Joint Committee consisting of the Parole Board's chairman and vice-chairman, a psychiatrist member of the Board, together with two senior Home Office officials. The Committee either fixed a date for the first formal review or, if it was clear that release was several years away, asked for the case to be brought before it again at a specified date.

THE SCHEME IN PRACTICE

From 1968 until 1983 the parole system operated without any significant alteration to its fundamental principles. But, as the Carlisle Committee

for considering suitability for release on licence. The Court of Appeal in *Payne* held that the rules of natural justice did not require the code to be supplemented by any further implied duties or rights.

[16] *The Parole System in England and Wales*, Report of the Review Committee Chaired by Lord Carlisle of Bucklow QC, Cmnd. 532 (London, 1988).

[17] In R. v. *Chairman of the Parole Board ex parte Gunnell* unreported (CA) 30 Oct. 1984 the Court of Appeal held that anything short of a recommendation for immediate release meant that release remained a matter for the Secretary of State's discretion. Accordingly a recommendation for release subject to a satisfactory release plan did not bind the Home Secretary to order immediate release.

commented, it was not without its critics. The publication of the May Committee Report in October 1979 marked an important change away from the belief that prison could ever truly be rehabilitative.[18] The parole scheme had been conceived at a time when optimism was high that prison could change people for the better and that 'experts' could identify and select those prisoners best suited for early release. If prison could not reform people, what was the justification for a selective system of early release? A second factor was the twin assault from different wings of the political spectrum which attacked parole as either being 'soft on criminals' at a time when the crime rate continued to rise or, alternatively, too arbitrary and lacking in basic procedural fairness safeguards necessary to ensure that all prisoners were considered equally and on the basis of reliable, relevant information. As more and more prisoners were released on licence it became harder to resist the argument that parole could not be considered as a mere privilege, undeserving of the full panoply of the rules of natural justice.[19] Finally, the existence of the MQP for release on licence led to the anomaly that the great majority of prisoners serving less than two years had to stay in prison until they had served two-thirds of their sentence whereas those serving three or four years (or even longer) were gaining freedom after serving a third of their sentence. Why should prisoners sentenced for more serious crimes be eligible for release at an earlier stage than those sentenced for relatively minor ones?

This latter criticism was met by the passage of s. 33 of the Criminal Justice Act 1982 which enabled the Home Secretary to reduce by order the twelve-months minimum qualifying period for parole. The Act also gave the Home Secretary a power to order the release of certain categories of prisoners up to six months early where the consent of both Houses of Parliament was obtained. This 'Executive release' power was a measure of last resort designed to avert any potential breakdown in the prison system threatened by overcrowding in local prisons.[20]

1983: THE BRITTAN POLICY

In 1983 major changes to the parole system were announced by the new Home Secretary, Leon Brittan, in the course of his first speech to the Conservative Party Conference in Blackpool. Responding to what he described as 'growing public criticism about the gap between the length of

[18] *Report of the Committee of Inquiry into the United Kingdom Prison Services* Cmnd. 7673, (London, 1980).

[19] For further argument in favour of the extension of natural justice principles, see the evidence called in support of the applicant in *Payne* v. *Lord Harris of Greenwich*.

[20] s. 33 of the 1982 Act remains on the statute book after the passage of the Criminal Justice Act 1991, though it has yet to be used.

sentence passed and the length of the sentence actually served' and a feeling that the public 'do want to know with certainty what will actually happen to the most serious offenders and . . . [that] what happens will reflect the gravity of the offences that have been committed', he indicated that he would in due course make a statement to Parliament setting out the new policy. This was done on 30 November 1983 when Mr Brittan announced three changes to the existing policy.[21] The first of these measures affected all prisoners serving more than five years for sexual offences or those involving drug trafficking, arson or violence. In future, the Home Secretary said that he would use his discretion to ensure that such prisoners would be granted parole 'only when release under supervision for a few months before the end of a sentence is likely to reduce the long-term risk to the public, or in circumstances which are genuinely exceptional'. Only compelling reasons would lead to release on licence in the future, he said. The second change affected lifers convicted of the murder of police or prison officers, terrorist murderers, sexual or sadistic murderers of children, and murderers by firearm in the course of robbery. Such persons would now expect to serve at least twenty years in custody, and other murderers, outside these categories, might merit equal treatment to mark the seriousness of the offence. The Home Secretary's statement made it clear that, although he would consult with the Parole Board about how these new policy objectives would be achieved in practice so that the Board's role was maintained, he had not consulted the Board in advance of the announcement. It was also made clear that the policy would be retrospective, and thus prisoners with release dates already identified were suddenly faced with the devastating destruction of their expectation of freedom. The policy changes were promptly challenged in the courts by four prisoners (two lifers and two serving fixed sentences) affected in this way, but the Home Secretary's right to alter his policy was upheld, finally, in the House of Lords.[22]

The third and final change announced by Mr Brittan was a liberalizing one. Using the new power contained in s. 33 of the Criminal Justice Act 1982, the Home Secretary reduced the minimum qualifying period for parole from twelve to six months with effect from 1 June 1984.[23] The effect of this was that whereas before only those serving a minimum of nineteen-and-a-half months could in practice attract parole, now those serving as little as ten and a half months were able to benefit. It meant that 2,500 more prisoners were out on parole at any one time.

[21] Hansard, Sixth Series, xlix (1983), Written Answers, cols. 505–8.
[22] *Re Findlay* [1985] 1 AC 318. The case reached the House of Lords after both the Divisional Court and the Court of Appeal had been divided over the lawfulness of the policy and its mode of implementation.
[23] Eligibility for Release on Licence Order 1983, SI 1959.

THE CARLISLE COMMITTEE AND THE 1990 WHITE PAPER

Though it was the least controversial aspect of the Brittan policy changes at the time of their announcement, it was the extension of parole to those serving six months which ultimately produced the greatest pressure for yet more reform and led to the setting up of a Review Committee under Lord Carlisle, QC, in July 1987. Only a year after the more liberal policy had come into force, the judiciary had begun to express its disquiet about the way in which parole was operating for those serving less than two years' imprisonment.[24] There were four main concerns. First, too many prisoners serving less than two years were being paroled. A rate of 75–80 per cent had turned parole into an entitlement rather than a privilege to be earned. Secondly, the lower six-month minimum qualifying period meant that prisoners serving anything between nine and eighteen months came out in practice on the same day just as before those serving between eighteen months and three years were freed at the same time. Since there were far more prisoners serving between nine and eighteen months and far more were granted parole, the anomaly was simply exacerbated. Thirdly, the expansion of parole had undermined the partly suspended sentence introduced by the Criminal Justice Act 1982 in that prisoners who had received an unsuspended sentence frequently emerged earlier from prison than those subject to partial suspension. Finally, the rule that time spent on remand counted towards both the total sentence and the one-third to be served for parole eligibility but not towards the MQP created anomalies which were made the greater by the reduction in 1983 of the MQP. There were also significant practical consequences for the work of the Parole Board and the LRCs, whose case-load increased dramatically.[25] It was widely felt that a system of selective release on licence for short-term prisoners had become unworkable.

---

[24] In the autumn of 1985 the Lord Chief Justice drew the attention of the Home Secretary to mounting concern among members of the judiciary about the way in which parole was operating for those sentenced to less than two years' imprisonment. A working group proposed a number of administrative changes to ensure that more information was available to LRCs when considering cases. It also proposed that the presumption in favour of parole which the Home Office and the Parole Board had asked LRCs to apply in under two-year cases should be withdrawn and all cases, irrespective of sentence length, be considered on the basis of relative risks and benefits.

[25] In 1983 only 265 prisoners serving less than two years were considered for parole whereas in 1986 it was 10,603. By 1991 the Board and LRCs were considering 33,000 cases in all and, though parole was more sparingly granted to long-term prisoners as a result of the Brittan policy, the entitlement to be reviewed after a third of the sentence remained unchanged. As a result many prisoners stayed in prison for longer and had more reviews than before.

It was against this background of judicial concern and administrative strain that the Carlisle Committee was appointed with a remit 'to examine the operation of the parole scheme in England and Wales, its relationship with the current arrangements for remission, time spent in custody on remand, and partly and fully suspended sentences and their effect on the time which offenders sentenced to imprisonment spend in custody'. With the announcement in July 1987 that remission was to be increased to 50 per cent for those serving twelve months or less, there was an ever increasing discrepancy between the prospects of early release for prisoners serving relatively short sentences and those long-term prisoners subject to the strictures of the 1983 Brittan policy. The Carlisle Committee identified the problem as one which had gone to the heart of the parole system as a whole, pointing out that 'there has in some quarters been the feeling that parole for those serving 2 years or more (something loosely referred to as "section 60 parole") is the real thing and parole for those serving less than 2 years ("section 33 parole") something different'.[26] It concluded that the introduction of 50 per cent remission for those serving twelve months or less and the lowering of the minimum qualifying period for parole had created an unacceptable disparity between what sentences say and what they mean. The effect of parole and remission, taken together, had over the years 'created an increasing unreality in the criminal justice system and handed to the executive too much control over the length of custodial sentences served'.[27]

Quite apart from the anomalies resulting from the different way in which parole and remission affected sentences of different length, there was an increasing volume of criticism from a wide range of opinion which questioned the obsessive secrecy surrounding the process of parole decision-making, its arbitrariness and lack of fairness. There was also fierce criticism of the effect of the Brittan policy on long-term and life sentence prisoners. The Carlisle Committee summed this up in the following terms:

[B]y making the change administratively, rather than by inviting Parliament to amend the law, the Home Secretary had modified the criteria for parole without being able to adjust the actual eligibility rules. Thus everyone still had to be reviewed after only a third of the sentence and to preserve the legality of the policy the Home Secretary had to make it clear that parole would still be granted in exceptional cases. This meant that prisoners, prison staff, probation staff, the Parole Board, the LRCs and indeed the Home Office itself were all required to go through a process which in the vast majority of cases was entirely nugatory. Not only was

[26] 'section 60' was a reference to s. 60 of the Criminal Justice Act 1967, which provided the power for prisoners serving determinate sentences to be released on licence by the Home Secretary. Strictly all parole cases were s. 60 cases but the term was used loosely to describe those cases which were not handled under the streamlined s. 33 procedure introduced by the Criminal Justice Act 1982, which was applicable to all sentences below two years.

[27] Carlisle Report, para. 194.

this wasteful of everyone's time but for the prisoner and his family it was positively harmful. The prisoner, though knowing his chances were slim, normally felt unable to get out of the process for fear of prompting his family to conclude that he did not really want to come home. His family were inevitably encouraged by the mere fact of a visit from the home probation officer to think that release was a real possibility. As a result, when parole was refused, as it almost invariably was, the relationship between inmates and their families was frequently damaged.[28]

Against this background the Carlisle Committee made a number of wide-ranging recommendations, the most important of which were that:

1. the MQP should be abolished and replaced with a minimum qualifying sentence for parole. Parole should be available only to those sentenced to more than four years with the parole eligibility date fixed at one-half of sentence rather than one-third. Misconduct in prison should delay both the parole eligibility date and the release date if parole is not granted. Those not granted parole should continue to be released after two-thirds of sentence, subject to any disciplinary awards for misconduct;

2. all prisoners sentenced to more than four years should be subject to a period of supervision on release until the three-quarters point of their sentence;

3. all prisoners sentenced to four years' imprisonment or less should be required to serve half of the sentence in custody, plus any additional days for misconduct in prison, and half in the community;

4. remission should be abolished;

5. all released prisoners, whatever the length of their sentences, should be liable, if convicted of another offence punishable with imprisonment committed before the end of the original sentence, to be ordered by the court to serve part or all of the portion of that sentence outstanding in addition to any new custodial sentence imposed;

6. the parole decision should be based on a more specific test with the criteria for parole prescribed by statute. The decision-making process should be more open, with disclosure of reports to prisoners (save in highly exceptional cases) and the giving of meaningful reasons for a refusal to grant parole;

7. The Home Secretary's responsibility for releasing determinate sentence prisoners on licence should cease and the decision to release prisoners on parole should be taken by the Parole Board alone. With the greatly reduced work-load of the Parole Board, LRCs should be abolished.

The 1990 White Paper *Crime, Justice and Protecting the Public* had at its core the aim of ensuring that convicted criminals are punished according to

[28] Carlisle Report, para. 172.

the seriousness of the offence so that 'they get their just deserts'.[29] It generally accepted the recommendations of the Carlisle Report save in two important respects.[30] It did not agree that the Home Secretary should withdraw completely from the responsibility for decisions to release all determinate sentenced prisoners but instead expressed the view that he should continue to consider the release of those serving very long determinate sentences of seven years or more who will necessarily have committed very serious crimes. It also rejected the proposal that reports before the Parole Board should be disclosed to prisoners. It did so on the somewhat dubious grounds that 'the likely benefits of greater openness outweigh its disadvantages' since 'there is a risk that openness could lead to less full and telling reports and so to less well informed decisions'.[31]

Neither the Carlisle Committee nor the White Paper dealt specifically with the position of life sentence prisoners. But two important decisions by the European Court of Human Rights ensured that the procedure for dealing with the release and subsequent recall of those prisoners given a discretionary life sentence would have to be amended in the light of a finding that English law was in breach of Article 5 (4) of the Convention.[32] The pressure for widespread reform of the system governing the early release of prisoners culminated in the passage of the Criminal Justice Act 1991. Below we examine its provisions as well as the case-law which, though pre-dating it, continues to be relevant to its interpretation.

THE CRIMINAL JUSTICE ACT 1991: A SUMMARY

The 1991 Act impacts in different ways on four categories of prisoner: those serving less than four years; those serving four years or more; discretionary lifers; and, finally, mandatory lifers. Transitional provisions apply to deal with the position of prisoners sentenced prior to 1 October 1992, when the Act came into force. Below we summarize the changes brought about by the Act before explaining in more detail how the new scheme works in practice.

SHORT-TERM PRISONERS (LESS THAN FOUR YEARS)

1. They must be released after serving half their sentence unconditionally if the sentence is for less than twelve months (s. 33 (1) (*a*)) and on licence if the sentence is for twelve months or more (s. 33 (1) (*b*));

---

[29] Cmnd. 965, (London, 1990).   [30] Ibid., para. 6.11.   [31] Ibid., para. 6.26.
[32] *Weeks* v. *United Kingdom* (1987) 10 EHRR 293 ECHR, and *Thynne, Wilson and Gunnell* v. *United Kingdom* (1990) 13 EHRR 666 ECHR.

2. a short-term prisoner may be returned to prison if he commits any offence punishable with imprisonment before the date on which he would (but for his release) have served his sentence in full (s. 40);

3. a short-term prisoner's licence shall (unless revoked or suspended) remain in force until the date on which he would (but for his release) have served three-quarters of his sentence (s.3 7(1));

4. a short-term prisoner who fails to comply with the conditions of his licence shall be guilty of an offence (s. 38 (1)) and punished by a fine. He may also be recalled to prison for as long as six months (s. 38 (2)) but must in any event be released unconditionally after he has served three-quarters of the sentence originally imposed (s. 33 (3)) unless the offence is a sexual one and s. 44 applies, in which case he may be recalled for the whole of the remainder of the sentence;

5. a short-term prisoner may be released on licence on compassionate grounds at any time and the Parole Board need not be consulted by the Secretary of State (s. 36 (1) and (2));

6. a short-term prisoner who is guilty of a disciplinary offence under the Prison Rules may be awarded additional days in prison as a punishment (s. 42);

7. special provisions apply to young offenders (s. 43) and fine defaulters/contemnors (s. 45).

LONG-TERM PRISONERS (FOUR YEARS OR MORE)

1. A long-term prisoner must be released on licence after he has served two-thirds of his sentence (s. 33 (2));

2. a long-term prisoner serving between four and under seven years imposed after 1 October 1992 shall, if recommended by the Parole Board, be released on licence when he has served one-half of his sentence (s. 35 (1), s. 50 (2), and Parole Board (Transfer of Functions) Order 1992) and any conditions attached to or inserted in his licence must be in accordance with the recommendation(s) of the Parole Board (s. 37 (5), s. 50 (3) and the Parole Board (Transfer of Functions) Order 1992);

3. a long-term prisoner serving more than seven years may be released on licence by the Secretary of State, if so recommended by the Parole Board, after serving half his sentence (s. 35 (1));

4. any long-term prisoner released on licence shall remain subject to the conditions of the licence until the date on which he would (but for his release) have served three-quarters of his sentence (s. 37 (1));

5. a long-term prisoner serving seven years or more may be recalled by the Secretary of State either upon the recommendation of the Parole Board (s. 39 (1)) or without its recommendation if the conditions of s. 39 (2) are fulfilled;

6. a long-term prisoner serving between four and under seven years imposed after 1 October 1992 may only be recalled to prison by the Secretary of State if the Parole Board so recommend (s. 39 (1), s. 50 (4) and the Parole Board (Transfer of Functions) Order 1992);

7. a long-term prisoner who is recalled under s. 39 (1) must, in any event, be released unconditionally after he would have served three-quarters of his sentence (but for his release) (s. 33 (3)), unless the offence is a sexual one and s. 44 applies, in which case he may be recalled until the expiry of his original sentence;

8. any long-term prisoner who is recalled to prison may make representations with respect to his recall and must be informed of the reasons for his recall (s. 39 (3));

9. the Home Secretary must refer to the Parole Board the case of a long-term prisoner who is either recalled under s. 39 (2) or who is recalled under s. 39 (1) and who makes representations about it (s. 39 (4)) and where the Board recommends his immediate release on licence the Home Secretary must give effect to the recommendation;

10. a long-term prisoner may be released at any time by the Secretary of State on compassionate grounds after consultation with the Parole Board (s. 36);

11. a long-term prisoner may be returned to prison if he commits an offence punishable with imprisonment before the date on which he would have served his sentence in full (s. 40 (1));

12. additional days in prison may be awarded to any long-term prisoner who commits a disciplinary offence during the course of his imprisonment (s. 42).

DISCRETIONARY LIFERS

1. A discretionary lifer must be released once he has completed the punitive portion ('the tariff') of his sentence and the Parole Board has directed his release on licence (s. 34 (3));

2. the licence of any discretionary lifer shall, unless revoked under s. 39 (1) or (2), remain in force until his death (s. 37 (3));

3. a discretionary lifer who has been released on licence may be recalled to prison by the Home Secretary on the recommendation of the Parole Board (s. 39 (1)) or without such a recommendation if the requirements of s. 39 (2) are fulfilled;

4. any discretionary lifer who is recalled to prison either under s. 39 (1) or (2) must be informed of the reason for his recall and told of his right to make representations with respect to it (s. 39 (3));

5. the Secretary of State must refer to the Parole Board the case of any discretionary lifer who is recalled under s. 39 (2) or who is recalled under s.

39 (1) and who makes representations about it (s. 39 (4)) and where the Board directs his immediate release the Home Secretary shall give effect to the recommendation (s. 39 (5));

6. discretionary lifers may be released on licence on compassionate grounds by the Secretary of State after consultation with the Parole Board, if practicable (s. 36).

MANDATORY LIFERS

1. The Home Secretary may release on licence a mandatory lifer where the Parole Board so recommends and after he has consulted with the Lord Chief Justice and the trial judge (if available) (s. 35 (2));

2. the licence of any mandatory lifer so released remains in force until his death unless revoked under s. 39 (s. 37 (3));

3. a mandatory lifer may be recalled to prison by the Home Secretary if the Parole Board so recommends (s. 39 (1)) or without such a recommendation where the requirements of s. 39 (2) are fulfilled;

4. a mandatory lifer who is recalled to prison must be informed of the reasons for his recall and his right to make representations with respect to it (s. 39 (3));

5. the Home Secretary must refer to the Parole Board the case of any mandatory lifer who is recalled either under s. 39 (2) or under s. 39 (1) and who makes representations about it (s. 39 (4)) and where the Board recommends his immediate release on licence the Home Secretary must give effect to the recommendation (s. 39 (5));

6. a mandatory lifer may be released at any time on compassionate grounds in consultation with the Parole Board, if practicable (s. 36 (1)).

THE 1991 ACT IN MORE DETAIL

DETERMINATE SENTENCE PRISONERS: THREE RELEASE SCHEMES

The 1991 Act introduced radically new arrangements for the early release of prisoners sentenced on or after 1 October 1992, wholly replacing the previous scheme of automatic remission and selective release on parole. The new arrangements apply to all prisoners although the Act makes special provisions to deal with those sentenced before 1 October 1992, defined as 'existing prisoners'.[33] Regrettably, but perhaps inevitably, the Act has spawned a plethora of new terms and jargon which will have to

---

[33] For a useful summary of the implications of the 1991 Act see Standing Order 3C and Circular Instructions 29/92 and 39/1992.

be adopted in this account, if only to acquaint the reader with the terminology used by the Home Office in its literature. In the hope that it will enlighten rather than confuse, a glossary of new terms is to be found in Appendix 6.

The 1991 Act creates three different routes to freedom for determinate sentence prisoners based on the length of their sentence, as follows:

1. *Automatic Unconditional Release.* This covers short-term prisoners (including young offenders serving sentences of up to twelve months) serving sentences of up to twelve months who are released automatically at the half-way stage of their sentence without statutory supervision. Released prisoners are 'at risk' of being returned to custody until the expiry of their full sentence if they commit a further imprisonable offence during this time.

2. *Automatic Conditional Release Scheme.* This covers short-term prisoners (including young offenders aged 18 and over) serving sentences of twelve months to under four years who will be released automatically at the half-way point of the sentence under licence and supervised by the probation service to the three-quarter stage of the sentence. They are also 'at risk' from the date of release until the expiry of their full sentence.

3. *Discretionary Release Scheme.* This covers long-term prisoners serving sentences of four years or more who will be eligible for discretionary release on licence by the Parole Board (in the case of those serving under seven years) or the Home Secretary (in the case of those serving seven years or more) from the half-way point of sentence to the two-thirds point. Those not released before two-thirds will be released automatically at that stage under licence and supervised by the probation service to the three-quarters point. The 'at risk' provisions apply from the date of release until the expiry of the full sentence.

EXISTING PRISONERS: TRANSITIONAL PROVISIONS

Schedule 12 of the 1991 Act contains important provisions governing the application of the new release schemes to prisoners sentenced before 1 October 1992 but who remain in custody beyond that date. These prisoners are referred to as 'existing prisoners' and, in the case of prisoners sentenced to long terms of imprisonment before 1 October 1992, the transitional provisions are likely to be relevant for some years to come. Paragraph 8 (1) of Schedule 12 also defines an 'existing licensee' as someone who at commencement has already been released on licence under s. 60 of the Criminal Justice Act 1967 and whose licence remains in force at commencement. The aim of the transitional arrangements is to ensure that existing prisoners/licensees are treated, at worst, no less favourably than they would have been under the old parole scheme.

Existing prisoners will not be subject to the supervision or 'at risk' provisions of the 1991 Act and will continue to be treated as 'existing prisoners' throughout the currency of their sentence even if they subsequently receive a concurrent or consecutive sentence or a revised sentence on appeal after 1 October 1992.[34] It follows that an existing prisoner who is released on parole and subsequently has his licence revoked will be treated as an existing prisoner on return to prison. An existing prisoner at commencement who has been released and fulfilled all the liabilities of the sentence but who is subsequently returned to prison on a fresh sentence will obviously be liable to the 1991 Act's provisions in full.

An existing prisoner or licensee who received a sentence of more than twelve months but less than four years would ordinarily be treated as a short-term prisoner but is in fact treated as a long-term prisoner.[35] This ensures that such prisoners fall to be dealt with under the Discretionary Release System and paragraph 8 (6) of Schedule 12 makes provision to adjust the normal points for eligibility for release on licence and unconditional release so that they reproduce the old system of release on licence at one-third and entitlement to unconditional release after two-thirds.

Schedule 12 also contains important provisions governing the position of existing prisoners/licensees who are discretionary lifers but these are dealt with below when we examine the scheme of the Act for discretionary lifers in general.

THE SINGLE TERM

Before a prisoner's release date is calculated it is essential to identify the release scheme applicable to his case. For this purpose the 1991 Act provides for all sentences of imprisonment to be considered as a 'single term'.[36] A single term can be made up of a single sentence, consecutive sentences, or a combination of multiple concurrent and overlapping sentences. The single term runs from the date of the prisoner's first entry into custody following sentence for the full term of the sentence(s) imposed by the court. For the purpose of determining the single term, sentence length should not be reduced by time on remand or police custody time. The single term simply determines the applicable release scheme. Calculation of release dates can only take place after the Sentence Expiry Date (SED) of each sentence within the term has been adjusted to take account of remand and police custody time. Standing Order 3C sets out in great detail the calculations which must be made to determine release dates and this is reproduced in part in Appendix 6.

---

[34] Schedule 12, para. 8 (3). [35] Schedule 12, para. 8 (5). [36] CJA 1991, s. 51 (2).

DECISION MAKING AND PROCEDURAL FAIRNESS UNDER THE DISCRETIONARY RELEASE SCHEME

The Discretionary Release Scheme applies to two categories of determinate sentence prisoners—those serving between four years and under seven years and those serving seven years or more. The discretionary release of the former is decided exclusively by the Parole Board whereas the release of the latter is still governed by a 'dual key' system of responsibility. The Home Secretary may release if the Parole Board so recommend but he is not obliged to release upon its recommendation. Save in respect of existing prisoners and licensees, Local Review Committees no longer have any role in the decision-making process of the DRS.[37]

What is the structure for decision-making under the DRS? The White Paper stated that the arrangements for making the parole decision should be 'equitable and fair and be seen to be so; they should be simple and easy to understand for all parties; delays should be avoided'. Under the old parole system the main criticisms were that prisoners had no right to a hearing before the Parole Board deciding their case, there was no right to have access to the written material before the Board, the criteria for release were unclear (and unpublished), and reasons were not given for a decision to refuse parole. In what ways, if any, do the new provisions meet these criticisms?

THE RIGHT TO BE HEARD

The 1991 Act does not provide a right to an oral hearing before the Parole Board. Indeed, it was one of the disappointments of the Carlisle Committee that the majority did not favour such a right in cases where the Board did not feel able to grant parole after considering any written material before it. Instead, as the majority of the Carlisle Committee recommended, the 1991 Act gives the Board a discretion to arrange for one of its members to interview the prisoner before it reaches a decision .[38] During the passage of the 1991 Act through the House of Lords, Earl Ferrers expressed the government's view that:

it is important for an inmate to have an interview with someone involved in the decision-making process. We intend, by virtue of [s. 32 (5)] to issue rules which would ensure that all inmates receive such an interview. The advantage of this approach is that it allows for flexibility and changes to the system in the light of

---

[37] Schedule 12, para. 8 (7). It is estimated that the 1991 Act will reduce the Parole Board's case-load from 24,000 to 4,000 cases per year.
[38] CJA 1991, s. 32 (3).

experience . . . the interview should be an administrative arrangement designed to assist the Board and not a legal right from which other rights could be inferred.[39]

It is anticipated that the first set of rules issued under s. 32 (5) of the 1991 Act will be published in about November 1993 and it remains to be seen at what level a prisoner will have a right of face-to-face contact with 'a person involved in the decision-making process'. The White Paper proposed that the experience of existing members of Local Review Committees should be utilized by formally appointing them as local Parole Assessors whose task would be to act as agents of the Parole Board. It described the kind of role they would perform in the following terms:

The Parole Assessor would see the prisoner, hear and record any further points he wishes to make to the Board, probe any points of doubt or difficulty and check the completeness of the parole dossiers forwarded by the prison to the Board. The hearings would not be adversarial and no lawyers or other advisors would attend.[40]

Under the old parole system a prisoner had a right to an interview with a single member of the Local Review Committee considering his case.[41] It seems likely that the Rules to be issued under s. 32 (5) of the 1991 Act will reproduce such an arrangement and simply give a right to an interview with a single Parole Assessor who would be regarded, to use Earl Ferrers's words, as 'someone involved in the decision-making process'. Of course this would leave prisoners in exactly the same position as before in that they will have no legal right to appear before, and make oral representations to, the actual decision-making body deciding on their release. Under the old parole system the courts were unwilling to imply any right to an oral hearing before the Parole Board, even in the case of a discretionary lifer who was recalled to prison after his release on licence.[42] Accordingly it seems unlikely that different principles will be applied to the construction of the new provisions in the 1991 Act.

RIGHT OF ACCESS TO MATERIAL BEFORE THE BOARD

The Carlisle Committee was strongly in favour of the view that a system of open reporting should be introduced into the parole system, giving prisoners a right to have access to all material/reports before the Parole Board save in exceptional cases where the Board is satisfied that disclosure would

---

[39] HL Debates, 21 May 1991, vol dxxix, col. 115.   [40] White Paper, para. 6.27.
[41] Local Review Committee Rules 1967 (SI 1462), para. 3.
[42] In *R.* v. *Chairman of the Parole Board ex parte Gunnell* (DC) 2 Nov. 1983 the Divisional Court rejected the argument that a single member of an LRC could not act as an agent of the Home Secretary for the purposes of conducting an interview with a recalled lifer. The single member was directly answerable to the Home Secretary and not to the Parole Board under the 1967 Act and Rules made thereunder.

cause 'specific harm' to a third party or the prisoner himself.[43] It stressed that one of the vices of the old system was that:

secrecy breeds rumour and suspicion. Inmates suspect that a prison officer or probation officer has spoiled their chances of parole even when in fact he has submitted a fair and favourable report. Sometimes an officer may write something which is unintentionally misleading or inaccurate but the prisoner will have no opportunity to correct it or put his side of the story.

The idea that open reporting might somehow inhibit the production of reliable, honest reports had also received short shrift from the Divisional Court in the case of *R. v. Home Secretary ex parte Benson.*[44] Mr Benson was a discretionary lifer who, having been granted leave to apply for judicial review of the Home Secretary's decision not to release him despite three favourable recommendations from the Parole Board, sought discovery of certain medical reports which had been available to both the Parole Board and the Home Secretary. The Home Secretary resisted the application and relied on the 'candour' argument to the effect that those who prepare medical reports for parole decisions might feel unable to give as full and candid an assessment of a prisoner as they would wish if they knew the report might not remain confidential. Lord Justice Lloyd was unimpressed with this reasoning, saying:

I find myself unconvinced by that argument. I cannot believe that a professional man, such as a doctor, would trim his views or be less than candid just because he thought that his report might one day see the light of day. If that is true of medical men in general, then I cannot follow why it should not be true of a prison medical officer just because he is working in a prison context.

Lloyd LJ went on to hold that he was not deciding that a prisoner is entitled to have his medical reports produced in every case where the Parole Board declines to recommend his release on parole. The reason why Mr Benson was entitled to see the reports was that he had succeeded in obtaining leave to apply for judicial review and he had established that the medical reports were not only relevant but indeed central to his application. Nevertheless the reasoning of the court was an important rejection of the most commonly cited argument against a duty of general disclosure of adverse material to prisoners. And in a subsequent case under the old parole scheme (albeit decided after the passage of the 1991 Act) it was held that a determinate sentence prisoner who had been released in 1976 on a licence due to expire in 1977, but whose licence was revoked in 1976 when

---

[43] Carlisle Report, para. 336.

[44] The original hearing was an interlocutory application for discovery before the Divisional Court (unreported 1 Nov. 1988). The substantive hearing of Mr Benson's application for judicial review was reported in The Times 21 Nov. 1988. See also *R. v. Lancashire County Council ex parte Huddleston* [1986] 2 All ER 941.

he left the country without permission, was entitled to be told of any adverse material before the Parole Board which was considering whether to release him after he was finally arrested in 1992![45]

Despite the weight of opinion in favour of disclosing reports to prisoners, the White Paper rejected the creation of a general duty of disclosure. Having stated that 'the Government will give further consideration to the detailed arrangements for making the parole decisions in the future, with the aim of moving towards disclosing reports made to the Board', it concluded that 'there is a risk that openness could lead to less full and telling reports and so to less well informed decisions'.[46] In fact the very next sentence in the White Paper almost certainly reveals the government's true concern when it warned that 'disclosing reports is likely to fuel demands for even more disclosure and the information could be used to contest adverse decisions through judicial review and any other available means'. It remains unclear, therefore, what exactly the rules to be made under s. 32 (5) of the 1991 Act will have to say on this subject. If, as the White Paper states, the government believes that 'on balance . . . the likely benefits of greater openness outweigh its disadvantages', why move towards disclosing reports as a long-term aim?

The principle in *Ex parte Benson* will still apply to determinate sentence prisoners under the 1991 Act and so a prisoner who can obtain leave to move for judicial review of a decision not to release him on parole, on a basis other than non-disclosure of adverse material, will be able to obtain discovery of such material if it can be shown to be of central relevance to the application itself.

CRITERIA GOVERNING DISCRETIONARY RELEASE

The Carlisle Committee recommended statutory criteria for parole and that the parole decision should not be a form of resentencing. The White Paper endorsed the view that clear and published criteria for parole should become the linchpin of an integrated parole system 'linking the decision to release, arrangements for taking that decision, the conditions of the licence and the arrangements for supervision in the community'. Section 32 (6) of the 1991 Act gives the Home Secretary a discretion to give directions to the Parole Board as to matters to be taken into account in discharging any of its functions under the Act. In August 1992 the Home Secretary published his policy directions and associated training guidance for the Parole Board to take into account when considering the release and recall of prisoners.

---

[45] R. v. *Parole Board ex parte Georghiades* Independent 27 May 1992.
[46] White Paper, para. 6.26.

So far as the release of determinate sentence prisoners is concerned, the directions are as follows:

1.1 The decision whether or not to recommend parole should focus primarily on the risk to the public of a further offence being committed at a time when the offender would otherwise be in prison. This should be balanced against the benefit, both to the public and the offender, of early release back into the community under a degree of supervision which might help rehabilitation and so lessen the risk of re-offending in the future.

   1.2 Each case should be considered on its individual merits, without discrimination on any grounds.

   1.3 Before recommending parole, the Parole Board should be satisfied that:

> *a.* the longer period of supervision that parole would provide is likely to reduce the risk of further imprisonable offences being committed. In assessing the risk to the community, a small risk of violent offending is to be treated as more serious than a larger risk of non-violent offending;
>
> *b.* the offender has shown by his attitude and behaviour in custody that he is willing to address his offending and has made positive efforts and progress in doing so;
>
> *c.* the resettlement plan will help secure the offender's rehabilitation.

   1.4 Before deciding whether or not to recommend parole, the Parole Board must take into account the supervising officer's recommendation as to suitability for release including co-operation with a programme of supervision and adherence to the conditions of the licence.[47]

The Home Secretary's training guidance requires twelve factors to be taken into account by the Parole Board when making a recommendation about parole. Plainly, any prisoner making representations to the Board (either in writing or orally via the Parole Assessor) should seek to address each factor as it applies to his individual case:

> *a.* the offender's background, including any previous convictions and their pattern, and responses to any previous periods of supervision;
>
> *b.* the nature and circumstances of the original offence;
>
> *c.* where available, the sentencing judge's comments and probation and medical reports prepared for the Court;
>
> *d.* any risk to the victim or possibility of retaliation by the victim, victim's family or local community;
>
> *e.* any risk to other persons, including persons outside the jurisdiction;
>
> *f.* any available statistical indicators as to the likelihood of re-offending;
>
> *g.* attitude and behaviour in custody including offences against prison discipline;
>
> *h.* attitude to other inmates and positive contributions made to prison life;
>
> *i.* remorse, insight into offending behaviour, attitude to the victim and steps

---

[47] The criteria are available from the Home Office Parole Unit, Abell House, John Islip Street, London SW1P 4LH, and are set out in a letter from the Home Secretary to the chairman of the Parole Board.

taken, within available resources, to achieve any treatment or training objectives set out in a sentence plan;

 j. realism of the release plan and resettlement prospects, including home circumstances and the likelihood of co-operation with supervision, relationship with the home probation officer, attitude of the victim and local community, extent to which the release plan continues rehabilitative work started in the prison and the extent to which it lessens or removes the occurrence of circumstances which led to the original offence;

 k. any medical or psychiatric considerations;

 l. any other information, including representations by or on behalf of the offender, which may have a bearing on risk assessment.

The White Paper expressed the view that the criteria it intended to publish would be sufficiently stringent to replace the 1983 Brittan policy as it affected prisoners serving more than five years for offences involving drug trafficking, sex, arson, or violence. Of course it is only prisoners serving seven years or more who require both the approval of both the Parole Board and the Home Secretary before they are released. It remains to be seen to what extent the Brittan policy is truly left behind by Home Office officials who rarely identified the 'exceptional cases' deserving of early release under the apparently rigid restrictions of the Brittan scheme. In fact, it is hard to see how the 'new' criteria differ from those taken into account by the Parole Board under the old scheme. A decision to release on parole will always be a balancing exercise between the risks of further offences and the benefits of supervised early release in rehabilitative terms. It is certainly the government's intention that parole should, somehow, be harder to obtain than it used to be even for those long-term prisoners not covered by the Brittan policy. We will have to wait and see if this proves to be the case.

GIVING REASONS FOR A PAROLE 'KNOCK-BACK'

Nothing caused more bitterness and frustration under the old parole system than the refusal to give any reasons for a 'knock-back'—a decision to refuse parole. The decision of the Court of Appeal in *Payne* v. *Lord Harris of Greenwich* to the effect that the general duty of fairness imposed on the Parole Board did not include a duty to disclose to a mandatory life sentence prisoner the reasons for a refusal to release him on licence was, until recently, a road-block in the path of any attempt to enhance the procedural rights of all prisoners under the parole system.[48] The Court of Appeal refused to follow *Payne* in the later case of *R.* v. *Parole Board ex parte Wilson,*[49] a case concerning a discretionary lifer, and in *R.* v. *Parole Board*

[48] [1981] 1 WLR 754. But see now *R.* v. *Home Secretary ex parte Doody*, 24 June 1993, in which the House of Lords finally held that Payne is no longer good law.
[49] [1992] QB 740.

ex parte Creamer and Scholey Rose, J., indicated his personal view that *Payne* was no longer good law even for mandatory lifers.[50] Nonetheless, *Payne* remains the law so far as determinate sentence prisoners are concerned. There is no general principle in administrative law that reasons must be given for a decision, even one as significant as a refusal to release on parole.[51]

Despite the unanimous recommendation of the Carlisle Committee that the Parole Board should be required to give prisoners meaningful reasons for refusing to grant parole, the 1991 Act falls short of ensuring such a statutory right. Unsatisfactorily, the matter is left in the discretion of the Home Secretary, who may, but need not, create such a duty by rules to be issued under s. 32 (5) of the 1991 Act. The arguments in favour of such a right are overwhelming and it is hard to interpret the reluctance to enshrine it in a statutory provision as anything more than the fear of numerous challenges in the courts should reasons be given.

As under the old scheme, there is no right of appeal against a refusal to grant parole though, again, it is intended to provide for a right to complain to the chairman of the Parole Board on procedural grounds. Judicial review will obviously remain a vital remedy in this area of law and the publication of policy directions/training guidance for the Board will make it easier to pursue such an application, especially if, in the end, the rules do provide for disclosure of reasons.

ADDITIONAL DAYS FOR DISCIPLINARY OFFENCES

The power to punish a determinate sentence prisoner by an award of 'additional days' is dealt with in Chapter 7. Under the old parole system a prisoner could be awarded loss of remission as a punishment, but though this had the effect of delaying his 'automatic' release date, it did not formally affect his eligibility for release on parole. (In practice, of course, a prisoner with a bad disciplinary record would be unlikely to be released on parole.) The 1991 Act provides that, where additional days are awarded to either a short- or long-term prisoner or to a remand prisoner who is subsequently convicted, then both the period which he must serve before he is entitled to release on licence and the period during which the licence remains in force shall be extended by the aggregate of the additional days awarded.[52] To take a practical example, P is serving three years but receives twenty-one additional days for a disciplinary offence. He would normally be released

[50] DC 21 Oct. 1992.
[51] See Ch. 3, pp. 34, 36–7, where the extent of the obligation to give reasons for administrative decisions is discussed.
[52] CJA 1991, s. 42.

on licence after eighteen months but instead is now released twenty-one days later. Further, when he is released his licence will not expire at the normal twenty-seven-month stage (three-quarters of his sentence) but will extend for twenty-one days beyond that date. Any award of additional days may be remitted under Rule 56, and if it is then obviously a prisoner's release date will be adjusted accordingly.

ORDER TO RETURN TO PRISON FOR OFFENCE COMMITTED DURING CURRENCY OF ORIGINAL SENTENCE

The White Paper accepted the Carlisle Committee's recommendation that all released prisoners should be liable, if convicted of another offence punishable with imprisonment committed before the end of the original sentence, to be ordered by the court to serve part or all of that portion of the sentence outstanding at the time the fresh offence was committed. This would be in addition to any new custodial sentence which might be imposed concurrently or consecutively. Thus a prisoner may spend up to half his sentence in the community but he will still be under sentence for the whole period of that sentence and 'at risk' of being returned to prison if he offends again. The Carlisle Committee was concerned to ensure that early release should no longer mean that the sentence ends prematurely (as happened under the old scheme of remission), thereby restoring some meaning to the full sentence originally imposed by the court.

This aim is achieved by s. 40 of the 1991 Act. It applies to a short- or long-term prisoner (but not a lifer) who is released on licence and who, before the date on which he would (but for his release) have served his sentence in full, commits an offence punishable with imprisonment. Where a court convicts the prisoner of the fresh offence it may order him to be returned to prison for the whole or part of the period which remained between the date of the fresh offence and the expiry of the full original sentence (s. 40 (1) (*a*) and (*b*)). It does not matter if the prisoner is not convicted of the fresh offence until after the expiry of the full term of his original sentence—the key issue is whether the fresh offence was committed before the original sentence came to an end. A magistrates' court has the power to order a prisoner to return to prison under s. 40 for a maximum of six months but may commit any prisoner in custody or on bail to the Crown Court for sentence pursuant to s. 42 of the Powers of Criminal Courts Act 1973. The order to return to prison is to be regarded as a sentence of imprisonment and may be served consecutively to or concurrently with any sentence imposed for the fresh offence (s. 40 (4)). Further, the sentence for the fresh offence shall not be influenced by the fact of the order to return to prison. These provisions may have implications for whether, upon return to prison, the prisoner is treated as a short- or long-term prisoner. For example, a court may decide that the fresh offence merits a three-year sentence and, if so, must impose that irrespective of the fact that it may also be ordering a return to prison for twelve months. It then goes on to decide whether the order to return to prison should be made to run concurrently with or consecutively to the fresh sentence. If it decides to make it run consecutively, the prisoner's single term becomes four years and he is liable to be dealt with under the Discretionary Release Scheme.

Section 40 represents a major change from the old system because it applies even where the fresh offence is committed after the expiry of the licence period and where release is 'unconditional' (i.e. in the last quarter of the sentence). Under the Criminal Justice Act 1967 the sanction of revocation of a licence only applied during the unexpired part of the licence, not the original sentence, and was limited to offences punishable on indictment with imprisonment.[53]

BREACH OF LICENCE CONDITIONS

Quite separately from the power to order a return to prison under s. 40, the 1991 Act creates a summary offence applicable to short-term prisoners of failing to comply with the conditions of a licence (s. 38 (1)). It is punishable only with a fine and hence cannot be invoked as a basis for a s. 40 order. However, the convicting court may, in addition to a fine, suspend the licence and recall the prisoner to prison for a period not exceeding six months but not, of course, if this exceeds the remaining period of the licence (s. 38 (2)). On the suspension of the licence the prisoner is deemed to be detained pursuant to the original sentence (s. 38 (3)).

Section 38 of the 1991 Act does not apply to long-term prisoners. They fall to be dealt with under s. 39, which gives the Home Secretary the power to revoke a long-term prisoner's licence, whether or not there is a technical breach of the licence conditions, and recall him to prison either on the recommendation of the Parole Board (s. 39 (1)) or without such a recommendation where it appears expedient in the public interest to recall before it is practicable to consult the Board (s. 39 (2)). This general power existed under the old parole system (and applied of course to lifers as well as all paroled prisoners). Any long-term prisoner recalled under either s. 39 (1) or (2) must be told the reasons for it and of his right to make written representations to the Parole Board, which has an absolute power to order

---

[53] s. 40 has the potential greatly to increase the prison population. Estimates have been made that with a return to prison rate of 20%, the overall prison population may increase by as much as 2,000.

immediate release (as it did under the old system).[54] The 1991 Act gives no statutory right of access to material before the Parole Board when it considers a revocation case under s. 39 but the case of *Ex parte Georghiades* suggests that any determinate sentence prisoner whose licence is revoked is entitled to access to any adverse reports/allegations before the Parole Board considering his release as a matter of natural justice.[55]

SEXUAL OFFENCES

The 1991 Act treats sex offenders whom the original sentencing court identified as coming under the provisions of s. 44 CJA 1991 differently from other offenders. Such persons will have been labelled as 's. 44 sex offenders' in the light of the criteria in s. 32 (6) of the 1991 Act, which refers to the need to protect the public from serious harm and the desirability of preventing the commission of further offences as well as securing rehabilitation. Such persons will be required to remain under supervision following their release right up to the end of their sentence and not just till the three-quarters stage. This is not meant to be punitive, rather it is intended to assist both the public and the offender by greater supervision. However, there is a punitive sting to s. 44 in that where a sex offender is recalled to prison for breach of his licence conditions he may be liable to remain there until the expiry of his original sentence and not simply until the three-quarters stage (s. 44 (*b*)).

YOUNG OFFENDERS (UNDER 21)

The 1991 Act provides that (except in very grave cases dealt with separately under s. 53 (2) of the Children and Young Persons Act 1933) no offender under the age of 15 will be committed to the custody of the Prison Service. Those offenders aged 15 and over may receive a sentence of detention in a young offenders' institution. The most serious offences may be punished by a sentence of life imprisonment.

The rules for calculating the release dates of young offenders serving determinate sentences are the same as those for adult prisoners. There are however limitations on the powers of the courts to impose detention on a young offender based on age. For offenders aged 15, 16, or 17 the maximum term of detention is twelve months. This cannot be increased by imposing a consecutive or concurrent and overlapping sentence. Section 43 (4) of the 1991 Act has the effect that offenders under the age of 18 who

---

[54] The recommendation must be for immediate release (see note 17) above.
[55] *Independent* 27 May 1992.

are sentenced to the maximum twelve months' detention will be released unconditionally half-way through their sentence rather than on licence as would apply to those aged 18 and over.

For offenders aged 18–21 there is no upper limit on the length of sentence other than that specified for the particular offence. The minimum period of detention is twenty-one days. Section 43 (2) of the 1991 Act deals with young offenders sentenced to indeterminate sentences under either s. 53 (2) of the CYPA 1933 or s. 8 of the Criminal Justice Act 1982. Such persons are to be treated for the purposes of early release as if they were life sentence prisoners.

LIFE SENTENCE PRISONERS

A combination of decisions by the domestic courts and the European Court of Human Rights together with the provisions of the 1991 Act has transformed the rights of all life sentence prisoners since 1987 but especially those of discretionary lifers. The root of the lifer 'problem' has been the continued existence of the mandatory life sentence for murder. A reluctance to distinguish between the rights of prisoners given a discretionary as opposed to a mandatory life sentence meant that a large number of prisoners were subject to the unique disadvantages and vulnerabilities which go with being on licence, literally for the whole of one's life. Eventually the domestic courts (and finally the government) accepted decisions of the European Court which made a clear distinction between the rights of prisoners detained or recalled to prison pursuant to a discretionary life sentence and those for whom the life sentence was mandatory. In the course of developing this argument, mandatory lifers were necessarily left behind as standards of procedural fairness as well as substantive rights for discretionary lifers were improved and extended. Recently, there has been an extension of some of these better standards to mandatory lifers but they still remain subject to grave disadvantages in comparison with discretionary lifers. Below we examine in some detail the development of the law applicable to lifers and describe the current rights which they enjoy.

DISCRETIONARY LIFERS

The rationale for the discretionary life sentence

While murder is the only offence which attracts a mandatory life sentence, there are several others where it may be imposed in the discretion of the sentencing judge. The use of a discretionary life sentence dates from the 1950s, when it was developed as a form of preventive detention for unstable, dangerous offenders.[56] It tends to be imposed for manslaughter, rape, buggery, or arson. The rationale for passing such a sentence is a desire to protect the public from an offender who, at the date of sentencing, has some quality of instability which makes it unsafe, in the opinion of the judge, to pass a determinate sentence. At the same time, the fact that release is discretionary gives rise to the possibility that an offender who progresses rapidly, so that he ceases to be regarded as a danger, may achieve freedom earlier than he would if a lengthy determinate sentence were imposed.[57] In 1991 it was estimated that there were 330 prisoners serving discretionary life sentences in the prison population.[58]

The Court of Appeal identified the criteria which had to be met before a judge could pass a discretionary life sentence in the criminal appeal of *R. v. Hodgson.*[59] They are:

1. when the offence or offences in themselves are grave enough to require a very long sentence;[60]
2. when it appears from the nature of the offences or from the defendant's history that he is a person of unstable character likely to commit such offences in the future;
3. when, if the offences are committed, the consequences to others may be specially injurious, as in the case of sexual offenders or cases of violence.

In a later appeal, Lord Lane, LCJ, emphasized the most exceptional circumstances which must exist before a discretionary life sentence may be imposed, saying that:

it is reserved broadly speaking for offenders who for one reason or another cannot be dealt with under the provisions of the Mental Health Act yet who are in a mental state which makes them dangerous to the life or limb of the public. It is sometimes impossible to say when the danger will subside, therefore an indeterminate sentence is required so that the prisoner's progress may be monitored and so that he

---

[56] See David Thomas, *Current Sentencing Practice* (London, 1982).

[57] The discretionary life sentence involves a departure from the basic principle of proportionality in sentencing since, of necessity, judges were imposing a more draconian sentence than the individual offence otherwise merited.

[58] This figure was based on the assumption that adult prisoners who had been convicted of murder while under the age of 18, and who had been ordered to be detained during Her Majesty's Pleasure pursuant to s. 53 of the Children & Young Persons Act 1933, were mandatory lifers. In. *R. v. Home Secretary ex parte Singh* (DC) The Times 27 Apr. 1993, the Divisional Court (Evans, LJ, and Morland, J.) held that such prisoners should be regarded as discretionary lifers at least for the purposes of deciding whether the decision in *Payne* v. *Lord Harris of Greenwich* [1981] 1 WLR 754 remained binding authority to deny such a lifer the right to have access to all reports before the Parole Board following the revocation of his licence and his recall to prison.

[59] (1968) 53 Cr. App. Rep. 13.

[60] The Court of Appeal was envisaging that only where the fixed sentence would otherwise have been fifteen years or more would a discretionary life sentence be imposed.

will be kept in custody only so long as public safety may be jeopardised by his being let loose at large.[61]

Lord Lane's reference to the need to monitor the prisoner and release him once public safety no longer demands his incarceration is of course a vital consideration, since it makes clear that punishment is not the purpose of a resort to the life sentence. Indeed frequently (though almost certainly wrongly according to the *Hodgson* criteria) judges pass discretionary life sentences in circumstances where only a relatively short sentence of three or four years would have been imposed had a purely punitive, determinate sentence been passed. This fact was of great significance when the domestic courts and the ECHR came to review the practice whereby prisoners serving discretionary life sentences were considered for release on licence on exactly the same basis as those serving mandatory life sentences for murder.

The development of the domestic case law

In four cases decided between 1987 and 1992 the English courts acknowledged the special position of discretionary lifers and identified a number of rights which could be derived from the discrete rationale applicable to their sentences. The decision of the Divisional Court in *R. v. Home Secretary ex parte Handscomb and Others* was the first breakthrough .[62] At the centre of the applications for judicial review brought by four discretionary lifers was the Home Secretary's policy of delaying for as long as three or four years the first reference of every lifer's case to the Local Review Committee and the Parole Board for the initial review of his suitability for release to be conducted. A similar period elapsed before the Home Secretary consulted the relevant judge (if available) and the Lord Chief Justice about the 'tariff' applicable to a lifer, that is, that period of time necessary to satisfy the requirements of punishment and deterrence. Since the process of review leading to eventual release on parole lasted, on average, some three years it followed that the effect of the policy inevitably meant that all lifers were detained for a minimum of six to seven years. The court held this policy to be *Wednesbury* unreasonable in so far as it applied to discretionary lifers because it ignored the rationale of their sentence, which demanded that they should not be detained as punishment for any longer than they would have been detained under a fixed-term sentence (with full remission) had a life sentence not been imposed in order to protect the public. At the point where his tariff expired, a discretionary lifer was entitled to be released if, upon review, it was decided that he was no longer dangerous. Accordingly, the court said it was essential that the judiciary were consulted immediately

---

[61] R. v. *Wilkinson* (1983) 5 Cr. App. Rep. 105.

[62] (1988) Cr. App. Rep. 59. In the earlier case of *Ex parte Gunnell* unreported CA 30 Oct. 1984 no argument had been advanced about the special rationale applicable to the discretionary life sentence.

after a life sentence was passed so that, at the outset of custody, the Home Secretary could fix the date of first review by the Parole Board and efforts could be made to monitor the prisoner's progress with that date in mind. The court also held that in determining how long a discretionary lifer ought to be detained for punitive purposes the Home Secretary should fix the first review date strictly in accordance with the judicial view on the appropriate tariff. In other words the Home Secretary had no discretion to create his own tariff by a form of Executive resentencing. This latter point was eventually overruled by the Court of Appeal,[63] but *Handscomb* was important in that, for the first time, a distinction was clearly made between discretionary and mandatory lifers which required different procedures to be adopted to ensure that the process of consideration for release on parole did not frustrate the purpose of the sentence.[64]

The following year, in the case of *R. v. Home Secretary ex parte Benson*, the Divisional Court examined the test which the Parole Board and the Home Secretary had to apply when considering whether further detention was justified after the expiry of a discretionary lifer's tariff.[65] Mr Benson had been sentenced to life imprisonment in 1972 for attempted grievous bodily harm because he was considered to have a dangerous, psychopathic personality. By 1988, however, the Parole Board had recommended his release on three occasions and a 1985 medical report (obtained on discovery) described him as 'a manipulative and immature fantasiser with a damaged personality [but] he is not dangerous'. The Home Secretary's refusal to order his release was accordingly challenged on the basis that he must have misdirected himself as to the purpose of the discretionary life sentence or taken into account irrelevant matters when addressing the dangerousness issue. Lord Justice Lloyd, commenting that dangerousness is an 'elusive quality and difficult to forecast', went on to identify what kind of risk had to be addressed by the Home Secretary:

If risk to the public is the test, risk must mean risk of dangerousness. Nothing less will suffice. It must mean there is a risk of Mr Benson repeating the sort of offence for which the life sentence was originally imposed; in other words risk to life or limb.

---

[63] R. v. *Home Secretary ex parte Doody and Others* [1992] 3 WLR 956.

[64] Following the *Handscomb* decision the Home Secretary had explained that, so far as discretionary life sentences were concerned, the first review would begin three years before the time when the prisoner would have served two-thirds of the tariff advised by the Lord Chief Justice and the trial judge. Thus if the judicial tariff was twelve years, the review would begin after five years. In *Ex parte Doody*, the Court of Appeal held that *Handscomb* had been wrongly decided on the issue of whether the Home Secretary was obliged to accept the judicial view on tariff in the case of a discretionary lifer. In view however of the acceptance of the decision by the Home Secretary in a policy statement in July 1987, it was unclear to what extent he had created a legitimate expectation that he would accept the judicial view on tariff in future.

[65] The Times 21 Nov. 1988.

On the facts, the court declined to find that the Home Secretary's refusal to accept the Parole Board's recommendation was perverse and pointed out that a decision whether or not to release on parole a discretionary lifer was a grave responsibility which the courts should be slow to characterize as irrational. Nevertheless it did find that two matters which the Home Secretary had relied on as counting against release were irrelevant to the dangerousness test and accordingly quashed the decision to refuse parole. It was the first time a lifer had successfully challenged the Home Secretary's exercise of his discretion to refuse to accept a Parole Board recommendation for release.

In *R. v. Parole Board ex parte Bradley*, the *Benson* test was approved and this time the Divisional Court explored the degree of risk which must exist to justify the continued detention of a discretionary lifer in the post-tariff period.[66] It was submitted that, because a discretionary lifer is effectively serving 'extra time' after his tariff date has been passed, the test for release should be no lower than that which had to be satisfied in the first place when the life sentence was passed. In other words, the Parole Board had to be satisfied that there was a 'likelihood' of the prisoner committing offences dangerous to life or limb in the future. The Divisional Court rejected this argument and distinguished the test which the sentencing judge had to apply before taking the drastic step of imposing a sentence which may well cause a defendant to serve longer in prison than his 'just deserts' from that which the Parole Board applies, often years later, which would have the effect of endangering public safety to an extent by enabling the prisoner's release back into society. In the circumstances, it was perfectly appropriate for the Board to formulate a lower test of dangerousness—one less favourable to the prisoner—and the court held that it was sufficient if the risk was 'substantial', which meant 'no more than that it is not merely perceptible or minimal'.

It was also argued on Mr Benson's behalf that the Court of Appeal decision in *Payne v. Lord Harris of Greenwich*[67] did not bind the court to rule that reasons need not be given for a decision to refuse parole nor to hold that he should be denied access to the reports before the Parole Board. Both these submissions were rejected by the court, but two years later, in *R. v. Parole Board and Home Secretary ex parte Wilson*[68] the Court of Appeal finally did overrule *Payne* (and the later decision of the Court of Appeal in *R. v. Parole Board and Home Secretary ex parte Gunnell*[69]) in so far as it applied to deny a discretionary lifer a right of access to any reports before the Parole Board considering his release on licence. Lord Justice Taylor characterized *Payne* as having been decided in 1981 'when established views on prisoners' rights were very different from those of today'

[66] [1991] 1 WLR 134.   [67] [1981] 1 WLR 754.   [68] [1992] QB 740.
[69] Unreported (CA) 30 Oct. 1984.

and listed six factors which had arisen since then which, in combination with the facts of Mr Wilson's case, 'constitute a formidable case for disclosure'.[70] Not least among these factors were the impending changes in the law brought about by the Criminal Justice Act 1991 and the decision of the European Court of Human Rights in the case pursued by Mr Wilson himself, *Thynne, Wilson and Gunnell*,[71] to which we turn next.

The jurisprudence of the European Court of Human Rights

The two cases to reach the ECHR, and which together transformed the rights of discretionary lifers, centred on Articles 5 (1) and 5 (4) of the European Convention on Human Rights which state that:

5 (1) Everyone has the right to liberty and security of person. No one shall be deprived of his liberty save in the following cases and in accordance with a procedure prescribed by law:

   (a) the lawful detention of a person after conviction by a competent court . . .

   (4) Everyone who is deprived of his liberty by arrest or detention shall be entitled to take proceedings by which the lawfulness of his detention shall be decided speedily by a court and his release ordered if the detention is not lawful.

In the case of *Weeks* it was argued that both these provisions had been breached.[72] Robert Weeks had been sentenced to life imprisonment in December 1966 at the age of 17 when he pleaded guilty to a bungled armed robbery during which no one was injured.[73] He was released on licence for the first time in 1976 but recalled to prison the following year. Between 1977 and 1986 he was released and re-detained several times and spent a further six years in custody. In his application to the ECHR Mr Weeks argued, first, that his recalls to prison from 1977 onwards and consequent detention were in breach of Article 5 (1) (a) of the Convention because there was not a sufficient causal connection between the original conviction

---

[70] The six factors were (1) the existence since 1983 of a parallel system of review by Mental Health Tribunals under s. 72 of the Mental Health Act 1983, in which disclosure of medical reports takes place; (2) the publication of the Carlisle Report unanimously affirming the advantages of open reporting and the giving of reasons for parole refusals; (3) the publication of the House of Lords Select Committee Report on Murder and Life Imprisonment (a committee including three former Chairmen of the Parole Board), which made similar recommendations; (4) the government's White Paper generally accepting the Carlisle Report; (5) the decision of the ECHR in *Thynne, Wilson and Gunnell*; and (6) the government's acceptance of *Thynne, Wilson and Gunnell* and the consequent enactment of the 1991 Act.

[71] (1990) 13 EHRR 666.   [72] (1987) 10 EHRR 293.

[73] The facts of Mr Weeks's original case were startling in the light of subsequent events. He had entered a pet shop in Gosport, Hampshire, with a starting pistol loaded with blanks, pointed it at the owner and told her to hand over the till. He stole 35 pence which were later found on the shop floor. Later he phoned the police and said he would give himself up. He was apprehended later by two police officers and, when he took out his starting pistol, it went off. Two blanks were fired causing no injury. It emerged that Weeks had committed the offence to pay back his mother the sum of £3 as she had told him that morning that he would have to find lodgings elsewhere.

and sentence in 1966 and the later deprivations of liberty from 1977 onwards to satisfy the requirements of Article 5 (1). In other words, it could not be said that his re-detention following his original release in 1976 was in accordance with 'a procedure prescribed by law' and undergone 'after conviction by a competent court'. The Court rejected this argument. After analysing the nature of a discretionary life sentence and the specific remarks made by the sentencing judge in Mr Weeks's case which placed it in a special category,[74] the Court found that it was inherent in the life sentence that, whether Mr Weeks was inside or outside prison, his liberty was at the discretion of the Executive for the rest of his life subject to the controls introduced by the Criminal Justice Act 1967. It was not for the Court to review the appropriateness of the imposition of the life sentence in the first place and, on the special facts of the case, the Court found there was a sufficient causal connection for the purposes of Article 5 (1) (a) between the 1966 conviction and the 1977 recall to prison. The Court inferred, however, that if a decision not to release a discretionary lifer or to re-detain him were based on grounds which were inconsistent with the objectives of the sentencing court, then the causal link may be broken and a detention lawful at the outset would be transformed into a deprivation of liberty that was arbitrary and hence incompatible with Article 5.

The second and alternative argument advanced by Mr Weeks was that on his recall to prison in 1977, or at reasonable intervals throughout his detention, he had not been able to take proceedings to challenge his imprisonment which complied with the requirements of Article 5 (4). The Court held that the stated purpose of social protection and rehabilitation which underlay the imposition of the discretionary life sentence and the grounds relied upon by the sentencing judge in Mr Weeks's case were, by their very nature, susceptible of change with the passage of time. Accordingly, Mr Weeks was entitled to apply to a 'court' having jurisdiction to decide speedily whether or not his deprivation of liberty had become unlawful at the moment of his recall to prison and also at reasonable intervals during the course of his imprisonment. The Parole Board did not satisfy this requirement because to be a 'court' the body in question must not have merely advisory functions but must be competent to decide the lawfulness of detention and order release if the detention has become unlawful. Save in a

---

[74] The trial judge, Thesiger, J., said that 'an indeterminate sentence is the right sentence for somebody of this age, of this character and disposition, who is attracted to this form of conduct. That leaves the matter with the Secretary of State who can release him if and when those who have been watching him and examining him believe that with the passage of years he has become responsible. It may not take long. Or the change may not occur for a long time—I do not know how it will work out. So far as the first count of the indictment is concerned, I think the right conclusion, terrible though it may seem, is that I pass the sentence that the law authorises me to pass for robbery and for assault with intent to rob with arms, that is life imprisonment. The Secretary of State can act if and when he thinks it is safe to act.'

recall case the Board had no power to order release. Furthermore, the lack of any right of access to the reports and other material before the Parole Board meant that the prisoner affected was not able properly to participate in the decision-making process, which was one of the principal guarantees of a judicial procedure for the purposes of the Convention. In these two important respects it followed that, in relation to both his original recall to prison and the periodic examination of his detention, the Parole Board could not be regarded as having satisfied the requirements of Article 5 (4) in Mr Weeks's case.

The *Weeks* judgment did not lead to a change in the procedures for reviewing the detention of all discretionary lifers. That came about as a result of the next case to reach the ECHR, *Thynne, Wilson and Gunnell*.[75] Whereas Mr Weeks's case was highly unusual in that the original offence which led to the life sentence was not a particularly serious one on its facts, the three applicants in *Thynne* had all been convicted of very grave offences of rape and buggery. In response to the submission that their detention was also in breach of Article 5 (4), the British government sought to argue that *Weeks* had been decided on its own very special facts and that its reasoning could not apply to all discretionary lifers, especially those where punishment was a significant aspect of the original life sentence.[76] In short, it was submitted that it was impossible to disentangle the punitive and security components in the vast majority of discretionary life sentences.

The Court rejected this argument and pointed out that the discretionary life sentence had a clear lineage and purpose. It had developed as a measure to deal with mentally unstable and dangerous offenders and:

although the dividing line may be difficult to draw in particular cases it seems clear that the principles underlying such sentences, unlike mandatory life sentences, have developed in the sense that they are composed of a punitive element and subsequently a security element designed to confer on the Home Secretary the responsibility for determining when the public interest permits the prisoner's release. This view is confirmed by the judicial description of the 'tariff' as denoting the period of detention considered necessary to meet the requirements of retribution and deterrence.

In the case of all three applicants, though they had committed grave offences deserving long sentences, they had remained in prison beyond their tariff dates and accordingly, so the Court found, they were in exactly the same position as Mr Weeks. The factors of mental instability and dangerousness which had led to the imposition of the life sentence were susceptible to change over time and new issues of lawfulness may thus arise in the

[75] (1990) 13 EHRR 666.
[76] When sentencing Gunnell Roskill J., had said 'These must be amongst the worst cases of rape or attempted rape ever to come before a court in this country . . . Punishment must be an element in this case and that punishment can only be achieved by imprisonment.'

course of their detention. It followed that they too were entitled to have their detention reviewed by a 'court' which satisfied the requirements of Article 5 (4) in ways which the Parole Board and the machinery of the Criminal Justice Act 1967 did not. The significance of the *Thynne* decision was clear. The process of reviewing the detention of discretionary lifers once they had completed the tariff portion of their sentences and on any recall to prison had to be changed in order to comply with the European Convention.[77] The 1991 Act has achieved that change though the government was strict in its interpretation that a clear distinction had been drawn between discretionary and mandatory lifers. Below we examine the rights of discretionary lifers under the 1991 Act

Discretionary Lifer Panels: The 1991 Act

Section 34 of the 1991 Act gives statutory recognition to the two parts of a discretionary life sentence—the punitive and protective phases—and requires the Home Secretary to refer the case of all discretionary lifers to the Parole Board as soon as the punitive phase has elapsed for consideration to be given as to whether they remain a danger to the public. If release on licence is not ordered, the Act gives a right to further biennial reviews by the Board so long as detention continues. The reviews are conducted by three-person panels of the Parole Board, known as Discretionary Lifer Panels (DLPs) and chaired by a judge, whose procedural rules are set out in the Parole Board Rules 1992. Before analysing the practice and procedures of DLPs however it is necessary to examine the concept of the 'penal term', since this is a key determinant of the rights and protections under the 1991 Act.

Since 1 October 1992 every judge who imposes a discretionary life sentence has been required to consider whether or not s. 34(1) should apply to it and, if it does, to specify the 'penal term' (or tariff) applicable to it. This will be the length of time which the judge would have imposed as a fixed sentence but for the 'dangerousness' element which justifies a life sentence. By stating that s. 34 (1) does apply and defining the penal term, the judge is making it clear that lifelong punishment is not the sentencing purpose. (Of course it is possible under the Act for a judge to impose a purely punitive, discretionary life sentence but it is hard to conceive of such a situation in practice)[78].

---

[77] In March 1993 the European Commission ruled as admissible an application by a mandatory lifer, Edward Wynne, who sought to extend the principles in *Thynne, Wilson and Gunnell* to mandatory lifers. Until Wynne's application, all previous applications had been ruled inadmissible. A decision from the full Court is not expected until 1994.

[78] Even in the most serious murder convictions where trial judges may specify a minimum term to be served it is rare for this to exceed thirty years. Accordingly, lifelong punishment is alien to basic sentencing principles. For the practice governing the making of an order under s. 34 see Practice Direction (Crime: Life Sentences) [1993] 1 WLR 233.

The position of discretionary lifers who were sentenced before 1 October 1992 (existing lifers) is dealt with under Schedule 12 of the 1991 Act. Paragraph 9 provides for the Home Secretary to certify as to whether, if s. 34 had been in force at the date of sentence, the judge would have ordered that it should apply to the sentence. This leaves the Home Secretary with a discretion to exclude existing discretionary lifers from the scope of s. 34 on the basis that the sentence was wholly punitive. It is a discretion susceptible to challenge by way of judicial review on normal principles. Paragraph 9 (2) of Schedule 12 gives the Home Secretary the power to specify the penal term in the certificate he issues in respect of existing discretionary lifers. Where this greatly exceeds the fixed-term sentence which would be imposed, then this would also be amenable to judicial review.[79] Once an existing lifer is confirmed as covered by s. 34, exactly the same procedures apply as to those sentenced after 1 October 1992. It was estimated that some 330 existing discretionary lifers fell to be dealt with at the coming into force of the 1991 Act and it was hoped that all of them would receive a hearing from a DLP by the end of 1993. Since the principle which underlies the *Thynne* decision and the 1991 Act itself is that all discretionary lifers have a right to be reviewed as soon as they complete their penal terms, and a right to be released if no longer dangerous, this fifteen-month delay is of dubious validity. Regrettably, the lengthy delay in obtaining a full hearing of an application for judicial review is likely to render this an ineffective remedy in most cases.

Procedural rules for DLPs

Section 32 (5) empowers the Home Secretary to make rules 'with respect to the proceedings of the [Parole] Board, including provisions authorising cases to be dealt with by a prescribed number of its members or requiring cases to be dealt with at prescribed times'. The Parole Board Rules 1992, issued pursuant to s. 32 (5), provide a comprehensive code for the conduct of DLPs in respect of both the initial release of post-tariff discretionary lifers under s. 34 and the recall to prison of such prisoners under s. 39.[80]

The appointment of the Panel

The chairman of the Parole Board appoints DLPs, each one of which must be chaired by a judge (Rule 3). The constitution of DLPs is not set out in the Rules but has been the subject of correspondence between the chairman of the Board and the Home Secretary. A High Court judge will chair each DLP which involves a prisoner convicted of terrorist offences, attempted murder/wounding of a police or prison officer, the sexual assault/mutilation

---

[79] See *Handscomb* and *R. v. Home Secretary ex parte Walsh* The Times, 18 Dec. 1991.

[80] See Appendix 6. The Parole Board Rules do not have the *status* of a Statutory Instrument and do not require parliamentary approval.

and killing of a child, serial rape, manslaughter following release on a previous manslaughter sentence, and offences involving multiple life sentences. A Circuit judge will preside in other cases. The second member of the DLP will generally be a psychiatrist but, if there is clear evidence that no real concern exists about the prisoner's state of mind, a psychologist or probation officer may be appointed. The third member will be a lay member, criminologist, or psychologist/probation officer.

Representation

Each 'party' to the hearing may be represented (Rule 6). 'Parties' means the prisoner and the Secretary of State. A prisoner may not be represented by a person liable to detention under the Mental Health Act 1983, a serving prisoner, a prisoner currently on release on licence, or any person with an unspent conviction. The Parole Board may appoint a representative where the prisoner does not authorize someone to act on his behalf. Legal aid is available for legal representation before a DLP under the ABWOR scheme.[81]

Submission of Evidence

Within eight weeks of the listing of a case before a DLP, the Home Secretary must provide the Parole Board with certain specified information set out in Schedule 1 of the Rules. This is vital material since it covers not only the background to the original offence and the sentencing judge's remarks but all pre-trial reports together with current reports on the prisoner's performance and behaviour in prison. Most importantly it includes assessment(s) of his suitability for release on licence and an up-to-date home circumstances report prepared by a probation officer. There will frequently be as many as eight to ten individual reports prepared by a wide range of people who have had contact with the prisoner during his sentence. All these reports must be served on the prisoner or his representative at the same time as they are served on the Parole Board (Rule 5 (1)). There is a discretion to withhold information from the prisoner where the Home Secretary believes it should be withheld because it might adversely affect the health or welfare of the prisoner or others (Rule 5 (2)). Information falling into this category will be kept separately and disclosed to the Board, but not the prisoner, with reasons for the decision to withhold. The chairman of the DLP will then consider the matter and make a ruling on whether to uphold the decision to withhold (Rule 9 (1) (d)). Where information is withheld, it will nevertheless be disclosed by the Home Secretary

---

[81] Assistance By Way Of Representation is a special scheme within the legal aid system which covers, for example, representation before mental health review tribunals and, formerly, disciplinary proceedings before Boards of Visitors. It will extend to covering the preparation of reports and representation by either counsel or solicitor.

to the prisoner's representative if he is a barrister or solicitor, a registered medical practitioner, or, in the view of the chairman, a suitably experienced or qualified person (Rule 5 (3)). Plainly this can give rise to difficulties for a representative who may be privy to vital information of crucial or even decisive relevance to the issue of dangerousness but upon which no instructions can be taken from the prisoner. Indeed when the information is dealt with by the DLP, the prisoner will be excluded from that part of the hearing. It is anticipated that this power will be sparingly exercised and, if it is not, it is almost certain that a challenge could be mounted on the basis that the ECHR in *Weeks* found a breach of Article 5 (4) precisely because the old procedure did not permit access to all reports before the Board.

The prisoner's initial representation about his case must be served on the Board and the Home Secretary within fifteen weeks of the case being listed for hearing (Rule 8 (1)). Any other documentary evidence which he wishes to adduce must then be served at least fourteen days before the date of the hearing. By this time the prisoner will have had access to the material disclosed by the Home Secretary.

Each party must apply in writing to the Board for permission to call witnesses before the DLP within twelve weeks of the case being listed. The identity of the witness and the substance of his evidence must be disclosed. The chairman of the DLP will then decide if the witness may be called and give reasons in writing for any refusal (Rule 7). If a prisoner wishes any other person, such as a relative, to attend the hearing as an observer/ support this too must be agreed to by the Board who will consult the governor of the prison where the DLP will sit (Rule 6 (5)).

Preliminary hearing

The DLP chairman will give preliminary consideration to the case papers and give directions on any request for attendance of witnesses, withholding of information, the timetable of proceedings, and the service of documents (Rule 9 (1)). Where necessary he may conduct an oral hearing in private and the prisoner will not attend unless he is unrepresented. Within fourteen days of being notified of any direction given by the chairman, an aggrieved party may appeal to the chairman of the Board whose decision shall be final (subject to a challenge by way of judicial review).

The hearing

An oral hearing will be held unless all parties agree otherwise (Rule 10 (1)). At least twenty-one days' notice of the date, time, and location of the hearing will be given (Rule 11 (2)). Hearings take place in the prison where the prisoner is detained and shall be held in private (Rule 12). Save in so far as the chairman of the panel directs, information about the proceedings and the names of any persons concerned shall not be made public.[82] There is accordingly a presumption in favour of confidentiality unless an express direction to the contrary is given. There is a discretion to admit persons to the hearing on appropriate terms (Rule 12 (3)). The rules for the conduct of the hearing are intended to be flexible and to provide for informality (Rule 13). Parties are entitled to appear before the DLP and be heard, hear and question each other's evidence (unless it involves withheld information), and put questions to witnesses called before the DLP. Hearsay evidence is admissible (Rule 13(5)). At conclusion of the evidence, the prisoner or his representative has a right to address the DLP (Rule 13 (7)). There is a general discretion to adjourn (Rule 14).

The DLP's decision

The decision of the DLP shall be communicated in writing to the prisoner not more than seven days after the end of the hearing (Rule 15). This is a welcome obligation since the tension of awaiting decisions concerning release can be a source of extreme anxiety. Under s. 34 (3) it is only a direction to release a discretionary lifer which binds the Home Secretary to release him on licence. Where something less than an order for release is given, then the prisoner remains liable to further detention. The DLP may indicate that release should only take place after a period of further testing in, say, open prison conditions or after a period in a pre-release employment scheme hostel (PRES). Alternatively, the DLP may direct a further hearing in, say, six months so that progress can be closely monitored. The Home Secretary is likely to comply with such recommendations and any unreasonable failure to follow a DLP's recommendation would be susceptible to judicial review. Where release is not ordered, the prisoner has a right to a further hearing no more than two years after the disposal of his last hearing.

The test to be applied by the DLP

Section 34 (4) (*b*) sets out the test to be applied by the Board as being whether it is 'satisfied that it is no longer necessary for the protection of the public that the prisoner should be confined'. The Act does not place the burden of proof on the Home Secretary to establish that further detention is necessary nor does it identify the level of risk applicable. The previous case law set out in *Benson/Bradley*[83] required there to be at least a

---

[82] In *P* v. *Liverpool Daily Post PLC* [1991] 2 AC 370 the House of Lords held that a mental health review tribunal was a court to which the law of contempt applied but that a newspaper which published the date, time, and place of a hearing and the fact that a patient had been discharged was not guilty of contempt. It would however be a contempt to publish evidence and other material on which the tribunal's decision was based in breach of Rule 21 (5) of the MHRT Rules 1983.

[83] [1991] 1 WLR 134.

substantial risk of further offences dangerous to life or limb (including of a sexual nature) and this is still relevant to s. 34. Accordingly the DLP cannot authorize further detention simply to prevent a prisoner from being a social nuisance or from committing further non-violent offences. If it did so, it would be frustrating the purpose of the Act. It remains to be seen whether the *Benson/Bradley* test is itself contrary to Article 5 (4) by placing too low a standard for further detention—'possible' rather than 'probable' dangerousness.

Of course the route by which the DLP may conclude that there exists an unacceptable risk of future dangerousness can be circuitous. Representatives must be alert to the dangers of excessive subjectivity. Naturally the final judgement of 'risk' is necessarily a subjective one but it can go too far. For example, there is no rule that requires a DLP to direct release only where a prisoner has progressed satisfactorily to open prison conditions and demonstrated his reliability and trustworthiness in such conditions. The overriding duty is to apply the s. 34 (4) (*b*) test and decide on the risk of dangerousness, and this must be done irrespective of whether the lifer concerned is in an open prison or still subject to category A conditions.[84]

How should a representative and the DLP treat a prisoner who maintains his innocence of the original offence? Of course as a matter of law this has no relevance to the requirements of s. 34 but in practice it inevitably presents problems. Those assessing the prisoner for reports to be submitted to the DLP will often state that the refusal to accept guilt makes it hard (or sometimes impossible) to assess future dangerousness. In such circumstances the evidential basis will not exist for the DLP to conclude that release can properly be ordered. It is impossible to make a hard and fast rule for such cases. Representatives must make clear to the prisoner what the consequences are of maintaining innocence and seek to direct the DLP to the objective material which exists both before and after sentence. The DLP is not the Court of Appeal and will inevitably assume that the prisoner was properly convicted, but this need not prevent it from concluding that release presents an acceptable risk.

Recall cases

A discretionary lifer who is recalled to prison under s. 39 has identical rights to a hearing before a DLP save that the rules are more flexible and designed to ensure an early hearing following recall. As with decisions under s. 34, the DLP has a power to direct release following a hearing. In practice s. 39 hearings will give rise to different problems since the focus will often be on finding, as a matter of fact, whether the conduct which gave rise to the recall

---

[84] In Nov. 1992 a Discretionary Lifer Panel directed the immediate release of Ramar Tucker from HMP Blundeston, a category C prison. See also *R.* v. *Parole Board ex parte Telling* The Times 10 May 1993.

actually occurred. This may require the calling of witnesses from the community. The same test of 'dangerousness' applies. Naturally, where a lengthy period has elapsed before recall, the evidence will have to be all the more cogent before it can justify continued detention.

The government did not accept the conclusion of the 1989 House of Lords Select Committee on Murder and Life Imprisonment. Its central proposal was that the sentence for murder should be at the discretion of the court, so that it could determine the circumstances of the offence and pass a sentence commensurate with it. This may, but need not, be an indeterminate sentence.[85] During the debate in the House of Lords on the Bill which ultimately became the 1991 Act, an attempt was made to create a 'Life Sentences Review Tribunal' which would have jurisdiction to consider the suitability for release of all lifers whose tariffs were due to expire. This amendment was later defeated in the House of Commons and accordingly mandatory lifers remain subject to the principle that release is dependent on the Home Secretary's exercise of discretion. The procedures for the review of discretionary lifers set out in the 1991 Act and the Parole Board Rules do not apply to mandatory lifers, whose release is governed by s. 35 (2) of the 1991 Act. It states that: 'If recommended to do so by the Board, the Secretary of State may, after consultation with the Lord Chief Justice together with the trial judge, if available, release on licence a life prisoner who is not a discretionary life prisoner.'

The ECHR in both *Weeks* and *Thynne, Wilson and Gunnell* distinguished the purpose which lay behind a discretionary life sentence from that which underlies a mandatory one. It derived from this the principle that, at the expiry of the punitive phase, a discretionary lifer had an Article 5 (4) entitlement to review by a 'court' whereas, arguably, no such right existed for a mandatory lifer. This distinction is of dubious validity now that it is recognized that a mandatory life sentence is also divided into a punitive (tariff) phase and a protective (post-tariff) phase. The domestic courts have, until recently, given universal acceptance to this approach on the basis that in a discretionary sentence there is always a 'notional equivalent determinate sentence' whereas there is no equivalent determinate sentence for murder ('just how long a prisoner should remain in custody for murdering someone', per Watkins, LJ, in *Handscomb*). In R. v. *Home Secretary ex parte Doody and others* Glidewell, LJ, doubted whether this is a sufficient basis

---

[85] In R. v. *Howe and Others* [1987] AC 417 Lord Hailsham observed that 'murder consists in a whole bundle of offences of vastly differing degree of culpability ranging from brutal . . . to the . . . mercy killing of a beloved partner'. Yet in the House of Commons several years later, the Home Secretary, Kenneth Baker, described it as a 'crime of dreadful finality'.

for distinguishing between the two.[86] None the less it is a distinction which survives in practice.

The procedural rights and entitlements of mandatory lifers have not however remained unaltered since the decision of the Court of Appeal in *Payne*. In *Ex parte Doody* the Court of Appeal focused on what rights a mandatory lifer enjoyed at the outset of imprisonment when his tariff was determined. Furthermore, in *R. v. Home Secretary ex parte Creamer and Scholey* the Divisional Court examined the procedural rights of mandatory lifers in the post-tariff stage, when their cases are considered by the Parole Board. The judgment of Rose, LJ, accepted that *Payne* remained binding upon him in 1992 but pointed strongly to its demise if the case is reconsidered by the Court of Appeal or House of Lords.[87] In the light of these decisions, together with the policy statements made by successive Home Secretaries during the 1980s applicable to mandatory lifers, a clear framework of rights and entitlements can now be said to exist for mandatory lifers in both the tariff and post-tariff stages.[88]

The tariff stage

When a prisoner is sentenced to life imprisonment, the Home Secretary consults the trial judge and Lord Chief Justice within a short period of time

---

[86] In *Ex parte Doody* Glidewell, LJ, pointed out that when the trial judge and the Lord Chief Justice express their views on the length of time a particular murderer should serve for the purposes of retribution and deterrence they are involved in part of the same consideration as they would be when deciding a proper determinate sentence for a serious offence other than murder. He was not therefore convinced that the distinction was a sufficient basis for concluding that the Home Secretary is bound by the judicial view in the case of discretionary lifers but not in mandatory lifers. Staughton, LJ, however, pointed out that the difficulty faced by the Home Secretary in assessing the requirements of retribution and deterrence from the judges' tariff in a murder case was a sufficient ground for his reserving the decision in mandatory life sentence cases to himself. Both judges thought *Handscomb* was wrong in so far as it held that the Home Secretary was obliged to accept the judicial view on tariff in the case of discretionary lifers, but in view of the Home Secretary's July 1987 statement accepting the *Handscomb* decision, it remained unclear if this had created a legitimate expectation in discretionary lifers that the Home Secretary would not depart from the judicial view on tariff. On 24 June 1993 the House of Lords gave its ruling on the appeal and cross appeal in *Ex parte Doody*. See Preface for details.

[87] At the time of writing the appeal in *Ex parte Creamer and Scholey* stands adjourned. This is because the effect of the Home Secretary's announcement in December 1992 that mandatory lifers will have the same rights as discretionary lifers from 1 April 1993 in terms of access to material before the Parole Board and reasons for a refusal to grant parole has made it academic to argue whether or not *Payne* is indeed good law for mandatory lifers post-April 1993.

[88] The three policy statements are (1) the 'Brittan policy': this, contained in the then Home Secretary's statement to the House of Commons on 30 Nov. 1983 in which he identified certain categories of murderers who would normally expect to serve a minimum of twenty years, remains 'in force'; (2) the Mar. 1985 parliamentary answers given by Mr Brittan in which he said that lifers with a tariff of twenty years of more would have a local review committee in any event after seventeen years; (3) the July 1987 statement by Douglas Hurd in which he said that, with both discretionary and mandatory lifers, the date of the first formal review of their cases by the Parole Board would be fixed as soon as practicable after conviction and sentence and that accordingly the views of the trial judge and Lord Chief Justice on tariff would be obtained as soon as possible after conviction. Unlike discretionary lifers, however, the Home Secretary retained a discretion to fix a tariff different from that of the judiciary.

about the length of time they consider he should serve for the purposes of retribution and deterrence—the tariff. The form which the trial judge fills out requires him to describe the circumstances in which the offence was committed, the relative culpability of co-defendants, the issues before the court, medical considerations arising (if any), and the judge's comments on the degree of dangerousness presented by the defendant. Finally, the tariff is set out. In *Ex parte Doody*, Farquharson, LJ, expressed the view that a trial judge should state in open court, after hearing submissions from defence counsel, what advice he will give the Home Secretary about the tariff, and this is increasingly the practice.

It is theoretically possible that the trial judge and the Lord Chief Justice will differ on the appropriate tariff. Where this occurs, the Lord Chief Justice's opinion prevails. It is important to stress that the 'judicial tariff' is merely advice tendered to the Home Secretary, who retains the final say on what the tariff should be. The 'real' tariff is the Home Secretary's tariff. It is not possible to seek to challenge the 'judicial tariff' because, as mere advice, it is not a 'decision' which is susceptible to challenge.[89]

The Home Secretary can and frequently does increase the judicial tariff, a practice regarded as unlawful in *Handscomb* but upheld by the Court of Appeal in *Ex parte Doody*. The Home Secretary is entitled to take into account different factors from the judiciary when approaching the tariff, including such things as public confidence in the administration of criminal justice. If, however, he increases the tariff without good reason or by an excessive amount so that it is 'so out of touch with reality that it is irrational', his decision would be challengeable by an application for judicial review.[90] Statistics presented to the Court of Appeal in *Ex parte Doody* revealed that in the period 1 April to 30 September 1988 the cases of 106 mandatory lifers were considered and in sixty-three cases the Home Secretary set a higher tariff than the judges. In sixteen cases the Lord Chief Justice increased the trial judge's tariff.

Before he sets the tariff for a mandatory lifer, the Home Secretary must:

1. provide the prisoner with the judicial tariff (both the trial judge's and the Lord Chief Justice's where they differ); and
2. provide the prisoner with any other necessary, relevant material, e.g. any comments made by the judges which will affect the Home Secretary's decision but not any departmental advice or the details of any 'in-house tariff scale' or general matters of policy unrelated to the individual prisoner; and
3. give the prisoner an opportunity to make representations in writing on what the tariff should be.

---

[89] See per Farquharson, LJ, in *Ex parte Doody* [1992] 3 WLR 956, at 991 D–E.
[90] Ibid., at 991 A–B.

There is no right to an oral hearing before the tariff is set. Once it is determined, the current practice is to inform prisoners with a tariff of twenty years or less of their first review dates and that this is three years before the date their tariff expires. Prisoners with a tariff which exceeds twenty years are not currently told the precise level of their tariff but are merely informed that their cases will be formally considered after seventeen years in any event. In other words they cannot conclude from this that their tariff is twenty years. This bizarre practice was successfully challenged in the case of discretionary lifers in the case of *R. v. Home Secretary ex parte Walsh* but it did not extend to mandatory lifers.[91] Where the Home Secretary does fix the tariff at a level which exceeds the judicial tariff he is not obliged to give reasons for departing from it. However Staughton, LJ, expressed the view in *Ex parte Doody*[91a] that in complex cases it is better to give reasons to avoid a *Wednesbury* challenge.

The post-tariff stage

Section 35 (2) of the 1991 Act does not specify any procedure by which the Home Secretary should decide whether and when to release a mandatory lifer. The choice of an appropriate procedure is within his discretion. The stage at which the Parole Board considers whether to recommend release is therefore crucial since s. 35 (2) envisages that only when the Board recommends release will the Lord Chief Justice and the trial judge (if available) be re-consulted about the question of releasing a mandatory lifer.

The current practice, as described above, is for the first review date to be set three years before the expiry of the tariff or, at the latest, after seventeen years for prisoners with tariffs in excess of twenty years. The Board has a discretion to arrange for one of its members to interview a lifer before it makes a recommendation about his release but there is no right to an oral hearing before the Board itself or even before a single member. Until December 1992, the procedural rights of mandatory lifers in the post-tariff stage remained governed by the decision in *Payne* and thus there was no legal right of access to reports submitted to the Board nor to reasons for a refusal to grant parole. In *Creamer and Scholey* Rose, LJ, expressed his disquiet at the consequences of *Payne*, stating that:

A prisoner's right to make representations is valueless unless he knows the case against him and secret, unchallengeable reports which may contain damaging inaccuracies and which result in continuing loss of liberty are, or should be, anathema in a civilised, democratic society.

---

[91] The Times 18 Dec. 1991. This decision was later upheld by the same Court of Appeal as decided *Ex parte Doody*—see (1993) ALR 138. The practice of refusing to disclose the tariff to a mandatory lifer where it exceeds twenty years is due to be challenged in the case of *R. v. Home Secretary ex parte Draper*, which, at the time of writing, is awaiting a first instance hearing before the Divisional Court.

[91a] On appeal the House of Lords held that reasons must always be given where the Secretary of State departs from the judicial view of the tariff/penal element.

In December 1992 the Home Secretary announced that mandatory lifers would from 1 April 1993 be able to have access to the reports before the Parole Board and would be given reasons for a parole 'knock-back'.

It is now clear that the test to be applied at the post-tariff stage is the same for both mandatory and discretionary lifers, namely a substantial risk of further offences dangerous to life or limb or of a serious sexual nature.[92]

Recall to prison of mandatory lifers

Where a mandatory lifer is recalled to prison under s. 39 he must be given reasons for it and be informed of his right to make written representations to the Parole Board. There is no duty to give reasons in writing but they must communicate in sufficiently clear terms what is alleged to justify the recall.[93] A lifer may be recalled by the Home Secretary without a prior recommendation of the Board, but even where the Board was consulted and recommended revocation of the licence, the prisoner's case must still be referred for further consideration by the Board where written representations are lodged.[94] There is no right to a hearing before the Board following a recall decision. Where the Board recommends immediate release, the Home Secretary must give effect to it. Otherwise, release is governed by the tripartite decision-making in s. 35 (2) of Parole Board, judiciary and Home Secretary, with the latter retaining the ultimate discretion. It is now clear that the rules of natural justice require that a mandatory lifer who is recalled to prison has a right of access to all the reports and statements which are to be presented to the Parole Board (subject to any claim for public interest immunity) before it makes its decision whether or not to uphold the decision to revoke the licence. In *R. v. Home Secretary ex parte Singh*, a case decided under the provisions of the Criminal Justice Act 1967 and before the Home Secretary's new policy for mandatory lifers came into force on 1 April 1993, the Divisional Court held that *Payne v. Lord Harris of Greenwich* did not apply to recall cases but only to the initial decision to release a mandatory lifer on licence.[95] The basis for the distinction was the fact that recalled lifers have (and always had under the 1967 Act) a statutory right to make representations to the Parole Board with respect to the reasons for their recall, a right which the Divisional Court held was meaningless without access to the material the Board will be considering. Of course, the new policy means that disclosure has happened in any event in all recall cases post-April 1993. To this extent, *Payne* has been overtaken by

[92] See *R. v. Home Secretary ex parte Cox* (1993) ALR 17, Popplewell, J., and *R. v. Home Secretary ex parte Singh* (DC) The Times 27 Apr. 1993, both cases involving mandatory lifers who were recalled to prison following their release on licence. The *Benson/Bradley* test was applied as the appropriate way of assessing the rationality of the decision further to detain them.

[93] *R. v. Home Secretary ex parte Gunnell* CA 30 Nov. 1984.

[94] CJA 1991, s. 39 (4).

[95] The Times 27 Apr. 1993.

events. The decision in *Ex parte Singh* means, however, that all mandatory lifers who were recalled to prison before 1 April 1993, and who were denied access to the material before the Parole Board, could apply for leave to move for judicial review out of time on the basis that their original revocation decision was taken in breach of the rules of natural justice. Finally, as the case of *R.* v. *Home Secretary ex parte Cox* demonstrated, a decision to cancel a provisional release date of a mandatory lifer is susceptible to judicial review on *Wednesbury* grounds. Any conduct said to justify cancellation must be related to 'dangerousness', not merely 'disobedience' or the fact that a person has breached his licence conditions by committing minor offences of dishonesty or possession of a small quantity of cannabis. The prisoner's proximity to his provisional release date, any previous periods of time at liberty and the further time to be spent in custody will also be relevant, on proportionality principles, to the decision to cancel the release plan. The Home Secretary is not however obliged to consult the Parole Board before cancelling a release date, but a lengthy delay in referring such a case back for reconsideration would be capable of challenge.[96]

---

[96] In *Ex parte Cox* (1993) ALR 17 Popplewell, J., declined to accept the submission that before cancelling Mr Cox's release date the Home Secretary was obliged to consult the Parole Board on the basis that its view on risk was so fundamental to the decision that it was irrational to fail to consult it. He commented, however, that in his view that Home Secretary would have avoided falling into error if he had consulted the Board!

10

Conclusions: Prisons and the Law

INTRODUCTION

In this final chapter we attempt an assessment of the relationship between law and prisons that has developed since the demise of the judicial 'hands off' approach, both in the United Kingdom and in Strasbourg, in the 1970s. We will consider two issues in particular. The first is the extent to which a coherent approach by the legal system and in particular the judiciary towards prisons and prisoners' rights has developed in that period. The second is what actual impact the courts' willingness to extend the rule of law to prisons has made on conditions and relationships within prisons. Neither of these matters, especially the second, has as yet been the object of much study in the United Kingdom. Legal writing has tended to emphasize the development of the law in relation to particular aspects of prison administration, whether it be discipline, access, or release. Sociologists of the prison have not paid much attention to the impact of law on prison regimes. Therefore our conclusions must be tentative. Both lawyers and sociologists in the United States have examined these issues in greater detail and we will be examining some aspects of their work to provide possible models for study of the experience in the United Kingdom.[1]

Despite the tentative nature of our conclusions we have included this chapter as we believe it is the next stage for prison law in the United Kingdom. There is a need now to consider in a more systematic way how legal regulation of prisons has developed and whether we can now speak of a distinct subject of prison law as opposed to a series of decisions and regulations on particular areas of prison life. There is also a need to consider what impact legal intervention has had on the lives of prisoners and prison staff, whether it has produced change, and what form this change has taken. Without asking, let alone answering these questions, the future legal regulation of prisons may fail to respond adequately to the problems of all who live and work in prisons. This is not to say that law can provide all the answers to the problems of prisons, just that we need to consider what

---

[1] A study of the literature on control in prisons for the Home Office observes that there is an absence of material on prison management and the impact of legal regulation in the United Kingdom. More studies are available in the American context and although one must be wary of too easily transferring ideas and observations this American literature offers useful insights, see J. Ditchfield, *Control in Prisons: A Review of the Literature* (London, 1990), 5.

answers it currently provides and what problems it could address. Before examining these two issues however it is necessary to appraise the relationship that existed between prisons and the law before the present period of judicial intervention if we are to see how judicial oversight of the prison may have led to changes.

THE PARADOX OF PRISONS AND THE LAW

In one sense there is no more legal institution than the prison. Liberal political theory posits an opposition of the public (the state) and the private (variously seen as the individual, family, or company). Whereas the various forms of the private are seen as somehow 'natural' the state is seen as 'artificial' and any intervention of the public into the private realm is seen as in need of justification.[2] The normal form of such justification is the existence of a public interest that outweighs the private interest subject to regulation, a public interest whose genuineness and legitimacy, in the democratic version of liberal theory, can ultimately be traced back to the expression of public will in elections and the delegation of that will to elected representatives. The expression of that public interest takes the form of law and in this sense the state is the law (hence the German term of *Rechtstaat* often used by political theorists); if officials detain someone or remove their property without legal authority they are no longer acting as officials of the state. That many political theorists would argue that the state precedes the law, in the sense that a monopoly on force in a society creates the right to make law, rather than the other way round, need not concern us here. We are interested in the claims to legitimate authority the state makes.

There can be few more significant interventions by the public into the private than imprisoning someone; only the imposition of the death penalty comes to mind. The decision to imprison a person, to take away their capacity to act in private society and to subject them constantly and totally to the supervision of the state, stands therefore in need of particularly clear justification by law. It is no accident then that decisions about what grounds justify imprisoning people are amongst the most fiercely contested in political debate and that the legal system has developed its most extensive procedural safeguards in the area of criminal law. Such safeguards can be seen as driven by the need to ensure that only people who do actually come within the scope of the grounds specified as justifying imprisonment actually find themselves subject to it.[3] Prisons therefore are entirely the creation of law.

---

[2] For a general discussion of the relation of public and private in liberal political theory see R. Unger, *Knowledge and Politics* (New York, 1975), Ch. 2.

[3] Public trial and fair hearing rights can also be seen as fulfilling other functions in addition to the avoidance of mistakes about who has committed an offence, such as participation in law enforcement and legitimization of punishment.

Without law they cannot be justified (hence the tort of false imprisonment) nor arguably would they be necessary. When the American radical George Jackson commented, 'the ultimate expression of law is not order, it's prison'[4] there was at least one sense in which he was clearly correct.

However, while much public and legal discussion is occupied by the question of who is to be sent to prison it is often almost assumed that the state's coercive involvement ends with the decision to imprison. This is untrue. The state does not at present (nor has it ever really) simply lock people up and throw away the key. Instead people in prison are usually required to work (if convicted), are required to take baths and submit to medical examinations, and find that they cannot take exercise, eat meals, or even go to the toilet whenever they want. Prison is therefore a continuing experience of coercion, of having nearly every aspect of one's life regulated by the state and of being dependent on the state for nearly every necessity of life. To some extent this experience of near total coercion/dependency is the inevitable outcome of imprisonment and exclusion from the outside world. To a greater extent though it is a consequence of the particular regime operating in prisons. As we observed in Chapter 6, the separation between prisons and the outside world was much less before the late eighteenth century than it is today, goods and services flowed into and out of the prison at the behest of prisoners (at least those who had money) more freely than they do now. Even in the late twentieth century the experience of the 'compound system' in Northern Irish prisons shows that prisons can exist with a much lesser degree of daily regulation of inmates' lives than is normally seen as necessary.[5] In the compounds, which were open to all prisoners convicted of terrorist-related offences between 1973 and 1976, prisoners were not required to get up or go to sleep at particular times, were not required to do prison work, and had no set times for association. In most twentieth-century prisons however the commitment to ensuring security, order, and rehabilitation of prisoners means that there is significant regulation of inmates' daily lives (often more so in medium than in maximum security environments).

We do not intend to imply that eighteenth-century gaols or the Northern Irish compounds were 'better' prison environments than that which prevails in most contemporary prisons in England and Wales. More extensive regulation may well be justified in the interests of the public, staff, and prisoners themselves. It must be said, however, that such regulation stands in need of justification; that the state cannot regulate people's lives without justification for each element of regulation (that is, at least if it is accepted that prisoners remain legal subjects and human beings, a position which

---

[4] G. Jackson, *Blood in my Eye* (London, 1972), 119.
[5] For discussion of the compound regime see C. Crawford, 'Long Kesh: An Alternative Perspective', M.Sc. thesis, Cranfield Institute of Technology (1979).

was not unequivocally held when a prisoner was subject to 'civil death' by the fact of imprisonment).[6]

It can be argued though that this extensive intervention is constantly justified through the myriad of Prison Rules, Standing Orders, and Circular Instructions. Prison is an extensively rule-bound institution where the prison authorities can almost always point to a rule at some level of the hierarchy as the basis for any action that they take. Indeed prisoners frequently list as one of their main complaints about their experience of prison that 'there are too many silly rules'.[7] Many prison officers take a pride in doing things 'by the book' and look with disfavour on decisions to ignore breaches of the rules or reach informal arrangements with prisoners. Given this density of rules in prison, this extensive literature of legal justification for intervention in prisoners' lives, why is it that some commentators have referred to prison as a 'lawless agency'.[8]

The commentators who made this judgement have argued that prisons formed islands of lawless discretion in a society guided by the values and often the practice of the rule of law. They claimed that in prisons the authorities, especially those at the lowest level,[9] exercised arbitrary power over prisoners' lives. One must make allowances for the fact that these comments have been made in respect of state prison systems in the United States, whose rule-making structures have always been more informal and decentralized than in the United Kingdom, and that they were made before extensive federal judicial oversight of prison regimes became the norm. Nevertheless such comments do strike a chord in the British prison system, certainly in its pre-*St Germain* days, and point towards the paradox of prisons, the creatures of law and highly rule-bound institutions, being places where often the rule of law is absent.[10]

When prisoners claim that there are too many silly rules they are obviously expressing a view that too many of these rules work against them. There is no necessary reason why this should always be the case. Even the most coercive system of rules is capable of being a double-edged sword, offering rights as well as burdens, constraining as well as legitimizing the exercise of power.[11] A reading of the Prison Rules 1964 shows them hardly

---

[6] Until the Forfeiture Act 1870 convicted felons forfeited all their land and chattels.

[7] This view has recently been expressed again in the *National Prison Survey 1991* (London, 1992).

[8] See D. Greenberg and F. Stender, 'The Prison as a Lawless Agency' (1972) 21 *Buffalo LR* 799.

[9] See B. Hirshkop and D. Millemann, 'The Unconstitutionality of Prison Life' (1969) 55 *Va. LR* 795.

[10] See e.g. the discussion of prison discipline in M. Fitzgerald and J. Sim, *British Prisons* (2nd edn., Oxford, 1982), 74–82, after which the authors conclude 'As in the other aspects of imprisonment we have discussed discretion, arbitrariness, expediency and unfairness are the hallmarks of discipline in British prisons.'

[11] For an exposition of this view see E. P. Thompson, *Whigs and Hunters* (London, 1977), 258–69.

to be the most coercive system of rules ever devised; many of their provisions offer protection to prisoners against excessive action by the authorities.[12] Indeed they compare favourably with many consumer contracts or housing agreements. Yet many prisoners clearly do not perceive them in this way and feel that the authorities have unlicensed power to do what they want. In terms of everyday life in prisons this view may well be exaggerated but arguably it has more than a grain of truth because of three related factors.

First, as has been pointed out several times in this book, the Prison Rules themselves are mostly vague and confer discretionary powers on the authorities. They do not set out a clear and comprehensive scheme of rights and duties on behalf of prisoners, with the result that it is often difficult to say exactly whether a particular act by the prison authorities or direction given to a prisoner is within the scope of the Rules or not. Indeed the survival of the disciplinary offence of 'disobeying any lawful order' indicates that there remains a scope of legitimate power for the authorities which is not clearly captured by the content of the Rules. Standing Orders and Circular Instructions do narrow down the extent of this discretion somewhat but invocation of them by prisoners is inhibited by the second factor, the lack of information about the Rules. Despite the requirement of Prison Rule 7 that a copy of the Prison Rules be displayed and made available to any prisoner who wishes to see them (though there is no positive duty to acquaint prisoners with them), researchers have frequently found that prisoners experience difficulty in informing themselves as to the content of the Rules.[13] So much greater then is the difficulty in becoming acquainted with the content of Standing Orders and Circular Instructions, where no obligation to publish exists. Prodded originally by some European Court decisions the government has now undertaken to make Standing Orders available but has made no similar commitment in respect of Circular Instructions. This is despite the latter's importance to decisions such as Rule 43 segregation, transfer between prisons, and dealing with prisoners diagnosed HIV positive. Without knowledge of the contents of these secondary documents prisoners are in a fairly weak position to challenge claims that there is pre-existing authority for decisions taken by officers and the prison administration.

Yet even when prisoners did challenge decisions and did have good grounds for believing that such decisions were not authorized by the Rules they found, at least in pre-*St Germain* days, that such challenges fell on

---

[12] See e.g. Prison Rule 44, limiting the use of force to what is reasonably necessary, Rule 25, on food quality, and Rule 28, limiting work to a maximum of ten hours a day.

[13] See, for example, the discussion in Loucks, N., (ed) 'Prison Rules: A Working Guide', Prison Reform Trust (London 1993), pp. 15–20. Researchers found the prisoners were still not given copies of the Prison Rules on reception at prison and that prison officers still frequently denied requests to see Standing Orders and Circular Instructions.

deaf ears. Breaches by the authorities, even of their own rules and directions, generally attracted no sanction within the prison system and until *St Germain* met no reproach in the courts. The combination of these three factors, lack of specificity in the Rules, lack of information as to their content, and lack of impartial applicability of them to prisoner and prison officer alike, deprived the extensive code of rules in prison of the character of being an expression of the rule of law.[14] Instead they became the rather threadbare cloak of a highly discretionary and frequently arbitrary regime.

Yet it was also a regime that sociologists of prison life have observed was well suited to (or indeed developed into) a particular style of control system within prisons. In his review of the literature on management regimes in prisons Ditchfield outlines a typology of four basic management models. The first of these is the 'authoritarian' model. In this all power is vested in either the prison governor or prison officers. Prisoners have no rights, although privileges (such as the more attractive prison jobs or a blind eye turned to rule infractions) may be extended to prisoners perceived by both prisoners and prison officers as leaders of the inmate population. The selective granting of privileges is aimed at ensuring the conformity of prisoners who might otherwise prove troublesome. Control, in this model, is maintained by 'a combination of terror, often brutal corporal punishment, some rudimentary incentives and favouritism'. The second model he terms 'bureaucratic-lawful'. In this model formal rules and standards exist to reduce discretion and make those who exercise power accountable, at least to the level above them in the official hierarchy. In this regime control is maintained less by coercion than by providing clear and consistent treatment of prisoners, by assuring them that they are not subject to arbitrary power. Prison officers often react to this loss of total power by becoming increasingly bureaucratized themselves, through the formation of unions, and lobby for clearer job assignments and higher staffing levels. The third he describes as 'shared powers'. This he sees as arising with the increasing emphasis on 'rehabilitation' and 'treatment' of prisoners (often of a medical or psychological character) in the prison systems of the 1950s and 1960s. This model provided for a greater involvement of civilian professional staff (such as teachers or psychologists) in prisons which in turn led to conflicts in approaches and objectives between 'treatment' and 'custody' staff. Inmate pressure groups evolved to play their part in this dispute and the prison administration found itself needing to respond to such groups to maintain control. However inmate influence is greatest in the fourth model, that of 'inmate control'. This is the view some writers take of gang influence in several American prisons. In these gangs have effectively gained

[14] This type of requirement is often seen as essential by what might be described as 'procedural' approaches to the rule of law, approaches which do not include any substantive rights in their definition; see e.g. J. Raz, 'The Rule of Law and its Virtue' (1977) 93 *LQR* 195.

control of the prison and it is only through negotiation with gang leaders, and indeed between gang leaders, that control is tentatively maintained.[15]

As observed earlier most of the examples around which such theories are constructed come from the United States. Not since the late nineteenth century does it seem that anything like a pure form of the authoritarian model has operated in British prisons. Nor has anything like the inmate control model ever prevailed in prisons in England and Wales. Yet for most of this century, despite the rehabilitative rhetoric of the Prison Commissioners,[16] the regime in most English prisons seems to have been closer to the authoritarian than the bureaucratic lawful model. The metaphor of prison as a military regime is often employed in early prison discipline cases and like an army the authoritarian model needs rules but not law. Breaches of rules are a matter of censure from the next level above, not of rights for the level below.

The adoption of more formal discipline and complaints procedures, tribunals in release decisions, and the move towards giving reasons for decisions show the prison system in the United Kingdom moving more towards a bureaucratic model in the last third of the twentieth century. Making the prison service an agency and introducing contracted-out prisons and prisoners' compacts all accelerate this trend, given their emphasis on targets, performance indicators, and auditing of results. The extent to which legal intervention is responsible for the moves in this direction is difficult to ascertain. The growth in the inmate population, a perceived rise in the number of difficult and dangerous prisoners, and rising staff costs have probably played a greater role in the feeling that there was a need for change. The authoritarian model emphasized 'getting by' or 'doing your time' with the minimum of friction, but lacked the capacity to develop priorities and objectives that are needed when 'getting by' is clearly no longer enough. It is also important not to overemphasize the move from one model to another. Most studies of prison regimes have noted how even the change in a governor can have an enormous impact on a prison, it is probably safe to assume that, even if the ideology of prison management is moving towards the bureaucratic, aspects of a more authoritarian regime continue to prevail in many prisons.

Nevertheless the courts have clearly had an impact and the next two sections examine what this has been. Before doing so it is worth returning to the notions of liberal political theory and the notion of government under law that we considered at the beginning of this section. To many

---

[15] See Ditchfield, *Control in Prisons*, 9.

[16] In the inter-war years especially the English Prison Commissioners stressed the reformative potential of prison. Such an approach, which aimed to 'treat' prisoners, also emphasizes the notion of prisoners as objects rather than the subjects of rights. See W. Forsythe, *Penal Discipline, Reformatory Projects and the English Prison Commissioners 1895–1939* (Exeter, 1990).

commentators the idea that these notions describe the British constitution has always been something of a joke.[17] Pointing to the extensive survival of prerogative powers, the power of Parliament to pass any legislation it chooses unrestrained by any constitutional norms, the ability of the governing party (which often represents a minority of the electorate) to control that Parliament, the vague power-conferring nature of many of the statutes passed, extensive government secrecy, and the weakness of other parliamentary checks such as the committee system, they have suggested the rule of law has a rather weak grip on British politics. Despite the constitutional battles of the seventeenth and eighteenth centuries they see the British constitution as remaining a 'top down' affair where the sovereign has the fact of power but delegates some limitations on it to citizens, rather than a 'bottom up' arrangement where the sovereign's power is granted, conditionally and on limited terms, entirely by the will of the populace. That the person of sovereign has changed from the monarch to the governing political party and civil service does not in the end alter the character of the fundamental relationship.

Viewed from this perspective the legal nature of the prison, the most total and coercive embodiment of the state, does not appear such an aberration but rather a heightening of the normal relation of state and citizen in the United Kingdom. Bringing the values and processes of the rule of law to bear on prisons may therefore turn out to be part of a more general move of strengthening such values in the British constitution.

JUDICIAL INTERVENTION IN PRISON LIFE

In the 1970s the judiciary, at both the domestic and European level, moved away from the 'hands off' approach to legal claims by prisoners. The *St Germain* case in England and the *Golder* case in Strasbourg were major breakthroughs in the willingness of courts to apply ordinary legal principles to the exercise of power in prisons. In the 1980s prisoners' cases began to feature ever more frequently in the judicial review and European Convention case-load. Courts also began to appear more comfortable with the role of reviewing rules and practices in prisons. Whereas in the early 1970s judges had commented that it was 'intolerable' that prison officers should have to go about their work with the threat of judicial review hanging over them,[18] by the late 1980s predictions that 'the tentacles of the law' would reach into every aspect of prison life were not enough to discourage the House of Lords from holding that it had power to review the decisions

---

[17] See M. Loughlin, *Public Law and Political Theory* (Oxford, 1992), esp. 184–230.
[18] See Lord Denning, MR, in *Becker* v. *Home Office* [1972] 2 QB 407.

of governors.[19] The judicial approach to prisons in the 1980s has often been referred to by commentators as an example of the judiciary at its most progressive, as one casebook observed: 'Generally, and as part of a process that is not yet complete, the courts have displayed a remarkable and quite unexpected willingness to involve themselves in the control of prison administration.'[20]

Yet the view that the courts have proved themselves to be firm guardians of prisoners' rights has not gone unchallenged. Other writers have argued that judicial intervention has been limited and that its impact on prisoners' lives has been far less significant than many lawyers assume.[21] This more critical view is one that we largely share. We believe that judicial intervention, while very welcome and important, has been partial and that its focus has primarily been on establishing the authority of courts over the actions of prison administrators rather than on defining and protecting the rights of prisoners. It has been at its most extensive when dealing with matters which lawyers are generally most comfortable with: issues of jurisdiction, access to courts, and trial procedures. It has been much less extensive when dealing with matters that involve the control and management of prisoners, such as transfer, segregation, safety, and living conditions. Yet decisions relating to these matters arguably have at least as great an impact on prisoners' lives and are in just as great a need of minimum standards and process values. In the words of one experienced American prison litigator, while courts have ensured that prisoners are no longer treated as slaves of the state, they have yet to recognize them as citizens behind bars.[22]

To support our argument it is worth looking again at the record of the courts in dealing with a variety of prisoners' claims. We believe that this record shows that the willingness of the judiciary to intervene varies with the type of issue posed. It also demonstrates that generally the European Commission and Court of Human Rights have proved more protective of prisoners' rights than those in England, perhaps because the legal standards they apply explicitly require them to conceptualize questions in terms of rights and legitimate restrictions on them.

---

[19] See the speech of Lord Bridge in *Leech* v. *Deputy Governor Parkhurst Prison* [1988] AC 533.

[20] See S. Bailey, D. Harris, and B. Jones, *Civil Liberties: Cases and Materials* (3rd edn.), London, 1991), 684.

[21] See e.g. the views of G. Richardson, 'Judicial Intervention in Prison Life', in M. Maguire, R. Morgan, and J. Vagg, *Accountability and Prisons: Opening up a Closed World* (London, 1985), 46–60; C. Gearty, 'The Prisons and the Courts', in J. Muncie and R. Sparks, *Imprisonment: European Perspectives* (London, 1991), 219–42; S. Livingstone, 'Prisoners Have Rights, But What Rights' (1988) 51 *MLR* 525.

[22] See A. Bronstein, 'Criminal Justice: Prisons and Penology', in N. Dorsen, *Our Endangered Rights* (New York, 1984), 221–34.

1. ESTABLISHING JURISDICTION OVER PRISON ADMINISTRATION

Clearly the issue on which judges have been most activist is in establishing their own jurisdiction over actions of the prison administration. The initial steps regarding this in domestic law were taken in the first *St Germain* case, where the Court of Appeal rejected the idea that any judicial oversight of the actions of prison officers and administrators would rapidly render the prisons unmanageable. However, this was still intervention on a fairly narrow front. Shaw, LJ, stressed the idea that 'the courts are in general the ultimate custodians of the rights and liberties of the subject whatever his status and however attenuated those rights and liberties may be as a result of some punitive or other process',[23] but his brethren based their decisions more on the quasi-judicial powers of a Board of Visitors hearing disciplinary charges. This gave the impression that the hearing of serious disciplinary charges could in some way be hived off from the general running of the prison and that judicial intervention would not affect the day-to-day actions of prison officers and governors. If this was the impression that the majority of the Court of Appeal wanted to give it was not entirely successful, as in the next nine years courts did show themselves willing at least to consider complaints by prisoners which did not relate purely to the exercise of disciplinary powers by Boards of Visitors.[24] Whether they had authority to do so remained unclear and indeed the Court of Appeal decision in *Ex parte King*,[25] suggested that judicial control did not even extend to the exercise of disciplinary powers by governors. This view was of course to be rejected by the House of Lords in the *Leech* case.[26]

The decision in *Leech* not only clearly established that governors exercising disciplinary powers were subject to judicial review, it also hinted that judicial supervision of governors' powers might extend beyond the discipline sphere. In rhetoric that echoed Shaw, LJ, in *St Germain* Lord Bridge observed that historically development of the courts' jurisdiction had been impeded 'by the court's fear that unless an arbitrary boundary is drawn it will be inundated by a flood of unmeritorious claims'.[27] Disciplinary powers of governors were always exercised more frequently than those of Boards. They were much more clearly seen by prison administrators as one of the governor's resources to maintain order in a prison, resources which included the power to segregate or transfer. Hence it was always likely to be more difficult to claim that opening their exercise to judicial review still left the 'administration' of a prison outside judicial scrutiny. Any lingering

---

[23] [1979] 1 All ER 701, 716.

[24] See e.g. R. v. *Home Secretary ex parte McAvoy* [1984] 1 WLR 1408 (transfer); R. v. *Home Secretary ex parte Hickling* The Times 7 Nov. 1985 (separation of mother and baby); R. v. *Home Secretary ex parte Herbage (No. 2)* [1987] QB 872 (prison conditions).

[25] R. v. *Deputy Governor, Camphill Prison ex parte King* [1985] QB 375.

[26] *Leech* v. *Deputy Governor Parkhurst Prison* [1988] AC 533.   [27] Ibid.

doubts as to the courts' willingness to extend jurisdiction to prison administration were removed by the House of Lords decision in Hague.[28] This decision unequivocally brought the exercise of discretionary powers in prisons within the ambit of judicial review. When combined with the courts' recognition of prisons owing prisoners a duty of care to ensure that they are not injured during the course of their imprisonment[29] it marks a clear victory for the views of Shaw LJ in *St Germain* and a declaration that the rule of law applies to prisons. The courts have indicated that the prison authorities must act within their legal powers, that indeed their power to make decisions affecting prisoners' lives comes entirely from the law, and that the courts stand ready to police breaches of those legal powers. They do not however give any real endorsement to a notion of prisoners having any rights which they might assert against the authorities, or rights which might shape or constrain the exercise of official power. Indeed in *Hague* the House of Lords rejected the idea that even some of the Prison Rules might confer rights on a prisoner to seek compensation in respect of breaches of the Rules, even though such breaches might have had a very adverse effect on a prisoner's life.

The impression the courts give is primarily of a desire to ensure that the bureaucracy functions correctly according to its defined goals. There is less concern with shaping what those goals might be, something which goes some way towards explaining the greater degree of caution courts have displayed when it comes to ruling on the content of the prison authorities' powers. Nevertheless decisions establishing jurisdiction also play a valuable role by ensuring that the public light of litigation will be shone on what is often a closed world in prisons. Even if prisoners cases prove unsuccessful, public hearings and often media coverage of them can lead to debate and reappraisal of a particular policy pursued in prisons.

At the European level the Commission and Court cast off any ideas that prisons might be beyond the scope of the Convention in the *Golder* case.[30] Therein the Court rejected any notion that prisoners' were subject to 'inherent limitations' on their rights by the fact of their imprisonment. Instead the Court indicated that the application of the qualifying clauses contained in many articles of the Convention would have to take account of the fact that the applicant was imprisoned and that a particular regime obtained in prisons which justified some restrictions which would not be acceptable in the outside world. This approach subjects prison administrations to the obligation to protect prisoners' human rights and enables

---

[28] See *Hague* v. *Deputy Governor Parkhurst Prison* [1992] 1 AC 58.

[29] See *Pullen* v. *The Prison Commissioners* [1957] 3 All ER 470.

[30] See *Golder* v. *United Kingdom* [1975] 1 EHRR 524, Series A No. 18; an earlier Commission decision to the same effect can be seen in (1961) 5 Yearbook of the European Convention on Human Rights 126.

prisoners to assert those rights. Moreover it justifies restrictions on rights by reference to the need to maintain security and safety in prison establishments as opposed to a moral judgement that prisoners should forfeit certain rights as part of their punishment, a view that was always hard to sustain when it came to restrictions on the rights of remand prisoners. As we shall see later, this basis has not prevented the European institutions from often giving a fairly limited reading to the content of prisoners' rights, but it may have encouraged a more serious examination of the interests to be considered in explaining that content than has always been evident in the English courts.

2. ACCESS TO LAWYERS AND THE COURTS

A second area of extensive judicial activity has been that of prisoners' access to legal advice and judicial proceedings. In many ways this follows as a natural consequence of the assertion of jurisdiction. If prisoners are unable to bring their claims before the courts then it is unlikely that the courts will have the opportunity to ensure that the prison administrators are acting within their powers. Hence courts in England have effectively dismantled barriers to prisoners petitioning the courts and even have gone so far as to speak in terms of such restrictions being 'unconstitutional', something rarely alluded to in British jurisprudence.[31] As a result of the decisions in *Raymond* v. *Honey* and *Ex parte Anderson* the prison authorities may not stop correspondence which is sent directly to the courts. In order to ensure effective access to the courts it has increasingly been recognized that prisoners should have unimpeded access to legal advice with respect to claims or potential claims. Here the European bodies have been more to the fore than the domestic judiciary with the decisions in *Golder*, *Boyle and Rice*, *McComb*, and *Campbell* effectively establishing that prison authorities may not stop or even read any correspondence between prisoners and their lawyers regarding ongoing or even potential litigation. Indeed the very different decision reached by the European Court in *Campbell*[32] from that of the English Divisional Court in *Leech*,[33] on the question of the confidentiality of a prisoner's correspondence with a lawyer, illustrates the impact of different judicial approaches. The European decision starts from the premiss of a prisoner having certain rights, which may be restricted in limited circumstances. The English decision starts from the idea of the administration having certain powers, whose exercise may be subject to limited constraints. The outcome of the *Campbell* decision, which may necessitate a change in the Prison Rules, ensures that prisoners should have

[31] See the observations of Lord Bridge in *Raymond* v. *Honey* [1983] 1 AC 1.
[32] See *Campbell* v. *United Kingdom* (1992) 15 EHRR 137, Series A No. 233-A.
[33] R. v. *Secretary of State ex parte Leech* unreported 22 Oct. 1991. But see now the decision of the Court of Appeal reversing the Div. Ct., The Times 20 May 1993.

unrestricted access to legal advice and this in turn completes the journey domestic and European courts have taken towards ensuring a prisoner's unrestricted access to the courts.

3. DISCIPLINARY HEARINGS

As we pointed out in Chapter 7 of this book prison discipline is arguably the area where the courts have been busiest. After jurisdiction over board hearings was established in the first *St Germain* case the courts spent much of the 1980s developing the scope of procedural requirements to ensure a fair hearing of disciplinary charges by Boards. Guidance on calling witnesses, cross-examination, availability of statements, the burden of proof, interpretation of charges and legal representation has all been forthcoming from the English courts. Though the Prison Rules remained unchanged by this litigation much of the guidance given by it was incorporated into Circular Instructions and the adjudications manual. The European Court played a less prominent role here than in the issue of access to lawyers and the courts. However its one major intervention, in the *Campbell and Fell* case,[34] was arguably a particularly influential one as without this decision (or at least without the Commission's decision) the *Tarrant*[35] court in England might not have reached its conclusion that legal representation was required in at least some circumstances in discipline hearings.

Questions of what is required to ensure a fair hearing when someone is facing a charge that might lead to punishment on conviction are of course questions that most lawyers are very comfortable with, even if they may disagree sharply on answers. They are matters which lawyers who are involved in criminal trials deal with every day.[36] Indeed when Boards of Visitors had powers to order unlimited loss of remission on conviction for certain disciplinary offences they had powers considerably in excess of most criminal courts (at least where a prisoner's length of sentence resulted in him being entitled to over six months' remission).[37] It may not be entirely fanciful to suggest that at least in the early discipline cases judicial willingness to intervene resulted from a feeling that a non-judicial body (or at least non-explicitly judicial) should not be entitled to hand out more severe punishments and with fewer constraints than a judicial body. As those punishment powers were reduced the courts showed a decreasing willingness to intervene in the conduct of discipline hearings and stopped short of apply-

[34] (1984) 7 EHRR 165, Series A No. 80.

[35] R. v. *Home Secretary ex parte Tarrant* [1985] QB 251.

[36] By this we mean that lawyers constantly work within the framework of these concepts, not that they are necessarily raising issues of the legal definition of fair trial every day. See, generally, P. Leith and J. Morison, *The Barristers World* (London, 1991).

[37] S. 31 (1) of the Magistrates' Courts Act 1980; magistrates may not impose a sentence of imprisonment in excess of six months.

ing the full procedural protections of the criminal law to Board hearings by refusing to recognize a right to legal representation in the *Hone* case.[38]

As we have argued in Chapter 7, in the end the courts appeared happy to allow board hearings to continue with what was effectively a medium level of process requirements. However political and administrative pressures led to their demise, with the criticisms in the Woolf Report being their death-knell. The removal of more serious offences to outside courts leaves only governors with limited powers to convict and order punishment within the prison system. Such powers are much less extensive than those of courts and much more clearly integrated into issues of prisoner management and control. The Northern Irish cases discussed in Chapter 7 suggest that the courts will approach these in a manner essentially similar to that adopted with regard to clearly discretionary administrative powers. That is, while there are guide-lines for their exercise and while courts will be prepared to intervene when administrators have clearly exceeded them, in general they will show deference towards the interpretation of such powers adopted by administrators. Judicial supervision is thus likely to remain at the margins and the courts will not see themselves as having an active role to protect prisoners' rights and ensure that such rights are reflected in the content and application of disciplinary powers.

4. RELEASE PROCEDURES

The fastest growing area of judicial intervention in recent years has been in reviewing the procedures governing the release of life sentence prisoners. Again the European Court has taken the lead by indicating in the *Weeks* case[39] that decisions to recall life sentence prisoners released on licence should be subject to some sort of judicial procedure and subsequently in *Thynne, Wilson and Gunnell*[40] that decisions not to release discretionary lifers purely on the grounds of their continuing dangerousness also required a judicial hearing. These decisions have prompted the legislative changes described in Chapter 9 of this book which have had the effect of substantially judicializing release procedures for discretionary lifers.

Domestic decisions also played a part in prompting such changes. In respect of discretionary lifers such decisions established that prisoners had a right to know the judicial tariff,[41] to see information considered by the Home Secretary in making such release decisions,[42] to be assured that the Home Secretary only takes account of relevant factors in reaching decisions

---

[38] *Hone* v. *Board of Visitors, Maze Prison* [1988] 1 All ER 321.
[39] *Weeks* v. *United Kingdom* (1987) 10 EHRR 293, Series A No. 143-A.
[40] *Thynne, Wilson and Gunnell* v. *United Kingdom* (1990) 13 EHRR 666, Series A No. 190.
[41] See *R.* v. *Home Secretary ex parte Handscomb and Others* (1987) 86 Cr. App. R. 59.
[42] *R.* v. *Home Secretary ex parte Benson* The Times 21 Nov.1988.

on whether a prisoner is too dangerous to be released,[43] and to make representations as regards both the tariff set by the Home Secretary and his judgement of their dangerousness.[44] However they have stopped short of indicating clearly that a prisoner has a right to be given reasons when he is refused release[45] or even that prisoners sentenced under one formulation of release policy have any rights to be consulted when that policy is changed.[46]

The life sentence cases have been one of the most significant aspects of judicial intervention in respect of prisons and have resulted in substantial changes in law and practice. In doing so the domestic and European courts have subjected what was previously a highly discretionary, secretive, and frequently arbitrary aspect of decision making within the prison system to a significant level of regulation. This regulation emphasizes the values of a fair process and hence has given prisoners an element of participation in decision-making that affects their lives which is, formally at least, generally absent in the rest of the prison system. Yet there is also a way in which judicial activity in this area can be seen as another facet of the struggle between the administration and the courts and of the courts taking back a function, control of sentencing, which they see as naturally belonging to them rather than to the administration. Judicial evidence to the House of Lords Select Committee on the sentence for murder was largely in favour of determinate sentencing.[47] With faith in the rehabilitative ideals that inspired the use of indeterminate sentences waning (though arguably this has been replaced by protecting the public from 'dangerous' people as the guiding principle), those who argue for non-judicial control of sentencing appear to be on weaker ground. Yet because life sentence release decisions can be seen as an aspect of sentencing rather than prison administration the judiciary may be more comfortable with exercising a substantial degree of control over how such decisions are reached.

5. DISCRETIONARY POWERS OF PRISON ADMINISTRATION

When it comes to the discretionary powers that prison administrators are granted over prisoners' everyday lives, powers to segregate, classify, or transfer, we witness a much more limited level of judicial intervention. As was noted earlier the courts have shown a willingness to extend their

---

[43] Ibid.

[44] See *R. v. Home Secretary ex parte Doody and Others* [1993] 1 All ER 151.

[45] See *Payne v. Lord Harris of Greenwich* [1981] 1 WLR 754. Some doubt has been cast on this position by decisions such as *R. v. Parole Board ex parte Wilson* [1992] QB 740, at least as regards discretionary lifers.

[46] See *Findlay v. Home Secretary* [1985] 1 AC 318.

[47] See *Report of the Select Committee on Murder and Life Imprisonment* (1988–9) HL Paper 78I-III.

jurisdiction over such powers, notably in the *Hague* case. However, when it comes to defining the legitimate scope of such powers, and by extension the legitimate scope of prisoners' rights to challenge their exercise, there has been a reluctance to import any procedural requirements, let alone any substantive guide-lines. Thus it has been indicated that prisoners have no right to any form of hearing in respect of decisions to classify,[48] transfer either between prisons[49] or between jurisdictions,[50] segregate,[51] or remove their baby from the prison.[52] All of these decisions may have a major impact on the lives of prisoners and often also on the lives of their families. Yet the courts seem to have accepted the view that there is a need for swift, decisive, and sometimes secret action to maintain security and order within the prison and that a greater level of process would unduly inhibit this. That prisoners may not have a hearing in respect of such decisions does not mean that the administration may act arbitrarily in making them. The courts have indicated that they will strike down exercises of discretionary power where they go beyond the scope of the powers granted or result in a decision which is clearly unreasonable. However a decision such as that in *R. v. Secretary of State for the Home Department ex parte McAvoy*[53] indicates how unlikely it is that such a conclusion will ever be reached. There the Divisional Court indicated that it could look at the Home Secretary's decision to transfer a prisoner to another prison and decide whether it was one no reasonable Secretary of State would have made. Once the Home Secretary indicated that the decision was made for 'operational and security' reasons the court swiftly ended its inquiry in his favour. Yet, as one commentator has noted, 'given the nature of the subject matter, it will not be difficult for the Home Office, quite fairly, to characterise all movement of prisoners as being for either operational or security reasons. Indeed, so broad are these criteria that it is hard to visualise such a move being made on any other basis.'[54]

These decisions suggest a strong reluctance to interfere with the exercise of discretionary administrative powers in the prison context. As we have seen throughout this book the statutory framework currently confers broad discretion on prison administrators. There is no countervailing code of constitutional or statutory rights of prisoners. As we will argue later in this chapter, judges anxious to give effect to a notion of prisoners having rights might have imported ideas of rights, perhaps drawing on the European Convention, to define and structure the exercise of discretion. This they

---

[48] See *Payne* v. *Home Office* unreported 2 May 1977.
[49] See *R.* v. *Home Secretary ex parte McAvoy* [1984] 1 WLR 1408.
[50] See *R.* v. *Home Secretary ex parte McComb* The Times 15 Apr. 1991.
[51] See *Williams* v. *Home Office (No. 2)* [1982] 1 All ER 1211.
[52] See *R.* v. *Home Secretary ex parte Hickling* The Times 7 Nov. 1985.
[53] [1984] 1 WLR 1408.   [54] See Gearty, 'The Prisons and the Courts', at 228.

have not done.[55] Instead a deference has generally been exhibited to the interpretation of the authorities. After the decision in *Hague* that none of the Prison Rules may give rise to an enforceable right to sue the need for a new Prison Act is even clearer. The passing of such a statute would be an occasion to examine how much discretion is necessary or desirable and such an Act could set enforceable limits on administrative power within prisons.

6. PRISONERS' LIVING CONDITIONS

While the cases in the categories discussed above refer to situations where the state makes decisions which affect prisoners' lives, those in this category reflect situations where the state has failed to act. Complaints relating to overcrowding or insanitary conditions, inadequate medical treatment, or a lack of safety often allege neglect and lack of action by the state. Perhaps because of a traditional reluctance to impose affirmative duties on the state prisoners' claims in this area have met with little success. Hence claims relating to medical treatment in *Freeman* and *Knight*[56] were unsuccessful and suggest a lack of willingness in the courts to concern themselves with issues of medical treatment in prison, which have been the subject of extensive discussion among doctors and prison reform groups. The negligence cases relating to prisoner assaults[57] largely demonstrate that the courts' sympathies lie initially with the difficult task of the authorities. Only where clear warning signs have been ignored or where the authorities have been grossly neglectful have prisoners been able to recover in respect of assaults committed by other prisoners. The courts do not appear to see themselves as having a role of encouraging prison administrators to review their approach to ensuring prisoner safety.[58]

Perhaps most disturbing in this area has been judicial reluctance to respond to prisoner claims of inhuman and degrading living conditions. Given the regularity with which reports by the Chief Inspector of Prisons throughout the 1980s have drawn attention to the dirty, overcrowded, and

[55] In *McAvoy*, the court did note that there were both prisoner (closeness to family and better preparation for trial) and administration (security and managerial) interests to be balanced. They then however struck the balance essentially by adopting the view of the administration as no attempt was made to weigh the operational or security interests. Aleinikoff, looking at American courts where balancing rhetoric is more often employed, notes that such deferring balancing is not uncommon, see T. Aleinikoff, 'Constitutional Law in the Age of Balancing', (1987) 96 *Yale LJ* 943, 976.

[56] See *Freeman* v. *Home Office* [1984] 1 All ER 1036 and *Knight* v. *Home Office* [1990] 3 All ER 237.

[57] See e.g. *Ellis* v. *Home Office* [1953] 2 QB 135, *Egerton* v. *Home Office* [1978] Crim. LR 494 and *Palmer* v. *Home Office*, The Guardian 31 March 1988.

[58] Courts in the United States, by including issues of safety and staffing levels in injunctions relating to Eighth Amendment prison conditions claims, have addressed these issues. Arguably they have also been faced by higher levels of violence in American prisons.

insanitary conditions in which many prisoners spend their lives, and given the apparent failure of government to do anything about this, the area might seem ripe for some sort of judicial intervention. We observed in Chapters 3 and 5 that there are significant procedural hurdles to bringing a conditions case before the British courts. Nevertheless ways have been found to present conditions issues to the judiciary, which has effectively discouraged further efforts. Perhaps the worst example of such judicial passivity occurred in the Williams[59] case and exhibited deference to actions of the authorities which even they came to understand as indefensible. The narrow, possibly very narrow, scope for judicial intervention in respect of 'intolerable' conditions that was sketched out by the House of Lords in the *Hague* case does not hold out too much hope for those seeking to change inhuman and degrading prison conditions by way of court action. If the views of Lord Goff in that decision prevail then conditions suits will probably only be available where the treatment of the prisoner is such as to justify a finding of negligence, an approach which will require some evidence of physical injury. Lord Bridge suggested a broader basis for claims of intolerable conditions but in seeming to permit negligence claims without proof of injury it lacks much support in precedent. Only in the *Herbage* case[60] has there been clear judicial endorsement for the idea that inhuman conditions, however they have occurred, are a matter of judicial concern. Even then this was in a case primarily concerned with discovery and the Court had no need to give any indication of what steps it would be prepared to take to remedy a set of conditions that amounted to cruel and unusual punishment.

Given the existence of an Article 3 prohibition on torture and inhuman or degrading treatment and punishment, one might expect the Strasbourg institutions to be rather more forthright in tackling unsatisfactory living conditions in prisons. They certainly have indicated that things like the toleration of overcrowding can amount to degrading treatment[61] and have produced rather more sympathetic rhetoric on these issues than in the English courts, for example in the *McFeeley* case.[62] In the end however decisions such as *Hilton*[63] indicate that the Commission and Court are rather more happy with tackling whether specific state practices are prohibited by Article 3 than with developing the kind of 'totality of conditions' approach that can be seen in the jurisprudence of the United States courts. The Strasbourg institutions also seem unwilling to involve themselves too deeply in second guessing European prison administrations on how prisons

---

[59] [1981] 1 All ER 1211.   [60] [1987] QB 872, 1077.
[61] See *Denmark and Sweden* v. *Greece* (1969) 12 Yearbook of the European Convention on Human Rights.
[62] *McFeeley* v. *United Kingdom* (1980) 3 EHRR 161.
[63] *Hilton* v. *United Kingdom* (1978) 3 EHRR 104.

should be managed or resources allocated. They too appear to feel that such matters are within the competence of the prison authorities and that only instances of clearly harsh treatment justify intervention.

Judges therefore have not intervened in a uniform way into prison life. Rather the more the issue looks like sentencing or an adjunct to judging the more willing they have been to lay down procedures and standards. The more it looks like an issue of prison management the more they have exhibited deference to the administration and a reluctance to uphold prisoners' claims. Such caution about reaching too deeply into the internal administration of prisons is very understandable. Judges may claim that they lack the expertise and information to set standards and monitor performance as regards prison administration. Freed from the context of litigation, though, members of the judiciary, such as Judge Tumim and Lord Woolf, have made significant contributions to the debate on how prisons should be treating prisoners.

Judges in the United Kingdom might also argue that the experience of the United States, where the judiciary has intervened on a more extensive scale, has not been a universally happy one. The federal judiciary in the United States has played a prominent role in establishing procedures and standards to govern broad areas of prison life. At its most extensive such intervention has placed individual prisons or entire state prison systems under judicial supervision in the name of protecting prisoners' constitutional rights. However, a fierce debate now rages as to whether such intervention has crossed the boundary that separates judicial from legislative action and whether the legitimacy of judicial action has thereby been undermined. Commentators also debate whether such intervention has produced more humane prison conditions or whether it has led to a breakdown in authority in prisons and an increase in violence perpetrated against both prison officers and other prisoners.[64]

In the end British judges may argue that they go as far as the law allows them. They can point out that Parliament in the 1952 Prison Act has given the responsibility for running the prison system to the Home Secretary and that he is answerable ultimately to Parliament for how this is done. The Home Secretary is within the powers granted to him when he lays before Parliament rules conferring broad discretionary powers on prison staff and administrators. All the courts can do is ensure that such officials act within their powers and do not abuse their discretion. There is no independent legal basis under which courts may intervene to challenge such powers or structure the way in which they are exercised. Such an approach lays bare the 'top down' nature of the United Kingdom's constitution. It also shows the ambiguous nature of Lord Wilberforce's remark that 'In English law a

[64] See the summary in J. DiIulio, 'Conclusion: What Judges can do to Improve Jails', in J. DiIulio, *Courts, Corrections and the Constitution* (New York, 1990), 287.

convicted prisoner retains all civil rights which are not taken away expressly or by necessary implication', for it is not clear in English law what civil rights any of us have. Nor is it clear whether the courts see themselves as having the power to determine independently when it is necessary to remove them, as opposed to deferring to the views of prison administrators on the issue of necessity.

We would argue strongly against such ultimate passivity and suggest that there are legitimate reasons why courts should be prepared to intervene to protect prisoners' rights. These lie in the fact that prisoners are, in the words of Chief Justice Stone of the United States Supreme Court, a 'discrete and insular minority'.[65] Prisoners are not a popular political cause and it is therefore unlikely that their interests will receive much of a hearing in the political sphere. Politicians, in the United Kingdom as elsewhere, spend a fair amount of time discussing who should go to prison, and for what, but much less time on what happens once they get there. The one time that prisons do force their way on to the political agenda is when a major disturbance occurs, a fact not lost on some prisoners. This suggests that leaving everything to the accountability of the Home Secretary to Parliament is a particularly weak safeguard of prisoners' interests. Courts are well placed to correct this deficit in accountability. They can do this without taking over the running of prisons by instead assisting the development of clearer standards and procedures within the prison that allow for effective prison management without sacrificing the principle that deprivation of liberty is sufficient punishment for an offender. An American writer, Susan Sturm, reviewing the American experience of judicial intervention, argues that judges who take on the function of being a 'catalyst' for change which enlists the participation of staff and prisoners in making changes are more likely to produce healthy and lasting changes than those who seek to 'direct' all changes themselves.[66]

If judicial reluctance to intervene more broadly results from a view that English law supplies no standards of rights on which such intervention could be based it can be argued that such standards can be supplied by European human rights law. Anthony Lester and Jeffrey Jowell have argued that courts can give more specific content to the application of the 'reasonableness' test of the exercise of discretionary power by drawing on the rights protected by the Convention. Where administrative decisions have been taken which infringe rights granted by the Convention they argue that the courts should find such actions unreasonable.[67] Although such an approach is unlikely to impinge on

---

[65] See footnote 4 to *United States* v. *Carolene Products* (1938) 304 US 144, 152–3.

[66] S. Sturm, 'Resolving the Remedial Dilemma: Strategies of Judicial Intervention in Prisons' (1990) 138 *University of Pennsylvania LR* 805.

[67] J. Jowell and A. Lester, 'Beyond Wednesbury: Substantive Principles of Administrative Law' [1987] *Public Law* 368.

actions of alleged neglect by the state it could, in the prison context, lead to a more activist judicial approach towards the widespread exercise of discretionary powers which affect the lives of prisoners. Unfortunately for advocates of such a way forward the House of Lords has rejected it in the *Brind*[68] case as incorporating the Convention by the back door.

In the absence of European standards to give judges a helping hand towards a more sure-footed intervention attention is likely to focus again on a domestic code of standards. Such standards, which are likely to be significantly more specific and comprehensive than the European Convention provisions, could provide judges with a basis for subjecting the actions of prison staff to a greater degree of regulation than has hitherto been the case. If prisoners were given a right of action in respect of breaches of the code this could be done directly. On the other hand if a code were given the status of being something that had to be taken account of, though not in itself legally enforceable, along the lines of the PACE or equal opportunities codes, then it could still help define the authorities' duty of care or what would constitute 'unreasonable action by them'. Making them enforceable would obviously seem the preferable solution since it would be less likely to lead to a period of confusion and uncertainty. The provisions in prisoners' contracts or 'compacts', or the standards set out in the contracts relating to contracted-out prisons, might fulfil a similar legal function to a code of standards. However, since compacts will not be legally binding and since prisoners are not parties to the contracts between the Prison Service and the contractor it seems likely that prisoners will only be regarded as having a 'legitimate expectation' to the entitlements contained in them. This characterization would limit prisoners' claims in respect of deprivation of one of these entitlements to a claim that they should have been given some sort of a hearing before its deprivation.[69] This is a much less satisfactory way of ensuring that the judiciary safeguard the interests of this discrete and insular minority. As we have argued earlier a new statutory framework is probably needed to achieve this objective.

THE IMPACT OF JUDICIAL INTERVENTION ON PRISON LIFE

At a formal level it is clear that the willingness of the courts to examine the lawfulness of actions by the prison authorities has had a significant impact. Earlier chapters of this book detail the legal changes that have come about

[68] See *Brind* v. *Secretary of State for the Home Dept.* [1991] 2 WLR 588. That the Convention is not dead in English law is however shown by the decision of the Court of Appeal in *Derbyshire CC* v. *Times Newspapers* [1992] 2 WLR 28.

[69] See the views of G. Richardson, 'From Rights to Legitimate Expectations', in M. Jenkins and E. Player, *Prisons After Woolf* (London, forthcoming).

since the courts renounced their 'hands off' policy with respect to prisons. There has been a complete overhaul of the prison disciplinary system. The procedures for deciding on the release of discretionary life sentence prisoners have been comprehensively altered by the Criminal Justice Act 1991. Both the Prison Rules and Standing Orders regarding correspondence and, in particular, communications with lawyers have been rewritten on several occasions. The procedures regarding transfer of prisoners for reasons of good order and discipline have been substantially altered. Prisoners are now allowed to marry.

All of these changes can be traced to court decisions. However, we would not want to suggest that court intervention has been the only cause of change within the prison system in the last fifteen or twenty years. Clearly factors such as government policy objectives, sentencing policy, staffing disputes, and prisoner disturbances have all played a role. In dealing with a complex and dynamic entity such as a prison system it is often difficult and perhaps unwise to point to one factor which has brought about change.[70] It may be worth pointing out though that writers on the prison system in the United Kingdom, even those with legal backgrounds, do not tend to focus on the impact of the courts as a major influence on developments within prisons.[71] Despite the impact that the courts have had the government has not seen the need yet to alter the Prison Act 1952, although in its response to Woolf it indicated that this was being considered and with the introduction of prisoners' contracts and a code of standards it may be needed.

The courts and the law may thus be an influence at the margin rather than the centre of prison life. However the shadow they cast has grown considerably in recent years. Apart from the formal changes in rules which were referred to above it is worth asking what impact judicial intervention has had on the character of imprisonment. In the first part of this chapter we suggested that the absence of judicial oversight helped foster an attitude that, while prison was densely rule governed, the rules only cut one way: that in reality they served as a cloak for a mixture of authoritarianism and selective accommodation. Unfortunately there does not appear as yet to have been an attempt to examine to what extent the courts' limited willingness to intervene over the last ten to fifteen years has brought about a change in this culture. Again studies of the more extensive judicial intervention into the management of prisons in the United States may provide some

[70] Sturm, 'Resolving the Remedial Dilemma', notes that failure to see prisons as dynamic entities wherein simple direction from the top is likely to find itself resisted and deflected at a variety of levels has impeded judicial efforts at prison reform.

[71] Books such as A. Rutherford, *Prisons and the Process of Justice* (Oxford, 1986), V. Stern, *Bricks of Shame: Britain's Prisons* (London, 1987), and P. Cavadino and J. Dignan, *The Penal System: An Introduction* (London, 1992) tend to stress sentencing decisions, concern with security, and staffing problems as the key determinants in prison policy-making.

pointers. They are pointers, however, which remain a matter of considerable debate among prisons scholars in the United States.[72]

Such studies suggest that judicial oversight does lead to a more formalized and bureaucratic style of prison management. Judicial demands for standards and the introduction of process values (such as reasons, record-keeping, and hearings) tend to undercut informal deal-making and the selective exercise of power. For a time prison staff and administrators find this new regime difficult to deal with and the prison may pass through a period of uncertainty, which in turn may foster a sense of tension and an increase in unrest. Most studies suggest though that this period passes and that prison staff (often new prison staff) accommodate themselves to the new type of regime and develop different techniques for maintaining control and order in the prison.[73] What may then tend to become a problem is that fulfilling bureaucratic requirements becomes the main objective of actors in the prison system and that the prison may lose a sense of purpose. Liberal 'treatment-orientated' regimes can just as easily fall foul of judicial imperatives as harsh, authoritarian regimes. This is because treatment-orientated regimes also require individualized and frequently unequal treatment of prisoners. This American experience does however reinforce the idea that maintaining control in the prison environment is compatible with a number of different styles of regime. Excessive judicial deference to claims by prison administrators that certain actions are necessary to maintain good order and discipline, that control cannot be maintained otherwise, may therefore be misplaced.

With contracting out, prisoner contracts, and the move to agency status of the Prison Service, prisons in the United Kingdom may be entering a new bureaucratic phase, even if this is primarily in response to financial rather than legal influences. These changes should lead to clearer lines of authority, to more explicit policy-making, and to more detailed standards, but whether they lead to the better protection of the human rights of prisoners may depend on the extent to which the courts are prepared to ensure such standards are kept to. Failure to intervene, for example to ensure that accommodation or sanitation standards contained in a code are adhered to, or too great a willingness to accept inadequacy of resources arguments may lead to prisoners having rights in form but not in substance. The experience of the past fifteen years has shown that prisons are a matter for the law. The issue in the years ahead is just what the nature of that relationship is.

---

[72] For a flavour of this debate see the books reviewed in S. Rhodes, 'Prison Reform and Prison Life: Four Books on the Process of Court Ordered Change' (1992) 26 *Law and Society Review* 189.

[73] See E. Alexander, 'New Directions in Prison Law' (1978) 56 *Texas LR* 963, arguing that prison administrators in the United States reacted to adverse court decisions by rewriting rules to provide a more formal basis for the exercise of discretionary power without altering the substance.

Appendix 1
Prison Act 1952

(15 & 16 Geo 6 and 1 Eliz 2 c 52)

An Act to consolidate certain enactments relating to prisons and other institutions for offenders and related matters with corrections and improvements made under the Consolidation of Enactments (Procedure) Act 1949 [1 August 1952]

Central administration

1 General control over prisons

All powers and jurisdiction in relation to prisons and prisoners which before the commencement of the Prison Act 1877 were exercisable by any other authority shall, subject to the provisions of this Act, be exercisable by the Secretary of State.

2 *(Repealed by the Prison Commissioners Dissolution Order 1963, SI 1963/597.)*

3 Officers and servants of Prison Commissioners

(1) The Secretary of State [may, for the purposes of this Act, appoint such officers

and servants as he] may, with the sanction of the Treasury as to number, determine.

(2) There shall be paid out of moneys provided by Parliament to [the officers and servants appointment under this section] such salaries as the Secretary of State may with the consent of the Treasury determine.

4 General duties of Prison Commissioners

(1) [The Secretary of State] shall have the general superintendence of prisons and shall make the contracts and do the other acts necessary for the maintenance of prisons and the maintenance of prisoners.

(2) [Officers of the Secretary of State duly authorised in that behalf] shall visit all prisons and examine the state of buildings, the conduct of officers, the treatment and conduct of prisoners and all other matters concerning the management of prisons and shall ensure that the provisions of this Act and of any rules made under this Act are duly complied with.

(3) [The Secretary of State and his officers] may exercise all powers and jurisdiction exercisable at common law, by Act of Parliament, or by charter by visiting justices of a prison.

5 Annual report of Prison Commissioners

[(1) The Secretary of State shall issue an annual report on every prison and shall lay every such report before Parliament.]

(2) The report shall contain—

(*a*) a statement of the accommodation of each prison and the daily average and highest number of prisoners confined therein;

(*b*) such particulars of the work done by prisoners in each prison, including the kind and quantities of articles produced and the number of prisoners employed, as may in the opinion of the Secretary of State give the best information by Parliament;

(*c*) a statement of the punishments inflicted in each prison and of the offences for which they were inflicted . . .

[5A Appointment and functions of Her Majesty's Chief Inspector of Prisons

(1) Her Majesty may appoint a person to be Chief Inspector of Prisons.

(2) It shall be the duty of the Chief Inspector to inspect or arrange for the inspection of prisons in England and Wales and to report to the Secretary of State on them.

(3) The Chief Inspector shall in particular report to the Secretary of State on the treatment of prisoners and conditions in prisons.

(4) The Secretary of State may refer specific matters connected with prisons in England and Wales and prisoners in them to the Chief Inspector and direct him to report on them.

(5) The Chief Inspector shall in each year submit to the Secretary of State a report in such form as the Secretary of State may direct, and the Secretary of State shall lay a copy of that report before Parliament.

(6) The Chief Inspector shall be paid such salary and allowances as the Secretary of State may with the consent of the Treasury determine.]

Visiting committees and boards of visitors

6 Visiting committees and boards of visitors
(1) . . .

(2) The Secretary of State shall appoint for every prison . . . a board of visitors of whom not less than two shall be justices of the peace.

(3) Rules made as aforesaid shall prescribe the functions of . . . boards of visitors and shall among other things require members to pay frequent visits to the prison and hear any complaints which may be made by the prisoners and report to the Secretary of State any matter which they consider it expedient to report; and any member of a . . . board of visitors may at any time enter the prison and shall have free access to every part of it and to every prisoner,

(4) . . .

Prison officers

7 Prison Officers
(1) Every prison shall have a governor, a chaplain and a medical officer and such other officers as may be necessary.

(2) Every prison in which women are received shall have a sufficient number of women officers; . . .

(3) A prison which in the opinion of the Secretary of State is large enough to require it may have a deputy governor or an assistant chaplain or both.

(4) The chaplain and any assistant chaplain shall be a clergyman of the Church of England and the medical officer shall be duly registered under the Medical Acts.

(5) . . .

8 Powers of prison officers
Every prison officer while acting as such shall have all the powers, authority, protection and privileges of a constable.

9 Exercise of office of chaplain
(1) A person shall not officiate as chaplain of two prisons unless the prisons are within convenient distance of each other and are together designed to receive not more than one hundred prisoners.

(2) Notice of the nomination of a chaplain or assistant chaplain to a prison shall, within one month after it is made, be given to the bishop of the diocese in which the prison is situate; and the chaplain or assistant chaplain shall not officiate in the prison except under the authority of a licence from the bishop.

10 Appointment of prison ministers
(1) Where in any prison the number of prisoners who belong to a religious denomination other than the Church of England is such as in the opinion of the Secretary of State to require the appointment of a minister of that denomination, the Secretary of State may appoint such a minister to that prison.

(2) The Secretary of State may pay a minister appointed under the preceding subsection such remuneration as he thinks reasonable.

(3) [The Secretary of State] may allow a minister of any denomination other than

the Church of England to visit prisoners of his denomination in a prison to which no minister of that denomination has ben appointed under this section.

(4) No prisoner shall be visited against his will by such a minister as is mentioned in the last preceding subsection; but every prisoner not belonging to the Church of England shall be allowed, in accordance with the arrangements in force in the prison in which he is confined, to attend chapel or to be visited by the chaplain.

(5) The governor of a prison shall on the reception of each prisoner record the religious denomination to which the prisoner declares himself to belong, and shall give to any minister who under this section is appointed to the prison or permitted to visit prisoners therein a list of the prisoners who have declared themselves to belong to his denomination; and the minister shall not be permitted to visit any other prisoners.

11 Ejectment of prison officers and their families refusing to quit
(1) Where any living accommodation is provided for a prison officer or his family by virtue of his office, then, if he ceased to be a prison officer or is suspended from office or dies, he, or, as the case may be, his family, shall quit the accommodation when required to do so by notice of [the Secretary of State].

(2) Where a prison officer or the family of a prison officer refuses or neglects to quit the accommodation forty-eight hours after the giving of such a notice as aforesaid, any two justices of the peace, on proof made to them of the facts authorising the giving of the notice and of the service of the notice and of the neglect or refusal to comply therewith, may, by warrant under their hands and seals, direct any constable, within a period specific in the warrant, to enter by force, if necessary, into the accommodation and deliver possession of it to [a person acting on behalf of the Secretary of State].

Confinement and treatment of prisoners

12 Place of confinement of prisoners
(1) A prisoner, whether sentenced to imprisonment or committed to prison on remand or pending trial or otherwise, may be lawfully confined in any prison.

(2) Prisoners shall be committed to such prisons as the Secretary of State may from time to time direct; and may by direction of the Secretary of State be removed during the term of their imprisonment from the prison in which they are confined to any other prison.

(3) A writ, warrant or other legal instrument addressed to the governor of a prison and identifying that prison by its situation or by any other sufficient description shall not be invalidated by reason only that the prison is usually known by a different description.

13 Legal custody of prisoner
(1) Every prisoner shall be deemed to be in the legal custody of the governor of the prison.

(2) A prison shall be deemed to be in legal custody while he is confined in, or is being taken to or from, any prison and while he is working, or is for any other reason, outside the prison in the custody or under the control of an officer of the

prison [and while he is being taken to any place to which he is required or authorised by or under this Act [or the Criminal Justice Act 1982] to be taken or is kept in custody in pursuance of any such requirement or authorisation.]

14 Cells

(1) The Secretary of State shall satisfy himself from time to time that in every prison sufficient accommodation is provided for all prisoners,

(2) No cell shall be used for the confinement of a prisoner unless it is certified by an inspector that its size, lighting, heating, ventilation and fittings are adequate for health and that it allows the prisoner to communicate at any time with a prison officer.

(3) A certificate given under this section in respect of any cell may limit the period for which a prisoner may be separately confined in the cell and the number of hours a day during which a prisoner may be employed therein.

(4) The certificate shall identify the cell to which it relates by a number or mark and the cell shall be marked by that number or mark placed in a conspicuous position; and if the number or mark is changed without the consent of an inspector the certificate shall cease to have effect.

(5) An inspector may withdraw a certificate given under this section in respect of any cell if in his opinion the conditions of the cell are no longer as stated in the certificate.

(6) In every prison special cells shall be provided for the temporary confinement of refractory or violent prisoners.

15 *(Repealed by the Criminal Justice Act 1967, ss 66 (2), 103 (2), Sch 7, Pt I.)*

16 Photographing and measuring of prisoners

The Secretary of State may make regulations as to the measuring and photographing of prisoners and such regulations may prescribe the time or times at which and the manner and dress in which prisoners shall be measured and photographed and the number of copies of the measurements and photographs of each prisoner which shall be made and the persons to whom they shall be sent.

17 Painful tests

The medical officer of a prison shall not apply any painful tests to a prisoner for the purpose of detecting malingering or for any other purpose except with the permission of [the Secretary of State] or the visiting committee or, as the case may be, board of visitors.

18 *(Repealed by the Criminal Justice Act 1967, ss 65, 103 (2), Sch 7, Pt I.)*

19 Right of justice to visit prison

(1) A justice of the peace for any county . . . may at any time visit any prison in that county . . . and any prison in which a prisoner is confined in respect of an offence committed in that county . . . , and may examine the condition of the prison and of the prisoners and enter in the visitors' book, to be kept by the governor of the prison, any observations on the condition of the prison or any abuses.

(2) Nothing in the preceding subsection shall authorise a justice of the peace to communicate with any prisoner except on the subject of his treatment in the prison, or to visit any prisoner under sentence of death.

(3) The governor of every prison shall bring any entry in the visitors' book to the attention of the visiting committee or the board of visitors at their next visit.

20 *(Repealed by the Courts Act 1971, s 56, Sch 11, Pt IV.)*

21 Expenses of conveyance to prison
A prisoner shall not in any case be liable to pay the cost of his conveyance to prison.

22 Removal of prisoners for judicial and other purposes
(1) Rules made under section forty-seven of this Act may provide in what manner an appellant within the meaning of [Part I of the Criminal Appeal Act 1968], when in custody, is to be taken to, kept in custody at, or brought back from, any place which he is entitled to be present for the purposes of that Act, or any place to which the Court of Criminal Appeal or any judge thereof may order him to be taken for the purpose of any proceedings of that court.

(2) The Secretary of State may—

    (*a*) . . .
    (*b*) if he is satisfied that a person so detained requires [medical investigation or observation or] medical or surgical treatment of any description, direct him to be taken to a hospital or other suitable place for the purpose of the [investigation, observation or] treatment;

and where any person is directed under this subsection to be taken to any place he shall, unless the Secretary of State otherwise directs, be kept in custody while being so taken, while at that place, and while being taken back to the prison in which he is required in accordance with the law to be detained.

23 Power of constable etc to act outside his jurisdiction
For the purpose of taking a person to or from any prison under the order of any authority competent to give the order a constable or other officer may act outside the area of his jurisdiction and shall notwithstanding that he is so acting have all the powers, authority, protection and privileges of his office.

Length of sentence, release on licence and temporary discharge

24 Calculation of term of sentence
(1) In any sentence of imprisonment the word 'month' shall, unless the contrary is expressed, be construed as meaning calendar month.

    (2) . . .

25 Remission for good conduct and release on licence of persons sentenced to terms of imprisonment
(1) Rules made under section forty-seven of this Act may make provision whereby, in such circumstances as may be prescribed by the rules, a person serving a sentence of imprisonment for such a term as may be so prescribed may be granted remission of such part of that sentence as may be so prescribed on the ground of his industry and good conduct, and on the discharge of a person from prison in pursuance of any such remission as aforesaid his sentence shall expire.

    (2)–(6) . . .

    [(7) A person who is committed to prison in default of payment of a sum adjudged to be paid by a conviction shall be treated for the purposes of subsection

(1) of this section, . . . , as undergoing a sentence of imprisonment for the term for which he is committed, and consecutive terms of imprisonment shall be treated for all the purposes of this section as one term.]

26, 27 *(Repealed by the Criminal Justice Act 1967, s 103 (2), Sch 7, Pt I.)*

28 Power of Secretary of State to discharge prisoners temporarily on account of ill health

(1) If the Secretary of State is satisfied that by reason of the condition of a prisoner's health it is undesirable to detain him in prison, but that, such condition of health being due in whole or in part of the prisoner's own conduct in prison, it is desirable that his release should be temporary and conditional only, the Secretary of State may, if he thinks fit, having regard to all the circumstances of the case, by order authorise the temporary discharge of the prisoner for such period and subject to such conditions as may be stated in the order.

(2) Where an order of temporary discharge is made in the case of a prisoner not under sentence, the order shall contain conditions requiring the attendance of the prisoner at any further proceedings on his case at which his presence may be required.

(3) Any prisoner discharged under this section shall comply with any condition stated in the order of temporary discharge, and shall return to prison at the expiration of the period stated in the order, or of such extended period as may be fixed by any subsequent order of the Secretary of State, and if the prisoner fails so to comply or return, he may be arrested without warrant and taken back to prison.

(4) Where a prisoner under sentence is discharged in pursuance of an order of temporary discharge, the currency of the sentence shall be suspended from the day on which he is discharged from prison under the order of the day on which he is received back into prison, so that the former day shall be reckoned and the latter shall not be reckoned as part of the sentence.

(5) Nothing in this section shall affect the duties of the medical officer of a prison in respect of a prisoner whom the Secretary of State does not think fit to discharge under this section.

Discharged prisoners

29 *(Repealed by the Criminal Justice Act 1961, ss 21, 41 (2), Sch 5.)*

[30 Payments for discharged prisoners
The Secretary of State may make such payments to or in respect of persons released or about to be released from prison as he may with the consent of the Treasury determine.]

Provision, maintenance and closing of prisons

33 Power to provide prisons, etc
(1) The Secretary of State may with the approval of the Treasury alter, enlarge or rebuild any prison and build new prisons.

(2) The Secretary of State may provide new prisons by declaring to be a prison any building or part of a building built for the purpose or vested in him or under his control.

(3) A declaration under this section may with respect to the building or part of a building declared to be a prison make the same provisions as an order under the next following section may make with respect to an existing prison.

(4) A declaration under this section may at any time be revoked by the Secretary of State.

(5) A declaration under this section shall not be sufficient to vest the legal estate of any building in the [Secretary of State].

34 Jurisdiction of sheriff, etc

(1) The transfer under the Prison Act 1877 of prisons and of the powers and jurisdiction of prison authorities and of justices in sessions assembled and visiting justices shall not be deemed to have affected the jurisdiction of any sheriff or coroner or, except to the extent of that transfer, of any justice of the peace or other officer.

(2) The Secretary of State may by order direct that, for the purposes of any enactment, rule of law or custom dependent on a prison being the prison of any county or place, any prison situated in that county or in the county in which that place is situated, or any prison provided by him in pursuance of this Act, shall be deemed to be the prison of that county or place.

35 Prison property

(1) Every prison and all real and personal property belonging to a prison shall be vested in the Secretary of State and may be disposed of in such manner as the Secretary of State, with the consent of the Treasury, may determine.

(2) For the purposes of this section the Secretary of State shall be deemed to be a corporation sole.

(3) Any instrument in connection with the acquisition, management or disposal of any property to which this section applies may be executed on behalf of the Secretary of State by an Under-Secretary of State or any other person authorised by the Secretary of State in that behalf; and any instrument purporting to have been so executed on behalf of the Secretary of State shall be deemed, until the contrary is proved, to have been so executed on his behalf.

(4) The last foregoing subsection shall be without prejudice to the execution of any such instrument as aforesaid, or of any other instrument, on behalf of the Secretary of State in any other manner authorised by law.]

36 Acquisition of land for prisons

(1) [The Secretary of State may purchase by agreement or] compulsorily, any land required for the alteration, enlargement or rebuilding of a prison or for establishing a new prison or for any other purpose connected with the management of a prison (including the provision of accommodation for officers or servants employed in a prison).

[(2) The [Acquisition of Land Act 1981] shall apply to the compulsory purchase of land by the Secretary of State under this section . . .]

(3) In relation to the purchase of land by agreement under this section, [the provisions of Part I of the Compulsory Purchase Act 1965 (so far as applicable) other than sections 4 to 8, section 10, and section 31, shall apply].

37 Closing of prisons

(1) Subject to the next following subsection, the Secretary of State may by order close any prison.

(2) Where a prison is the only prison in the county, the Secretary of State shall not make an order under this section in respect of it except for special reasons, which shall be stated in the order.

(3) In this section the expression 'county' means a county at large.

(4) For the purposes of this and the next following section a prison shall not be deemed to be closed by reason only of its appropriation for use as a remand centre, detention centre or [youth custody centre].

38 *(Repealed with a saving by the Criminal Justice Act 1972, ss 59, 64 (2), Sch 65, Pt II, in this title post and Vol 12, title* Criminal Law.*)*

Offences

39 Assisting prisoner to escape

Any person who aids any prisoner in escaping or attempting to escape from a prison or who, with intent to facilitate the escape of any prisoner, conveys any thing into a prison or to a prisoner or places any thing anywhere outside a prison with a view to its coming into possession of a prisoner, shall be guilty of felony and liable to imprisonment for a term not exceeding [five years].

40 Unlawful conveyance of spirits or tobacco into prison, etc

Any person who contrary to the regulations of a prison brings or attempts to bring into the prison or to a prisoner any spirituous or fermented liquor or tobacco, or places any such liquor or any tobacco anywhere outside the prison with intent that it shall come into the possession of a prisoner, and any officer who contrary to those regulations allows any such liquor or any tobacco to be sold or used in the prison, shall be liable on summary conviction to imprisonment for a term not exceeding six months or a fine not exceeding [level 3 on the standard scale] or both.

41 Unlawful introduction of other articles

Any person who contrary to the regulations of a prison conveys or attempts to convey any letter or any other thing into or out of the prison or to a prisoner or places it anywhere outside the prison with intent that it shall come into the possession of a prisoner shall, where he is not thereby guilty of an offence under either of the two last preceding sections, be liable on summary conviction to a fine not exceeding [level 3 on the standard scale].

42 Display of notice of penalties

The Prison Commissioners shall cause to be affixed in a conspicuous place outside every prison a notice of the penalties to which persons committing offences under the three last preceding sections are liable.

Remand centres, detention centres and Borstal institutions

The reference in the heading above to borstal institutions should be construed as a reference to Young Offender Institutions which are now provided under s 43 post in place of borstals.

[43 Remand centres, detention centres and Young Offender Institutions

(1) The Secretary of State may provide—

 (*a*) remand centres, that is to say places for the detention of persons not less than 14 but under 21 years of age who are remanded or committed in custody for trial or sentence;

 (*b*) detention centres, that is to say places in which male offenders not less than 14 but under 21 years of age who are ordered to be detained in such centres under the Criminal Justice Act 1982 may be kept for short periods under discipline suitable to persons of their age and description; and

 (*c*) Young Offender Institutions, that is to say places in which offenders not less than 15 but under 21 years of age may be detained and given training, instruction and work and prepared for their release.

(2) The Secretary of State may from time to time direct—

 (*a*) that a woman aged 21 years or over who is serving a sentence of imprisonment or who has been committed to prison for default shall be detained in a remand centre or a youth custody centre instead of a prison;

 (*b*) that a woman aged 21 years or over who is remanded in custody or committed in custody for trial or sentence shall be detained in a remand centre instead of a prison;

 (*c*) that a person under 21 but not less than 17 years of age who is remanded in custody or committed in custody for trial or sentence shall be detained in a prison instead of a remand centre or a remand centre instead of a prison, notwithstanding anything in section 27 of the Criminal Justice Act 1948 or section 23 (3) of the Children and Young Persons Act 1969.

(3) Notwithstanding subsection (1) above, any person required to be detained in an institution to which this Act applies may be detained in a remand centre for any temporary purpose or for the purpose of providing maintenance and domestic services for that centre.

(4) Sections 5A, 6 (2) and (3), 16, 22, 25 and 36 of this Act shall apply to remand centres, detention centres and Young Offender Institutions and to persons detained in them as they apply to prisons and prisoners.

(5) The other provisions of this Act preceding this section, except sections 28 and 37 (2) above, shall apply to such centres and to persons detained in them as they apply to prisons and prisoners, but subject to such adaptations and modifications as may be specified in rules made by the Secretary of State.

(6) References in the preceding provisions of this Act to imprisonment shall, so far as those provisions apply to institutions provided under this section, be construed as including references to detention in those institutions.

(7) Nothing in this section shall be taken to prejudice the operation of section 12 of the Criminal Justice Act 1982.]

44–46 (*Repealed by the Criminal Justice Act 1982, s 78, Sch 16.*)

Rules for the management of prisons and other institutions

47 Rules for the management of prisons, remand centres, detention centres and Borstal institutions
(1) The Secretary of State may make rules for the regulation and management of prisons, remand centres, detention centres and [youth custody centres] respectively, and for the classification, treatment, employment, discipline and control of persons required to be detained therein.

(2) Rules made under this section shall make provision for ensuring that a person who is charged with any offence under the rules shall be given a proper opportunity of presenting his case.

(3) Rules made under this section may provide for the training of particular classes of persons and their allocation for that purpose to any prison or other institution in which they may lawfully be detained.

(4) Rules made under this section shall provide for the special treatment of the following persons whilst required to be detained in a prison, that is to say—

   (a)–(c) . . .
   (d) any . . . person detained in a prison, not being a person serving a sentence or a person imprisoned in default of payment of a sum adjudged to be paid by him on his conviction [or a person committed to custody on his conviction].

(5) Rules made under this section may provide for the temporary release of persons [detained in a prison, [remand centre, youth custody centre], or detention centre, not being persons committed in custody for trial [before the Crown Court] or committed to be sentenced or otherwise dealt with by [the Crown Court] or remanded in custody by any court].

Miscellaneous

48 *(Repealed by the Criminal Justice Act 1961, s 41 (2), (3), Sch 5.)*

49 Persons unlawfully at large
(1) Any person who, having been sentenced to imprisonment, . . . [custody for life or youth custody] or ordered to be detained in a detention centre [or a young offenders institution], or having been committed to a prison or remand centre, is unlawfully at large, may be arrested by a constable without warrant and taken to the place in which he is required in accordance with law to be detained.

(2) Where any person sentenced to imprisonment, . . . or [youth custody], or ordered to be detained in a . . . detention centre, is unlawfully at large at any time during the period for which he is liable to be detained in pursuance of the sentence or order, then, unless the Secretary of State otherwise directs, no account shall be taken, in calculating the period for which he is liable to be so detained, of any time during which he is absent from the [place in which he is required in accordance with law to be detained].
   Provided that—

   (a) this subsection shall not apply to any period during which any such person as aforesaid is detained in pursuance of the sentence or order or in pursuance of any other sentence of any court [in the United Kingdom] in a prison, [youth custody centre, remand centre or detention centre];

(*b*), (*c*) . . .

(3) The provisions of the last preceding subsection shall apply to a person who is detained in custody in default of payment of any sum of money as if he were sentenced to imprisonment.

(4) For the purposes of this section a person who, after being temporarily released in pursuance of rules made under subsection (5) of section forty-seven of this Act, is at large at any time during the period for which he is liable to be detained in pursuance of his sentence shall be deemed to be unlawfully at large if the period for which he was temporarily released has expired or if an order recalling him has been made by the [Secretary of State] in pursuance of the rules.

50 *(Repealed in part by the Children and Young Persons Act 1969, s 72 (4), Sch 6; remainder spent upon the repeal of s 18 of this Act by the Criminal Justice Act 1967, ss 65, 103 (2), Sch 7, Pt I.)*

Supplemental

51 Payment of expenses out of moneys provided by Parliament

All expenses incurred in the maintenance of prisons and in the maintenance of prisoners and all other expenses of the Secretary of State . . . incurred under this Act shall be defrayed out of moneys provided by Parliament.

52 Exercise of power to make orders, rules and regulations

(1) Any power of the Secretary of State to make rules or regulations under this Act and the power of the Secretary of State to make an order under section thirty-four or section thirty-seven of this Act shall be exercisable by statutory instrument.

(2) Any statutory instrument containing regulations made under section sixteen or an order made under section thirty-seven of this Act, . . . shall be laid before Parliament.

(3) The power of the Secretary of State to make an order under section six or section thirty-four of this Act shall include power to revoke or vary such an order.

53 Interpretation

(1) In this Act the following expressions have the following meanings:—

'Attendance centre' means a centre provided by the Secretary of State under [section 16 of the Criminal Justice Act 1982];

'Prison' does not include a naval, military or air force prison;

. . . . .

(2) For the purposes of this Act the maintenance of a prisoner shall include all necessary expenses incurred in respect of the prisoner for food, clothing, custody and removal from one place to another, from the period of his committal to prison until his death or discharge from prison.

(3) References in this Act to the Church of England shall be construed as including references to the Church in Wales.

(4) References in this Act to any enactment shall be construed as references to that enactment as amended by any other enactment.

54 Consequential amendments, repeals and savings

(1) The enactments mentioned in the Third Schedule to this Act shall have effect subject to the amendments specified therein, being amendments consequential on the provisions of this Act.

(2) . . .

(3) Nothing in this repeal shall affect any rule, order, regulation or declaration made, direction or certificate given or thing done under any enactment repealed by this Act and every such rule, order, regulation, direction, certificate or thing shall, if in force at the commencement of this Act, continue in force and be deemed to have been made, given or done under the corresponding provision of this Act.

(4) Any document referring to any Act or enactment repealed by this Act shall be construed as referring to this Act or to the corresponding enactment in this Act.

(5) The mention of particular matters in this section shall not be taken to affect the general application to this Act of section thirty-eight of the Interpretation Act 1889 (which relates to the effect of repeals).

55 Short title, commencement and extent

(1) This Act may be cited as the Prison Act 1952.

(2) This Act shall come into operation on the first day of October, nineteen hundred and fifty-two.

(3) . . . Part II of the Fourth Schedule to this Act shall extend to Scotland, . . .

(4) Except as provided in the last preceding subsection or [the Criminal Justice Act 1961], this Act shall not extend to Scotland.

(5) This Act shall not extend to Northern Ireland.

(Schs 1, 2 repealed by the Criminal Justice Act 1961, s 41 (2), (3), Sch 5; Sch 3 repealed in part by the Mental Health Act 1959, s 149 (2), Sch 7, Pt I, the Criminal Justice Act 1961, s 41 (2), (3), Sch 5, and the Firearms Act 1968, s 59, Sch 7; the residue is spent upon the repeal of the Courts-Martial (Appeals) Act 1951, s 17, by the Courts-Martial (Appeals) Act 1968, s 60, Sch 6; Sch 4 repealed by the SL(R) Act 1974.)

Appendix 2
Prison Rules 1964

ENGLAND AND WALES

The Prison Rules 1964
as amended by
The Prison (Amendment) Rules of
1968, 1971, 1972, 1974, 1976, 1981, 1982, 1983, 1987, 1988, 1989, 1990, 1992

PART I

PRISONERS

General

Religion

Medical attention, &c

Physical welfare and work

101 Citation and commencement

In pursuance of section 47 of the Prison Act 1952, as amended by sections 23 (2) and 41 (1) of, and Schedule 4 to, the Criminal Justice Act 1961, I hereby make the following Rules:

PART I
PRISONERS

General

Purpose of prison training and treatment

1. The purpose of the training and treatment of convicted prisoners shall be to encourage and assist them to lead a good and useful life.

Maintenance of order and discipline

2. (1) Order and discipline shall be maintained with firmness, but with no more restriction than is required for safe custody and well ordered community life.

   (2) In the control of prisoners, officers shall seek to influence them through their own example and leadership, and to enlist their willing co-operation.

   (3) At all times the treatment of prisoners shall be such as to encourage their self-respect and a sense of personal responsibility, but a prisoner shall not be employed in any disciplinary capacity.

Classification of prisoners

3. (1) Prisoners shall be classified, in accordance with any directions of the Secretary of State, having regard to their age, temperament and record and with a view to maintaining good order and facilitating training and, in the case of convicted prisoners, of furthering the purpose of their training and treatment as provided by Rule 1 of these Rules.

   (2) Unconvicted prisoners shall be kept out of contact with convicted prisoners as far as this can reasonably be done.

   (3) Nothing in this Rule shall require a prisoner to be deprived unduly of the society of other persons.

Privileges

4. There shall be established at every prison systems of privileges approved by the Secretary of State and appropriate to the classes of prisoners there, which shall include arrangements under which money earned by 'prisoners in prison may be spent by them within the prison.

5. Not allocated

Temporary release

6. (1) A prisoner to whom this Rule applies may be temporarily released for any period or periods and subject to any conditions.

(2) A prisoner may be temporarily released under this Rule for any special purpose or to enable him to engage in employment, to receive instruction or training or to assist him in his transition from prison life to freedom.

(3) A prisoner released under this Rule may be recalled to prison at any time whether the conditions of his release have been broken or not.

(4) This Rule applies to prisoners other than persons committed in custody for trial or to be sentenced or otherwise dealt with before or by the Crown Court or remanded in custody by any court.

Information to prisoners

7. (1) Every prisoner shall be provided, as soon as possible after his reception into prison, and in any case within 24 hours, with information in writing about those provisions of these Rules and other matters which it is necessary that he should know, including earnings and privileges, and the proper method of making requests and complaints.

(2) In the case of a prisoner aged less than 18, or a prisoner aged 18 or over who cannot read or appears to have difficulty in understanding the information so provided, the governor, or an officer deputed by him, shall so explain it to him that he can understand his rights and obligations.

(3) A copy of these Rules shall be made available to any prisoner who requests it.

Requests and complaints

8. (1) A request or complaint to the Governor or board of visitors relating to a prisoner's imprisonment shall be made orally or in writing by the prisoner.

(2) On every day the governor shall hear any requests or complaints that are made to him under paragraph (1) above.

(3) A written request or complaint under paragraph (1) above may be made in confidence.

Women prisoners

9. (1) Women prisoners shall be kept entirely separate from male prisoners.

(2) Not allocated.

(3) The Secretary of State may, subject to any conditions he thinks fit, permit a woman prisoner to have her baby with her in prison, and everything necessary for the baby's maintenance and care may be provided there.

Religion

Religious denomination

10. A prisoner shall be treated as being of the religious denomination stated in the record made in pursuance of section 10 (5) of the Prisons Act 1952 but the governor may, in a proper case and after due enquiry, direct that record to be amended.

Special duties of chaplains and prison ministers

11. (1) The chaplain or prison minister of a prison shall—

  (*a*) interview every prisoner of his denomination individually soon after the prisoner's reception into that prison and shortly before his release; and

  (*b*) if no other arrangements are made, read the burial service at the funeral of any prisoner of his denomination who dies in that prison.

  (2) The chaplain shall visit daily all prisoners belonging to the Church of England who are sick, under restraint or undergoing cellular confinement; and a prison minister shall do the same, as far as he reasonably can, for prisoners of his own denomination.

  (3) If the prisoner is willing, the chaplain shall visit any prisoner not of the Church of England who is sick, under restraint or undergoing cellular confinement, and is not regularly visited by a minister of his own denomination.

Regular visits by ministers of religion

12. (1) The chaplain shall visit the prisoners belonging to the Church of England.

  (2) A prison minister shall visit the prisoners of his denomination as regularly as he reasonably can.

  (3) Where a prisoner belongs to a denomination for which no prison minister has been appointed, the governor shall do what he reasonably can, if so requested by the prisoner, to arrange for him to be visited regularly by a minister of that denomination.

Religious services

13. (1) The chaplain shall conduct Divine Service for prisoners belonging to the Church of England at least once every Sunday, Christmas Day and Good Friday, and such celebrations of Holy Communion and weekday services as may be arranged.

  (2) Prison ministers shall conduct Divine Service for prisoners of their denominations at such times as may be arranged.

Substitute for chaplain or prison minister

14. (1) A person approved by the Secretary of State may act for the chaplain in his absence.

  (2) A prison minister may, with the leave of the Secretary of State, appoint A substitute to act for him in his absence.

Sunday work

15. Arrangements shall be made so as not to require prisoners of the Christian religion to do any unnecessary work on Sunday, Christmas Day or Good Friday, or prisoners of other religions on their recognised days of religious observance.

Religious books

16. There shall, so far as reasonably practicable, be available for the personal use of

every prisoner such religious books recognized by his denomination as are approved by the Secretary of State for use in prisons.

Medical attention, &c

Medical attendance

17. (1) The medical officer of a prison shall have the care of the health, mental and physical, of the prisoners in that prison.
    (2) Every request by a prisoner to see the medical officer shall be recorded by the officer to whom it is made and promptly passed on to the medical officer.
    (3) The medical officer may call another medical practitioner into consultation at his discretion, and shall do so if time permits before performing any serious operation.
    (4) If an unconvicted prisoner desires the attendance of a registered medical practitioner or dentist, and will pay any expense incurred, the governor shall, if he is satisfied that there are reasonable grounds for the request and unless the Secretary of State otherwise directs, allow him to be visited and treated by that practitioner or dentist in consultation with the medical officer.

Special illnesses and conditions

18. (1) The medical officer shall report to the governor on the case of any prisoner whose health is likely to be injuriously affected by continued imprisonment or any conditions of imprisonment. The governor shall send the report to the Secretary of State without delay, together with his own recommendations.
    (2) The medical officer shall pay special attention to any prisoner whose mental condition appears to require it, and make any special arrangements which appear necessary for his supervision or care.
    (3) The medical officer shall inform the governor if he suspects any prisoner of having suicidal intentions, and the prisoner shall be placed under special observation.

Notification of illness or death

19. (1) If a prisoner dies, becomes seriously ill, sustains any severe injury or is removed to hospital on account of mental disorder, the governor shall, if he knows his or her address, at once inform the prisoner's spouse or next of kin, and also any person who the prisoner may reasonably have asked should be informed.
    (2) If a prisoner dies, the governor shall give notice immediately to the coroner having jurisdiction, to the board of visitors and to the Secretary of State.

Physical welfare and work

Clothing

20. (1) An unconvicted prisoner may wear clothing of his own if and insofar as it is suitable, tidy and clean, but, subject to that, the provisions of this Rule shall apply to him as a convicted prisoner. An unconvicted prisoner shall be permitted to arrange for the supply to him from outside prison of sufficient clean clothing

    (2) A convicted prisoner shall be provided with clothing adequate for warmth and health in accordance with a scale approved by the Secretary of State.

    (3) The clothing provided under this Rule shall include suitable protective clothing for use at work, where this is needed.

    (4) Subject to the provisions of Rule 38 (3) of these Rules, a convicted prisoner shall wear clothing provided under this Rule and no other, except on the directions of the Secretary of State.

    (5) A prisoner may be provided, where necessary, with suitable and adequate clothing on his release.

Food

21. (1) Not allocated

    (2) Subject to any directions of the Secretary of State, no prisoner shall be allowed, except as authorised by the medical officer, to have any food other than that ordinarily provided.

    (3) No prisoners shall be given less food than is ordinarily provided, except upon the written recommendation of the medical officer.

    (4) The food provided shall be wholesome, nutritious, well prepared and served, reasonably varied and sufficient in quantity.

    (5) The medical officer shall regularly inspect the food both before and after it is cooked and shall report any deficiency or defect to the governor.

    (6) In this Rule 'food' includes drink.

Alcohol and tobacco

22. (1) No prisoner shall be allowed to have any intoxicating liquor except under a written order of the medical officer specifying the quantity and the name of the prisoner.

    (2) No prisoner shall be allowed to smoke or to have any tobacco except as a privilege under Rule 4 of these Rules and in accordance with any orders of the governor.

Sleeping accommodation

23. (1) No room or cell shall be used as sleeping accommodation for a prisoner unless it has been certified in the manner required by section 14 of the Prison Act 1952 in the case of a cell used for the confinement of a prisoner.

    (2) A certificate given under the section or this Rule shall specify the maximum number of prisoners who may sleep or be confined at one time in the room

or cell to which it relates, and the number so specified shall not be exceeded without the leave of the Secretary of State.

Beds and bedding

24. Each prisoner shall be provided with a separate bed and with separate bedding adequate for warmth and health.

Special accommodation

25. The governor or board of visitors may, on application by an unconvicted prisoner, permit him on payment of a sum fixed by the Secretary of State—

   (*a*) to occupy a room or cell specially fitted for such prisoners and provided with suitable bedding and other articles in addition to, or different from, those ordinarily provided, and to have at his own expense the use of private furniture and utensils approved by the governor; and

   (*b*) to be relieved of the duty of cleaning his room or cell and similar duties.

Hygiene

26. (1) Every prisoner shall be provided with toilet articles necessary for his health and cleanliness, which shall be replaced as necessary.

   (2) Every prisoner shall be required to wash at proper times, have a hot bath on reception and thereafter at least once a week and, in the case of a man not excused or excepted by the governor or medical officer, to shave or be shaved daily, and to have his hair cut as may be necessary for neatness:

   Provided that an unconvicted prisoner or a convicted prisoner who has not yet been sentenced shall not be required to have his hair cut or any beard or moustache usually worn by him shaved off except where the medical officer directs this to be done for the sake of health or cleanliness.

   (3) A woman prisoner's hair shall not be cut without her consent except where the medical officer certifies in writing that this is necessary for the sake of health or cleanliness.

Daily exercise

27. (1) A prisoner not engaged in outdoor work, or detained in an open prison, shall be given exercise in the open air for not less than one hour in all, each day, if weather permits:

   Provided that exercise consisting of physical training may be given indoors instead of in the open air.

   (2) The Secretary of State may in special circumstances authorise the reduction of the period aforesaid to half an hour a day.

   (3) The medical officer shall decide upon the fitness of every prisoner for exercise and physical training, and may excuse a prisoner from, or modify, any activity on medical grounds.

Work

28. (1) A convicted prisoner shall be required to do useful work for not more than

10 hours a day, and arrangements shall be made to allow prisoners to work, where possible, outside the cells and in association with one another.

(2) The medical officer may excuse a prisoner from work on medical grounds, and no prisoner shall be set to do work which is not of a class for which he has been passed by the medical officer as being fit.

(3) No prisoner shall be set to do work of a kind not authorised by the Secretary of State.

(4) No prisoner shall work in the service of another prisoner or an officer, or for the private benefit of any person, without the authority of the Secretary of State.

(5) An unconvicted prisoner shall be permitted, if he wishes, to work as if he were a convicted prisoner.

(6) Prisoners may be paid for their work at rates approved by the Secretary of State, either generally or in relation to particular cases.

Education and social welfare

Education

29. (1) Every prisoner able to profit from the education facilities provided at a prison shall be encouraged to do so.

(2) Programmes of evening educational classes shall be arranged at every prison and, subject to any directions of the Secretary of State, reasonable facilities shall be afforded to prisoners who wish to do so to improve their education by correspondence courses or private study, or to practice handicrafts, in their spare time.

(3) Special attention shall be paid to the education of illiterate prisoners, and if necessary they shall be taught within the hours normally allotted to work.

Library books

30. A library shall be provided in every prison and, subject to any directions of the Secretary of State, every prisoner shall be allowed to have library books and to exchange them.

Outside contacts

31. (1) Special attention shall be paid to the maintenance of such relationships between a prisoner and his family as are desirable in the best interests of both.

(2) A prisoner shall be encouraged and assisted to establish and maintain such relations with persons and agencies outside prison as may, in the opinion of the governor, best promote the interests of his family and his own social rehabilitation.

After-care

32. From the beginning of a prisoner's sentence, consideration shall be given, in consultation with the appropriate after-care organisation, to the prisoner's future and the assistance to be given him on and after his release.

Letters and visits

Letters and visits generally

33. (1) The Secretary of State may, with a view to securing discipline and good order or the prevention of crime or in the interests of any persons, impose restrictions, either generally or in a particular case, upon the communications to be permitted between a prisoner and other persons.

(2) Except as provided by statute or these Rules, a prisoner shall not be permitted to communicate with any outside person, or that person with him, without the leave of the Secretary of State.

(3) Except as provided by these Rules, every letter or communication to or from a prisoner may be read or examined by the governor or an officer deputed by him, and the governor may, at his discretion, stop any letter or communication on the ground that its contents are objectionable or that it is of inordinate length.

(4) Every visit to a prisoner shall take place within the sight of an officer, unless the Secretary of State otherwise directs.

(5) Except as provided by these Rules, every visit to a prisoner shall take place within the hearing of an officer, unless the Secretary of State otherwise directs.

(6) The Secretary of State may give directions, generally or in relation to any visit or class of visits, concerning the days and time when prisoners may be visited.

Personal letters and visits

34. (1) An unconvicted prisoner may send and receive as many letters and may receive as many visits as he wishes within such limits and subject to such conditions as the Secretary of State may direct, either generally or in a particular case.

(2) A convicted prisoner shall be entitled—

(*a*) to send and to receive a letter on his reception into a prison and thereafter once a week; and

(*b*) to receive a visit twice in every period of four weeks, but only once in every such period if the Secretary of State so directs.

(3) The governor may allow a prisoner an additional letter or visit where necessary for his welfare or that of his family.

(4) The governor may allow a prisoner entitled to a visit to send and to receive a letter instead.

(5) The governor may defer the right of a prisoner to a visit until the expiration of any period of cellular confinement.

(6) The board of visitors may allow a prisoner an additional letter or visit in special circumstances, and may direct that a visit may extend beyond the normal duration.

(7) The Secretary of State may allow additional letters and visits in relation to any prisoner or class of prisoners.

(8) A prisoner shall not be entitled under this Rule to receive a visit from any person other than a relative or friend, except with the leave of the Secretary of State.

(9) Any letter or visit under the succeeding provisions of these Rules shall not be counted as a letter or visit for the purposes of this Rule.

Police interviews

35. A police officer may, on production of an order issued by or on behalf of a chief officer of police, interview any prisoner willing to see him.

Securing release

36. A person detained in prison in default of finding a surety, or of payment of a sum of money, may communicate with and be visited at any reasonable time on a weekday by, any relative or friend to arrange for a surety or payment in order to secure his release from prison.

Legal advisers

37. (1) The legal adviser of a prisoner in any legal proceedings, civil or criminal, to which the prisoner is a party shall be afforded reasonable facilities for interviewing him in connection with those proceedings, and may do so out of hearing but in the sight of an officer.
    (2) A prisoner's legal adviser may, subject to any directions given by the Secretary of State, interview the prisoner in connection with any other legal business out of hearing but in the sight of an officer.

Further facilities in connection with legal proceedings

37A (1) A prisoner who is a party to any legal proceedings may correspond with his legal adviser in connection with the proceedings and unless the governor has reason to suppose that any such correspondence contains matter not relating to the proceedings it shall not be read or stopped under Rule 33 (3) of these Rules.
    (2) A prisoner shall on request be provided with any writing materials necessary for the purposes of paragraph (1) of this Rule.
    (3) Subject to any directions given in the particular case by the Secretary of State, a registered medical practitioner selected by or on behalf of such a prisoner as aforesaid shall be afforded reasonable facilities for examining him in connection with the proceedings, and may do so out of hearing but in the sight of an officer.
    (4) Subject to any directions of the Secretary of State, a prisoner may correspond with a solicitor for the purpose of obtaining legal advice concerning any cause of action in relation to which the prisoner may become a party to civil proceedings or for the purpose of instructing the solicitor to issue such proceedings.

Removal, record and property

Custody outside prison

38. (1) A person being taken to or from a prison in custody shall be exposed as little as possible to public observation, and proper care shall be taken to protect him from curiosity and insult.

   (2) A prisoner required to be taken in custody anywhere outside the prison shall be kept in the custody of an officer appointed or a police officer.

   (3) A prisoner requiring to be taken in custody to any court shall wear his own clothing or clothing different from the dress worn at any institution to which the Prison Act 1952 applies.

Search

39. (1) Every prisoner shall be searched when taken into custody by an officer, on his reception into a prison and subsequently as the governor thinks necessary.

   (2) A prisoner shall be searched in as seemly a manner as is consistent with discovering anything concealed.

   (3) No prisoner shall be stripped and searched in the sight of another prisoner, or in the sight or presence of an officer not of the same sex.

   (4) Not allocated.

Record and photograph

40. (1) A personal record of each prisoner shall be prepared and maintained in such manner as the Secretary of State may direct.

   (2) Every prisoner may be photographed on reception and subsequently, but no copy of the photograph shall be given to any person not authorised to receive it.

Prisoners' property

41. (1) Subject to any directions of the Secretary of State, an unconvicted prisoner may have supplied to him at his expense and retain for his own use books, newspapers, writing materials and other means of occupation, except any that appears objectionable to the board of visitors or, pending consideration by them, to the governor.

   (2) Anything, other than cash, which a prisoner has at a prison and which he is not allowed to retain for his own use shall be taken into the governor's custody. An inventory of a prisoner's property shall be kept, and he shall be required to sign it, after having a proper opportunity to see that it is correct.

   (3) Any cash which a prisoner has at a prison shall be paid into an account under the control of the governor and the prisoner shall be credited with the amount in the books of the prison.

   (3A) Any article belonging to a prisoner which remains unclaimed for a period of more than three years after he leaves prison, or dies, may be sold or otherwise disposed of; and the net proceeds of any sale shall be paid to the

National Association for the Care and Resettlement of Offenders, for its general purposes.

(4) The governor may confiscate any unauthorised article found in the possession of a prisoner after his reception into prison, or concealed or deposited anywhere within a prison.

Money and articles received by post

42. (1) Any money or other article (other than a letter or other communication) sent to a convicted prisoner through the post office shall be dealt with in accordance with the provisions of this Rule, and the prisoner shall be informed of the manner in which it is dealt with.

(2) Any cash shall, at the discretion of the governor be—

(*a*) dealt with in accordance with Rule 41 (3) of these Rules; or

(*b*) returned to the sender; or

(*c*) in a case where the sender's name and address are not known, paid to the National Association for the Care and Resettlement of Offenders, for its general purposes:

Provided that in relation to a prisoner committed to a prison in default of payment of any sum of money, the prisoner shall be informed of the receipt of the cash and, unless he objects to its being so applied, it shall be applied in or towards the satisfaction of the amount due from him.

(3) Any security for money shall, at the discretion of the governor, be—

(*a*) delivered to the prisoner or placed with his property at the prison; or

(*b*) returned to the sender; or

(*c*) encashed and the cash dealt with in accordance with paragraph (2) of this Rule.

(4) Any other article to which this Rule applies shall, at the discretion of the governor, be—

(*a*) delivered to the prisoner or placed with his property at the prison; or

(*b*) returned to the sender; or

(*c*) in the case where the sender's name and address are not known or the article is of such a nature that it would be unreasonable to return it, sold or otherwise disposed of, and the net proceeds of any sale applied in accordance with paragraph (2) of this Rule.

Special control and restraint

Removal from association

43. (1) Where it appears desirable, for the maintenance of good order or discipline or in his own interests, that a prisoner should not associate with other prisoners, either generally or for particular purposes, the governor may arrange for the prisoner's removal from association accordingly.

(2) A prisoner shall not be removed under this Rule for a period of more than 3 days without the authority of a member of the board of visitors or of the Secretary of State. An authority given under this paragraph shall be for a

period not exceeding one month, but may be renewed from month to month except that, in the case of a person aged less than 21 years who is detained in prison such an authority shall be for a period not exceeding 14 days, but may be renewed from time to time for a like period.

(3) The governor may arrange at his discretion for such a prisoner as aforesaid to resume association with other prisoners, and shall do so if in any case the medical officer so advises on medical grounds.

Use of force

44. (1) An officer in dealing with a prisoner shall not use force unnecessarily and, when the application of force to a prisoner is necessary, no more force than is necessary shall be used.

(2) No officer shall act deliberately in a manner calculated to provoke a prisoner.

Temporary confinement

45. The governor may order a refractory or violent prisoner to be confined temporarily in a special cell, but a prisoner shall not be so confined as a punishment, or after he has ceased to be refractory or violent.

Restraints

46 (1) The governor may order a prisoner to be put under restraint where this is necessary to prevent the prisoner from injuring himself or others, damaging property or creating a disturbance.

(2) Notice of such an order shall be given without delay to a member of the board of visitors, and to the medical officer.

(3) On receipt of the notice the medical officer shall inform the governor whether he concurs in the order. The governor shall give effect to any recommendation which the medical officer may make.

(4) A prisoner shall not be kept under restraint longer than necessary, nor shall he be so kept for longer than 24 hours without a direction in writing given by a member of the board of visitors or by an officer of the Secretary of State (not being an officer of a prison). Such a direction shall state the grounds for the restraint and the time during which it may continue.

(5) Particulars of every case of restraint under the foregoing provisions of this Rule shall be forthwith recorded.

(6) Except as provided by this Rule no prisoner shall be put under restraint otherwise than for safe custody during removal, or on medical grounds by direction of the medical officer. No prisoner shall be put under restraint as a punishment.

(7) Any means of restraint shall be of a pattern authorised by the Secretary of State, and shall be used in such manner and under such conditions as the Secretary of State may direct.

Offences against discipline

47. A prisoner is guilty of an offence against discipline if he—

(1) commits any assault;

(2) detains any person against his will;

(3) denies access to any part of the prison to any officer;

(4) fights with any person;

(5) intentionally endangers the health or personal safety of others or, by his conduct, is reckless whether such health or personal safety is endangered;

(6) intentionally obstructs an officer in the execution of his duty;

(7) escapes or absconds from prison or from legal custody;

(8) fails to comply with any condition upon which he is temporarily released under Rule 6 of these Rules.

(9) has in his possession—

    (*a*) any unauthorised article, or

    (*b*) a greater quantity of any article than he is authorised to have;

(10) sells or delivers to any person any unauthorised article;

(11) sells or, without permission, delivers to any person any article which he is allowed to have only for his own use;

(12) takes improperly any article belonging to another person or to a prison;

(13) intentionally or recklessly sets fire to any part of a prison or any other property, whether or not his own;

(14) destroys or damages any part of a prison or any other property, other than his own;

(15) absents himself from any place where he is required to be or is present at any place where he is not authorised to be;

(16) is disrespectful to any officer or any person visiting a prison;

(17) uses threatening, abusive or insulting words or behaviour;

(18) intentionally fails to work properly or, being required to work, refuses to do so;

(19) disobeys any lawful order;

(20) disobeys or fails to comply with any rule or regulation applying to him;

(21) in any way offends against good order and discipline;

(22) (*a*) attempts to commit

    (*b*) incites another prisoner to commit, or

    (*c*) assists another prisoner to commit or to attempt to commit,

    any of the foregoing offences.

Disciplinary charges

48. (1) Where a prisoner is to be charged with an offence against discipline, the charge shall be laid as soon as possible and, save in exceptional circumstances, within 48 hours of the discovery of the offence.

(2) Not allocated

(3) Every charge shall be inquired into, by the governor.

(4) Every charge shall be first inquired into not later, save in exceptional circumstances, than the next day, not being a Sunday or public holiday, after it is laid.

(5) A prisoner who is to be charged with an offence against discipline may be kept apart from other prisoners pending the governor's first inquiry.

Rights of prisoners charged

49. (1) Where a prisoner is charged with an offence against discipline, he shall be informed of the charge as soon as possible and, in any case, before the time when it is inquired into by the governor.

   (2) At an inquiry into a charge against a prisoner he shall be given a full opportunity of hearing what is alleged against him and of presenting his own case.

Governor's punishments

50. (1) If he find a prisoner guilty of an offence against discipline the governor may, subject to rule 52 of these Rules, impose one or more of the following punishments:

   (*a*) caution;

   (*b*) forfeiture for a period not exceeding 28 days or any of the privileges under Rule 4 of these Rules;

   (*c*) exclusion from associated work for a period not exceeding 14 days;

   (*d*) stoppage of or deduction from earnings for a period not exceeding 56 days and of an amount not exceeding 28 days earnings;

   (*e*) cellular confinement for a period not exceeding 3 days;

   (*f*) in the case of a short-term or long-term prisoner, an award of additional days not exceeding 28 days;

   (*g*) in the case of a prisoner otherwise entitled to them, forfeiture for any period of the right, under rule 41 (1) of these Rules, to have the articles there mentioned;

   (*h*) in the case of a prisoner guilty of escaping or attempting to escape and who is entitled to it, forfeiture of the right to wear his own clothing under rule 20 (1) of these Rules.

   (2) If a prisoner is found guilty of more than one charge arising out of an incident, punishments under this rule may be ordered to run consecutively but, in the case of an award of additional days, the total period added shall not exceed 28 days.

Forfeiture of remission to be treated as an award of additional days

51. (1) In this rule, 'existing prisoner' and 'existing licensee' have the meanings assigned to them by paragraph 8 (1) of Schedule 12 to the Criminal Justice Act 1991 (*a*).

   (2) In relation to any existing prisoner or existing licensee who has forfeited any remission of his sentence, the provisions of Part II of the Criminal Justice Act 1991 shall apply as if he had been awarded such number of additional days as equals the number of days of remission which he has forfeited.

Offences committed by young persons

52. (1) In the case of an offence against discipline committed by an inmate who was under the age of 21 when the offence was committed (other than an

offender in relation to whom the Secretary of State has given a direction under section 13 (1) of the Criminal Justice Act 1982 that he shall be treated as if he had been sentenced to imprisonment)—

(*a*) rule 50 of these Rules shall have effect, but—

  (i) the maximum period of forfeiture of privileges under rule 4 of these Rules shall be 14 days; and

  (ii) the maximum period of stoppage of earnings shall be 14 days;

(2) In the case of an inmate who has been sentenced to a term of youth custody or detention in a young offender institution, and by virtue of a direction of the Secretary of State under section 13 of the Criminal Justice Act 1982, is treated as if he had been sentenced to imprisonment for that term, any punishment imposed on him for an offence against discipline before the said direction was given shall, if it has not been exhausted or remitted, continue to have effect as if made pursuant to rule 50 of these Rules.

(3) In the case of an inmate detained in a prison who, by virtue of paragraph 12 of the Schedule 8 to the Criminal Justice Act 1988, on 1st October 1988 fell to be treated for all purposes of detention, release and supervision as if his sentence had been a sentence of detention in a young offender institution, any award for an offence against discipline made in respect of him before that date under rule 50, 51 or 52 of the Prison Rules 1964, which were then in force, or treated by virtue of rule 5 (4A) as having been imposed under those Rules, shall, if it has not been exhausted or remitted, continue to have effect as if it were a punishment imposed pursuant to rule 50 or 51 of these Rules.

Particular punishments

53. (1) Not allocated

  (2) No punishment of cellular confinement shall be imposed unless the medical officer has certified that the prisoner is in a fit state of health to be so dealt with.

Prospective award of additional days

54. (1) Subject to paragraph (2), where an offence against discipline is committed by a prisoner who is detained only on remand, additional days may be awarded notwithstanding that the prisoner has not (or had not at the time of the offence) been sentenced.

  (2) An award of additional days under paragraph (1) shall have effect only if the prisoner in question subsequently becomes a short-term or long-term prisoner whose sentence is reduced, under section 67 of the Criminal Justice Act 1967 (*b*), by a period which includes the time when the offence against discipline was committed.

Suspended Punishments

55. (1) Subject to any directions given by the Secretary of State, the power to impose a disciplinary punishment (other than a caution) shall include power

to direct that the punishment is not to take effect unless, during a period specified in the direction (not being more than six months from the date of the direction) the prisoner commits another offence against discipline and a direction is given under paragraph (2) below.

(2) Where a prisoner commits an offence against discipline during the period specified in a direction given under paragraph (1) above the person dealing with that offence may—

    (*a*) direct that the suspended punishment shall take effect,

    (*b*) reduce the period or amount of the suspended punishment and direct that it shall take effect as so reduced,

    (*c*) vary the original direction by substituting for the period specified a period expiring not later than six months from the date of variation, or

    (*d*) give no direction with respect to the suspended punishment.

Remission and mitigation of punishments and quashing of findings of guilt

56. (1) The Secretary of State may quash any finding of guilt and may remit any punishment or mitigate it either by reducing it or by substituting another award which is, in his opinion, less severe.

  (2) Subject to any directions given by the Secretary of State, the Governor may remit or mitigate any punishment imposed by a governor or the Board of Visitors.

57–62. Not allocated.

Other particular classes

Prisoners committed for contempt, &c

63. (1) A prisoner committed or attached for contempt of court, or for failing to do or abstain from doing anything required to be done or left undone, shall have the same privileges as an unconvicted prisoner under Rules 20 (1) and 34 (1) of these rules.

  (2) Such prisoners shall be treated as a separate class for the purposes of Rule 3 of these Rules but, notwithstanding anything in that Rule, such prisoners may be permitted to associated with any other class of prisoners if they are willing to do so.

  (3) Not allocated.

64–71. Not allocated

Prisoners under sentence of death

Application of foregoing Rules

72. The foregoing provisions of these Rules shall apply in relation to a prisoner under sentence of death only in so far as they are compatible with that sentence and with Rules 73 to 76 of these Rules.

Search

73. A prisoner under sentence of death shall be searched with special care and every article shall be taken from him which it might be dangerous or inexpedient to leave in his possession.

Confinement

74. (1) A prisoner under sentence of death shall be confined in a separate cell and shall be kept apart from all other prisoners.
   (2) He shall be kept by day and night in the constant charge of 2 officers.
   (3) He shall not be required to work, but shall, if he wishes, be given work to do in his cell.
   (4) Subject to the provisions of Rule 75 of these Rules, no person other than a member of the board of visitors or an officer shall have access to a prisoner under sentence of death without the leave of the Secretary of State.

Visits

75. (1) Every visit to a prisoner under sentence of death, other than a visit by the chaplain or a prison minister, shall take place in the sight and hearing of an officer
   (2) Such a prisoner may be visited by any relation, friend or legal adviser whom he wishes to see, and who is authorised to visit him by an order in writing of a member of the board of visitors or the Secretary of State.
   (3) The chaplain shall have free access to every such prisoner belonging to the Church of England, and to every other such prisoner who wishes to see him.
   (4) Where such a prisoner belongs to a denomination other than the Church of England, a minister of that denomination shall have free access to him.

Correspondence

76. A prisoner under sentence of death shall be given all necessary facilities to enable him to correspond with his legal advisers, relatives and friends.

PART II
OFFICERS OF PRISONS

General duty of officers

77. (1) It shall be the duty of every officer to conform to these Rules and the rules and regulations of the prison, to assist and support the governor in their maintenance and to obey his lawful instructions.
   (2) An officer shall inform the governor promptly of any abuse or impropriety which comes to his knowledge.

Gratuities forbidden

78. No officer shall receive any unauthorised fee, gratuity or other consideration in connection with his office.

Search of officers

79. An officer shall submit himself to be searched in the prison if the governor so directs.

Transactions with prisoners

80. (1) No officer shall take part in any business or pecuniary transaction with or on behalf of a prisoner without the leave of the Secretary of State.

    (2) No officer shall without authority bring in or take out, or attempt to bring in or take out, or knowingly allow to be brought in or taken out, to or for a prisoner, or deposit in any place with intent that it shall come into the possession of a prisoner, any article whatsoever.

Contact with former prisoners, &c

81. No officer shall, without the knowledge of the governor, communicate with any person whom he knows to be a former prisoner or a relative or friend of a prisoner or former prisoner.

Communications to the press, &c

82. (1) No officer shall make, directly or indirectly, any unauthorised communication to a representative of the press or any other person concerning matters which have become known to him in the course of his duty.

    (2) No officer shall, without authority, publish any matter or make any public pronouncement relating to the administration of any institution to which the Prison Act 1952 applies or to any of its inmates.

Quarters

83. An officer shall occupy any quarters which may be assigned to him.

Code of discipline

84. The Secretary of State may approve a code of discipline to have effect in relation to officers, or such classes of officers as it may specify, setting out the offences against discipline, the awards which may be made in respect of them and the procedure for dealing with charges.

PART III
PERSONS HAVING ACCESS TO A PRISON

Prohibited articles

85. No person shall, without authority, convey into or throw into or deposit in a prison, or convey or throw out of a prison, or convey to a prisoner, or deposit in any place with intent that it shall come into the possession of a prisoner, any money, clothing food, drink, tobacco, letter, paper, book, tool or other article whatever. Anything so conveyed, thrown or deposited may be confiscated by the governor.

Control of persons and vehicles

86. (1) Any person or vehicle entering or leaving a prison may be stopped, examined and searched.

(2) The governor may direct the removal from a prison of any person who does not leave on being required to do so.

Viewing of prisons

87. (1) No outside person shall be permitted to view a prison unless authorised by statute or the Secretary of State.

(2) No person viewing the prison shall be permitted to take a photograph, make a sketch or communicate with a prisoner unless authorised by statute or the Secretary of State.

PART IV
BOARDS OF VISITORS

Disqualification for membership

88. Any person interested in any contract for the supply of goods or services to a prison shall not be a member of the board of visitors for that prison.

89–91 Not allocated

Board of visitors

92. (1) A member of the board of visitors for a prison appointed by the Secretary of State under section 6 (2) of the Prison Act 1952 shall subject to paragraph (1A below) hold office for three years, or such less period as the Secretary of State may appoint.

(1A) The Secretary of State may terminate the appointment of a member if he is satisfied that—

(*a*) he has failed satisfactorily to perform his duties,

(*b*) he is by reason of physical or mental illness, or for any other reason, incapable of carrying out his duties, or

(*c*) he has been convicted of such a criminal offence, or his conduct has been such, that it is not in the Secretary of State's opinion fitting that he should remain a member.

(2) When a board is first constituted, the Secretary of State shall appoint one of its members to be chairman for a period not exceeding twelve months.

(3) Subject to paragraph (2) above, at their first meeting in any year of office the Board shall appoint one of their number to be chairman and one to be vice-chairman for that year and thereafter shall fill any casual vacancy in either office promptly.

(4) The vice-chairman's term of office shall come to an end when, for whatever reason, that of the chairman comes to an end.

Proceedings of boards

93. (1) The board of visitors for a prison shall meet at the prison once a month or,
    if they resolve for reasons specified in the resolution that less frequent meet-
    ings are sufficient, not fewer than eight times in twelve months.

    (2) The board may fix a quorum of not fewer than three members for proceed-
    ings.

    (3) The board shall keep minutes of their proceedings.

    (4) The proceedings of the board shall not be invalidated by any vacancy in the
    membership or any defect in the appointment of a member.

General duties of boards

94. (1) The board of visitors for a prison shall satisfy themselves as to the state of
    the prison premises, the administration of the prison and the treatment of
    the prisoners.

    (2) The board shall inquire into and report upon any matter into which the
    Secretary of State asks them to inquire.

    (3) The board shall direct the attention of the governor to any matter which
    calls for his attention, and shall report to the Secretary of State any matter
    which they feel it expedient to report.

    (4) The board shall inform the Secretary of State immediately of any abuse
    which comes to their knowledge.

    (5) Before exercising any power under these Rules the board and any member
    of the board shall consult the governor in relation to any matter which may
    affect discipline.

Particular duties

95. (1) The board of visitors for a prison and any member of the board shall hear
    any complaint or request which a prisoner wishes to make to them or him.

    (2) The board shall arrange for the food of prisoners to be inspected by a mem-
    ber of the board at frequent intervals.

    (3) The board shall inquire into any report made to them, whether or not by a
    member of the board, that a prisoner's health, mental or physical, is likely
    to be injuriously affected by any conditions of his imprisonment.

Members visiting prisons

96. (1) The members of the board of visitors for a prison shall visit the prison fre-
    quently, and the board shall arrange a rota whereby at least one of its mem-
    bers visits the prison between meetings of the board.

    (2) A member of the board shall have access at any time to every part of the
    prison and to every prisoner, and he may interview any prisoner out of the
    sight and hearing of officers.

    (3) A member of the board shall have access to the records of the prison.

Annual Report

97. The board of visitors for a prison shall make an annual report to the Secretary
of State at the end of each year concerning the state of the prison and its administration, including in it any advice and suggestions they consider appropriate.

PART V
SUPPLEMENTAL

Delegation by governor

98. The governor of a prison may, with the leave of the Secretary of State, delegate any of his powers and duties under these Rules to another officer of that prison.

Contracted out prisons

98A. (1) Where the Secretary of State has entered into a contract for the running of a prison under section 84 of the Criminal Justice Act 1991 (*a*) ('the 1991 Act') these rules shall apply to that prison with the following modifications—

   (*a*) references to an officer in the Rules shall include references to a prisoner custody officer certified as such under section 89 (1) of the 1991 Act and performing custodial duties;

   (*b*) references to a governor in the Rules shall be construed as references to a director approved by the Secretary of State for the purposes of section 85 (1) (*a*) of the 1991 Act except—

      (i) in rule 43, 45, 46, 48, 49, 50, 56 and 98 where references to a governor shall be construed as references to a controller appointed by the Secretary of State under section 85 (1) (*b*) of the 1991 Act, and

      (ii) in rules 77 (1), 81 and 94 where references to a governor shall be construed as references to the director and the controller;

   (*c*) Rule 84 shall not apply.

   (2) Where a director exercises the powers set out in section 85 (3) (*b*) of the 1991 Act (removal from association, temporary confinement and restraints) in cases of urgency, he shall notify the controller of that fact forthwith.

Interpretation

99. (1) In these Rules, where the context so admits, the expression—

   'convicted prisoner' means, subject to the provisions of Rule 63 of these Rules, a prisoner who has been convicted or found guilty of an offence or committed or attached for contempt of court or for failing to do or abstain from doing anything required to be done or left undone, and the expression 'unconvicted prisoner' shall be construed accordingly:

   'governor' includes an officer for the time being in charge of a prison;

   'legal adviser' means, in relation to a prisoner, his counsel or solicitor, and includes a clerk acting on behalf of his solicitor;

   'officer' means an officer of a prison;

   'prison minister' means, in relation to a prison, a minister appointed to that prison under section 10 of the Prison Act 1952;

   'short-term prisoner' and 'long-term prisoner' have the meanings assigned to them by section 33 (5) of the Criminal Justice Act 1991, as extended by sections 43 (1) and 45 (1) of that Act.

(2) In these Rules a reference to (*a*) an award of additional days means additional days awarded under these Rules by virtue of section 42 of the Criminal Justice Act 1991; and (*b*) the Church of England includes a reference to the Church in Wales.

(3) The Interpretation Act 1889 shall apply for the interpretation of these Rules as it applies for the interpretation of an Act of Parliament.

Revocations and savings

100. (1) The Rules specified in the Schedule to these Rules are hereby revoked.

(2) For the purposes of these Rules any appointment, approval, authority, certificate, condition, direction or restriction made, given or imposed under any provision of any of the Rules revoked by this Rule shall be treated as having been made, given or imposed under the corresponding provision of these Rules.

Citation and commencement

101. These Rules may be cited as the Prison Rules 1964 and shall come into operation on the fourteenth day after the day on which they are made.

HENRY BROOKE
One of Her Majesty's Principal
Secretaries of State

Home Office
WHITEHALL
11th March 1964
Consolidated November 1991
DIA1

SCHEDULE	Rule 100
RULES REVOKED	

Rules	References
The Prison Rules 1949	SI 1949/1073 (1949 I, p. 3470)
The Prison Rules 1951	SI 1951/1343 (1951 II, p. 289)
The Prison Rules 1952	SI 1952/1405 (1952 III, p. 2631)
The Prison Rules 1956	SI 1956/1986 (1956 II, p. 1897)
The Prison Rules 1962	SI 1962/1471 (1962 II, p. 1602)
The Prison Rules 1963	SI 1963/468 (1963 I, p. 530)

Young Offender Institution Rules 1988

1988 No. 1422

YOUNG OFFENDER INSTITUTIONS,

ENGLAND AND WALES

The Young Offender Institution Rules 1988

Made	5th August 1988
Laid before Parliament	18th August 1988
Coming into force	1st October 1988

ARRANGEMENT OF RULES

PART I

Preliminary

23.   Library books.

Medical attention

24.   Medical attendance.
25.   Special illnesses and conditions.
26.   Notification of illness or death.

Religion

27.   Religious denomination.
28.   Special duties of chaplains and appointed ministers.
29.   Regular visits by ministers of religion, etc.
30.   Religious services.
31.   Substitute for chaplain or appointed minister.
32.   Sunday work.
33.   Religious books.

Occupation and links with the community

34.   Regime activities.
35.   Education.
36.   Training courses.
37.   Work.
38.   Physical education.
39.   Outside contacts.
40.   After-care.

Discipline and control

41.   Maintenance of order and discipline.
42.   Custody outside a young offender institution.
43.   Search.
44.   Record and photograph.
45.   Inmates' property.
46.   Removal from association.
47.   Use of force.
48.   Temporary confinement.
49.   Restraints.
50.   Offences against discipline.
51.   Disciplinary charges.
52.   Rights of inmates charged.
53.   Governor's awards.
54.   Not allocated
55.   Not allocated
56.   Confinement to a cell or room
57.   Removal from wing or living unit.
58.   Suspended awards.
59.   Remission and mitigation of awards.
60.   Adult female inmates: disciplinary awards.
60A. Forfeiture of remission to be treated as an award of additional days.

PART III
Officers of Young Offender Institutions

61.   General duty of officers.
62.   Gratuities forbidden.

In pursuance of sections 25, 43 (5) and 47 of the Prison Act 1952ª, I hereby make the following Rules:—

ª 1952 c. 52; section 43 was substituted by the Criminal Justice Act 1982 (c. 48), section 11; section 47 was extended by the Criminal Justice Act 1961 (c. 39), section 23 (2), the Criminal Justice Act 1982, section 13 (5), and Schedule 17, paragraph 9 and the Criminal Justice Act 1988 (c. 33), Schedule 8, paragraph 14; and amended by the Criminal Justice Act 1961, Schedule 4, the Criminal Justice Act 1967 (c. 80), section 66 (5), the Courts Act 1971 (c. 23), Schedule 8, paragraph 33 and the Criminal Justice Act 1982, Schedule 14, paragraph 7. Section 47 of the 1952 Act was also affected by an amendment to section 52 (2) of that Act by the Criminal Justice Act 1967, section 66 (4). The Criminal Justice Act 1988, Schedule 8, paragraph 1, contains amendments affecting these provisions.

PART I
PRELIMINARY

Citation and commencement

1. These Rules may be cited as the Young Offender Institution Rules 1988 and shall come into force on 1st October 1988.

Interpretation

2.—(1) In these Rules, where the context so admits, the expression:—

'compulsory school age' has the same meaning as in the Education Act 1944;

'governor' includes an officer for the time being in charge of a young offender institution;

'inmate' means a person detained in a young offender institution;

'legal adviser' means, in relation to an inmate, his counsel or solicitor, and includes a clerk acting on behalf of his solicitor;

'minister appointed to a young offender institution' means a minister so appointed under section 10 of the Prison Act 1952;

'officer' means an officer of a young offender institution.

' "short-term prisoner" and "long-term prisoner" have the meanings assigned to them by section 33 (5) of the Criminal Justice Act 1991, as extended by sections 43 (1) and 45 (1) of that Act'.

(2) In these Rules a reference to (*a*) an award of additional days means additional days awarded under these Rules by virtue of section 42 of the Criminal Justice Act 1991; and (*b*) the Church of England includes a reference to the Church in Wales.

(3) The Rules set out in the Schedule to this Order are hereby revoked.

PART II
INMATES

General

Aims and general principles of young offender institutions

3.—(1) The aim of a young offender institution shall be to help offenders to prepare for their return to the outside community.

(2) The aim mentioned in paragraph (1) above shall be achieved, in particular, by—

   (*a*) providing a programme of activities, including education, training and work designed to assist offenders to acquire or develop personal responsibility, self-discipline, physical fitness, interests and skills and to obtain suitable employment after release;
   (*b*) fostering links between the offender and the outside community;
   (*c*) co-operating with the services responsible for the offender's supervision after release.

Classification of inmates

4, Inmates may be classified, in accordance with any directions of the Secretary of State, taking into account their ages, characters and circumstances.

Release

5.—Not allocated.

Temporary release

6.—(1) An inmate to whom this rule applies may be temporarily released for any period or periods and subject to any conditions.

(2) An inmate released under this rule may be recalled at any time whether any conditions of his release have been broken or not.

(3) This rule applies to inmates other than persons committed in custody for trial or to be sentenced or otherwise dealt with before or by the Crown Court or remanded in custody by any court.

Conditions

Privileges

7. There shall be established at every young offender institution systems of privileges approved by the Secretary of State and appropriate to the classes of inmates thereof and their ages, characters and circumstances, which shall include arrangements under which money earned by inmates may be spent by them within the young offender institution.

Information to inmates

8.—(1) Every inmate shall be provided, as soon as possible after his reception into the young offender institution, and in any case within 24 hours, with information in writing about those provisions of these Rules and other matters which it is necessary that he should know, including earnings and privileges, and the proper method of making requests and complaints.

(2) In the case of an inmate aged less than 18, or an inmate aged 18 or over who cannot read or appears to have difficulty in understanding the information so provided, the governor, or an officer deputed by him, shall so explain it to him that he can understand his rights and obligations.

(3) A copy of these Rules shall be made available to any inmate who requests it.

Requests and Complaints

9.—(1) A request or complaint to the governor or Board of Visitors relating to an inmate's detention shall be made orally or in writing by that inmate.

(2) On every day the governor shall hear any oral requests and complaints that are made to him under paragraph (1) above.

(3) A written request or complain under paragraph (1) above made be made in confidence.

Letters and visits generally

10.—(1) The Secretary of State may, with a view to securing discipline and good order or the prevention of crime or in the interests of any persons, impose restrictions, either generally or in a particular case, upon the communications to be permitted between an inmate and other persons.

(2) Except as provided by statute or these Rules, an inmate shall not be permitted to communicate with any outside person, or that person with him, without the leave of the Secretary of State.

(3) Except as provided by these Rules, every letter or communication to or from an inmate may be read or examined by the governor or an officer deputed by him, and the governor may, at his discretion, stop any communication on the ground that its contents are objectionable or that it is of inordinate length.

(4) Subject to the provisions of these Rules, the governor may give such directions as he thinks fit for the supervision of visits to inmates, either generally or in a particular case.

Personal letters and visits

11.—(1) An inmate shall be entitled—

> (*a*) to send and to receive a letter on his reception into a young offender institution and thereafter once a week; and
>
> (*b*) to receive a visit twice in every period of four weeks but only once in every such period if the secretary of State so directs.

(2) The governor may allow an inmate an additional letter or visit when necessary for his welfare or that of his family.

(3) The governor may allow an inmate entitled to a visit to send and to receive a letter instead.

(4) The governor may defer the right of an inmate to a visit until the expiration of any period of confinement to a cell or room.

(5) The board of visitors may allow an inmate an additional letter or visit in special circumstances, and may direct that a visit may extend beyond the normal duration.

(6) The Secretary of State may allow additional letters and visits in relation to any inmate or class of inmates.

(7) An inmate shall not be entitled under this rule to receive a visit from any person other than a relative or friend, except with the leave of the Secretary of State.

(8) Any letter or visit under the succeeding provisions of these Rules shall not be counted as a letter or visit for the purposes of this rule.

Police interviews

12.—A police officer may, on production of an order issued by or on behalf of a chief officer of police, interview any inmate willing to see him.

Legal advisers

13.—(1) The legal adviser of an inmate in any legal proceedings, civil or criminal, to which the inmate is a party shall be afforded reasonable facilities for interviewing him in connection with those proceedings, and may do so out of hearing of an officer.

(2) An inmate's legal adviser may, with the leave of the Secretary of State, interview the inmate in connection with any other legal business.

Further facilities in connection with legal proceedings

14.—(1) An inmate who is a party to any legal proceedings may correspond with his legal adviser in connection with the proceedings and unless the governor has reason to suppose that any such correspondence contains matter not relating to the proceedings it shall not be read or stopped under rule 10 (3) of these Rules.

(2) An inmate shall on request be provided with any writing materials necessary for the purposes of paragraph (1) of this rule.

(3) Subject to any directions given in the particular case by the Secretary of State, a registered medical practitioner selected by or on behalf of an inmate who is a party to any legal proceedings shall be afforded reasonable facilities for examining him in connection with the proceedings, and may do so out of hearing but in the sight of an officer.

(4) Subject to any directions of the Secretary of State, an inmate may correspond with a solicitor for the purpose of obtaining legal advice concerning any cause of action in relation to which the inmate may become a party to legal proceedings or for the purpose of instructing the solicitor to issue such proceedings.

Securing release of defaulters

15. Any inmate detained in a young offender institution in default of payment of a fine or any other sum of money may communicate with, and be visited at any reasonable time on a weekday by, any relative or friend to arrange for payment in order to secure his release.

Clothing

16.—(1) An inmate shall be provided with clothing adequate for warmth and health in accordance with a scale approved by the Secretary of State.

(2) The clothing provided under this rule shall include suitable protective clothing for use at work, where this is needed.

(3) Subject to the provisions of rule 42 (3) of these Rules, an inmate shall wear clothing provided under this rule and no other, except on the directions of the Secretary of State.

(4) An inmate shall where necessary be provided with suitable and adequate clothing on his release.

Food

17.—(1) Subject to any directions of the Secretary of State, no inmate shall be allowed, except as authorised by the medical officer, to have any food other than that ordinarily provided.

(2) The food provided shall be wholesome, nutritious, well prepared and served, reasonably varied and sufficient in quantity.

(3) The medical officer shall regularly inspect the food both before and after it is cooked, and shall report any deficiency or defect to the governor.

(4) In this rule, 'food' includes drink.

Alcohol and tobacco

18.—(1) No inmate shall be allowed to have any intoxicating liquor except under a written order of the medical officer specifying the quantity and the name of the inmate.

(2) No inmate shall be allowed to smoke or to have any tobacco except in accordance with any directions of the Secretary of State.

Sleeping accommodation

19.—(1) No room or cell shall be used as sleeping accommodation for an inmate unless it has been certified by an officer of the Secretary of State (not being an officer of a young offender institution) that its size, lighting, heating, ventilation and fittings are adequate for health, and that it allows the inmate to communicate at any time with an officer.

(2) A certificate given under this rule shall specify the maximum number of inmates who may sleep in the room or cell at one time, and the number so specified shall not be exceeded without the leave of the Secretary of State.

Beds and bedding

20. Each inmate shall be provided with a separate bed and with separate bedding adequate for warmth and health.

Hygiene

21.—(1) Every inmate shall be provided with toilet articles necessary for his health and cleanliness, which shall be replaced as necessary.

(2) Every inmate shall be required to wash at proper times, have a hot bath or shower on reception and thereafter at least once a week.

(3) Subject to any directions of the Secretary of State, a male inmate may be required by the governor to shave or be shaved and to have his hair cut as may be necessary for neatness or, as directed by the medical officer, for health or cleanliness.

(4) A female inmate's hair shall not be cut without her consent except where the medical officer directs that it is necessary for health or cleanliness.

Female inmates

22. The Secretary of State may, subject to any conditions he thinks fit, permit a female inmate to have her baby with her in a young offender institution, and everything necessary for the baby's maintenance and care may be provided there.

Library books

23. A library shall be provided in every young offender institution and, subject to any directions of the Secretary of State, every inmate shall be allowed to have library books and to exchange them.

Medical attention

Medical attendance

24.—(1) The medical officer of a young offender institution shall have the care of the health, mental and physical, of the inmates of that institution.

(2) Every request by an inmate to see the medical officer shall be recorded by the officer to whom it is made and promptly passed on to the medical officer.

(3) The medical officer may call another medical practitioner in consultation at his discretion, and shall do so if time permits before performing any serious operation.

Special illnesses and conditions

25.—(1) The medical officer shall report to the governor on the case of any inmate whose health is likely to be injuriously affected by continued detention or any conditions of detention. The governor shall send the report to the Secretary of State without delay, together with his own recommendations.

(2) The medical officer shall pay special attention to any inmate whose mental condition appears to require it, and make any special arrangements which appear necessary for his supervision or care.

(3) The medical officer shall inform the governor if he suspects any inmate of having suicidal intentions, and the inmate shall be placed under special observation.

Notification of illness or death

26.—(1) If an inmate dies, becomes seriously ill, sustains any severe injury or is removed to hospital on account of mental disorder, the governor shall, if he knows his or her address, at once inform the inmate's spouse or next of kin, and also any person who the inmate may reasonably have asked should be informed.

(2) If an inmate dies, the governor shall give notice immediately to the coroner having jurisdiction, to the board of visitors and to the Secretary of State.

Religion

Religious denomination

27. An inmate shall be treated as being of the religious denomination stated in the record made in pursuance of section 10 (5) of the Prison Act 1952[a], but the governor may, in a proper case after due inquiry, direct that record to be amended.

[a] 1952 c. 52.

Special duties of chaplains and appointed ministers

28.—(1) The chaplain or a minister appointed to a young offender institution shall—

(a) interview every inmate of his denomination individually as soon as he reasonably can after the inmate's reception into that institution and shortly before his release; and

(b) if no other arrangements are made, read the burial service at the funeral of any inmate of his denomination who dies in that institution.

(2) The chaplain shall visit daily all inmates belonging to the Church of England who are sick, under restraint or confined to a room or cell; and a minister appointed to a young offender institution shall do the same, as far as he reasonably can, for inmates of his own denomination.

(3) If the inmate is willing, the chaplain shall visit any inmate not of the Church of England who is sick, under restraint or confined to a room or cell, and is not regularly visited by a minister of his own denomination.

Regular visits by ministers of religion, etc.

29.—(1) The chaplain shall visit regularly the inmates belonging to the Church of England.

(2) A minister appointed to a young offender institution shall visit the inmates of his denomination as regularly as he reasonably can.

(3) The governor shall, if so requested by an inmate belonging to a denomination for which no minister has been appointed to a young offender institution, do what he reasonably can to arrange for that inmate to be visited regularly by a minister of that denomination.

(4) Every request by an inmate to see the chaplain or a minister appointed to a young offender institution shall be promptly passed on to the chaplain or minister.

Religious services

30.—(1) The chaplain shall conduct Divine Service for inmates belonging to the Church of England at least once every Sunday, Christmas Day and Good Friday, and such celebrations of Holy Communion and weekday services as may be arranged.

(2) A minister appointed to a young offender institution shall conduct Divine Service for inmates of his denomination at such times as may be arranged.

Substitute for chaplain or appointed minister

31.—(1) A person approved by the Secretary of State may act for the chaplain in his absence.

(2) A minister appointed to a young offender institution may, with the leave of the Secretary of State, appoint a substitute to act for him in his absence.

Sunday work

32. Arrangements shall be made so as not to require inmates to do any unnecessary work on Sunday, Christmas Day or Good Friday nor inmates of religions other than the Christian religion to do any unnecessary work on their recognised days of religious observance (as an alternative, but not in addition, to those days).

Religious books

33. There shall, so far as reasonably practicable, be available for the personal use of every inmate such religious books recognised by his denomination as are approved by the Secretary of State for use in young offender institutions.

Occupation and links with the community

Regime activities

34.—(1) An inmate shall be occupied in education, training courses, work and physical education provided in accordance with rule 3 of these Rules.

(2) In all such activities regard shall be paid to individual assessment and personal development.

(3) The medical officer may excuse an inmate from work or any other activity on medical grounds; and no inmate shall be set to participate in work or any other activity of a kind for which he is considered by the medical officer to be unfit.

(4) An inmate may be required to participate in regime activities for no more than 8 hours a day.

(5) Inmates may be paid for their work or participation in other activities at rates approved by the Secretary of State, either generally or in relation to particular cases.

Education

35.—(1) Provision shall be made at a young offender institution for the education of inmates by means of programmes of class teaching or private study within the normal working week and, so far as practicable, programmes of evening and weekend educational classes or private study. The educational activities shall, so far as practicable, be such as will foster personal responsibility and an inmate's interests and skills and help him to prepare for his return to the community.

(2) In the case of an inmate aged less than 17, arrangements shall be made for his participation in education or training courses for at least 15 hours a week within the normal working week.

(3) In the case of an inmate aged 17 or over who is illiterate or backward, arrangements shall be made for education appropriate to his needs, if necessary within the normal working week.

(4) In the case of a female inmate aged 21 or over who is serving a sentence of imprisonment or who has been committed to prison for default and who is detained in a young offender institution instead of a prison, reasonable facilities shall be afforded if she wishes to improve her education, by class teaching or private study.

Training courses

36.—(1) Provision shall be made at a young offender institution for the training of inmates by means of training courses, in accordance with directions of the Secretary of State.

(2) Training courses shall be such as will foster personal responsibility and an inmate's interests and skills and improve his prospects of finding suitable employment after release.

(3) Training courses shall, so far as practicable, be such as to enable inmates to acquire suitable qualifications.

Work

37.—(1) Work shall, so far as practicable, be such as will foster personal responsibility and an inmate's interests and skills and help him to prepare for his return to the community.

(2) No inmate shall be set to do work of a kind not authorised by the Secretary of State.

Physical education

38.—(1) Provision shall be made at a young offender institution for the physical education of inmates within the normal working week, as well as evening and weekend physical recreation. The physical education activities shall be such as will foster personal responsibility and an inmate's interests and skills and encourage him to make good use of his leisure on release.

(2) Arrangements shall be made for each inmate, other than one to whom paragraph (4) of this rule applies, to participate in physical education for at least two hours a week on average or, in the case of inmates detained in such institutions or parts of institutions as the Secretary of State may direct, for at least 1 hour each weekday on average, but outside the hours allotted to education under rule 35 (2) in the case of an inmate of compulsory school age.

(3) In the case of an inmate with a need for remedial physical activity, appropriate facilities shall be provided.

(4) A female inmate aged 21 years or other who is serving a sentence of imprisonment or who has been committed to prison for default and who is detained in a young offender institution instead of a prison shall, if not engaged in outdoor work or detained in an open institution, be given the opportunity of exercise in the open air for not less than one hour in all, each day, if weather permits; but the Secretary of State may in special circumstances authorise the reduction of the period aforesaid to half an hour a day:

Provided that exercise consisting of physical education may be given indoors instead of in the open air.

Outside contacts

39.—(1) The governor shall encourage links between the young offender institution and the community by taking steps to establish and maintain relations with suitable persons and agencies outside the institution.

(2) The governor shall ensure that special attention is paid to the maintenance of such relations between an inmate and his family as seem desirable in the best interests of both.

(3) Subject to any directions of the Secretary of State, an inmate shall be encouraged, as far as practicable, to participate in activities outside the young offender institution which will be of benefit to the community or of benefit to the inmate in helping him to prepare for his return to the community.

After-care

40.—(1) From the beginning of his sentence, consideration shall be given, in consultation with the appropriate supervising service, to an inmate's future and the help to be given him in preparation for and after his return to the community.

(2) Every inmate who is liable to supervision after release shall be given a careful explanation of his liability and the requirements to which he will be subject while under supervision.

Discipline and control

Maintenance of order and discipline

41.—(1) Order and discipline shall be maintained, but with no more restriction than is required in the interests of security and well-ordered community life.

(2) In the control of inmates, officers shall seek to influence them through their own example and leadership, and to enlist their willing co-operation.

Custody outside a young offender institution

42.—(1) A person being taken to or from a young offender institution in custody shall be exposed as little as possible to public observation and proper care shall be taken to protect him from curiosity and insult.

(2) An inmate required to be taken in custody anywhere outside a young offender institution shall be kept in the custody of an officer appointed under section 3 of the Prison Act 1952[a] or of a police officer.

(3) An inmate required to be taken in custody to any court shall wear his own clothing or clothing different from the dress worn at any institution to which the Prison Act 1952 applies.

[a] 1952 c. 52.

Search

43.—(1) Every inmate shall be searched when taken into custody by an officer, on his reception into a young offender institution and subsequently as the governor thinks necessary.

(2) An inmate shall be searched in as seemly a manner as is consistent with discovering anything concealed.

(3) No inmate shall be stripped and searched in the sight of another inmate or in the sight or presence of an officer not of the same sex.

(4) Not allocated.

Record and photograph

44.—(1) A personal record of each inmate shall be prepared and maintained in such manner as the Secretary of State may direct, but no part of the record shall be disclosed to any person not authorised to receive it.

(2) Every inmate may be photographed on reception and subsequently, but no copy of the photograph shall be given to any person not authorised to receive it.

Inmates' property

45.—(1) Anything, other than cash, which an inmate has at a young offender institution and which he is not allowed to retain for his own use shall be taken into the governor's custody.

(2) Any cash which an inmate has at a young offender institution shall be paid into an account under the control of the governor and the inmate shall be credited with the amount in the books of the institution.

(3) Any article belonging to an inmate which remain unclaimed for a period of more than 3 years after he is released, or dies, may be sold or otherwise disposed of; and the net proceeds of any sale shall be paid to the National Association for the Care and Resettlement of Offenders, for its general purposes.

(4) The governor may confiscate any unauthorised article found in the possession of an inmate after his reception into a young offender institution, or concealed or deposited within a young offender institution.

Removal from association

46.—(1) Where it appears desirable, for the maintenance of good order or discipline or in his own interests, than an inmate should not associate with other inmates, either generally or for particular purposes, the governor may arrange for the inmate's removal from association accordingly.

(2) An inmate shall not be removed under this rule for a period of more than 3 days without the authority of a member of the board of visitors or of the Secretary of State. An authority given under this paragraph shall, in the case of a female inmate aged 21 years or over, be for a period not exceeding one month and, in the case of any other inmate, be for a period not exceeding 14 days, but may be renewed from time to time for a like period.

(3) The governor may arrange at his discretion for such an inmate as aforesaid to resume association with other inmates, and shall do so if in any case the medical officer so advises on medical grounds.

Use of force

47.—(1) An officer in dealing with an inmate shall not use force unnecessarily and, when the application of force to an inmate is necessary, no more force than is necessary shall be used.

(2) No officer shall act deliberately in a manner calculated to provoke an inmate.

Temporary confinement

48.—(1) The governor may order an inmate who is refractory or violent to be confined temporarily in a special cell or room, but an inmate shall not be so confined as a punishment, or after he has ceased to be refractory or violent.

(2) A cell or room shall not be used for the purpose of this rule unless it has been certified by an officer of the Secretary of State (not being an officer of a young offender institution) that it is suitable for the purpose, that its size, lighting, heating, ventilation and fittings are adequate for health, and that it allows the inmate to communicate at any time with an officer.

(3) In relation to any young offender institution, section 14 (6) of the Prison Act 1952 shall have effect so as to enable the provision of special rooms instead of special cells for the temporary confinement of refractory or violent inmates.

Restraints

49.—(1) The governor may order an inmate, other than an inmate aged less than 17, to be put under restraint where this is necessary to prevent the inmate from injuring himself or others, damaging property or creating a disturbance.

(2) Notice of such an order shall be given without delay to a member of the board of visitors and to the medical officer.

(3) On receipt of the notice the medical officer shall inform the governor whether he concurs in the order. The governor shall give effect to any recommendation which the medical officer may make.

(4) An inmate shall not be kept under restraint longer than necessary, nor shall he be so kept for longer than 24 hours without a direction in writing given by a member of the board of visitors or by an officer of the Secretary of State (not being an officer of a young offender institution). Such a direction shall state the grounds for the restraint and the time during which it may continue.

(5) Particulars of every case of restraint under the foregoing provisions of this rule shall be forthwith recorded.

(6) Except as provided by this rule no inmate shall be put under restraint otherwise than for safe custody during removal, or on medical grounds by direction of the medical officer. No inmate shall be put under restraint as a punishment.

(7) Any means of restraint shall be of a pattern authorised by the Secretary of State, and shall be used in such manner and under such conditions as the Secretary of State may direct.

Offences against discipline

50. An inmate is guilt of an offence against discipline if he—

    (1)   commits any assault;

    (2)   detains any person against his will;

    (3)   denies access to any part of the young offender institution to any officer;

    (4)   fights with any person;

    (5)   intentionally endangers the health or personal safety of others or, by his conduct, is reckless whether such health or personal safety is endangered;

    (6)   intentionally obstructs an officer in the execution of his duty;

    (7)   escapes or absconds from a young offender institution or from legal custody;

    (8)   fails

        (*a*) not allocated

        (*b*) to comply with any condition upon which he was temporarily released under Rule 6 of these rules.

    (9)   has in his possession—

        (*a*) any unauthorised article, or

        (*b*) a greater quantity of any article than he is authorised to have;

  (10)   sells or delivers to any person any unauthorised article;

  (11)   sells or, without permission, delivers to any person any article which he is allowed to have only for his own use;

  (12)   takes improperly any article belonging to another person or to a young offender institution;

  (13)   intentionally or recklessly sets fire to any part of a young offender institution or any other property, whether or not his own;

  (14)   destroys or damages any part of a young offender institution or any other property other than his own;

(15) absents himself from any place where he is required to be or is present at any place where he is not authorised to be;

(16) is disrespectful to any officer or any person visiting a young offender institution;

(17) uses threatening, abusive or insulting words or behaviour;

(18) intentionally fails to work properly or, being required to work, refuses to do so;

(19) disobeys any lawful order;

(20) disobeys or fails to comply with any rule or regulation applying to him;

(21) on any way offends against good order and discipline;

(22) (*a*) attempts to commit,

(*b*) incites another inmate to commit, or

(*c*) assists another inmate to commit or to attempt to commit any of the foregoing offences.

Disciplinary charges

51.—(1) Where an inmate is to be charged with an offence against discipline, the charge shall be laid as soon as possible and, save in exceptional circumstances, within 48 hours of the discovery of the offence.

(2) Not allocated

(3) Every charge shall be inquired into, by the governor.

(4) Every charge shall be first inquired into not later, save in exceptional circumstances, than the next day, not being a Sunday or public holiday, after it is laid.

(5) An inmate who is to be charged with an offence against discipline may be kept apart from other inmates pending the Governor's first inquiry.

Rights of inmates charged

52.—(1) Where an inmate is charged with an offence against discipline, he shall be informed of the charge as soon as possible and, in any case, before the time when it is inquired into by the governor.

(2) At an inquiry into a charge against an inmate he shall be given a full opportunity of hearing what is alleged against him and of presenting his own case.

Governor's punishments

53.—(1) If he finds an inmate guilty of an offence against discipline the governor may, subject to rule 60 of these Rules, impose one or more of the following punishments:

(*a*) caution;

(*b*) forfeiture for a period not exceeding 14 days of any of the privileges under rule 7 of these Rules;

(*c*) removal for a period not exceeding 14 days from any particular activity or activities of the young offender institution, other than education, training courses, work and physical education in accordance with rules 34, 35, 36, 37 and 38 of these Rules;

(*d*) extra work outside the normal working week for a period not exceeding 14 days and for not more than 2 hours on any day;

(e) stoppage of or deduction from earnings for a period not exceeding 28 days and of an amount not exceeding 14 days earnings.

(f) confinement to a cell or room for a period not exceeding 3 days;

(g) removal from his wing or living unit for a period not exceeding 14 days;

(h) in the case of an inmate who is a short-term or long-term prisoner, an award of additional days not exceeding 28 days.

(2) If an inmate is found guilty of more than one charge arising out of an incident punishments under this rule may be ordered to run consecutively, but, in the case of an award of additional days, the total period added shall not exceed 28 days.

54.—Not allocated

55.—Not allocated

Confinement to a cell or room

56.—(1) No punishment of confinement to a cell or room shall be imposed unless the medical officer has certified that the inmate is in a fit state of health to be so dealt with.

(2) No cell or room shall be used as a detention cell or room for the purpose of a punishment of confinement to a cell or room unless it has been certified by an officer of the Secretary of State (not being an officer of a young offender institution) that it is suitable for the purpose; that its size, lighting, heating, ventilation and fittings are adequate for health; and that it allows the inmate to communicate at any time with an officer.

Removal from wing or living unit

57. Following the imposition of a punishment of removal from his wing or living unit, an inmate shall be accommodated in a separate part of the young offender institution under such restrictions of earnings and activities as the Secretary of State may direct.

Suspended punishments

58.—(1) Subject to any directions of the Secretary of State, the power to impose a disciplinary punishment (other than a caution) shall include the power to direct that the punishment is not to take effect unless, during a period specified in the direction (not being more than 6 months from the date of the direction), the inmate commits another offence against discipline and a direction is given under paragraph (2) below).

(2) Where an inmate commits an offence against discipline during the period specified in a direction given under paragraph (1) above, the person dealing with that offence may—

(a) direct that the suspended punishment shall take effect; or

(b) reduce the period or amount of the suspended punishment and direct that it shall take effect as so reduced; or

(c) vary the original direction by substituting for the period specified therein a period expiring not later than six months from the date of variation; or

(d) give no direction with respect to the suspended punishment.

Remission and mitigation of punishments and quashing of findings of guilt

59.—(1) The Secretary of State may quash any finding of guilt and may remit a disciplinary punishment or mitigate it either by reducing it or by substituting a punishment which is, in his opinion, less severe.

(2) Subject to any directions of the Secretary of State, the governor may remit or mitigate any punishment imposed by a governor or the board of visitors.

Adult female inmates: disciplinary punishments

60.—(1) In the case of a female inmate aged 21 years or over who is serving a sentence of imprisonment or who has been committed to prison for default—

  (i) rule 53 of these Rules shall not apply, but the governor may, if he finds the inmate guilt of an offence against discipline, impose one or more of the following punishments:

  (*a*) caution;

  (*b*) forfeiture for a period not exceeding 28 days of any of the privileges under rule 7 of these Rules;

  (*c*) removal for a period not exceeding 14 days from any particular activity or activities of the young offender institution, other than education, training courses, work and physical education in accordance with rules 34, 35, 36, 37 and 38 of these Rules;

  (*d*) extra work outside the normal working week for a period not exceeding 14 days and for not more than 2 hours on any day;

  (*e*) stoppage of or deduction from earnings for a period not exceeding 56 days and of an amount not exceeding 28 days earnings.

  (*f*) confinement to a cell or room for a period not exceeding 3 days;

  (*g*) in the case of an inmate who is a short-term or long-term prisoner, an award of additional days not exceeding 28 days;

  (2) Not allocated

  (3) If an inmate is found guilty of more than one charge arising out of an incident, punishments under this rule may be ordered to run consecutively, but in the case of an award of additional days, the total period added shall not exceed 28 days.

Forfeiture of remission to be treated as an award of additional days

60A.—(1) In this rule, 'existing prisoner' and 'existing licensee' have the meanings assigned to them by paragraph 8 (1) of Schedule 12 to the Criminal Justice Act 1991.

(2) In relation to any existing prisoner or existing licensee who has forfeited any remission of his sentence, the provisions of Part II of the Criminal Justice Act 1991 shall apply as if he had been awarded such number of additional days as equals the number of days of remission which he has forfeited.

PART III
OFFICERS OF YOUNG OFFENDER INSTITUTIONS

General duty of officers

61.—(1) It shall be the duty of every officer to conform to these Rules and the rules and regulations of the young offender institution, to assist and support the governor in their maintenance and to obey his lawful instructions.

(2) An officer shall inform the governor promptly of any abuse or impropriety which comes to his knowledge.

Gratuities forbidden

62. No officer shall receive any unauthorised fee, gratuity or other consideration in connection with his office.

Search of officers

63. An officer shall submit himself to be searched in a young offender institution if the governor so directs.

Transactions with inmates

64.—(1) No officer shall take part in any business or pecuniary transaction with or on behalf of an inmate without the leave of the Secretary of State.

(2) No officer shall, without authority, bring in or take out, or attempt to bring in or take out, or knowingly allow to be brought in or taken out, to or for an inmate, or deposit in any place with intent that it shall come into the possession of an inmate, any article whatsoever.

Contact with former inmates, etc.

65. No officer shall, without the knowledge of the governor, communicate with any person who he knows to be a former inmate or a relative or friend of an inmate or former inmate.

Communications to the press, etc.

66.—(1) No officer shall make, directly or indirectly, any unauthorised communication to a representative of the press or any other person concerning matters which have become known to him in the course of his duty.

(2) No officer shall, without authority, publish any matter or make any public pronouncement relating to the administration of any institution to which the Prison Act 1952 applies or to any of its inmates.

Quarters

67. An officer shall occupy any quarters which may be assigned to·him.

Code of discipline

68. The Secretary of State may approve a code of discipline to have effect in relation to officers, or such classes of officers as it may specify, setting out the offences against discipline, the awards which may be made in respect of them and the procedure for dealing with charges.

PART IV
PERSONS HAVING ACCESS TO A YOUNG OFFENDER INSTITUTION

Prohibited articles

69. No person shall, without authority, convey into or throw into or deposit in a young offender institution, or convey to an inmate, or deposit in any place with intent that it shall come into the possession of an inmate, any article whatsoever. Anything so conveyed, thrown or deposited may be confiscated by the governor.

Control of persons and vehicles

70.—(1) Any person or vehicle entering or leaving a young offender institution may be stopped, examined and searched.

(2) The Governor may direct the removal from a young offender institution of any person who does not leave on being required to do so.

Viewing of young offender institutions

71.—(1) No outside person shall be permitted to view a young person institution unless authorised by statute or the Secretary of State.

(2) No person viewing a young offender institution shall be permitted to take a photograph, make a sketch or communicate with an inmate unless authorised by statute or the Secretary of State.

PART V
BOARDS OF VISITORS

Disqualification for membership

72. Any person interested in any contract for the supply of goods or services to a young offender institution shall not be a member of the board of visitors for that institution.

Appointment

73.—(1) A member of the board of visitors for a young offender institution appointed by the Secretary of State under section 6 (2) of the Prison Act 1952 shall subject to paragraph (1A) below hold office for 3 years or such less period as the Secretary of State may appoint.

(1A) The Secretary of State may terminate the appointment of a member if satisfied that—

(a) he has failed satisfactorily to perform his duties,

(b) he is by reason of physical or mental illness, or for any other reason, incapable of carrying out his duties, or

(c) he has been convicted of such a criminal offence, or his conduct has been such, that it is not in the Secretary of State's opinion fitting that he should remain a member.

(2) When a board is first constituted, the Secretary of State shall appoint one of its members to be chairman for a period not exceeding twelve months.

(3) Subject to paragraph (2) of this rule, at their first meeting in any year of office the board shall appoint one of their members to be chairman and one to be vice-chairman for that year and thereafter shall fill any casual vacancy in either office promptly.

(4) The vice-chairman's term of office shall come to an end when, for whatever reason, that of the chairman comes to an end.

Proceedings of boards

74. (1) The board of visitors of a young offender institution shall meet at the institution at least once a month.

(2) The board may fix a quorum of not fewer than 3 members for proceedings.

(3) The board shall keep minutes of their proceedings.

(4) The proceedings of the board shall not be invalidated by any vacancy in the membership or any defect in the appointment of a member.

General duties of boards

75.—(1) The board of visitors for a young offender institution shall satisfy themselves as to the state of the premises, the administration of the institution and the treatment of the inmates.

(2) The board shall inquire into any report on any matter into which the Secretary of State asks them to inquire.

(3) The board shall direct the attention of the governor to any matter which calls for his attention, and shall report to the Secretary of State any matters which they consider it expedient to report.

(4) The board shall inform the Secretary of State immediately of any abuse which comes to their knowledge.

(5) Before exercising any power under these Rules, the board and any member of the board shall consult the governor in relation to any matter which may affect discipline.

Particular duties

76.—(1) The board of visitors for a young offender institution and any member of the board shall hear any complaint or request which an inmate wishes to make to them or him.

(2) The board shall arrange for the food of the inmates to be inspected by a member of the board at frequent intervals.

(3) The board shall inquire into any report made to them, whether or not by a member of the board, that an inmate's health, mental or physical, is likely to be injuriously affected by any conditions of his detention.

Members visiting young offender institutions

77.—(1) The members of the board of visitors for a young offender institution shall visit the institution frequently, and the board shall arrange a rota for the purpose.

(2) A member of the board shall have access at any time to every part of the institution and to every inmate, and he may interview any inmate out of sight and hearing of officers.

(3) A member of the board shall have access to the records of the young offender institution.

Annual report

78. The board of visitors for a young offender institution shall make an annual report to the Secretary of State at the end of each year concerning the state of the institution and its administration, including in it any advice and suggestions they consider appropriate.

PART VI
SUPPLEMENTAL

Delegation by governor

79. The governor of a young offender institution may, with the leave of the Secretary of State, delegate any of his powers and duties under these Rules to another officer of that institution.

Transitional

80. In the case of an inmate who, by virtue of paragraph 12 of Schedule 8 to the Criminal Justice Act 1988[a], falls to be treated for all purposes of detention, release and supervision as if he had been sentenced to detention in a young offender institution or who, under paragraph 13 of the said Schedule 8, is detained in such an institution, any award of an offence against discipline made in respect of him under rule 53 or 54 of the Detention Centre Rules 1983[b] or rule 53 or 54 of the Youth Custody Centre Rules 1983[c] shall, if it has not been exhausted or remitted, continue to have effect as if it had been made under rule 53 or 54, respectively, of those Rules.

One of Her Majesty's Principal
Secretaries of State

Consolidated Rules
DIA1 Division
November 1992

SCHEDULE
INSTRUMENTS REVOKED

The Detention Centre Rules 1983 (S.I. 1983/569)
The Youth Custody Centre Rules 19483 (S.I. 1983/570)
The Detention Centre (Amendment) Rules 1987 (S.I. 1987/1255)
The Youth Custody Centre (Amendment) Rules 1987 (S.I. 1987/1257)

---

[a] 1988 c. 33.
[b] S.I. 1983/569.
[c] S.I. 1983/570.

EXPLANATORY NOTE

(These notes are not part of the Rules)

These Rules make provision for the management of young offender institutions provided by section 43 of the Prison Act 1952 as amended by section 11 of the Criminal Justice Act 1982 and Schedules 8 and 15 of the Criminal Justice Act 1988 which comes into force on 1st October 1988 (S.I. 1988/1408). They replace the Detention Centre Rules 1983 (S.I. 1983/569) and the Youth Custody Centre Rules 1983 (S.I. 1983/570), as amended. They include provision for the treatment, occupation, discipline and control of inmates who may be detained therein, the conduct of officers of young offender institutions and the constitution, powers and duties of boards of visitors.

Rule 3 contains the aims and general principles of young offender institutions and rules 34 to 40 contain more detailed provisions regulating the regime. The provisions of the Rules are generally similar in substance to those of the Youth Custody Rules 1983, with some modifications to take account of the fact that the new sentence replaces both the detention centre order and the youth custody sentence. Rule 43 provides that an inmate may not be stripped and searched either in the sight of another inmate or in the sight or presence of an officer not of the same sex.

These Rules remove the disciplinary functions of boards of visitors. Transitional provision is made so that boards of visitors have continuing jurisdiction to deal with cases referred to them before the coming into force of these Rules.

The Rules increase the entitlement of a young offender to receive visits from once every four weeks to twice every four weeks.

The Rules abolish the separate disciplinary offence of failing to return to a young offender institution after a period of temporary release, which duplicates, in part, the offence of failing to comply with a condition of release.

These Rules make provision for the award of additional days which replaces forfeiture of remission as a punishment available to the governor of a young offender institution in respect of an inmate who is a short-term or long-term prisoner and who is guilty of a disciplinary offence.

The Rules also provide that the provisions of Part II of the Criminal Justice Act 1991 shall apply to an inmate who is an existing prisoner or licensee and who has forfeited any remission of his sentence as if he had been awarded the same number of additional days as the number of days of remission which he has forfeited.

In addition, the Rules prohibit smoking except in accordance with any directions of the Secretary of State and remove the requirement that inmates be searched only by an officer of the same sex.

The European Prison Rules 1987

REVISED EUROPEAN VERSION OF THE STANDARD MINIMUM RULES FOR THE TREATMENT OF PRISONERS

Preamble

The purposes of these rules are:

a. to establish a range of minimum standards for all those aspects of prison administration that are essential to human conditions and positive treatment in modern and progressive systems;

b. to serve as a stimulus to prison administrations to develop policies and management style and practice based on good contemporary principles of purpose and equity;

c. to encourage in prison staffs progressional attitudes that reflect the important social and moral qualities of their work and to create conditions in which they can optimise their own performance to the benefit of society in general, the prisoners in their care and their own vocational satisfaction;

d. to provide realistic basic criteria against which prison administrations and those responsible for inspecting the conditions and management of prisons can make valid judgments of performance and measure progress towards higher standards.

It is emphasised that the rules do not constitute a model system and that, in practice, many European prison services are already operating well above many of the standards set out in the rules and that others are striving, and will continue to strive, to do so. Wherever there are difficulties or practical problems to be overcome in the application of the rules, the Council of Europe has the machinery and the expertise available to assist with advice and the fruits of the experience of the various prison administrations within its sphere.

In these rules, renewed emphasis has been placed on the precepts of human dignity, the commitment of prison administrations to humane and positive treatment, the importance of staff roles and effective modern management approaches. They are set out to provide ready reference, encouragement and guidance to those who are working at all levels of prison administration. The explanatory memorandum that accompanies the rules is intended to ensure the understanding, acceptance and flexibility that are necessary to achieve the highest realistic level of implementation beyond the basic standards.

PART I
THE BASIC PRINCIPLES

1. The deprivation of liberty shall be effected in material and moral conditions which ensure respect for human dignity and are in conformity with these rules.

2. The rules shall be applied impartially. There shall be no discrimination on grounds of race, colour, sex, language, religion, political or other opinion, national or social origin, birth, economic or other status. The religious beliefs and moral precepts of the group to which a prisoner belongs shall be respected.

3. The purposes of the treatment of persons in custody shall be such as to sustain their health and self-respect and, so far as the length of sentence permits, to develop their sense of responsibility and encourage those attitudes and skills that will assist them to return to society with the best chance of leading law-abiding and self-supporting lives after their release.

4. There shall be regular inspections of penal institutions and services by qualified and experienced inspectors appointed by a competent authority. Their task shall be, in particular, to monitor whether and to what extent these institutions are administered in accordance with existing laws and regulations, the objectives of the prison services and the requirements of these rules.

5. The protection of the individual rights of prisoners with special regard to the legality of the execution of detention measures shall be secured by means of a control carried out, according to national rules, by a judicial authority or other duly constituted body authorised to visit the prisoners and not belonging to the prison administration.

6. (1) These rules shall be made readily available to staff in the national languages;
   (2) They shall also be available to prisoners in the same languages and in other languages so far as is reasonable and practicable.

PART II
THE MANAGEMENT OF PRISON SYSTEMS

Reception and registration

7. (1) No person shall be received in an institution without a valid commitment order.
   (2) The essential details of the commitment and reception shall immediately be recorded.

8. In every place where persons are imprisoned a complete and secure record of the following information shall be kept concerning each prisoner received:

   (a) information concerning the identity of the prisoner;
   (b) the reasons for commitment and the authority therefor;
   (c) the day and hour of admission and release.

9. Reception arrangements shall conform with the basic principles of the rules and shall assist prisoners to resolve their urgent personal problems.

10. (1) As soon as possible after reception, full reports and relevant information about the personal situation and training programme of each prisoner with a sentence of suitable length in preparation for ultimate release shall be drawn up and submitted to the director for information or approval as appropriate.

   (2) Such reports shall always include reports by a medical officer and the personnel in direct charge of the prisoner concerned.

   (3) The reports and information concerning prisoners shall be maintained with due regard to confidentiality on an individual basis, regularly kept up to date and only accessible to authorised persons.

The allocation and classification of prisoners

11. (1) In allocating prisoners to different institutions or régimes, due account shall be taken of their judicial and legal situation (untried or convicted prisoner, first offender or habitual offender, short sentence or long sentence), of the special requirements of their treatment, of their medical needs, their sex and age.

   (2) Males and females shall in principle be detained separately, although they may participate together in organised activities as part of an established treatment programme.

   (3) In principle, untried prisoners shall be detained separately from convicted prisoners unless they consent to being accommodated or involved together in organised activities beneficial to them.

   (4) Young prisoners shall be detained under conditions which as far as possible protect them from harmful influences and which take account of the needs peculiar to their age.

12. The purposes of classification or re-classification of prisoners shall be:

   (a) to separate from others those prisoners who, by reasons of their criminal records or their personality, are likely to benefit from that or who may exercise a bad influence; and

   (b) to assist in allocating prisoners to facilitate their treatment and social resettlement taking into account the management and security requirements.

13. So far as possible separate institutions or separate sections of an institution shall be used to facilitate the management of different treatment régimes or the allocation of specific categories of prisoners.

Accommodation

14. (1) Prisoners shall normally be lodged during the night in individual cells except in cases where it is considered that there are advantages in sharing accommodation with other prisoners.

   (2) Where accommodation is shared it shall be occupied by prisoners suitable to associate with others in those conditions. There shall be supervision by night, in keeping with the nature of the institution.

15. The accommodation provided for prisoners, and in particular all sleeping accommodation, shall meet the requirements of health and hygiene, due regard being paid to climatic conditions and especially the cubic content of air, a reasonable amount of space, lighting, heating and ventilation.

16. In all places where prisoners are required to live or work:

(*a*) the windows shall be large enough to enable the prisoners *inter alia*, to read or work by natural light in normal conditions. They shall be so constructed that they can allow the entrance of fresh air except where there is an adequate air conditioning system. Moreover, the windows shall, with due regard to security requirements, present in their size, location and construction as normal an appearance as possible;

(*b*) artificial light shall satisfy recognised technical standards.

17. The sanitary installations and arrangements for access shall be adequate to enable every prisoner to comply with the needs of nature when necessary and in clean and decent conditions.

18. Adequate bathing and showering installations shall be provided so that every prisoner may be enabled and required to have a bath or shower, at a temperature suitable to the climate, as frequently as necessary for general hygiene according to season and geographical region, but at least once a week. Wherever possible there should be free access at all reasonable times.

19. All parts of an institution shall be properly maintained and kept clean at all times.

Personal hygiene

20. Prisoners shall be required to keep their persons clean, and to this end they shall be provided with water and with such toilet articles as are necessary for health and cleanliness.

21. For reasons of health and in order that prisoners may maintain a good appearance and preserve their self-respect, facilities shall be provided for the proper care of the hair and beard, and men shall be enabled to shave regularly.

Clothing and bedding

22. (1) Prisoners who are not allowed to wear their own clothing shall be provided with an outfit of clothing suitable for the climate and adequate to keep them in good health. Such clothing shall in no manner be degrading or humiliating.

 (2) All clothing shall be clean and kept in proper condition. Underclothing shall be changed and washed as often as necessary for the maintenance of hygiene.

 (3) Whenever prisoners obtain permission to go outside the institution, they shall be allowed to wear their own clothing or other inconspicuous clothing.

23. On the admission of prisoners to an institution, adequate arrangements shall be made to ensure that their personal clothing is kept in good condition and fit for use.

24. Every prisoner shall be provided with a separate bed and separate and appropriate bedding which shall be kept in good order and changed often enough to ensure its cleanliness.

Food

25. (1) In accordance with the standards laid down by the health authorities, the administration shall provide the prisoners at the normal times with food

which is suitably prepared and presented, and which satisfies is quality and quantity the standards of dietetics and modern hygiene and takes into account their age, health, the nature of their work, and so far as possible, religious or cultural requirements.

(2) Drinking water shall be available to every prisoner.

Medical services

26. (1) At every institution there shall be available the services of at least one qualified general practitioner. The medical services should be organised in close relation with the general health administration of the community or nation. They shall include a psychiatric service for the diagnosis and, in proper cases, the treatment of states of mental abnormality.

(2) Sick prisoners who require specialist treatment shall be transferred to specialised institutions or to civil hospitals. Where hospital facilities are provided in an institution, their equipment, furnishings and pharmaceutical supplies shall be suitable for the medical care and treatment of sick prisoners, and there shall be a staff of suitably trained officers.

(3) The services of a qualified dental officer shall be available to every prisoner.

27. Prisoners may not be submitted to any experiments which may result in physical or moral injury.

28. (1) Arrangements shall be made wherever practicable for children to be born in a hospital outside the institution. However, unless special arrangements are made, there shall in penal institutions be the necessary staff and accommodation for the confinement and post-natal care of pregnant women. If a child is born in prison, this fact shall not be mentioned in the birth certificate.

(2) Where infants are allowed to remain in the institution with their mothers, special provision shall be made for a nursery staffed by qualified persons, where the infants shall be placed when they are not in the care of their mothers.

29. The medical officer shall see and examine every prisoner as soon as possible after admission and thereafter as necessary, with a view particularly to the discovery of physical or mental illness and the taking of all measures necessary for medical treatment; the segregation of prisoners suspected of infectious or contagious conditions; the noting of physical or mental defects which might impede resettlement after release; and the determination of the fitness of every prisoner to work.

30. (1) The medical officer shall have the care of the physical and mental health of the prisoners and shall see, under the conditions and with a frequency consistent with hospital standards, all sick prisoners, all who report illness or injury and any prisoner to whom attention is specially directed.

(2) The medical officer shall report to the director whenever it is considered that a prisoner's physical or mental health has been or will be adversely affected by continued imprisonment or by any condition of imprisonment.

31. (1) The medical officer or a competent authority shall regularly inspect and advise the director upon:

(*a*) The quantity, quality, preparation and serving of food and water;

(*b*) the hygiene and cleanliness of the institution and prisoners;

(*c*) The sanitation, heating, lighting and ventilation of the institution;

(*d*) the suitability and cleanliness of the prisoners' clothing and bedding.

(2) The director shall consider the reports and advice that the medical officer submits according to Rules 30, paragraph 2, and 31, paragraph 1, and, when in concurrence with the recommendations made, shall take immediate steps to give effect to those recommendations; if they are not within the director's competence or if the director does not concur with them, the director shall immediately submit a personal report and the advice of the medical officer to higher authority.

32. The medical services of the institution shall seek to detect and shall treat any physical or mental illness or defects which may impede a prisoner's resettlement after release. All necessary medical, surgical and psychiatric services including those available in the community shall be provided to the prisoner to that end.

Discipline and punishment

33. Discipline and order shall be maintained in the interests of safe custody, ordered community life and the treatment objectives of the institution.

34. (1) No prisoner shall be employed, in the service of the institution, in any disciplinary capacity.

(2) This rule shall not, however, impede the proper functioning of arrangements under which specified social, educational or sports activities or responsibilities are entrusted under supervision to prisoners who are formed into groups for the purposes of their participation in régime programmes.

35. The following shall be provided for and determined by the law or by the regulation of the competent authority:

(*a*) conduct constituting a disciplinary offence;

(*b*) the types and duration of punishment which may be imposed;

(*c*) the authority competent to impose such punishment;

(*d*) access to, and the authority of, the appellate process.

36. (1) No prisoner shall be punished except in accordance with the terms of such law or regulation, and never twice for the same act.

(2) Reports of misconduct shall be presented promptly to the competent authority who shall decide on them without undue delay.

(3) No prisoner shall be punished unless informed of the alleged offence and given a proper opportunity of presenting a defence.

(4) Where necessary and practicable prisoners shall be allowed to make their defence through an interpreter.

37. Collective punishments, corporal punishment, punishment by placing in a dark cell, and all cruel, inhuman or degrading punishment shall be completely prohibited as punishments for disciplinary offences.

38. (1) Punishment by disciplinary confinement and any other punishment which might have an adverse effect on the physical or mental health of the prisoner shall only be imposed if the medical officer, after examination, certifies in writing that the prisoner is fit to sustain it.

(2) In no case may such punishment be contrary to, or depart from, the principles stated in Rule 37.

(3) The medical officer shall visit daily prisoners undergoing such punishment and shall advise the director if the termination or alteration of the punishment is considered necessary on grounds of physical or mental health.

Instruments of restraint

39. The use of chains and irons shall be prohibited. Handcuffs, restraint-jackets and other body restraints shall never be applied as a punishment. They shall not be used except in the following circumstances:

(*a*) if necessary, as a precaution against escape during a transfer, provided that they shall be removed when the prisoner appears before a judicial or administrative authority unless that authority decides otherwise;

(*b*) on medical grounds by direction and under the supervision of the medical officer;

(*c*) by order of the director, if other methods of control fail, in order to protect a prisoner from self-injury, injury to others or to prevent serious damage to property; in such instances the director shall at once consult the medical officer and report to the higher administrative authority.

40. The patterns and manner of use of the instruments of restraint authorised in the preceding paragraph shall be decided by law or regulation. Such instruments must not be applied for any longer time than is strictly necessary.

Information to, and complaints by, prisoners

41. (1) Every prisoner shall on admission be provided with written information about the regulations governing the treatment of prisoners of the relevant category, the disciplinary requirements of the institution, the authorised methods of seeking information and making complaints, and all such other matters as are necessary to understand the rights and obligations of prisoners and to adapt to the life of the institution.

(2) If a prisoner cannot understand the written information provided, this information shall be explained orally.

42. (1) Every prisoner shall have the opportunity every day of making requests or complaints to the director of the institution or the officer authorised to act in that capacity.

(2) A prisoner shall have the opportunity to take to, or to make requests or complaints to, an inspector of prisons or to any other duly constituted authority entitled to visit the prison without the director or other members of the staff being present. However, appeals against formal decisions may be restricted to the authorised procedures.

(3) Every prisoner shall be allowed to make a request or complaint, under confidential cover, to the central prison administration, the judicial authority or other proper authorities.

(4) Every request or complaint addressed or referred to a prison authority shall be promptly dealt with and replied to by this authority without undue delay.

Contact with the outside world

43. (1) Prisoners shall be allowed to communicate with their families and, subject to the needs of treatment, security and good order, persons or representatives of outside organisations and to receive visits from these persons as often as possible.

(2) To encourage contact with the outside world there shall be a system of prison leave consistent with the treatment objectives in Part IV of these rules.

44. (1) Prisoners who are foreign nationals should be informed, without delay, of their right to request contact and be allowed reasonable facilities to communicate with the diplomatic or consular representative of the state to which they belong. The prison administration should co-operate fully with such representatives in the interests of foreign nationals in prison who may have special needs.

(2) Prisoners who are nationals of states without diplomatic or consular representation in the country and refugees or stateless persons shall be allowed similar facilities to communicate with the diplomatic representative of the state which takes charge of their interests or national or international authority whose task it is to serve the interests of such persons.

45. Prisoners shall be allowed to keep themselves informed regularly of the news by reading newspapers, periodicals and other publications, by radio or television transmissions, by lectures or by any similar means as authorised or controlled by the administration. Special arrangements should be made to meet the needs of foreign nationals with linguistic difficulties.

Religious and moral assistance

46. So far as practicable, every prisoner shall be allowed to satisfy the needs of his religious, spiritual and moral life by attending the services or meetings provided in the institution and having in his possession any necessary books or literature.

47. (1) If the institution contains a sufficient number of prisoners of the same religion, a qualified representative of that religion shall be appointed and approved. If the number of prisoners justifies it and conditions permit, the arrangement should be on a full-time basis.

(2) A qualified representative appointed or approved under paragraph 1 shall be allowed to hold regular services and activities and to pay pastoral visits in private to prisoners of his religion at proper times.

(3) Access to a qualified representative of any religion shall not be refused to any prisoner. If any prisoner should object to a visit of any religious representative, the prisoner shall be allowed to refuse it.

Retention of prisoners' property

48. (1) All money, valuables, and other effects belonging to prisoners which under the regulations of the institution they are not allowed to retain shall on admission to the institution be placed in safe custody. An inventory thereof shall be signed by the prisoner. Steps shall be taken to keep them in good condition. If it has been found necessary to destroy any article, this shall be recorded and the prisoner informed.

(2) On the release of the prisoner, all such articles and money shall be returned except insofar as there have been authorised withdrawals of money or the authorised sending of any such property out of the institution, or it has been found necessary on hygienic grounds to destroy any article. The prisoner shall sign a receipt for the articles and money returned.

(3) As far as practicable, any money or effects received for a prisoner from outside shall be treated in the same way unless they are intended for and permitted for use during imprisonment.

(4) If a prisoner brings in any medicines, the medical officer shall decide what use shall be made of them.

Notification of death, illness, transfer, etc.

49. (1) Upon the death or serious illness of or serious injury to a prisoner, or removal to an institution for the treatment of mental illness or abnormalities, the director shall at once inform the spouse, if the prisoner is married, or the nearest relative and shall in any event inform any other person previously designated by the prisoner.

(2) A prisoner shall be informed at once of the death or serious illness of any near relative. In these cases and wherever circumstances allow, the prisoner should be authorised to visit this sick relative or see the deceased either under escort or alone.

(3) All prisoners shall have the right to inform at once their families of imprisonment or transfer to another institution.

Removal of prisoners

50. (1) When prisoners are being removed to or from an institution, they shall be exposed to public view as little as possible, and proper safeguards shall be adopted to protect them from insult, curiosity and publicity in any form.

(2) The transport of prisoners in conveyances with inadequate ventilation or light, or in any way which would subject them to unnecessary physical hardship or indignity shall be prohibited.

(3) The transport of prisoners shall be carried out at the expense of the administration and in accordance with duly authorised regulations.

PART III
PERSONNEL

51. In view of the fundamental importance of the prison staff to the proper management of the institutions and the pursuit of their organisational and treatment objectives, prison administrations shall give high priority to the fulfilment of the rules concerning personnel.

52. Prison staff shall be continually encouraged through training, consultative procedures and a positive management style to aspire to humane standards, higher efficiency and a committed approach to their duties.

53. The prison administration shall regard it as an important task continually to

inform public opinion of the roles of the prison system and the work of the staff, so as to encourage public understanding of the importance of their contribution to society.

54. (1) The prison administration shall provide for the careful selection on recruitment or in subsequent appointments of all personnel. Special emphasis shall be given to their integrity, humanity, professional capacity and personal suitability for the work.

(2) Personnel shall normally be appointed on a permanent basis as professional prison staff and have civil service status with security of tenure subject only to good conduct, efficiency, good physical and mental health and an adequate standard of education. Salaries shall be adequate to attract and retain suitable men and women; employment benefits and conditions of service shall be favourable in view of the exacting nature of the work.

(3) Whenever it is necessary to employ part-time staff, these criteria should apply to them as far as that it is appropriate.

55. (1) On recruitment or after an appropriate period of practical experience, the personnel shall be given a course of training in their general and specific duties and be required to pass theoretical and practical tests unless their professional qualifications make that unnecessary.

(2) During their career, all personnel shall maintain and improve their knowledge and professional capacity by attending courses of in-service training to be organised by the administration at suitable intervals.

(3) Arrangements should be made for wider experience and training for personnel whose professional capacity would be improved by this.

(4) The training of all personnel should include instruction in the requirements and application of the European Prison Rules and the European Convention on Human Rights.

56. All members of the personnel shall be expected at all times so to conduct themselves and perform their duties as to influence the prisoners for good by their example and to command their respect.

57. (1) So far as possible the personnel shall include a sufficient number of specialists such as psychiatrists, psychologists, social workers, teachers, trade, physical education and sports instructors.

(2) These and other specialist staff shall normally be employed on a permanent basis. This shall not preclude part-time or voluntary workers when that is appropriate and beneficial to the level of support and training they can provide.

58. (1) The prison administration shall ensure that every institution is at all times in the full charge of the director, the deputy director or other authorised official.

(2) The director of an institution should be adequately qualified for that post by character, administrative ability, suitable professional training and experience.

(3) The director shall be appointed on a full-time basis and be available or accessible as required by the prison administration in its management instructions.

(4) When two or more institutions are under the authority of one director, each

shall be visited at frequent intervals. A responsible official shall be in charge of each of these institutions.

59. The administration shall introduce forms of organisation and management systems to facilitate communication between the different categories of staff in an institution with a view to ensuring co-operation between the various services, in particular, with respect to the treatment and re-socialisation of prisoners.

60. (1) The director, deputy, and the majority of the other personnel of the institution shall be able to speak the language of the greatest number of prisoners, or a language understood by the greatest number of them.

    (2) Whenever necessary and practicable the services of an interpreter shall be used.

61. (1) Arrangements shall be made to ensure at all times that a qualified and approved medical practitioner is able to attend without delay in cases of urgency.

    (2) In institutions not staffed by one or more full-time medical officers, a part-time medical officer or authorised staff of a health service shall visit regularly.

62. The appointment of staff in institutions or parts of institutions housing prisoners of the opposite sex is to be encouraged.

63. (1) Staff of the institutions shall not use force against prisoners except in self-defence or in cases of attempted escape or active or passive physical resistance to an order based on law or regulations. Staff who have recourse to force must use no more than is strictly necessary and must report the incident immediately to the director of the institution.

    (2) Staff shall as appropriate be given special technical training to enable them to restrain aggressive prisoners.

    (3) Except in special circumstances, staff performing duties which bring them into direct contact with prisoners should not be armed. Furthermore, staff should in no circumstances be provided with arms unless they have been fully trained in their use.

PART IV
TREATMENT OBJECTIVES AND RÉGIME

64. Imprisonment is by the deprivation of liberty a punishment in itself. The conditions of imprisonment and the prison and the prison régimes shall not, therefore, except as incidental to justifiable segregation or the maintenance of discipline, aggravate the suffering inherent in this.

65. Every effort shall be made to ensure that the régimes of the institutions are designed and managed so as:

    (a) to ensure that the conditions of life are compatible with human dignity and acceptable standards in the community;

    (b) to minimise the detrimental effects of imprisonment and the differences between prison life and life at liberty which tend to diminish the self-respect or sense of personal responsibility of prisoners;

(*c*) to sustain and strengthen those links with relatives and the outside community that will promote the best interests of prisoners and their families;

(*d*) to provide opportunities for prisoners to develop skills and aptitudes that will improve their prospects of successful resettlement after release.

66. To these ends all the remedial, educational, moral, spiritual and other resources that are appropriate should be made available and utilised in accordance with the individual treatment needs of prisoners. Thus the régimes should include:

(*a*) spiritual support and guidance and opportunities for relevant work, vocational guidance and training, education, physical education, the development of social skills, counselling, group and recreational activities;

(*b*) arrangements to ensure that these activities are organised, so far as possible, to increase contacts with and opportunities within the outside community so as to enhance the prospects for social resettlement after release;

(*c*) procedures for establishing and reviewing individual treatment and training programmes for prisoners after full consultations among the relevant staff and with individual prisoners who should be involved in these as far as practicable;

(*d*) communications systems and a management style that will encourage appropriate and positive relationships between staff and prisoners that will improve the prospects for effective and supportive régimes and treatment programmes.

67. (1) Since the fulfilment of these objectives requires individualisation of treatment and, for this purpose, a flexible system of allocation, prisoners should be placed in separate institutions or units where each can receive the appropriate treatment and training.

(2) The type, size, organisation and capacity of these institutions or units should be determined essentially by the nature of the treatment to be provided.

(3) It is necessary to ensure that prisoners are located with due regard to security and control but such measures should be the minimum compatible with safety and comprehend the special needs of the prisoner. Every effort should be made to place prisoners in institutions that are open in character or provide ample opportunities for contacts with the outside community. In the case of foreign nationals, links with people of their own nationality in the outside community are to be regarded as especially important.

68. As soon as possible after admission and after a study of the personality of each prisoner with a sentence of a suitable length, a programme of treatment in a suitable institution shall be prepared in the light of the knowledge obtained about individual needs, capacities and dispositions, especially proximity to relatives.

69. (1) Within the régimes, prisoners shall be given the opportunity to participate in activities of the institution likely to develop their sense of responsibility, self-reliance and to stimulate interest in their own treatment.

(2) Efforts should be made to develop methods of encouraging co-operation with and the participation of the prisoners in their treatment. To this end prisoners shall be encouraged to assume, within the limits specified in Rule 34, responsibilities in certain sectors of the institution's activity.

70. (1) The preparation of prisoners for release should begin as soon as possible

after reception in a penal institution. Thus, the treatment of prisoners should emphasise not their exclusion from the community but their continuing part in it. Community agencies and social workers should, therefore, be enlisted wherever possible to assist the staff of the institution in the task of social rehabilitation of the prisoners particular maintaining and improving the relationships with their families, with other persons and with the social agencies. Steps should be taken to safeguard, to the maximum extent compatible with the law and the sentence, the rights and other social benefits of prisoners.

(2) Treatment programmes should include provision for prison leave which should also be granted to the greatest extent possible on medical, educational, occupational, family and other social grounds.

(3) Foreign nationals should not be excluded from arrangements for prison leave solely on account of their nationality. Furthermore, every effort should be made to enable them to participate in régime activities together so as to alleviate their feelings of isolation.

Work

71. (1) Prison work should be seen as a positive element in treatment, training and institutional management.

(2) Prisoners under sentence may be required to work, subject to their physical and mental fitness as determined by the medical officer.

(3) Sufficient work of a useful nature, or if appropriate other purposeful activities shall be provided to keep prisoners actively employed for a normal working day.

(4) So far as possible the work provided shall be such as will maintain or increase the prisoner's ability to earn a normal living after release.

(5) Vocational training in useful trades shall be provided for prisoners able to profit thereby and especially for young prisoners.

(6) Within the limits compatible with proper vocational selection and with the requirements of institutional administration and discipline, the prisoners shall be able to choose the type of employment in which they wish to participate.

72. (1) The organisation and methods of work in the institutions shall resemble as closely as possible those of similar work in the community so as to prepare prisoners for the conditions of normal occupational life. It should thus be relevant to contemporary working standards and techniques and organised to function within modern management systems and production processes.

(2) Although the pursuit of financial profit from industries in the institutions can be valuable in raising standards and improving the quality and relevant of training, the interests of the prisoners and of their treatment must not be subordinary to that purpose.

73. Work for prisoners shall be assured by the prison administration:

(a) either on its own premises, workshops and farms; or

(b) in co-operation with private contractors inside or outside the institution in which case the full normal wages for such shall be paid by the persons to whom the labour is supplied, account being taken of the output of the prisoners.

74. (1) Safety and health precautions for prisoners shall be similar to those that apply to workers outside.

(2) Provision shall be made to indemnify prisoners against industrial injury, including occupational disease, on terms not less favourable than those extended by law to workers outside.

75. (1) The maximum daily and weekly working hours of the prisoners shall be fixed in conformity with local rules or custom in regard to the employment of free workmen.

(2) Prisoners should have at least one rest-day a week and sufficient time for education and other activities required as part of their treatment and training for social resettlement.

76. (1) There shall be a system of equitable remuneration of the work of prisoners.

(2) Under the system prisoners shall be allowed to spend at least a part of their earnings on approved articles for their own use and to allocate a part of their earnings to their family or for other approved purposes.

(3) The system may also provide that a part of the earnings be set aside by the administration so as to constitute a savings fund to be handed over to the prisoner on release.

Education

77. A comprehensive education programme shall be arranged in every institution to provide opportunities for all prisoners to pursue at least some of their individual needs and aspirations. Such programmes should have as their objectives the improvement of the prospects for for successful social resettlement, the morale and attitudes of prisoners and their self-respect.

78. Education should be regarded as a régime activity that attracts the same status and basic remuneration within the régime as work, provided that it takes place in normal working hours and is part of an authorised individual treatment programme.

79. Special attention should be given by prison administrations to the education of young prisoners, those of foreign origins or with particular cultural or ethnic needs.

80. Specific programmes of remedial education should be arranged for prisoners with special problems such as illiteracy or innumeracy.

81. So far as practicable, the education of prisoners shall:

(a) be integrated with the educational system of the country so that after their release they may continue their education without difficulty.

(b) take place in outside educational institutions.

82. Every institution shall have a library for the use of all categories of prisoners, adequately stocked with a wide range of both recreational and instructional books, and prisoners shall be encouraged to make full use of it. Wherever possible the prison library should be organised in co-operation with community library services.

Physical education, exercise, sport and recreation

83. The prison régimes shall recognise the importance to physical and mental health of properly organised activities to ensure physical fitness, adequate exercise and recreational opportunities.

84. Thus a properly organised programme of physical education, sport and other recreational activity should be arranged within the framework and objectives of the treatment and training régime. To this end space, installations and equipment should be provided.

85. Prison administrations should ensure that prisoners who participate in these programmes are physically fit to do so. Special arrangements should be made, under medical direction, for remedial physical education and therapy for those prisoners who need it.

86. Every prisoner who is not employed in outdoor work, or located in an open institution, shall be allowed, if the weather permits, at least one hour of walking or suitable exercise in the open air daily, as far as possible, sheltered from inclement weather.

Pre-release preparation

87. All prisoners should have the benefit of arrangements designed to assist them in returning to society, family life and employment after release. Procedures and special courses should be devised to this end.

88. In the case of those prisoners with longer sentences, steps should be taken to ensure a gradual return to life in society. This aim may be achieved, in particular, by a pre-release régime organised in the same institution or in another appropriate institution, or by conditional release under some kind of supervision combined with effective social support.

89. (1) Prison administration should work closely with the social services and agencies that assist released prisoners to reestablish themselves in society, in particular with regard to family life and employment.

   (2) Steps must be taken to ensure that on release prisoners are provided, as necessary, with appropriate documents and identification papers and assisted in finding suitable homes and work to go to. They should also be provided with immediate means of subsistence, be suitably and adequately clothed having regard to the climate and season, and have sufficient means to reach their destination.

   (3) The approved representatives of the social agencies or services should be afforded all necessary access to the institution and to prisoners with a view to making a full contribution to the preparation for release and after-care programme of the prisoner.

PART V
ADDITIONAL RULES FOR SPECIAL CATEGORIES

90. Prison administration should be guided by the provisions of the rules as a whole so far as they can appropriately and in practice be applied for the benefit of those special categories of prisoners for which additional rules are provided hereafter.

Untried prisoners

91. Without prejudice to legal rules for the protection of individual liberty or pre-scribing the procedure to be observed in respect of untried prisoners, these prison-ers, who are presumed to be innocent until they are found guilty, shall be afforded the benefits that may derive from Rule 90 and treated without restrictions other than those necessary for the penal procedure and the security of the institu-tion.

92. (1) Untried prisoners should be allowed to inform their families of their deten-tion immediately and given all reasonable facilities for communication with family and friends and persons with whom it is in their legitimate interest to enter into contact.

   (2) They shall also be allowed to receive visits from them under humane condi-tions subject only to such restrictions and supervision as are necessary in the administration of justice and of the security and good order of the insti-tution.

   (3) If an untried prisoner does not wish to inform any of these persons, the prison administration should not do so on its own initiative unless there are good overriding reasons as, for instance, the age, state of mind or any other incapacity of the prisoner.

93. Untried prisoners shall be entitled, as soon as imprisoned, to choose a legal representative, or shall be allowed to apply for free legal aid where such aid is available and to receive visits from that legal adviser with a view to their defence and to prepare and hand to the legal adviser, and to receive, confidential instruc-tions. On request they shall be given the free assistance of an interpreter for all essential contacts with the administration and for their defence. Interviews between prisoners and their legal advisers may be within sight but not within hearing, either direct or indirect, of the police or institution staff. The allocation of untried prison-ers shall be in conformity with the provisions of Rule 11, paragraph 3.

94. Except when there are circumstances that make it undesirable, untried prison-ers shall be given the opportunity of having separate rooms.

95. (1) Untried prisoners shall be given the opportunity of wearing their own clothing if it is clean and suitable.

   (2) Prisoners who do not avail themselves of this opportunity, shall be sup-plied with suitable dress.

   (3) If they have no suitable clothing of their own, untried prisoners shall be provided with civilian clothing in good condition in which to appear in court or on authorised outings.

96. Untried prisoners shall, whenever possible, be offered the opportunity to work but shall not be required to work. Those who choose to work shall be paid as other prisoners. If educational or trade training is available, untried prisoners shall be encouraged to avail themselves of these opportunities.

97. Untried prisoners shall be allowed to procure at their own expense or at the expense of a third part such books, newspapers, writing materials and other means of occupation as are compatible with the interests of the administration of justice and the security and good order of the institution.

98. Untried prisoners shall be given the opportunity of being visited and treated by their own doctor or dentist if there is reasonable ground for the application.

Reasons should be given if the application is refused. Such costs as are incurred shall not be the responsibility of the prison administration.

Civil prisoners

99. In countries where the law permits imprisonment by order of a court under any non-criminal process, persons so imprisoned shall not be subjected to any greater restriction or severity than is necessary to ensure safe custody and good order. Their treatment shall not be less favourable than that of untried prisoners, with the reservation, however, that they may be required to work.

Insane and mentally abnormal prisoners

100. (1) Persons who are found to be insane should not be detained in prisons and arrangements shall be made to remove them to appropriate establishments for the mentally ill as soon as possible.

(2) Specialised institutions or sections under medical management should be available for the observation and treatment of prisoners suffering gravely from other mental disease or abnormality.

(3) The medical or psychiatric service of the penal institutions shall provide for the psychiatric treatment of all prisoners who are in need of such treatment.

(4) Action should be taken, by arrangement with the appropriate community agencies, to ensure where necessary the continuation of psychiatric treatment after release and the provision of social psychiatric after-care.

Appendix 3
Standing Order 3D: Offences, Adjudications and Punishments*

Introduction

The purpose of the disciplinary system in Prison Service establishments is to preserve good order and to ensure that authority is exercised legitimately and fairly. It is based on:

(a) the statutory provisions of Rules 47–56 of the Prison Rules 1964 (as amended) and Rules 50–59 of the Young Offender Institution Rules 1988 (as amended) and

(b) the principles of natural justice.

The principles of natural justice require that:

(a) no person should be judge in his or her own cause

(b) proceedings should start afresh without reference to a previous hearing

(c) each party must have an opportunity to present his or her case and to call relevant witnesses

(d) each party must be given an opportunity to comment on all the material considered by the adjudicator

(e) the adjudicator should, wherever possible, give reasons for the decisions reached.

Many incidents of misconduct can be dealt with informally in the context of the general treatment approach of the regime, without recourse to formal measures of any kind. Where this is not possible, more formal methods are necessary, and this Standing Order and the Manual on the Conduct of Adjudications deal with those formal methods. The Standing Order provides a general outline of the procedure and details of administrative matters to do with charging and punishment. Detailed guidance on the conduct of hearings is in the Manual, copies of which are available for sale to the public from the Home Office Library, 50 Queen Anne's Gate, London SW1H 9AT (price £2.50); a copy is available in the library of every establishment for prisoners to refer to, or they may buy and retain their own copies if they wish.

---

* Standing Order 3D is currently undergoing revision. A revised version is expected to come into effect sometime after this work is published. Details can be obtained from the Prison Service.

CONTENTS

The Adjudicator

1. A governor has the task of deciding whether or not a prisoner is guilty of a disciplinary charge and, if guilty, what punishment should be awarded.

2. A governor acting in this role of adjudicator will inquire into the circumstances of the incident involving the alleged offence and should call and question witnesses in order to establish the facts.

Charges

Laying a charge

3. Prisoners are charged with an offence when they are handed a Notice of Report (Form 1127). The charge must be:

(*a*) of an offence described in the relevant Rules

(*b*) laid as soon as possible and, save in exceptional circumstances, within 48 hours of the alleged offence being discovered

(*c*) sufficiently detailed to leave prisoners in no doubt as to what is alleged against them

(*d*) explained to any prisoner who has difficulty in reading English.

The prisoner must be given at least two hours' notice of the adjudication.

4. Normally a charge will be laid by the officer against whom the offence was committed or who witnessed or dealt with the incident during which the alleged offence took place. Exceptions where the charge can be laid by another officer may for example include:

(*a*) where a member of staff was not present when the alleged offence was committed; or

(*b*) where a prisoner who was unlawfully at large is taken to another establishment (although it should be noted that young offenders should be deemed to have committed an offence under Young Offender Institution Rules, even if they have been taken to an adult establishment).

5. An officer wishing to lay a charge should seek the advice of a senior member of staff, normally at Principal Officer level or above. This officer will ensure that the charge is correctly laid. Consultation of this kind should also ensure that charges are not laid where knowledge of the prisoner and the circumstances of the incident suggest that informal advice to the prisoner would suitably dispose of the matter. In cases of doubt on the correct laying of a charge, Headquarters (Directorate of Inmate Administration) should be consulted before the charge is laid. Once a charge has been laid, the governor's initial hearing of it must commence not later, save in exceptional circumstances, than the next day (not being a Sunday or public holiday).

Serious criminal offences

6. The facts of any serious criminal offence, whether or not there is an identifiable suspect, should be reported immediately to the governor to decide whether the police should be informed. Details of the offence should also be reported to Headquarters (DOC1) under the normal incident reporting procedures.

7. In cases where the police are asked to investigate, a disciplinary charge should nevertheless be laid within 48 hours of discovery of the alleged offence. In such cases the governor should open the adjudication and satisfy himself or herself that there is a case to answer. Subject to this, the hearing should then be adjourned pending the outcome of the police investigation. Action will then take two courses depending on the result of the latter.

(*a*) If no prosecution results, or if the CPS decide that a prosecution cannot continue, the governor should consider whether to proceed with the disciplinary charge. If it is clear that the police or CPS have decided that a prosecution cannot be brought because the available evidence is insufficient, and the disciplinary charge is similar to and relies on the same evidence as the potential charge, the governor should dismiss the disciplinary charge. In addition, where the resumption of internal proceedings is likely to create an appearance of unfairness out of proportion to the seriousness of the alleged disciplinary offence, the governor should consider whether the charge should be dismissed. In other cases it is open to the governor to proceed with the charge.

(b) Of the CPS proceed to prosecute and present evidence in court, the governor should inform the prisoner that the disciplinary charge will not be proceeded with.

8. Where the police charge a prisoner with causing criminal damage to prison property, consideration should be given to applying to the court for a compensation order.

Offences at court

9. Before a disciplinary charge is laid for an offence committed at court it must be established that the prisoner was, at the material time, in the custody of the prison authorities. Headquarters (Directorate of Custody) should be consulted before any charge is laid.

Segregation

10. Where a governor considers that a prisoner suspected of a criminal offence should be kept apart from other prisoners while the police investigate the matter, the prisoner should be segregated under Prison Rule 43 or Young Offender Institution Rule 46. The provisions of Prison Rule 43 (2) and Young Offender Institution Rule 46 (2), which require the authority of a member of the Board of Visitors or the Secretary of State for more than 3 days' segregation, must be strictly observed.

11. A prisoner who is to be placed on report may be segregated under Prison Rule 48 (5) or Young Offender Institution Rule 51 (5), although the reasons for segregation should be connected with the adjudication. The prisoner may be segregated under these Rules until the initial hearing of the charge by the governor, but at any time thereafter prisoners should only be segregated under the provisions of Prison Rule 43 or Young Offender Institution Rule 46. The fact of, and reasons for, segregation following the governor's initial hearing should be recorded on the F256 and on the prisoner's record. Thus any segregation considered necessary during an adjournment should be under Prison Rule 43 or Young Offender Institution Rule 46, and where necessary the authority of the Board of Visitors must be sought. Prisoners segregated under Prison Rule 48 (5) and Young Offender Institution Rule 51 (5) will:

(a) keep their bed and bedding in the cell unless there is reason to suppose they will damage or destroy them

(b) not be deprived of any facilities (privileges) to which they are normally entitled except those that are incompatible with segregation.

12. A prisoner segregated as described in paragraphs 10 or 11 above should not be located in a segregation unit or any part of the establishment normally used for prisoners undergoing punishment, unless there is no suitable cell available elsewhere. However, where the adjudication room is in the punishment block, a prisoner may be lodged in the punishment block on the morning of the adjudication.

13. Opportunities for daily exercise as prescribed in the Rules will be given to segregated prisoners.

Medical examination

14. On the day of an adjudication or resumed hearing and before the hearing starts the prisoner charged will be examined by a medical officer, who will:

(*a*) note on the record of proceedings (F256) whether the prisoner is fit for adjudication and will in addition note whether the prisoner is in a fit state of health to undergo a punishment of cellular confinement

(*b*) report in writing to the adjudicator any matter affecting a prisoner's physical or mental condition which appears relevant to the adjudication (including, where appropriate, an opinion that their mental condition was such that they should not be held responsible for their actions at the time of the alleged offence).

15. If the prisoner is found to be unfit for adjudication, the medical officer should be asked for an opinion as to how long this condition is likely to last and, in the normal circumstances, a date should be arranged for a further examination.

16. Exceptionally, where there is no full-time medical cover available at the time and the part-time medical officer is unable to examine the prisoner charged as in paragraph 14 above, the examination must take place:

(*a*) within the 24 hours preceding the adjudication (or resumed hearing) or

(*b*) as soon as possible after the adjudication (normally within the next 24 hours), in which case the reasons for delay should be included in the F256 or in the medical report.

17. No punishment will be awarded to any prisoner about whose fitness for the punishment the adjudicator has any doubt.

18. No award of cellular confinement will be made unless the medical officer has certified that the prisoner is fit for it.

Hearing arrangements

General

19. The adjudicator has a responsibility to ensure that a record of the proceedings is kept on the F256.

20. The prisoner charged should be allowed to sit at a table and be provided with writing material at all adjudications.

21. Form 1145 (Explanation of Procedure at Disciplinary Charge Hearings) must be issued to the accused in sufficient time for it to be studied before the hearing, and should be available during the hearing.

22. Accused prisoners who have difficulty in hearing or understanding English should be assisted by the adjudicator, staff or, if necessary, an interpreter so that they can participate fully in the proceedings.

23. A prisoner who applies to consult a solicitor before or after an adjudication should be allowed to do so, although a hearing will not necessarily be adjourned for this reason.

24. Adjudicators must consider requests to allow prisoners to be legally represented at hearings. Prisoners may also with to be assisted by a friend of adviser

('McKenzie Friend'). The Manual on the Conduct of Adjudications sets out arrangements and guidance in both cases.

25. At an adjudication, all prisoners charged will be given an opportunity to:

(a) state whether they have had enough time to prepare their defence

(b) state whether they wish to apply for legal representation or other assistance

(c) plead guilty or not guilty to the charge

(d) call relevant witnesses (though it should be noted that a witness who is a prisoner cannot be compelled to give evidence)

(e) question any witnesses either directly or, if this is abused, through the adjudicator

(f) make their defence or explain their conduct

(g) give any reasons why, if found guilty, they should be treated leniently

(h) ask questions about the conduct report which is read out if they are found guilty.

26. If prisoners refuse to attend an adjudication, the adjudicator may proceed in their absence but must still ensure that a full enquiry is made. A prisoner must be informed of the result of the hearing and any punishment awarded as soon as the hearing is concluded, unless there are exceptional reasons which require a short delay before the prisoner is told.

Standard of proof

27. A prisoner will only be found guilty of a disciplinary charge if the adjudicator is satisfied beyond reasonable doubt that the prisoner committed the offence as charged.

Security

28. In determining the appropriate number and deployment of staff during a hearing the governor will ensure that proper account is taken of the general demeanour of the prisoner and the nature of the alleged offence.

Records

29. Immediately after an adjudication:

(a) details of the offence, the finding and any punishment should be entered on the prisoner's adjudication record (F2050E) in the same terms as those entered on the record of the adjudication

(b) the F2050E entry will be checked and initialled by the senior officer attending the adjudication and present when the punishment was imposed, who will also initial the relevant space provided on the F256

(c) where an earlier suspended punishment has been dealt with, the decision on that punishment should be recorded on the relevant previous F256 and against the relevant F2050E page 7 entry.

30. Written details of the punishment(s) imposed should be given to the Head of Custody whose duty it is to see that they are carried out.

31. The prisoner's record and the F256 should be made available to the appropriate member of staff who will:

(*a*) make any necessary amendments to the prisoner's earliest date of release (EDR)

(*b*) inform the prisoner of any EDR change (space is provided on the back of the F2052A Inmate History to record that this has been done)

(*c*) taken any other consequential action, eg notify the parole authorities.

32. The F256 and written statements that have been accepted as evidence should be filed in date order.

Punishment

General

33. Punishments must be within the range and terms of the Prison or Young Offender Institution Rules.

34. Any punishment other than forfeiture of remission will start immediately unless:

(*a*) it was ordered to be suspended or

(*b*) it was ordered to start at the end of a period of punishment already being served.

35. If more than one punishment of the same kind is imposed at the same time for separate offences, they may be ordered to run concurrently or consecutively to one another. Records should be clearly marked to show whether punishments are concurrent or consecutive.

36. Guidance on consistency in punishments is contained in the Manual, but adjudicators should note in particular that:

(*a*) cellular confinement should be reserved for serious offences of an anti-social kind

(*h*) forfeiture of remission should be reserved for serious or repeated offences, or for cases where the adjudicator is satisfied that an exemplary punishment is necessary.

37. The adjudicator should ensure that the prisoner fully understands the effect of any punishment imposed, particularly when it is suspended or perhaps expressed in alternative ways (see paragraphs 42 and 58 below).

Forfeiture of facilities (privileges)

38. Forfeiture of any facilities (privileges) (see Standing Order 4) may be given as a punishment by a governor for a period not exceeding 28 days. In young offender institutions, the maximum period is 14 days.

39. Educational notebooks, attendance at educational classes and correspondence courses should not normally be forfeited unless specifically included in the award. General notebooks, drawing books, radios and permission to purchase postage stamps should also not normally be forfeited.

Exclusion from associated work

40. Exclusion from associated work may be given as a punishment for a period not exceeding 14 days. In the case of young offenders the adjudicator may award a

maximum of 14 days removal from any particular activity or activities of the young offender institution, other than training course, work, education or physical education as provided under Young Offender Institution Rules 34, 35, 36, 37 and 38.

41. This punishment should be regarded as an alternative to cellular confinement, although it is possible to combine an award of exclusion from associated work with an award of one, two or three days' cellular confinement. In combining the punishments in this way the cellular confinement should be served first and the total period of punishment must not exceed 14 days. Exclusion from associated work should not in itself involve forfeiture of any facility (privilege) except where the exclusion arrangement necessarily affects its availability. Nor, so far as possible, should exclusions from associated work be awarded where exclusion from other forms of association will result.

Stoppage of earnings

42. The adjudicator may impose stoppage of all or part of a prisoner's daily pay, up to a maximum amount equivalent to 28 days' full pay for adult offenders and 14 days' full pay for young offenders. The maximum periods for which this punishment can run are 56 and 28 days respectively.

43. One day's pay should be treated for punishment purposes as one seventh of a week's pay (being gross earnings exclusive of overtime) regardless of the number of days actually worked each week.

44. The effect of any punishments awarded under paragraph 42 must be carefully explained to the prisoner at the hearing.

Cellular confinement

45. A governor may impose cellular confinement for a period not exceeding three days.

46. No award of cellular confinement will be imposed:

(a) consecutively to an award of exclusion from associated work (or, in the case of young offenders, of removal from any particular activity or activities) or

(b) if there is medical advice to the contrary (see paragraph 14).

47. No other punishment or punishments may be imposed which would be served in conditions amounting or virtually amounting to cellular confinement for a longer period than three days. Any punishment which is served in the segregation unit and has as its consequence the loss of association other than during exercise, must be treated as virtually amounting to cellular confinement and therefore not be awarded for longer than three days. In general, governors must seek to ensure that the minimum loss of facilities results when a punishment needs, for practical reasons, to be served in the segregation unit.

48. Prisoners serving a punishment of cellular confinement will:

(a) be located in an ordinary cell set aside for the purpose that will have removed from it the bed (except at night) and other such articles (except table, stool and chamber pot) as the governor directs

(b) be observed by an officer at least once an hour, and visited daily by the chaplain and by the medical officer

(c) be allowed to attend the main service of their religion unless prevented under Standing Order 7A

(d) be allowed to have books within the limits of general regulations

(e) except where an award of forfeiture of facilities (privileges) is also made, be allowed all normal facilities (privileges) except those which are incompatible with cellular confinement.

49. A governor's power under Prison Rule 34 (5) to defer the right of a prisoner to a visit until the expiration of any period of cellular confinement should not be exercised automatically. Visits should be allowed unless the prisoner's behaviour and attitude make removal from cellular confinement impracticable or undesirable, and should take place somewhere other than in the ordinary visiting room.

50. The governor should give the medical officer and chaplain a daily list of prisoners in cellular confinement.

Forfeiture of remission

51. A governor may give forfeiture of remission to a sentenced prisoner (including an appellant) for a period not exceeding 28 days. Two or more punishments imposed at the same time for separate offences will be treated as cumulative (ie consecutive) unless ordered to run concurrently, or unless for related offences in which case their total may not exceed 28 days. Punishments imposed at different times will be cumulative.

52. When a prisoner is serving overlapping concurrent sentences, great care must be taken to ensure that remission is only forfeited in respect of a sentence or sentences to which the offence is applicable.

53. A prisoner serving an indeterminate sentence cannot forfeit remission. However, when such a prisoner has at the time of the adjudication been given a provisional date of release, the adjudicator may inform the prisoner that a recommendation will be made to the Home Secretary that the date of release should be postponed.

54. If the parole dossier of a prisoner has been sent to the Parole Unit, the effect of a punishment of forfeiture of remission on the earliest date of release should be notified immediately to Headquarters (DSP2) by the governor.

Additional punishments available under Young Offender Institution Rule 53

55. These are:

(a) extra work outside the normal working week for a period not exceeding 14 days, with not more than two hours on any day

(b) removal from wing or living unit for periods not exceeding 14 days.

Unconvicted and unsentenced prisoners

56. Unconvicted prisoners may be punished by forfeiture for any period of their rights to have books, writing materials and other means of occupation and, if found guilty of escaping or attempting to escape, of the right to wear their own clothing.

57. An unsentenced prisoner may be awarded prospective forfeiture of remission, which will be taken into account in calculating the EDR if a custodial sentence is imposed by the court.

Suspended punishments

58. An adjudicator may order any punishment (other than a caution or prospective forfeiture of remission) to be suspended for up to six months so that it cannot take effect unless the prisoner is found guilty of another disciplinary offence in the suspension period. It should be noted that:

(*a*) suspended punishment should not be given where an unsuspected punishment of the same kind would be inappropriate

(*b*) a suspended punishment will not usually be given for an offence where violence has been used and injuries caused

(*c*) forfeiture of remission should be suspended only where there are special extenuating circumstances and, where a punishment of more than 14 days forfeiture of remission is suspended, the F256 should indicate what those circumstances are

(*d*) although any or all of the punishments given for a single offence may be suspended, it is preferable to suspend all or none

(*e*) an individual award may not be suspended in part.

59. If a prisoner is found guilty of a further offence against discipline during the period of suspension of an earlier punishment (and irrespective of the punishment given for the later offence), a governor may:

(*a*) direct the suspended punishment to be brought into effect fully

(*b*) direct the suspended punishment to be brought into effect in part, in which case the balance of the suspended punishment will lapse

(*c*) vary the supervision period by directing the punishment to remain suspended for up to 6 months from the date of variation or

(*d*) give no direction about the suspended award.

60. If action is taken under paragraphs 59 (*a*) or (*b*) above, the punishment can be directed to take effect immediately or to be consecutive to a punishment of the same kind imposed for the subsequent offence.

61. The activation of a suspended sentence should not be automatic and each case should be decided on its merits. The adjudicator should take into account such matters as a significant period of good conduct, or whether the subsequent offence is of a different character from the one for which the punishment was suspended. Such circumstances should also be considered when deciding whether or not to exercise options 59 (*b*) or (*c*).

Interruptions to punishment

62. If a prisoner under punishment commits an offence appropriate action will not be delayed.

63. Time spent in hospital will count as part of a punishment period.

64. Days on which a prisoner attends court will count as part of a punishment period. If the prisoner is likely to be sent from court to another Prison Service establishment, the establishment carrying out the punishment should check the outcome of the case with the court and notify any new establishment by telephone of any punishment still to be carried out.

65. When a punishment is interrupted by the prisoner being on bail or unlawfully

at large, the balance of the punishment will be served if the prisoner returns to custody in connection with the same legal proceedings or is recaptures. If the interruption is longer than three months the advice of Headquarters (Directorate of Inmate Administration) should be sought.

66. When the start of punishment is delayed more than 24 hours or is interrupted for any reason, the prisoner will be examined by the medical officer before the punishment is started or resumed.

Termination or mitigation of punishment

Powers of the Secretary of State

67. Prison Rule 56 (1) and Young Offender Institution Rule 59 (1) give the Secretary of State the power to terminate, mitigate or quash any disciplinary award. Prisoners who wish to complain about the conduct of an adjudication, a finding of guilt or a punishment imposed on the grounds of unfairness or error should be advised to submit a request/complaint form to the Area Manager giving their reasons. The governor should forward the form to the Area Manager with copies of all relevant documents (F256, statements of evidence) and a report from the adjudicator commenting on the prisoner's claims. Where the prisoner is serving the punishment imposed, this action should be taken without delay. Adjudicators may also ask the Area Manager to take action if they discover a reason to doubt the finding or punishment.

68. The Secretary of State has also given the governor authority under Prison Rule 56 (2) and Young Offender Institution Rule 59 (2) to remit or mitigate awards in certain circumstances (see paragraph 70 below).

Powers of the governor

69. A governor may not:

(a) remit a caution
(b) remit or mitigate a punishment of forfeiture of remission or
(c) remit or mitigate any punishment on the grounds of unfairness.

This action can only be taken by Headquarters on behalf of the Secretary of State (paragraph 67 above). However a governor may consider restoration of forfeited remission in the special circumstances described in paragraphs 72 and 73 below.

70. A governor may terminate or mitigate any punishment other than a caution or forfeited remission where it appears that the effect of the punishment already served has been such that the prisoner is unlikely to repeat the offence.

71. A record of any variation of a punishment should be made on the relevant F256 on the Actions/Events box relating to the charge on the prisoner's adjudication record (F2050E). Where the variation is the result of the Secretary of State exercising his power to remit or mitigate a punishment or forfeiture of remission, this should also be recorded on the Custodial Documents File (F20501) and the Release Dates box on page 3 of the Core Record (F2050).

Restoration of forfeited remission

72. In appropriate cases, forfeited remission may be restored by the governor in accordance with the directions made by the Secretary of State under Prison Rule 56 (2) or Young Offender Institution Rule 59 (2). In such cases:

(*a*) where an application has been made, the fact will be recorded on the Custodial Documents File (F2051) of the prisoner's record, with the outcome

(*b*) where restoration has been granted, that fact will be recorded on the Custodial Documents File (F2051) of the prisoner's record and in the Adjudication Record F2050E, and any action necessary under paragraph 31 above will be taken

(*c*) if remission forfeited has already been served, it may not be restored.

73. In order to be considered for restoration of lost remission, adult prisoners must spend six months, and young offenders four months, from the date of the offence for which they lost remission without losing any further remission. Any time spent unlawfully at large does not count towards that qualifying period. Prisoners can make their case in writing and in person, and will be given reasons for the decision that is taken.

Standing Order 3E: Management of Violent or Refractory Prisoners

This Standing Order deals with a range of matters concerning the management of violent or refractory prisoners. It sets out general principles on the use of force, the use of special accommodation and mechanical restraints on non-medical ground (including requirements for authorisation, recording, reporting and supervision) and instructions governing the use of ratchet handcuffs, protective clothing, staves, etc. Provisions on the use of special accommodation and mechanical restraints on medical grounds are contained in SO 13.

References to 'prisoner(s)' should be taken to include any inmate of a Prison Service establishment.

CONTENTS

SECTION A: THE USE OF FORCE

1. (1) In dealing with a prisoner, no officer shall use force unless this is absolutely necessary. When the use of force is necessary, no more force shall be used than is absolutely necessary to achieve the required objective. All reasonable efforts must be made to manage violent, refractory or disturbed behaviour by persuasion or other means which do not entail the use of force. The use of force must always be regarded as a matter of last resort.

   (2) Where the use of force is necessary, then, so far as practicable, only approved control and restraint techniques should be employed.

   (3) No member of staff shall intentionally act in a manner calculated to provoke a prisoner.

SECTION B: THE USE OF SPECIAL ACCOMMODATION AND MECHANICAL RESTRAINTS FOR NON-MEDICAL PURPOSES

Special accommodation

2. (1) In this section, 'special accommodation' means:

   (*a*) a cell which has been designated as a special cell; or

   (*b*) an unfurnished cell—that is, a cell in which a prisoner is temporarily confined under the terms of paragraph 3 (1) (being either a cell which is specifically designated as one to be used exclusively as an unfurnished cell for temporary confinement under the terms of paragraph 3 (1); or a cell which is designated and usually used for ordinary accommodation purposes but from which the usual furniture has been removed) and which is either totally unfurnished or does not contain basic items of furniture such as table and chair and is thus to be considered, for all practical purposes, to be unfurnished.

(2) So far as is practicable, cardboard furniture should be used in preference to none.

3. (1) Special accommodation may be used for the temporary confinement of a violent or refractory prisoner, but only if the use of such accommodation:

   (a) is necessary in order to prevent the prisoner causing self-injury, injuring another prisoner or staff, or damaging property, or creating a serious disturbance; and

   (b) has been properly approved in accordance with this Order.

   (2) No prisoner shall be confined in special accommodation as a punishment, and, as soon as the original justification for the use of special accommodation has ceased, the prisoner must be removed from that accommodation.

Mechanical restraints

4. In this section, 'mechanical restraints' means any of the following:

   (*a*) a body belt, with iron cuffs for a male prisoner; with leather cuffs for a female prisoner;
   (*b*) standard handcuffs, for a male prisoner;
   (*c*) leather wrist straps, for a female prisoner; or
   (*d*) ankle straps, for a male or female prisoner.

No other item shall be used as a mechanical restraint under this section.

5. (1) Subject to sub-paragraph (2), a mechanical restraint may be used in order to prevent a prisoner causing self-injury, injuring another prisoner or staff, or damaging property, but only if the use of a mechanical restraint:

   (*a*) is absolutely necessary to achieve the required objective; and
   (*b*) has been approved in accordance with this Order.

   The use of mechanical restraints is particularly undesirable and every effort must be made to avoid recourse to them. The use of special accommodation as an alternative to mechanical restraints should be considered before any decision is taken to use mechanical restraints. This is not intended to preclude the use of mechanical restraint and special accommodation in conjunction, where the circumstances of the case so demand.

   (2) No inmate below the age of 17 may be placed under mechanical restraint under the terms of this section.

   (3) A mechanical restraint shall never be used as a punishment, and, as soon as the original justification for the use of mechanical restraint has ceased, it must be removed.

6. Recourse to the use of a **body belt** must be regarded as an exceptional measure. A body belt should be applied only when the circumstances of the case necessitate the use of this particular item and the use of, for example, standard handcuffs or leather wrist straps would not achieve the necessary purpose. To enable the prisoner to take food and drink, the left hand (or, in the case of a left-handed prisoner, the right hand) will be released.

7. When used under this section, **handcuffs** will be applied with the prisoner's hands behind his body. But they will be changed to the front of the body for meals and, if they should have to be left on during the night, at bed-times.

8. **Leather wrist straps** will be applied with the prisoner's hands in front of her body.

9. In the case of both handcuffs and leather wrist straps, great care must be taken to ensure that the restraint is not too small for the prisoner's wrists. The Medical Officer will check on this point when visiting the prisoner (see paragraph 18 (3)).

10. In the case of a body belt, handcuffs or leather wrist straps, the prisoner will be temporarily released from the restraint as necessary in order to take food or drink, to attend to calls of nature, or for other essential purposes, unless the Governor considers that release would be dangerous.

11. **Ankle straps** may be used only where a prisoner resorts to kicking to resist removal from one part of the establishment to another and it is necessary to use ankle straps to prevent injury to the prisoner or to others. The restraint must be taken off as soon as the removal of the prisoner is completed.

The giving of authority

12. (1) No prisoner shall be placed in special accommodation or in a mechanical restraint under this section except on the prior authority of the Governor in charge. However, if the matter is urgent and the Governor in charge cannot be contacted, the decision may be taken by the most senior officer available, who must be at not less than Principal Officer level (save that, in exceptional circumstances in which no member of staff at Principal Officer level or above is available, the decision may be taken by a Senior Officer).

 (2) The Governor in charge must personally see the prisoner before a mechanical restraint is applied—unless the urgency of the case requires the restraint to be applied before the prisoner can be seen (in which case, the prisoner must be seen by the Governor in charge as soon as possible thereafter).

 (3) The decision will be recorded at once in F1980 and the form will be signed by the Governor in charge, or, if the Governor in charge cannot be contacted, the officer taking the decision. If the decision has been taken by other than the Governor in charge, the matter must be referred at the earliest opportunity to the Governor in charge who, if he or she agrees with the decision, will endorse and sign the F1980 to this effect.

Supervision

13. The removal of a prisoner to special accommodation, or the removal of furniture, or the application of a mechanical restraint will be supervised by an officer at not less than Principal Officer level—unless no such officer is on duty (in which case, the task will be performed by the most senior officer on duty).

Notification of the Medical Officer

14. (1) Where action has been taken under paragraph 12, two copies of the F1980 must be supplied by the quickest possible means to a Medical Officer, who will arrange an early medical examination.

 (2) Where the Medical Officer is satisfied on the basis of the medical examination that there are no clinical contra-indications, he or she will certify accordingly by signing and returning the F1980. The other copy will be

retained, and will be placed in the prisoner's medical record. See further paragraph 18 (3) below.

(3) Where the examining Medical Officer considers that there are clinical contra-indications, he or she will immediately inform the Governor in charge, or the other most senior officer available, of that opinion.

Notification of the Board of Visitors

15. A copy of every F1980 will be taken for the Board of Visitors, and the Board will be notified of the case as soon as practicable. In addition, the Governor in charge will report each such case (and supply a copy of the F1980) to the member of the Board who next visits the establishment. See further paragraph 16.

Authority for continued use

16 (1) If the use of special accommodation or of a mechanical restraint under this section needs to continue beyond 24 hours, the written authority of the Board of Visitors must be obtained in F2006.

(2) If the continued use is necessary beyond the period authorised in the F2006, the written authority of the Board must be obtained in F2007 for any further period.

(3) The period of authority given under sub-paragraph (1) or (2) is a matter to be determined by the member of the Board, according to the circumstances of the case. Application for such authority shall be made only by the Governor in charge. A first application should never, and a subsequent application should only very exceptionally, be made for a period in excess of 24 hours.

Recording

17. (1) A record must be maintained in F1980 during an incident involving the use under this section of special accommodation or a mechanical restraint (including any further approval given under paragraph 16 above); and the removal of the prisoner from special accommodation, or the removal of a mechanical restraint, must also be recorded.

(2) In addition, full particulars of every case in which special accommodation or a mechanical restraint is used will be entered in the Governor's journal.

Monitoring

18. (1) The need for the continued use of special accommodation or of a mechanical restraint must be monitored regularly so that such use may be discontinued as soon as it is no longer necessary.

(2) The Governor in charge will personally visit, at least twice in every 24 hours, any prisoner who is confined in special accommodation or is under mechanical restraint under the terms of this Order, or of SO 13 in relation to the use of special accommodation or mechanical restraint for medical reasons; and will record any remarks in (as appropriate) the wing/unit occurrence book or the hospital observation book, and in the Governor's journal.

(3) A Medical Officer will personally visit at least twice in every 24 hours, and more often as may be necessary, any prisoner who is confined in special accommodation or is under mechanical restraint; and will record any remarks in (as appropriate) the wing/unit occurrence book or the hospital observation book, and in the Medical Officer's journal.

(4) Any prisoner who is held in special accommodation or is under mechanical restraint will be observed by the officer in charge of the wing or unit where the prisoner is held, or by an officer specifically deputed to do so, at least once in every 15 minutes.

(5) Where the use of special accommodation or of a mechanical restraint has been authorised under this section, the Governor in charge will, in consultation as appropriate with the Medical Officer, test the effect of removing the prisoner from the special accommodation, or of removing the restraint, as soon as the initial outburst of violence appears to have passed, and it appears safe to do so.

(6) A prisoner who is confined in special accommodation will be taken out of the special accommodation for exercise as soon and as often as is considered desirable or practicable.

Reporting

19. (1) When a prisoner is finally removed from special accommodation, or when a mechanical restraint is finally removed, the record in F1980 will be completed and a copy will be sent to the Medical Officer by the quickest possible means; and the Board of Visitors will be informed as soon as possible and supplied with a copy of the endorsed F1980.

(2) A copy of every completed F1980 will be sent (with a compliments slip) to S2 Division, Abell House, for statistical purposes.

20. The originals of every F1980 (together with the copy signed by the Medical Officer—see paragraph 14) will be filed in a central place and be available for inspection, if required. A copy of every such form will also be kept on the prisoner's record.

SECTION C: OTHER MATTERS

Ratchet handcuffs

21. (1) Ratchet handcuffs may be applied temporarily if, following a violent outburst, it is necessary to remove a prisoner from one part of the establishment to another and

(a) some distance is involved, or narrow doorways or stairs need to be negotiated, and, in consequence, it is likely to be difficult to maintain approved holds; or

(b) the prisoner is particularly violent and powerful and failure to restrain the prisoner is likely to result in injury to the prisoner, another prisoner or staff, or in escalation of the incident.

(2) The use of ratchet handcuffs in such circumstances must be regarded as exceptional and must be authorised by an officer at not less than Principal Officer level—unless no such officer is on duty (in which case, the authority of the most senior officer available must be obtained).

22. Ratchet handcuffs will also be carried by dog handlers and may be used temporarily to restrain a prisoner caught attempting to escape.

23. (1) Ratchet handcuffs may not be used within an establishment other than under the terms of paragraphs 21 and 22. They will not be used under the terms of Section B.

(2) The use of ratchet handcuffs under paragraphs 21 and 22 is not subject to the requirements set out above for the recording, reporting, etc of the use of mechanical restraints.

The use of protective clothing

24. (1) Where a prisoner is placed in special accommodation or in a mechanical restraint in accordance with section B above, and, as part of the protective arrangements, it is necessary to deprive the prisoner of normal clothes, the prisoner should be provided with a suit of protective clothing of an approved type so that he or she can be decently dressed.

(2) Depriving a prisoner of his or her normal clothing and the use of protective clothing require the same level of authority as the use of special accommodation or of mechanical restraint under section B (see paragraph 12 (1)) and the authority must be recorded on the relevant F1980.

(3) A prisoner may be deprived of normal clothes only if, in the light of the circumstances of the individual case, this is considered essential to prevent self-injury or injury to others, and the prisoner's normal clothes must be returned at the earliest opportunity.

Staves, etc

25. (1) This paragraph concerns the standard stave.

(2) A stave may be carried only by a male officer grade.

(3) A stave shall not be carried within:

(a) a Category D establishment;

(b) a female establishment; or

(c) a young offender establishment which holds only young adults with not more than 6 months to serve or juveniles serving up to 12 months;

or by male hospital or nursing staff.

(4) The drawing and use of a stave must be regarded as an exceptional and extreme measure.

(5) A stave must never be regarded or used as other than an implement for defensive use. It may be drawn or used only when:

(a) it is necessary for the officer to defend himself or a third party from an attack threatening serious injury;

(b) no alternative action is practicable.

(6) A stave will not be drawn or used when more than one officer is present except on the instruction of the most senior officer who is present and in a position to give such an order.

(7) A stave must never be used to strike a prisoner unless the officer or another person is in grave danger and no other option is open to the officer to save himself or the other person; and then the stave may be directed only at the prisoner's arm and legs where serious injury is less likely to result.

26. (1) This paragraph concerns the mini-baton.

(2) A mini-baton may be carried only by a woman officer grade.

(3) The drawing and use of a mini-baton must be regarded as an exceptional measure. A mini-baton may be drawn and used:

  (*a*) only as a part of, and in accordance with, approved techniques; and then only

  (*b*) when it is judged that approved techniques are unlikely to succeed without the use of the mini-baton; and

  (*c*) its use is necessary in order to prevent injury to the officer or another person, or damage to property.

27. (1) This paragraph concerns the side-arm baton.

(2) A side-arm baton shall be carried only by selected staff forming part of a response force trained and deployed in accordance with standing instructions to deal with violent concerted acts of indiscipline, and never otherwise.

(3) The drawing and use of a side-arm baton must be regarded as an exceptional and extreme measure.

(4) A side-arm baton must never be regarded or used as other than a defensive implement. It may be drawn or used only:

  (*a*) on the instruction of a member of the response force authorised to give such an instruction, where such action is considered necessary to prevent serious injury to staff or others; or

  (*b*) on the judgement of an individual member of staff who becomes detached from the body of the response force and needs to exercise personal discretion, where such action is necessary to prevent serious injury;

and no alternative action is practicable.

(5) A side-arm baton may be used only as part of, and in accordance with, approved techniques. Where

  (*a*) in an extreme circumstance in which a member of staff or another person is in grave danger; and

  (*b*) there is no other option open to the member of staff to save himself or herself or the other person but to employ such a technique which entails striking a prisoner;

the side-arm baton may be aimed only at the prisoner's arms and legs where serious injury is less likely to result.

28. (1) This paragraph applies to the standard stave, mini-baton and side-arm baton.

(2) Whenever a stave, a mini-baton or a side-arm baton is drawn or used, a written report must be made to the Governor.

(3) Where

    (*a*) an officer acting alone, without the assistance or presence of other staff, draws or uses a stave or mini-baton; or

    (*b*) a member of staff draws or uses a side-arm baton under the terms of paragraph 27 (4) (b);

he or she will submit a report to the Governor giving reasons why it was necessary to draw or use the stave, mini-baton or side-arm baton.

(4) Where staves, mini-batons or side-arm batons are drawn or used by one or more of a number of officers on the same occasion, each member of staff will submit a report to the Governor through the senior member of staff present, who will submit a covering report.

29. No stave, mini-baton or side-arm baton (and no other implement) will be carried except as approved and issued by Headquarters.

Medical examination

30. (1) A Medical Officer will examine as soon as possible any prisoner injured as a result of violent behaviour by a prisoner.

(2) A Medical Officer will examine any prisoner who has been injured during efforts of staff to control him or her, and will make a written report on the prisoner's condition.

(3) A Medical Officer will examine at the earliest possible opportunity any prisoner against whom a stave, mini-baton or side-arm baton has been used, and will make a written report on the prisoner's condition. Such a report will be made even if no apparent injury has been sustained by the prisoner.

(4) A Medical Officer will at the earliest possible opportunity examine (with his or her consent) any member of staff who is injured as a result of violent behaviour by a prisoner.

(5) Medical reports made under this paragraph will be submitted to the Governor.

Standing Order 4: Facilities

This Standing Order describes the facilities that shall or may be made available to prisoners, and the circumstances under which they may be forfeited, withdrawn or withheld.

CONTENTS

STANDING ORDER 4 FACILITIES
(PRIVILEGES FOR PRISONERS)

1. Prison Rule 4 and Young Offender Institution Rule 7 require establishments to operate systems of privileges for prisoners that are approved by the Secretary of State. These will fall into two main categories:

(*a*) facilities which shall be available to all prisoners

(*b*) facilities which may be provided or withdrawn by the Governor (ie discretionary facilities).

2. This Standing Order describes the facilities that shall or may be made available, and the circumstances under which they may be forfeited, withdrawn or withheld. Unless specifically forfeited, the facilities available in establishments will, where practicable, apply to prisoners located other than in normal location (eg in a hospital or in segregation). Unconvicted prisoners are eligible for the facilities set out in this Order in the same way as convicted prisoners, subject to the variations indicated.

3. A 'facility' for the purpose of this Standing Order is taken to mean any of the following:

retention, acquisition and use of personal possessions, including smoking materials (except for sentenced young offenders in juvenile establishments or juvenile units)

use of the prison shop (canteen)

expenditure of private cash (ie all monies available to a prisoner that are not received as prison pay)

engaging in cell activities and hobbies

Entering public competitions

taking part in activities in association

wearing own clothes or footwear (other than where this is a right under Prison Rule 21). Female prisoners are allowed to wear own clothes as a facility: convicted male prisoners are not generally allowed to wear their own clothes, except where specially authorised in certain establishments. Clothes and footwear should conform to the health and safety requirements of the activity area in which they are worn.

4. Governors will publish a statement of the facilities at their establishments (including a list of items that prisoners may normally retain in their possession). Different statements may operate in different parts of an establishment if appropriate. The statement must normally include those items set out in paragraph 9 below. Where, exceptionally, after approval by the Area Manager, one or more of the facilities mentioned in paragraph 9 is not provided, or where the method of obtaining the items is restricted, the statement will explain the reasons. Arrangements must be made to review statements at least annually and prisoners should be consulted about possible changes in the facilities provided. The statement must be available for inspection by prisoners, their families and visitors.

5. Where, exceptionally, the Governor decides not to allow a prisoner one or more of the facilities in the published statement, he or she will give an explanation to the prisoner concerned, in writing if the prisoner so requests.

I. POSSESSIONS

A. Retention of personal possessions

Policy

6. Property in possession is for the prisoner's own use. It cannot be given, lent or sold to another prisoner without the permission of the Governor.

7. Prisoners shall be allowed to have sufficient property in possession to lead as normal and individual an existence as possible within the constraints of the prison environment and the limitations under this and other standing orders.

8. The number of items permitted in a cell should not be so great that searching is unduly hindered; that a cell cannot be kept tidy; or that possession cannot be carried by the prisoner on transfer or put with stored property until discharge. The number and type of possessions in a cell must not constitute a risk to health, safety or security or good order or discipline. The amount of property that can be stored shall be at the discretion of the Governor, but normally should not exceed the limits specified in paragraph 2 of Standing Order 1C (ie what could be carried without assistance).

In-possession items

9. The list of the items that prisoners shall be allowed to retain in their possession will include:

    at least six newspapers and periodicals
    at least three books
    a combined sound system or a radio and one of a record, cassette or compact disc player together with any number of records, compact discs or cassettes consistent with paragraphs 7 and 8 (see also paragraphs 12–15 below)
    smoking requisites (except where smoking is prohibited), including for convicted prisoners up to 62.5 grams of loose tobacco or 80 cigarettes, or cigars or a combination of both. (The limit for unconvicted prisoners is 137.5 grams of 180 cigarettes)
    hobbies materials
    games (including electronic games without data storage facilities)
    writing and drawing materials
    a wrist watch (including one with an alarm function)
    a manual typewriter (excluding electric or electronic typewriters or word processors)
    a battery shaver
    batteries for personal possessions (to a specific number and only of a type authorised by Headquarters)
    personal toiletries
    one wedding or other plain ring
    one medallion or locket
    religious articles, as authorised by the Governor
    photographs and pictures and unglazed frames

unpadded greeting cards
a calendar
a diary or personal organiser
an address book
phonecards (as permitted under SO5)
postage stamps.

Unless restrictions are stated elsewhere in this or other standing orders (or in the statement published under paragraph 4) these items may be sent or handed in by relatives or friends or purchased from earnings or private cash.

10. The Governor may add other items to the published statement, subject to the approval of the Area Manager.

11. The published statement should inform prisoners that all personal property brought into prison, particularly if held in possession is held at the prisoner's own risk and that the Department will not accept liability for the loss of or damage to that property unless resulting from the Department's negligence. They must be reminded in this statement that they will not be able to control or supervise such items at all times and that the Department cannot guarantee the security of items left in cells or other accommodation. They should also be advised not to keep in possession particularly valuable property. Prisoners must sign 'in possession' property card for all items retained in possession other than consumable/disposable items.

Personal radios, record players, cassette and compact disc players

12. A prisoner shall be allowed to have a battery-operated radio (which may include an internal mains adaptor but must not be run from any external power source) capable of receiving VHF/FM transmissions within the 88–108 MHz band, short wave transmissions between 1–18 MHz, and medium and long wave transmissions. An aerial (other than one which consists of long strong wires) is permitted but external leads are not. A prisoner may also have one of the following: a record player, a cassette player (including a digital audio tape player) or a compact disc player. Alternatively, a prisoner may have a combined sound system, comprising some or all of the aforementioned (ie radio, record player, cassette player and compact disc player).

13. The Governor may require prisoners to use earplugs or earphone attachments where not to do so would result in annoyance to others, or where broadcasts may be heard by prisoners who are not allowed access to a radio. If it appears necessary to do so in the interests of security or good order or discipline or safety, the Governor may impose restrictions on the use of record, cassette or compact disc players or combined sound systems or require modifications to be made to such equipment: the cost of which must be met by the prisoner.

14. Cassette players must take standard or digital audio cassettes. Prisoners may send out uncensored recorded messages on cassettes except where the routine censorship of letters occurs, when cassettes should be censored in line with the arrangements for letters under Standing Order 5B.

15. Prisoners in Category C and D establishments may obtain commercially pre-recorded cassettes from any source other than another prisoner (unless specifically permitted under paragraph 6 above). Other prisoners must obtain commercially pre-

recorded cassettes direct from established British retail suppliers, registered record clubs, or other approved suppliers. All prisoners may also obtain home recorded cassettes direct from their families, provided that the cassettes have transparent cases. The arrangements outlined in paragraph 14 extend to such home recorded cassettes.

B. Acquisition of personal possessions

16. The Governor, with the approval of the Area Manager, may place restrictions, in addition to those stated in this Standing Order, on those items that can be purchased from private cash, or be sent or handed in by relatives or friends. This information will be included in the statement of facilities issued under paragraph 4 above.

Private cash and prison pay

17. Unconvicted prisoners (except those subject to a deportation order: see SO 1 C 31) may spend as much private cash as they wish to purchase items allowed in possession, subject to any restrictions imposed in accordance with paragraph 5 above.

18. The Secretary of State may set a specified limit on the amount of private cash (which may include a hobbies materials allowance in prisons containing Category A or B prisoners: see paragraph 31) that convicted prisoners may spend in a 12-month period on items defined as facilities in this Order. The limit should be reduced on a pro rata basis for periods of less than 12 months. The Governor will inform prisoners of the current level of this financial limit (from which there shall be no local variation unless specifically authorised by the Area Manager) and when the 12 month period will start. Prisoners will be expected to control the use of their private cash so that they manage the rate of expenditure throughout the 12-month period. The Governor will, from time to time, or at the request of individual prisoners, inform prisoners of the amount of private cash that they have spent and the amount that is available for the remainder of the 12-month period. Prisoners will not be able to carry forward to, or anticipate from, the next 12-month period without the permission of the Governor.

19. All purchases may be made with either pay or private cash, subject to any restrictions in place in accordance with paragraph 16. Purchases of the following items shall not count against the private cash limit imposed by paragraph 18: battery-operated shaver: a radio: record player: cassette player: combination sound system: compact disc player: headphones: a wrist watch: writing materials: postage stamps or phonecards.

Purchases from the prison shop

20. Establishments shall run a shop from which prisoners may buy items including batteries, postage stamps, confectionery, toiletries, tobacco, phonecards, stationery and writing implements. Prisoners shall be informed of the times when they are able to use the shop. The governor should ensure that, as far as is practicable, the shop is stocked to meet the likely needs of prisoners, including those from the ethnic minorities.

21. The following items may not be handed or sent in and may only be purchased through the shop:

batteries for personal possessions   —(except in the case of young offender institutions, where batteries may be sent or handed in by a relative or friend)

food and confectionery
phonecards
tobacco   —(except for unconvicted prisoners)

Mail order purchases

22. The Governor shall permit (within the scope of paragraphs 7 and 8 above) the purchase by prisoners, at their own risk, of items permitted under this Standing Order through companies that belong to the Mail Order Protection Scheme (MOPS), provided that the goods are addressed to the prisoner at the full prison address. These items must remain the property of the prisoner who purchased them, and cannot be given, lent or sold to another prisoner without the permission of the Governor.

Smoking requisites for unconvicted prisoners

23. Unconvicted prisoners will be allowed to bring smoking requisites, including tobacco and cigarettes, into prison on reception and to have them sent in by relatives or friends. Those aged 16 or over may also purchase them from their private cash or prison pay. They may not order or receive in any week more than 112.5 grams of tobacco, whether in the form of loose tobacco, cigarettes or cigars. All tobacco, cigarettes and cigars sent in to unconvicted prisoners or purchased by them other than through the prison shop must be in sealed packets or tins: a record will be kept of such items received for individual unconvicted prisoners.

Newspapers, periodicals, magazines and books

24. Prisoners shall be allowed to receive newspapers, periodicals, magazines or books (see paragraph 9).

25. Except for prisoners in category C or D establishments, newspapers, periodicals and magazines must be received direct from publishers, newsagents or booksellers, either at the order of relatives or friends or by payment from prisoners' private cash or prison pay. The Governor may allow prisoners to receive books and, in the case of prisoners in category C or D establishments, newspapers, periodicals and magazines direct from relatives or friends, or from people not previously known to them.

26. When prisoners pay for newspapers, periodicals or magazines to be delivered they should normally do so in advance.

27. The Department will accept no responsibility if a newspaper, periodical or book is not received at the establishment.

Personal radios, record, cassette, compact disc players, or combined sound systems

28. Radios, record, cassette or compact disc players or combined sound systems may come from prisoners' stored property, by purchased from private cash or pay, or be sent or handed in by relatives or friends.

29. Radios and combined sound systems will be checked on arrival at establishments to ensure they do not contravene the restrictions on wave-bands. The governor may decide to place a seal and/or an identifying mark on any item as a condition of its issue, which may indicate that it has been modified to comply with any standard safety/security requirements in force at that time (see paragraph 13). The Governor will also arrange for these items and batteries to be examined at irregular intervals.

General notebooks and drawing books

30. A prisoner is entitled to be issued with a general notebook (in addition to notebooks issued by Education Officers) and a drawing book, which will normally be replaced when full.

Cell activities and hobbies

31. Items used for those cell activities and hobbies which have been allowed under paragraph 44 below may be purchased from pay or private cash or, in the case of unconvicted prisoners and prisoners in Category C and D establishments, may be sent or handed in by relatives or friends. In prisons containing Category A or B prisoners, material for hobbies may not be sent or handed in by relatives or friends to any prisoner other than an unconvicted prisoner. Instead, prisoners may be permitted to spend money on these materials from their private cash, subject to any limit on expenditure (see paragraph 18 above).

32. The Governor may permit prisoners who are members of handicraft classes to continue their handicrafts in their cells: see paragraph 45 below. Materials will be obtained under the arrangements for supplying those classes, and may be bought or disposed of—see paragraph 41 below—only under arrangements made by the Education Officer. Prisoners may be allowed to use their private cash (subject to the limits outlined in paragraph 18) to supplement the materials provided by the Education Officer.

C. Forfeiture, disposal etc. of personal possessions

Forfeiture

33. The circumstances in which facilities may be forfeited as a punishment are described in Standing Order 3D.

Enforced removal of possession

34. The Governor may require a prisoner, for the reasons set out in paragraph 8, to move personal possessions to stored property or to arrange to have them sent or handed to relatives or friends. He or she may refuse to allow the prisoner to receive any more property either for storage or to hold in possession, until the amount of property in prison (including that in store) has been reduced to the normal limits specified in that paragraph and Standing Order 1C. The Governor will inform the prisoner of the reason for this decision and record this reason in the prisoner's Inmate History record (F2052B or F2052A).

Withholding or withdrawal of newspapers, periodicals, magazines and books

35. The Governor may withhold or withdraw any newspaper, periodical or maga-zine or any particular issue, or any book, if he or she considers that the content presents a threat to good order or discipline or to the interests of prison or national security, or where the medical officer considers that possession of the material is likely to have an adverse effect on the prisoner's physical or mental condition. Any items so stopped will be put with the prisoner's stored property or may be sent or handed out to a relative or friend. The Governor will record the Governor's or the Medical Officer's reasons for this decision in the prisoner's Inmate History record (F2052B or F2052A) and that the prisoner has been informed of the reason.

36. Where a personal newspaper has been withdrawn or withheld from a pris-oner, either as part of a punishment for an offence against prison discipline or for any other reason, arrangements will be made, eg through the availability of newspa-pers provided at public expense (see paragraph 52), to ensure that the prisoner is not normally deprived of access to newspapers altogether.

Restrictions on display

37. The Governor may impose restrictions on the display of material which, he or she considers, is likely to cause offence by reason of its indecent or violent or racist content or is racially or sexually discriminatory.

Disposal of surplus books, records, compact discs and cassettes

38. Prisoners who acquire more books, records, compact discs or cassettes than they are allowed to retain in possession may reduce the number by placing them in store, by having them handed out on a visit, by having them sent out, by giving them to the public library collection held in the prison library or by giving them to another pris-oner (provided that the Governor has permitted the transfer and that the transfer is recorded). Where items are given away a prisoner will have no further claim on them.

39. The Governor may refuse a prisoner permission to receive another book, record, compact disc or cassettes until the latter has reduced the number to below the permitted total.

Disposal of completed notebooks etc.

40. Prisoners may send out completed books, diaries or personal organisers, or any other artistic or written material **that is not correspondence**, unless they contain any material of the kind prohibited in general correspondence under Standing Order 5B. If prisoners are not allowed to send out this material, they may be allowed to take it out on discharge, provided that it contains nothing relating to individually named staff or prisoners or matters relating to offences committed by others, or escape plans or other material that should be disallowed in the interests of prison security or good order and discipline, the prevention of crime, or national security.

41. If prisoners claim that the material that they have been prevented from taking out on discharge contains serious academic literary or artistic work, it must be pre-served and the Governor should make a report to Headquarters (Chief Education Officer's Branch). If there is no such claim, the material should be destroyed three months after the prisoner's final discharge.

Disposal of surplus items

42. Games that are no longer required, and completed works or spare material connected with hobbies unless sold through an approved charitable organisation (see paragraph 51), should be added to a prisoner's stored property, handed or sent out to relatives or friends, donated to the establishment for use by other prisoners, or given to a charity.

Withdrawal of notebooks

43. General notebooks are issued for writing, drawing or painting and will be withdrawn if they are misused in any way likely to affect the security, good order or discipline of the establishment, or if they contain material likely to jeopardise national security or encourage the commission of crime. Where a book is withdrawn a note of the reasons should be made in the prisoner's record, and the prisoner informed.

II. OTHER FACILITIES

Cell activities

44. The Governor may allow prisoners to play games or engage in hobbies in cells. These activities must not constitute a risk to health or safety or to security or good order or discipline or cause annoyance to other prisoners. To avoid the practical difficulties of transporting large quantities of materials or materials that are easily breakable, unconvicted prisoners who are regularly attending court may be restricted from engaging in hobbies that require such materials.

45. Prisoners who are members of handicraft classes may be permitted to continue their handicrafts in their cells under arrangements made by the Education Officer.

Smoking

46. Young persons aged under 16 (whether sentenced or not) may not be permitted to purchase tobacco.

47. Sentenced young offenders in YOIs will not be allowed to smoke or to have any tobacco except in accordance with the directions of the Secretary of State. These directions at present allow YOI Governors, in consultation with their Area Managers, to specify at what times and places smoking may be allowed, except in establishments or parts of establishments for juveniles. In juvenile establishments and juvenile units, neither smoking nor possession of tobacco will be allowed.

48. Sentenced prisoners and young offenders located elsewhere (i.e. not in YOIs) may smoke at times and in places specified by the Governor.

49. Prisoners will be allowed to take tobacco, cigarettes or cigars to court and to smoke them in the places where smoking is permitted.

50. Prisoners will not be allowed to smoke in health care centre consultation/ treatment rooms or waiting areas. Prisoners who have been admitted to the centre as in-patients may be allowed to smoke in such other parts of the centre at such times as the Managing Medical Officer directs.

Public Competitions and the sale of work through charitable organisations

51. Prisoners may enter public competitions, within their normal allowance for letters. They may also submit artwork or work of literary merit for sale or publication for profit through any charitable organisation (eg the Koestler Trust) approved for this purpose by the Chief Education Officer's Branch. The Governor's approval has to be obtained before a prisoner may enter a competition for which a fee is required. Money prizes or income from the sale of artwork or work of literary merit may be credited to private cash but not to the prisoner's pay. Other prizes may be kept, subject to the above provisions relating to the retention of personal possessions, placed in a prisoner's stored property, or disposed of in any way that the Governor approves.

Supply of newspapers at public expense

52. The Governor will make arrangements to supply at least one newspaper for every ten prisoners. In deciding which newspapers to provide the Governor should take into account the preference of prisoners, including those from ethnic minorities, and provide as full a range of newspapers as possible.

Association

53. All unconvicted and convicted prisoners, unless they are excluded under the provisions set out below, will, wherever practicable, be given the opportunity to participate in association with other prisoners. Such association can take various forms, including eating together, recreation (including physical recreation) and entertainment. (Work, religious services and education classes do **not** count as association under this Standing Order.)

54. Where opportunities for association are available only to a limited number of prisoners, young prisoners under 21, whether convicted or unconvicted, should generally have priority over adults in the same establishment. (See also Standing Order 8A on the supervised mixing of convicted and unconvicted prisoners.)

55. A prisoner may be excluded or removed from all association or from a specified group activity only under the circumstances set out below, as provided in the Prison Rules:

(a) in his or her own interest or for the maintenance of good order or discipline;
(b) if refractory or violent;
(c) to restrain the prisoner;
(d) pending the governor's initial hearing of a disciplinary charge;
(e) as a result of a punishment awarded at an adjudication.

Any decision to cancel a period of association must, on all occasions, be taken by an officer not below the rank of Governor 5.

Purchase of services

56. Prisoners may not purchase services (eg barber or hairdresser) or personal tuition. Governors may not make arrangements for such services to be provided commercially, except for distance learning facilities approved by the Education Department.

Standing Order 5: Communications

It is one of the roles of the Prison Service to ensure that the socially harmful effects of an inmate's removal from normal life are as far as possible minimised, and that contacts with the outside world are maintained. Outside contacts are therefore encouraged, especially between an inmate and his or her family and friends. At the same time, the Prison Service has an overriding duty to hold inmates in lawful custody in well ordered establishments, and to have regard to the prevention of crime and similar considerations; and some regulation of inmates' communications is therefore necessary.

CONTENTS

SECTION A: VISITS

See also:

Standing Order 5D on visits by Members of the United Kingdom Parliament and Members of the European Parliament.

Standing Order 5E on visits by consular officials.

Introduction

1. Inmates may receive visits in accordance with Rules 33, 34 and 37 of the Prison Rules and Rules 10, 11 and 13 of the Young Offender Institution Rules.

   2. In this Standing Order the term *statutory visit* refers to a visit to which an inmate is entitled within the basic minimum allowance as set out in paragraphs 3 and 4 below. Subject to paragraph 9 below, statutory visits may not be withdrawn or withheld as part of a punishment award. Any visits beyond these minimum entitlements which are routinely allowed to all inmates, or all inmates of a particular class, at a given establishment are referred to as *privilege visits*, and visits granted to individual inmates for particular reasons are referred to as *special visits*.

Allowances of visits

3. Visits to *convicted inmates* should be allowed as frequently as circumstances at establishments permit, subject to the following minimum entitlements:

   (1) adults—a visit after reception on conviction and every four weeks thereafter;
   (2) young offenders—a visit after reception on conviction and every two weeks thereafter.

Visits to convicted inmates should last at least thirty minutes, but where circumstances permit Governors should allow longer.

   4. Visits to *unconvicted inmates* should normally be permitted every day (Monday to Saturday) for a minimum of 15 minutes, and longer visits should be allowed wherever possible, and particularly where it is known that their close relatives are

not able to visit frequently. But where the number of inmates and staff resources make daily visiting impracticable at a particular establishment, the Regional Director may authorise a reduction in the frequency of visits, so long as the aggregated entitlement of one and a half hours per week is maintained.

5. In addition, Governors may allow one or more *special visits*:

(1) if an inmate's private or business affairs are not in order after conviction, to help him or her complete the necessary arrangements;

(2) if they are satisfied that it is necessary for the conduct of legal proceedings or the welfare of the inmate or the inmate's family;

(3) subject to the medical officer's advice, to an inmate who is seriously ill. Restrictions on the number of visitors or the time of visits should wherever practicable be waived in such cases.

6. The following visits should not normally count against an inmate's allowance:

(1) visits by legal advisers

(2) visits by probation officers

(3) pastoral visits by priests and ministers

(4) visits by MPs on special visiting orders (see SO 5D)

(5) visits by officers of the Parliamentary Commissioner for Administration (see paragraph 40 below)

(6) visits by consular officials (see SO 5E 7)

(7) visits by police officers or other public officials in discharge of their duty

(8) special visits granted under paragraph 5 above.

Letters in place of visits

7. An extra letter at public expense and a reply should be allowed in place of any statutory visit not paid which the inmate is not allowed or does not wish to accumulate under paragraphs 11–18 below.

Visits in advance

8. Governors have discretion to allow convicted inmates to receive visits in advance of their due date. The due dates of subsequent visits will remain unaltered unless also allowed in advance.

Deferment of visits during cellular confinement

9. A Governor may defer the right of an inmate to a visit until the expiration of any period of cellular confinement. This power should only be exercised when an inmate's behaviour and attitudes are such that removal from cellular confinement for the purpose of the visit is clearly impracticable or undesirable. Any such visits may take place in the ordinary visiting room or elsewhere at the Governor's discretion.

Sunday and Bank Holiday visits

10. Subject to staff availability, visits may be allowed on Sunday outside the times arranged for worship. Where Sunday visits are not allowed visiting orders should be so endorsed. Visits will not normally be allowed on Christmas Day, Boxing Day and Good Friday.

Accumulated visits

11. Subject to paragraphs 12–18 below, convicted inmates may be allowed to accumulate a minimum of three visits up to a maximum of 12 and apply to take their accumulated visits at the prison at which they are detained or by temporarily transferred to a local prison to take their visits. Inmates are entitled to accumulate only statutory visits, which for this purpose should be taken to be the first visit received in any month. Privilege visits may be accumulated at the Governor's discretion, but when an inmate is transferred to another establishment, either permanently or temporarily, permission to take accumulated privilege visits will be at the discretion of the receiving Governor.

12. An inmate is eligible for temporary transfer for accumulated visits 12 months after transfer from the local prison to which he or she was sent on conviction, though earlier transfer may be possible in appropriate cases with the consent of the Governor and Regional Director, or, in the case of Category A inmates, of Prison Department Headquarters. Once eligible for temporary transfer, inmates are eligible for further transfers in the same month in succeeding years even if transfer has been delayed in the previous year. Unless a previous transfer has been delayed inmates should not normally be transferred for accumulated visits more frequently than once a year, nor should they normally be transferred if they have less than six months to serve.

13. Convicted young offenders are eligible for accumulated visits after having served 12 months, provided that they have at least four months still to serve. A young offender who has become eligible for accumulated visits retains that eligibility if reclassified as an adult prisoner.

14. Inmates should be advised to petition for accumulated visits if:

   (*a*) they wish to be temporarily transferred to an establishment other than a local prison, or to a prison in Scotland, Northern Ireland, the Channel Islands or the Isle of Man;

   (*b*) they are in Category A.

15. Transfers will normally be for one month but may be extended for up to a further month with the agreement of the Governor of the receiving establishment, and, in the case of a Category A inmate, with Headquarters' approval.

16. The 12 accumulated visits may include visits in advance provided they do not exceed the number to which the inmate would be entitled during the remainder of the sentence, and provided that on return from temporary transfer he or she receives no more visits until any advance visits have been accounted for.

17. While at local prisons inmates should wear the clothing normally in use there, and should receive only such privileges as the facilities of the prison allow.

18. An inmate may take with him or her only such items of property currently in possession as are allowed at the receiving establishment.

Visits where both parties are inmates at the same establishment

19. Visits may, on application, be allowed between close relatives as defined in paragraph 30 below when both parties are inmates of the same establishment. Visiting Orders need not be surrendered and the frequency of such visits will be at the Governor's discretion. Each visit should be for as long as local circumstances permit.

Inter-prison visits

20. Visits may, on application, be allowed between close relatives as defined in paragraph 30 below when both parties are inmates in the same establishment. Arrangements may, subject to the requirements of security and the availability of transport and accommodation, be made for such visits to take place at three-monthly intervals, and each inmate must surrender one Visiting Order. Each visit should be for as long as local circumstances permit.

21. An inter-prison visit between unconvicted inmates may only be approved if the period of continuous custody up to the date of trial is expected to be at least a month, and where it has not been possible for the inmates to have a visit during any of their productions at court.

Visiting Orders

22. Visitors to convicted inmates, other than those covered by paragraph 6 above, must be in possession of a valid visiting order on which they are named.

23. Visiting orders should be issued sufficiently early to allow the visit to take place as soon as it becomes due. Should an order for which an inmate has applied be refused, or the issue of an order be liable to delay, the inmate should be told. Whenever possible visiting orders should be enclosed with inmates' letters.

Conditions for visits

24. (1) Visits should take place under the most humane conditions possible. Subject to sub-paragraph (2) below, they should take place in open visiting rooms, with both inmate and visitors seated at a table, and the inmate and visitors should be permitted to embrace each other.

   (2) Where security or control conditions so require, it may be stipulated that a special or closed visiting room or a closed visiting box is to be used, or that inmates and visitors are to be permitted no, or only limited, physical contact.

   (3) Up to three persons (not including any children under 10) should normally be allowed at each visit.

25. (1) Except where otherwise provided in Standing Orders of Circular Instructions, all visits will be in the sight of a prison officer and are liable to take place within the hearing of an officer. For the majority of social visits it should be sufficient for officers to be present in the room where visits are taking place, but where necessary it may be stipulated that visits are to be subject to closer supervision, including being within the direct hearing of an officer.

   (2) Information about a conversation during a visit may be disclosed outside the Home Office in a circumstance in which the contents of a letter may be disclosed under SO 5B 39.

   (3) Officers supervising visits will be responsible for ensuring that no unauthorised article is allowed to pass between a visitor and an inmate. Should a visitor try to pass anything to an inmate, or vice versa, the supervising officer should investigate the incident. If the article is unau-

thorised the visit should be stopped at once and the officer should report the matter immediately to the Governor or an appropriate senior officer.

26. Letters, notes, or other papers, like any other article, may not be taken into or out of an establishment without the authorisation of the Governor. It is for the Governor to decide how this principle should be enforced (for instance, whether a notice should be displayed requiring the specific permission of a supervising officer for any paper to be passed between an inmate and a visitor), but in general the degree of control should be commensurate with the level of censorship in force at the establishment. In any case where it is suspected that the rules governing correspondence are being evades the supervising officer should inspect the material in question or refuse permission for it to be handed over.

27. Subject to paragraph 29 below, inmates and their visitors may speak in the language of their choice, even though the supervising officer many not understand the language being spoken.

28. (1) Except in circumstances to which sub-paragraph (5) applies, no tape-recording may be made of a visit.

(2) Except in circumstances to which sub-paragraph (5) applies, visitors should not be allowed to retain cassette recorders, radios, cameras, cellular telephones or similar equipment in their possession at visits. Visitors should be advised that such items must be left at the main gate. A visitor who refuses to comply with this request should be advised that the visit cannot proceed and should be asked to leave.

(3) If, nevertheless, a visitor is discovered with a tape-recorder, radio, camera, or cellular telephone (or similar item of equipment) after entering the establishment, the visitor should be asked to surrender the item, which should be returned when the visitor leaves the establishment. In the case of a tape-recorder or camera, the visitor should be further advised that the tape or film is liable to confiscation (under Prison Rule 85 or Young Offender Institution Rule 69) and asked to surrender it. In case of refusal, the visitor should be advised that to attempt to take the tape or film out of the prison would constitute a criminal offence, under section 41 of the Prison Act 1952, and that if it is not handed over the police may be called.

(4) Any tape or film which is confiscated should be sealed in the presence of the visitor and forwarded to headquarters (P5 Division) with a full report of the circumstances. The visitor should be told that he or she may write to headquarters (P5 Division) about the return of the tape or film.

(5) The circumstances to which this paragraph applies are those described in paragraphs 29, 35 (legal advisers) and 39 (police officers) below, SO 5D 9 (MPs and peers), SO 5E 9 (consular officials), and visits by the European Committee Against Torture.

29. If it appears necessary to do so, in the interests of prison security, national security, public safety or the prevention or detection of crime

(*a*) an inmate and visitor may be required to speak English, if they are able to do so; or

(*b*) it may be required that a visit involving an inmate or a visitor who cannot speak English be overheard by a person who understands the language to be used; or

(c) a visit may be tape-recorded so that it can be heard later.

Visitors

Visits by close relatives

30. Subject to paragraph 31 below and, in particular, the special procedures which apply to applications to visit inmates in Category A, inmates may be visited by close relatives, who are defined as husband and wife (including a person with whom the inmate was living as man and wife immediately before reception), parent, child, brother, sister (including half- or step-brothers and sisters), fiance or fiancee (provided that the Governor is satisfied that a bona fide engagement to marry exists), or a person who has been in loco parentis to an inmate or to whom the inmate has been in loco parentis.

31. Governors have discretion to refuse visits from persons covered by paragraph 30 above on the grounds of security, good order or discipline, or in the interests of the prevention or discouragement of crime, but this should be done only in exceptional circumstances.

Social visits by other persons

32. Subject to paragraphs 33–39 below, inmates may be visited by other persons, but such visits may be disallowed on the grounds of security, good order and discipline, for the prevention of crime or because the Governor has reason to believe the visit would seriously impede the rehabilitation of the inmate concerned. Visitors to Category 'A' inmates are subject to special procedures.

Visits to and by minors

33. (1) The Governor has discretion to disallow any visit to an inmate under 18 years of age, if he or she considers that such a visit would not be in the inmate's best interests. In deciding whether to exercise this discretion the Governor will take account of any views which the inmate's parent or guardian may express.

(2) The Governor has discretion to disallow any visit to an inmate by any person under 18 years of age, if he or she considers that such a visit would not be in the best interests of the visitor.

(3) Any case in which the Governor proposes to disallow a visit by a close relative, as defined in paragraph 30 above, under the provisions of this paragraph should be reported to Headquarters (P3 Division).

Visits by legal advisers

34. An inmate's *legal adviser* may be the inmate's counsel, solicitor, or a clerk acting on behalf of the solicitor. Inmates may be visited by legal advisers acting in their professional capacity, and visiting orders need not be surrendered. All visits by legal advisers will take place in the sight of a prison officer, but will take place out of hearing if the purpose of the visit is:

(*a*) to discuss legal proceedings to which the inmate is a party;

(*b*) to allow an inmate to consult his or her legal adviser about possible legal proceedings. In this case the inmate must confirm that this is the purpose of the visit (but is not obliged to disclose what the contemplated proceedings are about);

(*c*) to allow an inmate to consult his or her legal adviser about other legal business, which does not involve possible proceedings (eg. selling a house or making a will); in such cases the purpose of the visit must be disclosed in advance, and must not offend against the restrictions in SO 5B 34;

(*d*) to allow an inmate to consult his or her legal adviser about a forthcoming adjudication; or

(*e*) to allow an inmate to consult his or her legal adviser about an application to the European Commission of Human Rights (or subsequent proceedings before the European Court) (see SO 5F 7).

Visits by a member of the legal profession other than in a professional capacity (eg. as a relative or friend) should take place in the sight and hearing of an officer.

35. An inmate's legal adviser may be allowed to use a cassette recorder when interviewing the inmate provided the legal adviser gives a written undertaking that the tape will be kept in his or her officer and will be used solely in connection with the proceedings or the legal business which the visit was to discuss.

36. Inmates who are taking or contemplating legal proceedings without using the services of a legal adviser as defined in paragraph 35 above should apply in writing to the Governor for any facilities such as special visits and letters that they may require, but such visits may be overheard by a prison officer See also SO 11.

Visits by journalists or writers

37. Visits to inmates by journalists or authors in their professional capacity should in general not be allowed and the governor has authority to refuse them without reference to headquarters. If a journalist or author who is a friend or relative wishes to visit an inmate in this capacity and not for professional purposes, the governor should inform the intending visitor that before the visit can take place he or she will be required to give a written undertaking that any material obtained at the interview will not be used for professional purposes and in particular for publication by the intending visitor or anyone else.

Visits by priests and ministers

38. An inmate may be allowed an occasional pastoral visit by his or her home minister of religion after consultation between a chaplain and the Governor. Such a visit may take place either in the sight but out of hearing of a prison officer (unless the Governor exceptionally considers it necessary for the conversation to be overheard) or in the Chaplain's office as is thought most desirable. Such visits may be paid without visiting orders being issued. visits by clergymen to whom inmates have sent visiting orders should take place in the visits area and are not affected by this paragraph.

Visits by police officers

39. (1) An inmate may be interviewed in an establishment by police officers. So far as practicable, such an interview shall be conducted in accordance with the terms of the Police and Criminal Evidence Act 1984, and of Code C of the Codes of Practice issued under that Act in relation to the conduct of interviews at police stations.

(2) Such an interview will take place within the sight and, where appropriate, within the hearing of a prison officer.

(3) If the police have reasonable grounds for suspecting that the inmate has committed an arrestable offence, the inmate may be required to remain with the police officers so that they may put their questions.

(4) If the police wish to interview the inmate in any other connection, the inmate may be required to remain with the police officers so that they can explain why they wish to conduct the interview. But if the inmate is then not prepared to be interviewed (or if at any stage during the interview the inmate wishes to terminate the interview), the interview will not continue.

(5) Before any interview in an establishment the inmate will be advised:

(*a*) of the right to consult a legal adviser;

(*b*) If the inmate has difficulty in understanding English, and the interviewing officer does not speak the inmate's own language, or if the inmate suffers from hearing difficulties, of the right to have an interpreter present during the interview.

An inmate who so wishes will be afforded the opportunity to consult with a legal adviser and to have a legal adviser and/or an interpreter present during an interview, unless the circumstances of the case are such that, under the terms of Code of Practice C, access to a legal adviser might be delayed, or an interview might proceed in the absence of a legal adviser and/or an interpreter if the interview were at a police station.

(6) If the inmate is under the age of 17, or is mentally ill or mentally handicapped, an appropriate adult shall be present during the interview. The appropriate adult shall be as defined in Code of Practice C. If no other more appropriate adult is available, a member of staff will undertake the role. An inmate who wishes to select a particular member of staff may do so if that member of staff is available.

(7) An interview may be tape-recorded by the police in accordance with section 60 of the Police and Criminal Evidence Act 1984, with Code of Practice E issued under the Act, and with police regulations.

(8) A copy of the Codes of Practice will be made available in every establishment for inmates' reference.

(9) This paragraph also applies to interviews by officers of other bodies, such as Customs & Excise.

Visits by officers of the Parliamentary Commissioner for Administration

40. Inmates may on request receive such visits, which Governors should arrange with the office of the PCA. Visits will be supervised as for visits by MPs (see Order

5D 7–8). The inmate will be allowed to have present at the interview a friend or adviser if that person would normally be allowed to visit the inmate.

Section B: Correspondence

On correspondence with Members of the United Kingdom or European Parliaments, see Order 5D below. On correspondence with consular or High Commission officials, see Order 5E below. On correspondence with the European Commission or Court of Human Rights, see Order 5F below.

Introduction

1. The examination and reading of correspondence to and from inmates is undertaken to prevent its use to plan escapes or disturbances or otherwise jeopardise the security or good order of establishments, to detect and prevent offences against prison discipline or the criminal law and to satisfy other ordinary and reasonable requirements of prison administration. Accordingly the extent to which correspondence needs to be read will vary according to the nature of the establishment, inmate and correspondent. Limits have to be placed on the amount of correspondence that inmates may send and receive, if such correspondence has to be examined and read, because the staff resources to supervise correspondence are limited; otherwise the diversion of staff to censorship duties will inevitably cause other aspects of the prison regime to suffer, to the detriment of all inmates.

2. The provisions affecting correspondence set out in this Order (which applies to all inmates of all prison service establishments) are necessary to meet the operational needs summarised above, while complying with Articles 8 and 10 of the European Convention of Human Rights. Article 8 provides that there shall be no interference by a public authority with the exercise of the right to respect for correspondence, except such as in accordance with the law and is necessary in a democratic society in the interests of national security, public safety or the economic well-being of the country, for the prevention of disorder or crime, for the protection of health or morals, or for the protection of the rights and freedoms of others. Article 10 provides that everyone has the right to freedom of expression, including the freedom to hold opinions and to receive and impart information and ideas. It goes on to provide that since the exercise of these freedoms carries with it duties and responsibilities, it may be subject to such formalities, conditions, restrictions and penalties as are prescribed by law and are necessary in a democratic society in the interests of national security, territorial integrity or public safety. They may also be directed at the prevention of disorder or crime, the protection of health or morals, the protection of the reputation or rights of others, preventing the disclosure of information received in confidence, or maintaining the authority and impartiality of the judiciary.

3. The restrictions on what may be included in correspondence apply irrespective of whether letters are examined or read by prison staff, and inmates and their correspondents should be aware that they remain personally responsible for the contents of their correspondence.

4. The examination and reading of correspondence will be done as quickly as possible. But letters which appear to include prohibited material are liable to delay while the prison authorities consider what action should be taken.

Allowances of letters

5. (1) A *statutory letter* is a letter to which an inmate is entitled under Prison Rule 34 or Young Offender Institution Rule 11, and which may not be withdrawn or withheld as part of a punishment award.

   (2) A *privilege letter* is a letter which inmates at a given establishment are regularly allowed to send over and above the entitlement to statutory letters.

   (3) A *special letter* is a letter not counted against an inmate's allocation of statutory or privilege letters which he or she is given permission to send for some special reason.

Allowances for convicted inmates

6. Convicted inmates may send:

   (1) one *statutory letter* a week, the first such letter to be issued immediately on reception;

   (2) at least one *privilege letter* a week in the case of adults and two in the case of young offenders, and as many more such letters as are practicable bearing in mind any need to examine and read correspondence and the staff resources available. (At establishments where routine censorship is not in force there is no limit on the number of privilege letters inmates may send);

   (3) *special letters*, in the circumstances set out in paragraph 7 below.

7. Convicted inmates should be issued with one or more special letters, according to their needs:

   (1) when they are about to be transferred to another establishment; or, if the inmate is not given a special letter before transfer, on reception at the new establishment. The number of letters issued should correspond to the number of visiting orders the inmate has outstanding;

   (2) immediately after conviction if he or she needs to settle business affairs;

   (3) where necessary for the welfare of the inmate or his or her family;

   (4) in connection with legal proceedings to which the inmate is a party, including an appeal against conviction or the issue of a counternotice to the discontinuance of proceedings;

   (5) where necessary to enable an inmate to write to a home probation officer or to an agency which is arranging employment or accommodation for him or her on release;

   (6) where necessary to enable an inmate to write to the Parliamentary Commissioner for Administration;

   (7) at Christmas. Governors have discretion to limit the numbers of such letters allowed.

Allowances for unconvicted inmates

8. Unconvicted inmates may send

(1) two *statutory letters* per week;

(2) as many *further privilege* letters as they wish;

(3) *special letters*

  (*a*) when about to be transferred to another establishment, or, if this has not been possible before transfer, on reception at the new establishment;

  (*b*) in connection with their defence, if they cannot afford the postage costs of a privilege letter for this purpose.

Length of outgoing letters

9. Except at establishments where all or most correspondence is not censored, letters written by convicted inmates should not exceed four sides of A5 paper.

Numbers and length of incoming letters

10. At establishments where all or most correspondence is not censored, there are no restrictions on the number or length of letters which inmates may receive.

11. At other establishments inmates are allowed to receive as many letters as they are allowed to send. If an inmate at such an establishment receives an excessive number of letters either habitually or on one occasion the governor has discretion to return excess letters to the sender(s); the inmate will be given the opportunity to select from the envelopes received those which he particularly wishes to read. Similarly anyone who makes a practice of sending overlong letters to inmates at such establishments may be asked to confine themselves to 4 sides of paper. If they ignore the request the governor may return subsequent overlong letters, in which case the inmate should be informed accordingly.

Letters instead of visits

12. In accordance with Prison Rule 34 (4), and Young Offenders Institution Rule 11 (3), a letter at public expense may be allowed in place of a statutory visit.

Letters in advance

13. The Governor has discretion to allow a letter in advance of the due date. The date of the next letter will be calculated from the due date.

Accumulated letters

14. A convicted inmate may, to the extent the Governor considers reasonable, accumulate his or her allowance of statutory and privilege letters.

Christmas cards

15. At Christmas an inmate may buy up to 12 cards, or more if the Governor considers this practicable, from the canteen.

Postage costs

16. (1) Statutory letters will be sent at public expense.

   (2) The postage costs of privilege letters may be met from prison earnings or private cash.

(3) The postage costs of special letters from convicted inmates should normally be met from prison earnings or private cash, except for letters on transfer (paragraph 7 (1)), which should be sent at public expense. Governors also have discretion to allow any other special letter which a convicted inmate is permitted to write under paragraph 7 above to be sent at public expense if they consider this justified in the circumstances of the case.

(4) Special letters from unconvicted inmates should be sent at public expense.

(5) The postage costs of Christmas cards may be met from private cash or prison earnings.

17. (1) If inmates pay the postage costs themselves they may choose between first or second class mail and, for overseas letters, air or sea mail.

(2) Subject to sub-paragraph (3) below, letters at public expense will normally be despatched at the cheapest possible rate but an inmate may opt for a higher class of postage provided he or she pays the differences from prison earnings or private cash.

(3) Letters at public expense will be sent first class or by air mail if:

(*a*) they are special letters sent on transfer, or

(*b*) they are in connection with an appeal, or

(*c*) exceptionally, postage at the higher rate has been approved by the Governor.

Format of letters

18. Letters may be written on the appropriate letter form and posted in the envelope provided. Inmates may also use envelopes and writing paper bought from the canteen, provided that they show the official address of the establishment at the top of the paper. Inmates may apply to omit this address in special circumstances.

Correspondents

19. An inmate may write to, and receive letters from, any person or organisation, subject only to the acceptability of the contents and to the restrictions set out in paragraphs 20–29 below. It does not necessarily follow from the fact that an inmate may correspond with a person or organisation that he or she may be visited by that person or a representative of that organisation.

20. If the recipient of correspondence from an inmate requests in writing to the prison authorities that further letters should not be sent, the inmate should be informed of the request, asked to co-operate by not writing and given the opportunity to discuss the matter with a member of staff. If the inmate then hands out a further letter for posting, the Governor may, at his or her discretion, give effect to the recipient's wishes and inform the inmate that the letter will not be sent. But the Governor should consult Headquarters (P3 Division) if he or she proposes to stop correspondence from an inmate to his or her spouse at the request of the spouse.

Correspondence with minors

21. A minor is a person under 18 years of age. The procedure set out in paragraph 20 above will apply:

(*a*) if the person or authority having paternal responsibility for the care of a minor requests the stopping of correspondence from an inmate to that minor; or

(b) if the person who normally has parental responsibility for the care of a minor who is in custody in a penal establishment requests the stopping of correspondence between that minor and any other person, except the minor's husband or wife.

22. If the Governor judges that correspondence between a minor in his or her custody and any person would not be in the inmate's best interest he or she has discretion to stop the correspondence. In exercising this discretion the Governor will take account of any views which the inmate's parent or guardian may express. If the Governor proposes to exercise this discretion so as to prevent correspondence with a close relative as defined in SO 5A 30, the circumstances should first be reported to Headquarters (P3 Division).

Correspondence with convicted inmates

23. Correspondence with another convicted inmate requires the approval of both governors, except where the inmates are close relatives as defined in SO 5A 30 or where they were co-defendants at their trial and the correspondence relates to their conviction or sentence. Subject to the provisions of paragraphs 21–29 approval should be given unless there is reason to believe that such correspondence will seriously impede the rehabilitation of either, or where it would be desirable, in the interests of security or good order and discipline, that the inmates should be prevented from communicating with each other. Any letter from one inmate to another should, if the governor of the writer's establishment has no objection, be sent to the governor of the recipient's establishment with a memorandum inviting him or her to consider whether it should be issued.

Ex-inmates

24. Correspondence with a person who has served a custodial sentence will be allowed, subject to paragraph 26 below and, if the ex-inmate is under supervision in the community, to the views of the supervisor, unless the governor believes that it would seriously impede the rehabilitation of either.

Victims of inmate's offences

25. Inmates wishing to write to a victim of their offences, or a victim's family, should first apply to the Governor for permission, which may be withheld if it is considered that the approach would add unduly to the victim's or family's distress. This restriction does not apply where:

(a) the victim is a close relative, as defined in SO 5A 30, or

(b) the victim has already written to the inmate since conviction, or

(c) the inmate concerned is unconvicted.

Threats to security and good order

26. The governor has discretion to disallow correspondence with a person or organisation if there is reason to believe that the person or organisation concerned is planning or engaged in activities which present a genuine and serious threat to the

security or good order of the establishment or other Prison Department establishments. The Governor should consult Headquarters (P3 Division) before disallowing correspondence between an inmate and a close relative, as defined in SO 5A 30, under the provisions of this paragraph.

27. When an inmate has been stopped from writing to any person or organisation, or would not be allowed to do so, in accordance with paragraph 26 above, it will follow that communication with any other person at the same address will be stopped unless the other person is a close relative as defined in SO 5A 30.

Penfriends

28. (1) Inmates who wish to advertise for penfriends should first submit the text of the proposed advertisement to the Governor for permission. Such permission should normally be granted, but should be refused if:

   (a) the advertisement invites respondents to write to a box number; or

   (b) the inmate is an adult and the publication concerned is aimed at, or read mainly by, children or young people;

   (c) the advertisement is to be placed in a periodical which caters for tastes or interests which may have motivated the inmate's offence; or

   (d) in the opinion of the Governor, correspondence arising from the advertisement might, having regard to the nature of the inmate's offence, place respondents in danger of harm from the inmate after release.

(2) It should be made clear to inmates at establishments where all or most correspondence is liable to censorship that Governors have discretion under paragraph 11 to withhold replies to such advertisements if an excessive number are received, without prior reference to the inmate.

Box numbers; anonymous letters

29. Inmates will not normally be allowed to write to a box number, but if the inmate does not know the person's private address, the governor may, if satisfied that security is not threatened, allow the letter to be sent. Similarly, inmates will not normally be allowed to receive anonymous letters, and the governor has discretion to withhold letters which do not show the sender's address.

Contents of correspondence

Choice of language

30. Inmates may write their letters in the language of their choice, but letters not written in English which are subject to censorship may be liable to delay.

Examination of correspondence for illicit enclosures

31. (1) Inmates may not without permission enclose any article or paper, other than a visiting order, with their correspondence. Inmates' correspondence are subject to the provisions of the Prisons Act 1952 on the importation of unauthorised articles into establishments. Accordingly, both incoming and

outgoing letters may be opened and examined for illicit enclosures. This procedure is to be distinguished from the reading of the correspondence.

(2) At open establishments, outgoing correspondence should only be examined if there is reason to suspect that it contains an illicit enclosure. At such establishments, correspondence may be handed in by inmates sealed.

(3) Special arrangements apply to the examination of incoming and outgoing correspondence marked 'SO 5B 32 (3)'. See paragraph 32 (3) below for details.

Reading of correspondence

32. (1) Subject to sub-paragraph (3) below and Order 5F7, at open establishments correspondence, both incoming and outgoing, should only be read where there is reason to believe:

(*a*) that the inmate concerned or his or her correspondence may attempt to infringe any of the restrictions on correspondence set out in this Order, or

(*b*) that the correspondence may contain information about criminal activities, or

(*c*) that reading may be in the inmate's own interest (for instance, if a severely depressed inmate is expecting to receive bad news which ought to be broken gently).

(2) Subject to sub-paragraph (3) below and Order 5F7, at other establishments all correspondence may be read. This is referred to in this Order as 'routine censorship'.

(3) Correspondence between an inmate and his or her legal adviser which relates only to legal proceedings to which the inmate is a party or to a forthcoming adjudication against the inmate carries special privileges under Prison Rule 37A (1) and Young Offender Institution Rule 14 (1). The enveloped carrying such correspondence should be marked 'SO 5B 32 (3)' and if outgoing may be handed in sealed by the inmate. Unless the Governor has reason to suppose that a letter purporting to be covered by this paragraph is not in fact covered, such a letter:

(*a*) may not be read;

(*b*) may not be stopped;

(*c*) may be opened for examination only in the presence of the inmate concerned (unless the inmate declines the opportunity).

Categories of correspondence

33. For the purpose of restrictions on content, inmates' correspondence, both incoming and outgoing, is divided into the following groups:

(1) General correspondence, that is to say correspondence with family, friends and other individuals and with organisations or public bodies not falling into any of the following categories;

(2) Correspondence with the Parliamentary Commissioner for Administration;

(3) Correspondence between an inmate and his or her legal adviser;

(4) An application to a court which constitutes the issue of proceedings. Such an application is not subject to the restrictions on correspondence and must be posed without delay.

Restrictions on general correspondence

34. General correspondence, as defined in paragraph 33 (1), may not contain the following:

    (1) Material which contains:

        (*a*) messages which are indecent or grossly offensive; or

        (*b*) a threat; or

        (*c*) information which is known or believed to be false;

        provided that it appears that the writers' intention is to cause distress or anxiety to the recipient or any other person.

    (2) Plans or material which would tend to assist or encourage the commission of any disciplinary offence or criminal offence (including attempts to defeat the ends of justice by suggesting the fabrication or suppression of evidence).

    (3) Escape plans, or material which if allowed would jeopardise the security of a prison establishment.

    (4) Material which would jeopardise national security.

    (5) Descriptions of the making or use of any weapon, explosive, poison or other destructive device.

    (6) Obscure or coded messages which are not readily intelligible or decipherable.

    (7) Material which is indecent and obscene under the Post Office Act 1963.

    (8) Material which would create a clear threat or present danger of violence or physical harm to any person, including incitement to racial hatred.

    (9) Material which is intended for publication or for use by radio or television (or which, if sent, would be likely to be published or broadcast) if it:

        (*a*) is for publication in return for payment (unless the inmate is unconvicted).

        (*b*) is likely to appear in a publication associated with a person or organisation to whom the inmate may not write as a result of the restriction on correspondence in paragraph 26 above.

        (*c*) is about the inmate's own crime or past offences or those of others, except where it consists of serious representations about conviction or sentence or forms part of serious comment about crime, the processes of justice or the penal system.

        (*d*) refers to individual inmates or members of staff in such a way that they might be identified.

        (*e*) contravenes any of the other restrictions on content applying to letters.

   (10) In the case of a convicted inmate, material constituting the conduct of a business activity: provided that for these purposes business activity does not include:

        (*a*) the conferment of a power of attorney or the making of other arrangements for a person outside prison to conduct a business activity on the inmate's behalf;

(*b*) the winding up of a business by an inmate as a result of his or her conviction;

(*c*) the sale, transfer or other disposal of the inmate's personal property, or the transfer of his or her personal funds;

(*d*) other personal financial transactions, provided that in a twelve month period such transactions do not exceed the current limit on the amount which may be spent from private cash within the establishment.

(11) In the case of an inmate against whom a deportation order is in force, material constituting or arranging any financial transaction unless the Governor is satisfied that there is a real need for such a transaction (eg. the support of near relatives or the seeking of advice to petition against deportation). This restriction does not apply to an inmate whose sentence includes a recommendation for deportation but in respect of whom a decision has not been made by the Secretary of State to act upon the recommendation.

(12) In the case of an inmate in respect of whom a receiving order has been made or who is an undischarged bankrupt, material constituting or arranging any financial transaction except:

(*a*) on the advice of the Official Receiver;

(*b*) to pay wholly or in part a fine or debt in order to secure the inmate's earlier release;

(*c*) to defend criminal proceedings brought against the inmate;

(*d*) to meet the cost of communicating with or instructing a solicitor to act on the inmate's behalf in bankruptcy proceedings;

(*e*) to meet the cost of the inmate's production in bankruptcy proceedings;

(*f*) to meet expenditure allowed under Prison Rules 17 (4), 20 (1), 25 and 41 (1).

Correspondence with legal advisers

35. Correspondence with a legal adviser, other than:

(*a*) correspondence about legal proceedings to which the inmate is already a party or about a forthcoming adjudication, on which see paragraph 32 (3) above, and

(*b*) correspondence about an application to the European Commission of Human Rights or proceedings resulting from it, on which see Order 5F 7

may be read and may not contain matter mentioned in paragraph 34 above.

Correspondence with the PCA

36. Correspondence with the Parliamentary Commissioner for Administration may only be stopped if it would be an offence to send it (e.g. if it contains matter which is indecent and obscene).

Evasion

37. An inmate may not ask, in writing or otherwise, another person to make on his or her behalf a communication which he or she would not be allowed to make direct, or which would contravene this Standing Order.

Copying and disclosure of correspondence

38. (1) Subject to sub-paragraph (3) below, no copy shall be taken of a letter which the Governor is precluded from reading under paragraph 32 (3) above.

(2) Subject to sub-paragraph (3) below, no copy shall be taken of any other letter to or from an inmate except

(i) where Headquarters needs to be consulted as to whether the letter should be stopped or withheld because it appears to be in contravention of this Order;

(ii) where an inmate is believed to have committed suicide and it is necessary to send Headquarters a copy of a note written by the deceased;

(iii) where, in connection with the censorship process, a letter written wholly or partly in a language other than English needs to be sent away for translation;

(iv) where only one copy of the formal notification of the outcome of an Appeal is received and it is necessary to take a copy to file in the inmate's record;

(v) where a letter contains material which suggests:

(a) that the inmate may be a suicide risk and the medical officer considers it necessary to retain a copy;

(b) that the inmate may be mentally disordered and a copy may be necessary to any psychiatric assessment which may be made; or

(c) that there is a medical history details of which the Medical Officer considers will materially contribute to the clinical management of the inmate's care.

(vi) where a letter contains material which seriously casts doubt upon an inmate's fitness for release on licence and, in particular, suggests that he or she would represent a risk if at large; and where the matter ought to be brought to the attention of

(a) the local Review Committee, or

(b) Headquarters to consider whether it should be laid before the Parole Board or Ministers;

(vii) where a letter contains information (for example, an admission by the inmate) which indicates that another person is innocent of an offence with which he or she has been charged or of which he or she has been convicted; or that an inmate proposes to admit to, or has admitted to, an offence and there is reason to believe he or she may be innocent;

(viii) where a letter contains information which may indicate that a person is, or has been involved in, the unlawful evasion or attempted evasion of immigration control, and it may be necessary to bring the matter to the attention of the Immigration authorities;

(ix) where a letter purportedly covered by paragraph 32 (3) above is found, on being read in accordance with that paragraph, to contain matter not relating to legal proceedings to which the inmate is a party or to an adjudication against him or her which is yet to be heard, and the matter needs to be referred to Headquarters;

(x) for the purpose of disclosure as set out in paragraph 39 below.

(3) Sub-paragraphs (1) and (2) do not apply

(a) to a letter which is addressed to the Home Office or which is referred to the Home Office by someone outside the Department (for example, a Member of Parliament); or

(b) to the copying of a letter at the request, or with the express consent, of an inmate.

39. (1) Subject to sub-paragraphs (4) and (5) below, no letter which the Governor is precluded from reading by paragraph 32 (3), no copy of such a letter, and no information about the contents of such a letter, may be disclosed to any person outside the Home Office.

(2) Subject to sub-paragraphs (4) and (5) below, no other letter to or from an inmate, no copy of such a letter, and no information about the contents of such a letter may be disclosed to any person outside the Home Office except:

(i) to a court;

(ii) to a coroner who is enquiring into the circumstances of a suicide or other death in custody;

(iii) to a person engaged to translate letters written in a foreign language;

(iv) where the letter forms part of the inmate's record, to another authority to whom the inmate, with the record, is transferred;

(v) in the case of an inmate who is, or may be, mentally disordered, to a medical practitioner charged with the care or assessment of the inmate, if the letter may be relevant to such care or assessment;

(vi) to the Local Review Committee in the circumstances set out in paragraph 38 (2) (vi) above;

(vii) to the Parole board in the circumstances set out in paragraph 38 (2) (vi) above;

(viii) to the police or some other relevant person or agency in the circumstances set out in paragraph 38 (2) (vii) above;

(ix) to another Government Department or to the Immigration Appellate authorities or to a person's legal representative in the circumstances set out in paragraph 38 (2) (viii) above;

(x) to an appropriate professional body or professionally qualified person where a letter from a legal adviser comes within the terms of paragraph 38 (2) (ix) above;

(xi) to the European Commission of Human Rights or the European Court of Human Rights where the letter is relevant to a matter which is at issue in an application to or proceeding before the Commission or Court; or

(xii) to the police or other relevant agency where information in the letter may (*a*) prevent an escape from an establishment;

(*b*) affect national security or public safety; or

(*c*) assist in the prevention or detection of crime or in the conviction of an offender, or in the recovery of the proceeds of crime or of articles used in the commission of crime.

(3) In the circumstances described in (2) (vii)–(xii) above, disclosure of a letter, or a copy of a letter, or information about the contents of a letter, may be made only by, or with the explicit authority of, Headquarters.

(4) Sub-paragraphs (1) and (2) do not apply to any letter addressed to the Home Office or which is referred to the Home Office by someone outside the Department (for example, a Member of Parliament).

(5) Notwithstanding sub-paragraphs (1) and (2), a letter or a copy of a letter, or information about the contents of a letter, may be communicated to the writer or intended recipient of the letter at the request or which the express consent of the writer, recipient or intended recipient of the letter.

Storage and disposal of stopped letters

40. (1) Letters to and from prohibited correspondents and letters containing prohibited material are liable to be stopped. When a letter is stopped the inmate should be informed and a note should be made of the decision in the inmate's record and the submitted letters book. Any letter stopped under paragraph 34 above should be placed on the inmate's F1150. With the exception of outgoing letters stopped under paragraphs 20–25 and 28 above, which are to be placed with the inmate's property and which may be used for reference for the purposes of a rewrite, outgoing stopped letters will not normally be returned to the inmate.

(2) Visiting orders enclosed with a stopped letter should normally be returned to the inmate.

ANNEX E TO CI 34/1990

REVISED STANDING ORDER 5c

SECTION C: REQUESTS/COMPLAINTS PROCEDURES

On applications to the European Commission of Human Rights, see Order 5F 4–6

Requests and complaints relating to a prisoner's imprisonment

Method of making a request or complaint

1. Inmates have the right to make requests or complaints to the Governor, the Board of visitors or to the Secretary of State. They may also raise their request or complaint outside the Prison Service or the Home Office with any individual or

organisation of their choice, subject to the provisions elsewhere in Standing Order 5.

2. Requests or complaints to the Governor or the Board of Visitors may be made either orally or in writing.

3. Prison Rule 8 (2) and YOI Rule 9 (2), as amended, require that on every day governors shall hear any oral requests or complaints made to them. The Secretary of State authorises governors, under Prison Rule 98, to delegate to any of their staff the task of hearing such complaints. Wing or landing applications shall be heard every day. Governor's applications shall be heard, at a minimum, on every day except Sundays and public holidays.

4. Requests or complaints to the Secretary of State must be made in writing. This enables inmates to communicate with headquarters about matters which local management cannot grant or to appeal against local decisions.

5. Inmates may make their requests or complaints under confidential cover to the Governor, the Chairman of the Board of Visitors or to the Area Manager.

6. Requests or complaints should normally be made within 3 months or the relevant facts coming to the prisoner's notice. However, there is discretion to consider requests or complaints made outside this period.

Inmates unable to write

7. Requests or complaints should normally be written by the inmate personally. The Governor should, however, allow an inmate who cannot write to dictate his or her request or complaint to an officer, who should make an appropriate endorsement on the form or letter and add his or her signature and rank below the inmate's signature or mark. Alternatively, an inmate should be allowed to dictate his or her request or complaint to another inmate, who should add his or her signature below the former's signature or mark.

Language

8. Inmates who are able to speak English will normally be required to write their requests or complaints in English, with the assistance of an officer or another inmate if necessary. An inmate who prefers to write in Welsh should be permitted to do so.

Replies

9. Every request or complaint shall be dealt with promptly and replied to without undue delay.

10. In the case of a written request or complaint to the Governor a reply shall be provided within 7 days of the form being received by the requests/complaints clerk. If a full reply cannot be given within 7 days an interim reply shall be provided. However, in the case of an allegation against a member of staff the relevant time limit shall be 14 days.

11. In the case of a written request or complaint to headquarters, including an appeal against a local decision, a reply shall be provided within 6 weeks of the receipt of the form in headquarters. If a full reply cannot be given within 6 weeks an interim reply shall be provided.

12. Written requests or complaints to the Governor or to headquarters shall normally receive a written reply, save that, in exceptional circumstances, the reply may be given orally. If a request is refused or a complaint rejected the reply shall give reasons.

Copies

13. Once the reply has been added, the request/complaint or appeal form shall be returned to the inmate. However, an inmate who wishes may be allowed a photocopy of his or her request or complaint prior to the reply being given, provided that the cost of copying is met from the inmate's prison earnings or private cash. An inmate who is unwilling or unable to pay for a photocopy should be advised to make a copy of the request or complaint in a notebook.

Records

14. The Governor shall keep a record of all oral requests or complaints and the action taken in each case.

15. Copies of a prisoner's written requests and complaints shall be kept in accordance with the instructions for maintaining prisoners' personal records.

Boards of Visitors

16. Every request to speak to the Board of Visitors shall be recorded by the officer to whom it is made and promptly passed on to the Chairman or another member of the Board of Visitors.

Petitions to the Queen

17. Every subject of The Queen has the constitutional right to petition Her Majesty. Others have no such right to do so, but their petitions are as a matter of grace treated in the same way as those of British subjects. Accordingly, all inmates should be allowed to petition The Queen if they wish and such petitions are not subject to the restrictions in Order 5B 34.

Petitions to Parliament

18. The right of petitioning Parliament for redress of grievances is acknowledged as a fundamental principle of the constitution and every inmate is entitled to petition Parliament.

Requests and complaints about conviction or sentence

19. (1) An inmate may make requests or complaints to the Secretary of State on matters connected with his or her conviction or sentence. The Secretary of State has power to intervene in a conviction or with regard to the sentence either by referring the whole of the case (if tried on indictment) to the Court of Appeal (section 17 of the Criminal Appeal Act 1968) or by recom-

mending to The Queen that she exercise the Royal Prerogative of Mercy to grant a Free Pardon or 'Special Remission' of the sentence.

(2) Although the Home Secretary has a duty to consider any representations about a conviction or sentence, he will normally consider intervening with regard to a conviction only where there has been a plea of not guilty; an appeal or application for leave to appeal against conviction; and where there has come to light some new evidence or new consideration of substance which was not before the courts and which appears to cast doubt on the rightness of the conviction.

(3) The Home Secretary will not normally consider recommending grant of a Free Pardon in any case tried on indictment, since he has power to refer it to the Court of Appeal. Nor will he normally consider doing so unless there has come to light some new evidence which shows that no offence was committed or that the person concerned is wholly innocent.

(4) The Home Secretary may recommend exercise of the Prerogative of Mercy to grant Special Remission of the sentence on the grounds of doubt about the conviction only in the most exceptional circumstances. The power is very rarely used in this way. The Secretary of State may also, in very exceptional circumstances, recommend grant of Special Remission for compassionate reasons, for example where an inmate is suffering from a terminal illness.

(5) It should be noted that the inmate is not required to submit a request or complaint in order to have a matter related to his or her conviction or sentence considered by the Secretary of State, to whom representation may be made direct by a third party.

(6) In order to raise a matter in respect of his or her conviction the inmate is not required to submit a request or complaint via a member of Parliament. Nor does he or she need to petition The Queen or Parliament, but nothing in this Order affects the inmate's right to do so.

Court martial prisoners

The Army Act 1955 and the Air Force Act 1955

20. (*a*) If the inmate wishes to petition after promulgation against the finding, otherwise than by means of an appeal petition, he or she must present a petition to a reviewing authority within 6 months of promulgation and in the appropriate form.

(*b*) If the inmate wishes to petition after promulgation against the sentence he or she must present a petition to a reviewing authority within 6 months of promulgation and in the appropriate form.

The Naval Discipline Act 1957

21. The finding and the sentence in respect of offences tried under the Naval Discipline Act 1957 are not subject to confirmation, but they may be reviewed by the Admiralty Board of the Defence Council at any time, and in the case of persons tried by court-martial must be so reviewed as soon as practicable, after the

Admiralty Board has received the record of the proceedings. Without prejudice to this procedure, however, an inmate sentenced under the Naval Discipline Act 1957 may present a petition to the Admiralty Board of the Defence Council at any time against the finding or the sentence or both.

Requests or complaints on other matters from court-martial prisoners

22. A court-martial prisoner also has the right to make requests or complaints to the Home Secretary about his or her case or about any other matter related to his or her imprisonment.

SECTION D. COMMUNICATIONS WITH MEMBERS OF THE UNITED KINGDOM AND EUROPEAN PARLIAMENTS

1. Inmates have the right to communicate with members of the United Kingdom and European Parliaments. It is a fundamental principle that contact between a Member of Parliament and those he or she has been elected to represent should be allowed freely, subject only to such conditions as are essential to the preservation of the security and good order of the establishment.

Members of the United Kingdom Parliament

2. The following paragraphs deal with correspondence with and visits to individual inmates by peers or Members of the House of Commons.

Letters

3. An inmate may write to any Member of Parliament, including the Home Secretary or a Home Office Minister. Inmates should be advised that a letter to a Member should show the inmate's home address if he/she has one, and should normally be addressed to the Member by name at the House of Commons or House of Lords, unless the Member has previously written to the inmate from another address.

4. Letters sent by inmates to, or received by inmates from, Members may be read in accordance with SO5B32; any matter which might prejudice the security, good order or discipline of the establishment should be brought to the attention of the governor, but the letter should not be stopped on these grounds.

5. A letter to a Member may only be stopped if it would be an offence to send it (e.g. if it contains matter which is indecent and obscene).

6. A letter to a Member should normally count against the inmate's allowance of letters, but an extra letter may be allowed if in the governor's view it is justified by the circumstances of the case. Postage will be met from prison earnings or private cash unless these are inadequate to meet the cost and the governor thinks the circumstances justify meeting the cost from public funds.

Visits

7. A Member of Parliament may visit a convicted or unconvicted inmate with the inmate's consent on production of a valid special visiting order issued by the governor; an order admits only the Member named thereon, during the visiting hours specified. The visit should take place within sight, but not within hearing, of prison staff, unless the Member or the inmate requests that it should be within hearing, or the governor so requires for reasons of security.

8. Visits by Members of Parliament to whom inmates have sent visiting orders should take place in the normal visits area and are not affected by paragraph 7 above.

9. A Member of Parliament may use a cassette or tape recorder when interviewing an inmate under paragraph 7 above.

Members of the European Parliament

10. Members of the European Parliament (MEPs) are directly elected and represent individual constituencies. This enables United Kingdom MEPs to represent their Euro constituents on issues involving Community matters and provides a basis for contact between them and the United Kingdom Government. MEPs have no standing in respect of exclusively domestic United Kingdom matters. MEPs will therefore not be afforded the special facilities available to Members of the United Kingdom Parliament described in paragraphs 2–9 above.

Letters

11. An inmate may write to and receive letters from an MEP. Such letters may be read and will be subject to the restrictions on content set out in Order 5B 34. A letter to a MEP should normally count against the inmate's allowance of letters; an extra letter may be allowed as in paragraph 6 above.

Visits

12. An inmate may send out visiting orders to a MEP. If a MEP asks to visit an inmate the inmate should be informed and offered an opportunity to send out a visiting order. The MEP should be informed of this. A visit with a MEP should be conducted and supervised in accordance with Order 5A 24–29.

SECTION E. COMMUNICATIONS WITH CONSULAR AND HIGH COMMISSION OFFICIALS

1. The Vienna Convention on Consular Relations, which has been supplemented by a number of bilateral agreements between the United Kingdom and other countries, guarantees freedom of communication between consular officers and their nationals. Consular officers have the right to correspond with and visit any of their nationals who may be in a penal establishment.

2. For the purpose of the following paragraphs an inmate's citizenship should be

accepted as that which he or she claims.

3. These paragraphs do not apply to an inmate who is known to be seeking asylum in this country or who is detained under the Immigration Act of 1971 and is representing on political, religious or ethnic grounds that he or she should not be deported or removed.

4. On first reception an inmate who claims citizenship of any foreign or Commonwealth country should be informed without delay of his or her right to communicate with the appropriate consular officer or High Commission. An extra letter at public expense should be allowed if the inmate wishes to notify the consular officer or High Commission of his or her reception into prison.

Letters and visits

5. Any inmate who is a citizen of any country other than the United Kingdom may at any time correspond with or be visited by the accredited representative of that country on any matter without having first to petition or otherwise seek permission. The letter should be posted or the visit arranged without delay.

6. A letter to a consular official may not be stopped unless it would be a criminal offence to post it (e.g. where the letter contains indecent or obscene matter, under the Post Office Act 1953).

7. The letters will normally count against the inmate's allowance of letters, but an extra letter may be allowed if in the governor's view it is justified by the circumstances of the case. Visits will not count against the inmate's allowance. Postage will be payable from prison earnings or private cash unless these are adequate to meet the cost and the Governor thinks the circumstances justify meeting the cost from public funds.

8. Visits should be in sight but out of the hearing of a prison officer, unless either party to a visit requests that it should be within hearing or the Governor so requires for reasons of security.

9. Consular and High Commission officials may use tape or cassette recorders when interviewing inmates.

Citizens of countries without consular representation and stateless persons

10. A citizen of a county without consular representation in the United Kingdom should be treated in the same way as a citizen of the country which looks after its interests here. If there is no such country or the inmate is stateless or a refugee, he or she should be given all reasonable facilities to communicate with any international authority charged with protecting the interests of such persons, such as the United Nations High Commissioner for Refugees.

SECTION F. THE EUROPEAN CONVENTION ON HUMAN RIGHTS

Introduction

1. The European Convention on Human Rights, which came into force on 3 September 1953, represents a collective guarantee at a European level of a number of principles set out in the Universal Declaration of Human Rights, which must be recognised and observed by contracting States. Since then, five protocols to the Convention have been concluded but only two of these (the First and Fourth) contain further substantive points and of these two only the First Protocol has been ratified by the United Kingdom. The United Kingdom is not bound by the Fourth Protocol. The Convention established 2 organs through which it is implemented. These are the European Commission of Human Rights and the European Court of Human Rights, which both have their seat in Strasbourg. The Committee of Ministers of the Council of Europe also has functions under the Convention including deciding whether there has been a breach of the Convention in cases which have not been brought before the European Court of Human Rights.

2. The Convention was drawn up within the Council of Europe. It is unconnected with the European Community (the Common Market) or the European Court of Justice.

3. Under the Convention a Contracting Party may allege violation of the Convention on the part of other Contracting Parties. There is also the procedure, accepted by the United Kingdom Government, under which individuals and non-government organisations may apply to the Commission alleging violations of the Convention by the United Kingdom. Thirteen other Contracting Parties have accepted this procedure in relation to complaints against their Governments. This is the so-called right of individual application. Representations to the Commission by inmates in this form, although sometimes referred to as petitions, should not be confused with petitions to the Secretary of State.

Applications

4. An inmate or his or her representative may submit an application to the European Commission of Human Rights without the need to petition the Secretary of State or otherwise seek permission, and may correspond with his or her legal adviser and with other persons in connection with the preparation of the application.

5. The cost of postage should be met by the inmate out of prison earnings or private cash. It should be paid from public funds only if the inmate has no money.

Censorship and contents of correspondence

6. Correspondence with the European Commission of Human Rights, or with the European Court of Human Rights, including applications to the former, may be read but should be stopped only if it contains:

(1) material which it is an offence to send through the post (e.g. material which is indecent and obscene (Post Office Act 1953));
(2) plans or material which would tend to assist or encourage the commission of any criminal offence (including attempts to defeat the ends of justice by suggesting the fabrication or suppression of evidence);
(3) escape plans, or material which if sent would jeopardise the security of a prison establishment;
(4) material which could jeopardise national security;
(5) obscure or coded messages which there is reason to suspect may contain any of the above.

7. An inmate may correspond with his or her legal adviser (who may be a barrister, a solicitor, or a clerk acting on behalf of a solicitor) in regard to an application or any proceedings resulting from it. Such correspondence should not be read unless the governor has reason to suspect that it contains other matter; if it is read it should only be stopped if found to contain matter described in paragraph 6 above.

8. An inmate should not be placed on report on account of anything contained in an application to the European Commission of Human Rights or in a letter to his or her legal adviser with a view to or in regard to an application to the Commission.

Visits by legal advisers

9. An inmate may consult a legal adviser within the sight of a prison officer but out of hearing in regard to an application or any proceedings resulting from it. A legal adviser may visit an inmate without first obtaining a visiting order and such visits should be in the sight, but not in the hearing, of a prison officer. An inmate's legal adviser may be allowed to use a cassette recorder when interviewing the inmate, provided that a written undertaking is given by the legal adviser that the cassette will be kept in his or her office and will be used solely in connection with the preparation of an application to the Commission or any resulting proceedings of the Commission or Court.

SECTION G. INMATES' USE OF TELEPHONES

1. At establishments where card-operated pay telephones have been installed for their use. inmates may make telephone calls during hours specified by the Governor. Where the facility is heavily used, the Governor has discretion to impose restrictions on the duration of calls allowed to each inmate.

2. Inmates are not permitted to make calls to or via the operator or receive incoming calls, or to communicate by telephone matters which they would not be allowed to include in correspondence under the terms of SO 5B 34.

3. If a member of the public requests in writing to the prison authorities that an inmate should not telephone him or her, the inmate should be ordered not to do so.

4. (1) Subject to sub-paragraph (2) below, if the person or authority having

parental responsibility for the care of a minor (i.e. a person under 18 years of age) requests in writing the stopping of telephone calls from an inmate to that minor, or if the person or authority normally having parental responsibility for the care of a minor who is in custody in a penal establishment requests the stopping of telephone calls from that minor to any other person, the inmate in question should be ordered not to telephone that person. In addition, where the Governor judges that telephone calls from a minor in his or her custody to any person would not be in the inmate's best interests, the Governor may order the inmate not to telephone that person. In exercising this discretion the Governor will take account of any views which the inmate's parent or guardian may express.

(2) The Governor should consult Headquarters (P3 Division) before ordering any inmate not to telephone a close relative, as defined in SO 5A 30, on the grounds set out in this paragraph.

5. The Governor should order an inmate not to telephone a person or organisation if there is reason to believe that the person or organisation is planning or engaged in activities which present a genuine and serious threat to the security or good order of the establishment or other Prison Department establishments. The Governor should consult Headquarters (P3 Division) before ordering an inmate not to telephone a close relative (as defined in SO 5A 30) under the provisions of this paragraph.

6. Inmates are not allowed to telephone a victim of their offences unless:

(1) he or she is a close relative, as defined in SO 5A 30; or

(2) the victim has first approached the inmate; or

(3) the Governor is satisfied that the call would not cause undue distress to the victim.

7. If an officer hears an inmate making a call which appears to contravene paragraphs 2–6 above the call may be terminated and the matter reported to the Governor, who has discretion to prohibit the inmate from using the telephone for a period, take appropriate disciplinary action, or, in the case of widespread abuse, to suspend the facility altogether for a period.

8. Inmates may consult an STD codebook but may not consult telephone directories. An inmate who wishes to obtain a number must apply to the Governor, explaining why he or she wishes to call the person in question.

9. Inmates may purchase Phone Cards from the canteen from their earnings, or, subject to any limit imposed by the Secretary of State, from private cash. Refunds will not be made under any circumstances for partially or wholly unused Phone Cards.

10. Inmates must not have more than two Phone Cards with units remaining on them at one time, unless this has been authorised by the Governor. They must not give Phone Cards to or receive them from other inmates. Phone Cards may not be sent in from outside.

Appendix 4
Circular Instruction 26/1990

To all Prison Service establishments

REMOVAL FROM ASSOCIATION UNDER PRISON RULE 43 AND YOUNG OFFENDER INSTITUTION RULE 46 AND ARRANGEMENTS FOR THE MANAGEMENT OF VULNERABLE PRISONERS

Introduction

This Instruction supersedes Circular Instructions 15/1974 and 5/1983. It contains consolidated guidance on the operation of arrangements for the removal of inmates from association under Prison Rule 43 and Young Offender Institution Rule 46 and implements recommendations of the Prison Department Working Group on the Management of Vulnerable Prisoners, whose report was circulated to Governors and Chairmen of Boards of Visitors on 10 August 1989. The Home Secretary announced on 31 October that the Working Group's recommendations had been accepted with a procedural modification in respect of one.

2. Circular Instruction 62/1989 announced that Orders amending Prison Rule 43 and YOI Rule 46 had been made and explained the changes which were to come into force on 18 December 1989. These were the increase from 24 hours to three days of the maximum period for which a Governor may authorise removal from association and the reduction from one month to 14 days of the maximum period for which the Board of Visitors may authorised the continued removal from association of unsentenced prisoners under 21 years of age. The Rules have not been amended to allow directions on their interpretation to be given by the Secretary of State, as recommended by the Working Group: this aim is achieved by guidance in this Circular Instruction (see paragraph 10 below and Annex B). The texts of the revised Prison Rule 43 and YOI 46 are set out in Annex A[1].

3. In order to avoid repetition of guidance set out in the Working Group's Report and to encourage familiarity with the issues considered, a copy of the report accompanies, and should be kept with, each copy of this Circular Instruction, which refers, as appropriate, to the text of the report.

4. In the rest of this Circular Instruction references to Rule 43 should be interpreted as applying to both Prison Rule 43 and YOI Rule 46, except where otherwise indicated. References to 'the report' are to the report of the Working Group on the Management of Vulnerable Prisoners.

---

[1] It should be noted that Rule 43 does not have parts A and B. The terms Rule 43A and 43B should accordingly not be used.

Policy

5. The purpose of Rule 43 is to provide statutory power for a Governor either to limit or to withdraw the opportunity for particular inmates to have contact with other inmates, for the maintenance of good order or discipline or in their own interests, in circumstances where such action is not justified under Rule 45 (temporary confinement of a refractory or violent prisoner), Rule 48 (2) (initial segregation pending Governor's adjudication) or Rules 50 or 51 (as a punishment for a disciplinary offence); and to provide related safeguards.

6. Segregation under Rule 43 is designed to assist Governors to prevent trouble and Governors may use it for this purpose in respect of known subversive inmates or inmates known to be at risk of assault either on initial reception or at any subsequent stage of their time in custody. It is not necessary to wait until an inmate has actually jeopardised control. It is sufficient that the inmate should have indicated the intention of doing so; and it is right to take into account a history of disruptive behaviour, either inside or outside the institutional setting. Similarly it is not necessary that an inmate should already have been assaulted or should have requested Rule 43 protection. It is sufficient for the Governor to be satisfied that the inmate is seriously at risk and that his or her safety cannot reasonably be assured by other means.

7. The power must not be used without good and sufficient reason and its use should be for the shortest possible period consistent with the need to maintain the good order or discipline of the establishment or to protect by that means a particular inmate. The aim should be to seek to manage the situation by means other than Rule 43 segregation wherever possible, but, where recourse to segregation is necessary, to return the inmate to full association with other inmates as soon as practicable.

8. Historically young offender establishments have not made frequent use of what is now YOI Rule 46. They have to a large extent succeeded in dealing with problems by providing staff support and jobs which ensure close supervision. such practices are, of course, to be encouraged, where possible, for all inmates, not only young offenders.

9. The report (paragraph 7.5) stressed the need for revision of the guidance in CI 32/1979 on the personal liability of officers for the protection of inmates from injury. Revised guidance is to be issued in CI 30/1990. This will make it clear that officers taking management decisions in good faith within the authority delegated to them can be assured of the support, assistance and financial backing of the Department in the event of a civil law action claiming damages for injuries sustained by an inmate.

Removal from association

10. Rule 43 also provides controls over managerial action. These are necessary because removing inmates from association with other inmates not only deprives them of the normal social contact which they could expect to have with other inmates in the course of daily life in custody, but may also result in their being unable to participate in some activities generally available to other inmates in the same establishment. Guidance on the factors which constitute removal from association, and on the circumstances in which Rule 43 controls may be dispensed with, is contained in Annex B.

Authorisation of removal from association

11. Under the revised Rule 43 the Governor may, on his or her own authority, remove an inmate from association for up to three days (ie 72 hours). If the Governor considers that the segregation needs to continue longer than three days, the authority of a member of the Board of Visitors (or, exceptionally, of the Secretary of State) must be sought, who may authorise the continuation of segregation for a further period of up to one month (ie. one calendar month) or, in the case of an inmate aged under 21, for a further period of up to 14 days. When prolonged removal from association is considered justified, the Board member may authorise continued segregation for further periods of up to one month or, in the case of an inmate aged under 21, up to 14 days.

12. The Governor should exercise his or her discretion to arrange for an inmate to resume normal association as soon as segregation is no longer considered necessary and is obliged to do so if the Medical Officer so advised on medical grounds. The Medical Officer should ensure that he or she is fully aware of the conditions in which inmates on Rule 43 are held and should keep a careful eye on the effect of segregation on individual inmates. To this end the Medical Officer should, wherever possible, see each inmate segregated under Rule 43 every day. If this is not practicable, the aim should be to see each such inmate at least once a week.

13. The Governor's powers under Rule 43 should, wherever possible, be exercised by the Governor or Head of Custody. These powers should not be delegated to an officer below the rank of Governor 5.

14. The procedure for authorisation and documentation of removal from association is set out in Annex c. This deals in particular with completion of F1299A and with the maintenance of a history sheet for each segregated prisoner.

Avoiding unnecessary recourse to segregation as a protective measure

15. Because of the stigma attached to Rule 43 it can be difficult to return inmates to normal location after a period of protective segregation. It is therefore in the interests of both the inmate and the Service to find, wherever possible, other ways of dealing with the problems which lead to requests for segregation. Section 8 of the report is concerned with the problem of outside influences which can lead to inmates requesting Rule 43 segregation on arrival in prison—often unnecessarily and before they understand what it involves. Section 9 refers to the effects of media publicity, of the identification of vulnerable inmates by other inmates at court, through unguarded remarks before and at the time of their reception into prison and through lack of privacy during reception. Officers involved in the escorting of inmates to and from court and in dealing with their reception into prison should be alive to these problems and take all possible precautions to prevent potential vulnerable inmates being identified to other inmates. As recommended by the Working Group, letters are being sent to the organisations concerned with the professional standards of barristers, solicitors, probation officers, court staff and police officers with a view to influencing the advice they may offer and the attitude they may adopt in dealing with potential vulnerable inmates.

16. While in some cases use of Rule 43 segregation will be inevitable, Governors and their staff should take all reasonable steps to avoid its use if other alternatives,

eg. change of cell, landing, wing, establishment, are available. Section 10 of the report provides guidelines on the handling of potential vulnerable inmates with a view to avoiding recourse to segregation wherever possible. These concern the procedure followed on reception, the need for a critical and questioning approach by the Governor in considering the justification for segregation and the role of prison officers in dealing with the anxieties of potential vulnerable inmates and helping them to cope on normal location. The measures referred to in Section 10 include transfer to another establishment. Such a transfer may be indicated in circumstances where the threat is presumptive, eg. when an inmate is expected to be a prosecution witness against others in the same (or another) prison. In these cases advance warning is usually given to the Governor or via Headquarters to the police. Action will be taken in the light of the police advice. The police do not recommend what the alternative establishment should be.

Getting inmates back to normal location

17. It should rarely be necessary for an inmate who has been removed from association for the maintenance of good order or discipline to remain segregated for a prolonged period. Such inmates can normally return to normal location after a short time (eg. once a potentially dangerous situation has been defused or following a delayed adjudication for which an inmate has needed to be kept in segregation). Transfer to another establishment of an inmate who has threatened another inmate should be necessary only in the most exceptional circumstances and should be sought only in those cases in which the Governor is satisfied that no other measures which it would be reasonable to take are likely to protect the threatened inmate effectively. Arrangements for the transfer of inmates in the interests of good order or discipline will be dealt with in separate Circular Instructions to be issued shortly.

18. Where the segregation is for the inmate's protection, the aim should be to bring it to an end as quickly as possible, so that the inmate is not deprived of normal social contact and access to normal regime activities for longer than is absolutely necessary. There will be a relatively small number of adult male inmates with substantial terms left to serve for whom all attempts at return to normal location have failed and who seem destined to remain segregated until their discharge. To meet the needs of such inmates Vulnerable Prisoner Units have been set up (see section 12 of the report and paragraphs 27–28 below). For the majority of inmates segregated for protection, however, a return to normal location should be achievable and Section 11 of the report provides guidance on management measures which may be deployed to this end.

Regimes

19. The principle which should guide Governors and their staffs in ameliorating the worst effects of segregation under Rule 43 is that the regime for segregated inmates should be as balanced and well integrated as for the rest of the population within the limits of affording them protection and maintaining good order or discipline, and nothing less than that.

20. Inmates segregated for maintenance of good order or discipline should not deliberately be subjected to a restricted regime by being denied facilities or privileges

available to others: their segregation is for management purposes, not as a punishment. Given that they are segregated in order to prevent trouble by removing their disruptive influence, it will not normally be appropriate for such inmates to share living accommodation. Some restriction of regime activities will normally be inevitable because of the inmates' separate location and limitations of accommodation, manpower and other resources. Such restrictions, however, should not go beyond what is necessary for these reasons. (It should be noted, in particular, that the former guidance in paragraph 3*a* of CI 5/1983, providing for a deliberately restrictive regime for highly disruptive young offenders, no longer applies.)

21. Similarly inmates segregated for their own protection should not be subject to any regime restriction which is not an unavoidable consequence of their separation from the rest of the population. Since inmates do not normally pose a threat to one another, there is no intrinsic objection to their sharing living accommodation, though care should obviously be taken to avoid unsuitable matches. Every effort should be made to enable them to associate together as a group and to provide them with normal regime activities so far as the need to protect them allows.

22. Even where accommodation is limited, much can be done to provide a decent standard of life for inmates segregated for protection, given the management will and some dedicated effort. Innovative approaches can ensure the provision of positive regimes without necessarily using more resources. Education, library, protection and PSIF resources can be extended or adapted to cater for them, though some activities might need to be re-designed or adjusted for the circumstances of the segregated inmates. Section 14 of the report contains noes on various regime activities which may be of assistance in the development of more positive regimes for inmates segregated under Rule 43 for their own protection. Much of this is relevant also to efforts to improve arrangements for those segregated for the maintenance of good order or discipline.

Counselling and help with offence-related behaviour

23. In addition to the need to seek to provide a range of normal regime activities among groups of segregated inmates, there are likely to be identifiable general or specific needs for counselling or assistance on a group or individual basis. Paragraphs 14.24–14.27 of the report refer to the types of programmes which have been developed from local initiatives at various establishments, involving staff of all disciplines and, in some cases, outside specialists. Governors should encourage and foster the development of appropriate activities wherever possible as in integral part of the overall regime. Guidelines on the general approach to the counselling of vulnerable inmates are being prepared and will be circulated as soon as possible.

24. Considerable knowledge and experience in this field is being built up at individual establishments and it is appropriate that this should be shared and made available to others who are engaged in similar work or who see scope for embarking upon programmes on similar lines at other establishments. P5 Division[2] will act as a repository for information about counselling programmes for vulnerable inmates, both on and off Rule 43, so that an establishment seeking information about a particular type of activity can receive up-to-date details of work at other

[2] From 25 September 1990, the Directorate of Inmate Programmes

establishments and be put in touch with colleagues who have acquired relevant knowledge and experience. In order for this reference service to operate effectively, however, it is important that establishments regularly supply P5 Division[2] with as much information as possible about relevant activities which they have developed. Notification should be as indicated in paragraph 32.

25. Research is planned with a view to assessing the validity and effectiveness of such counselling activities. Meanwhile Governors are encouraged to seek advice from DPMS or elsewhere, as appropriate, in setting up new counselling arrangements.

Allocation to training prisons of inmates segregated for protection

26. There are many sentenced prisoners on Rule 43 for their own protection who should eventually be able to return to normal location—eg. after publicity has died down or as a result of careful counselling—but who remain in segregated conditions at local prisons for substantial periods, long after they would have moved on to training prisons if they had not been on Rule 43. It is accepted that Rule 43 status should not be an obstacle to the normal process of allocating prisoners to training prisons, where they can benefit from better conditions. Future planning will therefore allow for certain training prisons to cater for Rule 43 prisoners as a segregated group, so that they may be allocated in the normal way from local prisons. This will involve some re-organisation of facilities at the training prisons concerned and may have manpower implications. To some extent, therefore, full implementation of this policy may need to await the availability of planned additional accommodation or staff resources. It is hoped that training prisons with very good facilities will be able to manage such prisoners without requiring Rule 43 controls. The aim, however, should be to continue efforts to move the prisoners out of segregated conditions to normal location as soon as this can safely be achieved.

Vulnerable Prisoner Units

27. There will continue to be a need for Vulnerable Prisoner Units (VPUs), as a last resort, for the separate accommodation of adult male prisoners who have failed all attempts to return them to normal location, who have a substantial term left to serve and who would otherwise remain in deprived segregated conditions until their discharge. Waiting lists for the national VPUs are at present maintained by P5 Division, who operate the criteria set out in Appendix I of the report in considering possible candidates for allocation to them. If a Governor believes that an inmate satisfies the criteria for transfer to a national VPU, a recommendation for transfer should be submitted promptly to the Regional Office or P2 Division, as appropriate, with full supporting details of the inmate's history and of the efforts made to return him to normal location.

28. With effect from 25 September 1990 recommendations for allocation to the national VPUs should be submitted to the Tactical Management and Planning Unit (TMPU) or, for prisoners serving life sentences, to the Directorate of Custody (Life Sentence Review Section). If a recommendation in respect of a life sentence prisoner is supported, it will be passed to the TMPU. The TMPU will be responsible for allocation of VPU places and also for allocations to the VPUs, Rule 43 Units and other special units at present operated by Regional Offices.

Monthly returns and monitoring

29. Individual cases of Rule 43 segregation and the operation of the arrangements at individual establishments are at present monitored by Regional Offices by scrutiny of F1299As (individual authorisations) and F1299s (monthly summary returns). From 25 September 1990, Area Managers will be notified of individual cases of segregation by sight of F1299As. The TMPU will monitor cases and the situation at each establishment by scrutiny of F1299s and report to Area Managers as appropriate. Guidance on the completion of F1299 and on monitoring arrangements is contained in Annex D.

Action to be taken at establishments

30. The guidance in this Circular Instruction and the Working Group report is concerned with the segregation of inmates both for maintenance of good order or discipline and for their own protection. It lays particular emphasis on the positive management of inmates identified as, or claiming to be, vulnerable to assault or intimidation by other inmates. The action to be taken will vary from one establishment to another, and, where significant changes are required, these will need to be introduced in a planned way and, where necessary, through Governors' annual contracts under Circular Instruction 55/1984. Where appropriate related management tasks and performance indicators should be developed for inclusion in Governors' contracts.

31. The Governor in charge of each establishment should assign to an officer of Governor 5 grade or above, who reports directly to him/her, responsibility for overseeing the operation and development of arrangements for the management of Rule 43 inmates (including groups segregated without Rule 43 controls where the aim is to achieve the inmates' return to normal location).

32. The officer given this responsibility should review current arrangements and take such steps as are necessary to ensure that the guidance in this Circular Instruction and the Working Group report is implemented as soon a possible. He or she should be responsible for maintaining momentum in efforts to avoid recourse to Rule 43, to remove inmates from Rule 43, to improve regimes and counselling facilities for segregated inmates and for encouraging staff to extend their knowledge and skills in dealing with Rule 43 inmates. A particular concern should be to seek to ensure that staff continuity is maintained in the management of these inmates and that local staff training includes appropriate emphasis on the contributions to be made by officers up to Principal Officer level. That officer should also see that information about any existing programmes for the counselling of Rule 43 and vulnerable inmates is sent to Mr A King, P5 Division, Cleland House, by 31 August 1990, and that information about any future initiatives, developments or modifications is similarly notified (to the Directorate of Inmate Programmes after 25 September 1990). The job description of the officer designated for this functional responsibility should adequately reflect these aspects of the management task.

Headquarters responsibilities

33. For the present P5 Division will continue to be responsible for general policy on the operation of Prison Rule 43, on arrangements for the management of vulnerable

inmates, and for the allocation of inmates to the national Vulnerable Prisoner Units. The Division will also act as a repository and reference point for information about counselling programmes for vulnerable inmates on and off Rule 43. P4 Division will similarly continue to be responsible for general policy on the operation of YOI Rule 46 and associated matters relating to young offenders.

34. From 25 September 1990 responsibility for policy and practice on the use of Prison Rule 43 and YOI Rule 46 for maintenance of good order or discipline will be exercised by the Directorate of Custody; the allocation of inmates to VPUs will be the responsibility of the Tactical Management and Planning Unit; and policy and practice on the segregation of inmates in their own interests, including the use of Prison Rule 43 and YOI Rule 46 for the protection of inmates and the collection of information on counselling programmes, will be the responsibility of the Directorate of Inmate Programmes.

Guidance for the Board of Visitors

35. Separate guidance on matters dealt with in this Circular Instruction is to be issued to Chairmen of Boards of Visitors and a copy will be sent to each Governor.

Standing Order amendment

36. In the second sentence of paragraph 12 of Standing Order 3D, '24 hours' should be amended to read 'three days'.

PDG/89 157/22/9
19 July 1990

Index under: Forms 1299 and 1299A
            Prison Rule 43
            Removal from association
            Segregation
            Vulnerable inmates
            Young Offender Institution Rule 46

ANNEX B

REMOVAL FROM ASSOCIATION

Factors constituting removal from association

The facts which normally constitute removal from association for the purposes of Rule 43 are detention in separate accommodation without an acceptable degree of social contact with other inmates and deprivation of the opportunity to participate in recreation (including physical education), entertainment, education classes or work (where generally available). Where work is not available generally at a particular establishment, it is not a relevant factor in considering the degree of deprivation arising from segregation. Exercise and religious services are not relevant because all inmates have rights to receive exercise and to participate in religious services and, subject to a few special exceptions, appropriate arrangements must be made for the exercise of these rights by all inmates. A requirement to take meals in cells is not regarded as a serious deprivation justifying Rule 43 controls.

Location

2. Inmates removed from association for maintenance of good order or discipline (including segregation pending adjudication—see CI 25/1989 and Standing Order 3D) are not segregated as a punishment and should not be located in punishment cells, unless there is no alternative. They should be allowed to keep their bed and bedding in the cell, unless there is reason to suppose they will damage or destroy them; and they should not be deprived of any privileges to which they are normally entitled, except those that are incompatible with segregation.

3. Inmates removed from association in their own interests (ie. for their own protection) should not be located in punishment cells nor in an area used to accommodate those segregated for maintenance of good order or discipline, unless there is no alternative, in which case they should be kept apart from the 'good order or discipline' inmates. They should, wherever possible, be located in normal accommodation set aside for the purpose, where they can retain their personal possessions.

4. The Rule allows for circumstances where an inmate may need to be removed from association for particular purposes only. Where the Rule is used to restrict an inmate's activities only partially (eg. stopping participation in physical education classes), because he or she is considered to be at risk only in a particular situation, it may be feasible for the inmate to continue to occupy normal accommodation.

5. If an inmate who is removed from association under Rule 43 is admitted to the establishment's hospital, formal consideration should be given to the continued application of the Rule. Rule 43 controls should be removed if the inmate is not to be held in conditions of segregation (ie. if he is to be located on a ward, or allowed to associate with inmates who are not segregated), or if the Medical Officer so directs.

'Own Protection' inmates held as a group

6. Where a group of inmates, for their own protection, are kept apart from the rest of the population in a separate wing, on a separate landing or in some other location providing such separation, it does not necessarily follow that Rule 43 controls must apply: it depends upon the degree of deprivation which their separation entails. Inmates should not need to be subject to Rule 43 controls if they have normal social contact with others in the group and are able to participate within the group in recreation (including physical education), entertainment, education and work to the extent that the rest of the population at that establishment can participate in such group activities. If they need to be escorted as a group to make use of the general facilities of the establishment, eg. to participate in religious services or film shows, where they are kept apart from other inmates, or to the exercise yard or gymnasium for separate exercise of PE sessions, this is not regarded as deprivation requiring Rule 43 controls.

7. If the Governor of an establishment considers that a group of inmates who have been segregated from the rest of the population under Rule 43 for their own

protection are not deprived of any activities or facilities available to other inmates of the same classification (eg. sentenced, unconvicted, adult, under 21) in that establishment, it is open to him or her to propose to the Board of Visitors that Rule 43 controls should be dispensed with. It is for the Board of Visitors, in the light of local conditions and in consultation with the Governor, to judge whether the size of the group is sufficient to provide an acceptable degree of social contact for the inmates in question. The sharing of a cell, however, by two or three inmates— which is permissible, subject to careful selection of those sharing—is not in itself adequate to satisfy this requirement. The Board will also need to be satisfied that the inmates in the group have the opportunity to take advantage of all the activities and facilities which are available to other inmates of the same classification in the establishment. The decision to dispense with Rule 43 controls in such a situation is for the Board of Visitors and, if they so decide, they should review the decision at least annually. This does not, however, affect the discretion of Headquarters management to designate a particular unit as a vulnerable prisoner unit which is not subject to Rule 43 controls.

ANNEX C

PROCEDURE FOR AUTHORISATION AND DOCUMENTATION OF REMOVAL FROM ASSOCIATION

1. The Governor's authority for removal of an inmate from association should be given on F1299A, which has been revised as indicated in the Appendix to this Annex.

2. It is particularly important in 'own protection' cases that the section of the form headed 'Detailed description of the circumstances' should explain fully the circumstances which have led the Governor to decide that it is not in the inmate's best interests for him or her to associate fully with other inmates. It is not sufficient to write simply 'Nature of offence'. The entry should make specific reference to any assault or intimidation which the inmate has suffered or any other grounds for believing that he or she will be at serious risk on normal location, eg. knowledge that the antagonistic feelings of other inmates have been aroused by local or national publicity of the case. Similarly, in 'good order or discipline' cases, including cases of segregation pending completion of adjudication proceedings, the entry should make it clear why the Governor considers that the inmate needs to be kept apart from other inmates, ie. precisely how the good order or discipline of the establishment would be threatened by continued association with other inmates.

3. If the Governor expects to return the inmate to normal location within three days, there is no need to inform the Board of Visitors of the segregation. If, however, the Governor considers at the outset that the inmate will need to remain segregated for more than three days, or comes to that conclusion on reviewing the case on or before the third day, a Board member should be informed of the circumstances and of the Governor's views and the Board's authority sought for the inmate's segregation beyond three days.

4. The Governor should suggest a specific period of further segregation in the light of the circumstances of the case and should not necessarily seek authority for continued segregation for the maximum period of one month (or 14 days) which the Board member has authority to approve. This is particularly important in 'good order or discipline' cases where authorisation for the maximum period should rarely be necessary. If the circumstances necessitate the Board member's authority being sought by telephone, the member should ensure that it is confirmed in writing on F1299A within 4 days (96 hours) of the Governor's initial authorisation.

5. The Board member will decide, in the light of the Governor's advice, for how long a period to authorise continued segregation and will not automatically give authority for the maximum of one month (or 14 days). The member may decide to give authority for only a few days initially, pending the opportunity to interview the inmate, or for any period less than the maximum which may seem sufficient in the circumstances of the case. Boards of Visitors have been asked to bear in mind that the removal of inmates from association should be for the minimum period necessary in each case and that automatic authorisation or renewal of authorisation for the maximum period should be avoided.

6. F1299A should normally be completed in duplicate, one copy being sent to the Regional Office[3] and the other filed in the inmate's record (F1150 or, for IPRS[4] records, next to the relevant F2050A card). Whenever a Category A inmate is segregated under Rule 43, a copy of the F1299A should be sent also to the Directorate of Custody (Category A Section). Where the inmate is returned to normal location or otherwise taken off Rule 43 within three days, the F1299A should be suitably endorsed before submission and filing.

7. The initial removal of an inmate from association should be recorded also on page 24 of the F1150 or IPRS card F2050A, as should the Board member's authorisation of continued segregation. Any subsequent renewal of that authorisation should be recorded similarly and the entry should be countersigned by the authorising Board member. The ending of the inmate's segregation should also be recorded on page 24/F2050A.

8. In the rare event that authorisation of continued segregation is given by Headquarters on behalf of the Secretary of State, rather than by the Board of Visitors, the name and location of the authorising officer and the period of segregation authorised should be recorded at the bottom of F1299A. Where renewed authorisation is given by Headquarters, similar details should be recorded on page 24/F2050A and the entry countersigned by the Governor.

9. The removal of an inmate from association is managerial measure taken on a decision by the Governor. Even when it is for an inmate's protection, it does not require a request from the inmate. If, however, an inmate does request segregation for his own protection, the inmate should be encouraged to explain the reasons fully and to put the request in writing. The decision to return an inmate to normal location is similarly one for the Governor and does not require the inmate's agreement. An inmate should not be required to sign any form of consent when being returned to normal location.

---

[3] From 25 September 1990, Area Manager
[4] Inmate Personal Record System (CI 22/1990 refers).

10. Inmates have no entitlement to be given reasons for their segregation under Rule 43. Whilst discretion rests with Governors and Boards of Visitors, however, the giving of reasons is encouraged as a sensible practice. It is, of course, recognised that considerations of security or good order or discipline may on occasion prevent complete openness. Similarly, though there is no obligation to give reasons in writing, it is considered sensible practice to do so where an inmate so requests.

Addendum to Circular Instruction 26/1990

To all Prison Service establishments
(with an additional copy for the Chairman of the Board of Visitors)

REMOVAL FROM ASSOCIATION UNDER PRISON RULE 43 AND YOUNG OFFENDER INSTITUTION RULE 46 IN THE INTERESTS OF GOOD ORDER OR DISCIPLINE

1. The report of the Woolf Inquiry dealt with the management of disruptive offenders in paragraphs 12.221–305. Paragraphs 12.264–271 were concerned specifically with segregation in the interests of good order and discipline. Paragraphs 12.292–296 set out the Inquiry's general conclusions on segregation and certain other options for the management of disruptive inmates.

2. The report concluded that segregation and the other options all have shortcomings. It commented that segregation 'involves a restriction on association and living conditions which should not be other than very short term'. The exercise of this option 'should therefore be as restricted as possible, and, we propose, more restricted than at present'. Nevertheless, the report concluded, the Prison Service is right to have in place options such as segregation.

The use of segregation

3. The activities of violent, disruptive or subversive inmates endanger control within establishments, and impede or undermine management's efforts to provide positive regimes for the benefit of the inmate population generally. The problems which such inmates pose must not be under-estimated. The activities of a mere handful of inmates can generate great difficulties. Nor must local management be fettered in its ability to take prompt and positive steps to deal with difficult situations and thereby prevent escalation. Moreover, as the report acknowledged, segregation within the same establishment may well be a better option—for management and the inmate alike—than transfer to another establishment.

4. The Department welcomes the report's endorsement of the necessity for the option of removing an inmate from normal association with the general inmate population where this is required in the interests of good order or discipline. But the thrust of the report's conclusion that segregation should not be used unless it is absolutely necessary is also fully accepted. It must not be regarded as the instant remedy for problems of a control nature. As CI 26/1990 already makes clear, the aim must be to seek to manage a situation by means other than segregation

wherever possible. And a decision to resort to the segregation of an inmate must be reasoned and carefully recorded.

5. It is particularly important in the case of inmates who are persistently disruptive that efforts should be made, through establishing good staff/inmate relationships, to seek by means such as counselling, persuasion and reasoning with the inmate to change the inmate's attitude and behaviour. The problems of this approach are fully recognised. The process may be slow and frustrating and in some cases will prove unsuccessful. But staff must be encouraged to make the effort. Notable successes have been achieved in this way through the sheer professionalism and dedication of staff.

6. Nor should an inmate be retained in conditions of segregation for longer than is absolutely necessary. As will be readily appreciated, the fact that authority for segregation has been given for a period of time does not mean that an inmate should continue to be segregated throughout that period. The case must be kept under continuous review and, as soon as segregation is no longer necessary in the interests of good order or discipline, the inmate must be returned to normal location.

7. Inmates who are under segregation are more likely to be at risk of suicide. In 1990, for example, 5 inmates (or 10% of the total figure) killed themselves while in segregation units. It is important, therefore, that feelings of isolation are eased as much as possible by regular face-to-face contact and out-of-cell activity. Wherever possible the staff in the unit should have received suicide awareness training, and a high state of alertness should be maintained. When talking to inmates, staff should try to encourage them to be open about how they are feeling, and should be supportive of those who are depressed. Inmates should be made aware of other sources of help and counselling which they may request, such as the Chaplaincy, Probation and local Samaritans. Where there is serious concern about an inmate's mood, Form 1997 should be completed and the medical officer should visit the inmate so that a fuller assessment can be made.

8. Inmates who are under segregation for reasons of good order or discipline should be visited at least daily by the Governor or by another governor grade specifically deputed to carry out the task. The Governor or other governor grade concerned should not only see, but also speak with, the inmate. This will be helpful in assessing the inmate's demeanour, and thus the need for the inmate's continued segregation (which should be specifically addressed on each occasion); may help in establishing a rapport with the inmate; and will be supportive of staff in the segregation unit.

9. The Managing Medical Officer or a doctor deputed by him must be informed as soon as possible when an inmate has been placed on segregation; and must visit the inmate as soon as possible after removal from association and thereafter as frequently as possible and as is thought appropriate in the circumstances of the individual case, and in no case less than once in every three days (which must be regarded as the bare minimum requirement: a regime of daily visits should be regarded as the norm). Governors are asked to bring this provision to the attention of the Managing Medical Officer.

10. A member of the Board of Visitors whose authority is being sought for continued segregation should be encouraged to see and speak with the inmate, if neces-

sary at some length, before deciding whether to give such authority. In addition, Board members should be invited to see and speak with any inmates under segregation in the course of regular rota visits, and other visits to the establishment, even when cases are not due for formal review. Board members will use the day book provided in the segregation unit to record interview details, to confirm that the inmate has been given the opportunity to complain, the nature of any complaint raised, and any action being taken to deal with it.

11. If an inmate complains about the actions of a Board of Visitors member, the Board will wish to give a written reply in accordance with their standard procedure for handling inmates' requests and complaints. This is likely to include also giving a reply orally in most instances.

The giving of reasons

12. The report of the Woolf Inquiry proposed that an inmate who is segregated in the interests of good order or discipline should be given reasons in writing. This was in line with one of the report's main recommendations (recommendation 12) that, to improve standards of justice within establishments (to which the report attached much importance), inmates should be given reasons for any decision which materially and adversely affects them.

13. As paragraph 10 of annex C to CI 26/1990 stated, inmates have no absolute entitlement to be given reasons for their segregation. Nevertheless, it encouraged the giving of reasons as a sensible practice; and further encouraged the giving of reasons in writing where the inmate so requests. CI 37/1990 went further and required that in all cases in which a prisoner is temporarily transferred from a dispersal prison to a local prison (under the type of arrangement previously covered by CI 10/74) and segregated at the local prison, the prisoner should be given reasons for his segregation (as well as for his transfer). A recent survey of a sample of establishments showed that inmates segregated in other circumstances generally are given reasons, at least orally. It has been decided that practice in relation to all cases of segregation should now be taken a stage further, as the report of the Woolf Inquiry recommended.

14. Therefore, in future, any inmate who is segregated must be advised in writing of the reasons as far as is practicable, and as soon as possible. In general, this should be done before, or at the time that, the inmate is placed in conditions of segregation. But an inmate does not have an absolute right to be informed of the reasons *before* the decision is taken or *before or at the time* that the inmate is placed in conditions of segregation, if the interests of good order or discipline dictate otherwise.

15. An inmate who is not given reasons before or at the time must be given reasons as soon as possible thereafter (and at the latest within 24 hours unless, in a most exceptional case, the inmate's demeanour necessitates delay—in which case at the earliest opportune time).

16. The giving of reasons must be undertaken by a member of staff not below governor 5 level who is familiar with the facts of the case. The reasons should be as full as is reasonably possible. But they should not include any material which should be withheld in the interests of good order or discipline (in particular, any material which identifies, or may tend to identify—and thereby put at risk—another

inmate), or in the interests of security. Where there are reasons which cannot be given, or where it is impossible to elaborate on reasons which are given, this should be stated. A copy of the written reasons given to the inmate should be held in the segregation unit (or other place where the segregated inmate may be held) so that it is readily available for the information of staff working there, and also governor grades and others (eg, members of the Board of Visitors) when visiting. A copy should also be placed on the inmate's record.

17. Where an inmate has not yet been given reasons, exceptionally authority may nevertheless be given for continued segregation if the inmate's demeanour makes the giving of reasons impossible; and, where appropriate, such authority may be for a period extending beyond the time by which the inmate should be given reasons.

18. Notwithstanding the requirement to give reasons in writing, inmates should also be given reasons orally—normally again by a member of staff not below governor 5 level who is familiar with the facts of the case. The governor grade concerned may find it helpful to speak to the inmate together with, say, a Principal Officer who knows the inmate well. This occasion will provide an early opportunity for the inmate to make representations. The requirement to give reasons in writing should not delay reasons being given. Where appropriate, reasons should first be given orally, and then followed up in writing.

19. The Governor in charge has overall responsibility for ensuring the quality of reasons which are given; and should personally, or through a member of his or her senior management team specifically deputed to carry out the task, check the statements of reasons which are given to inmates.

Circular Instruction 37/1990

To all Prison Service establishments
(with an additional copy of the Chairman of the Board of Visitors)

THE TRANSFER OF INMATES IN THE INTERESTS OF GOOD ORDER OR DISCIPLINE

1. Introduction

The purposes of this Instruction are:
  (i) to set out general principles governing the transfer of any inmate (which, for the avoidance of any doubt, includes female inmates and young offenders); and
  (ii) to lay down revised guidance governing the temporary transfer of an adult male prisoner under two specific sets of arrangements.

2. Under the Prison Act 1952, an inmate may be confined in any establishment of

an appropriate type, and may be transferred between establishments of appropriate types. Transfers may be necessary for a variety of reasons connected with the regulation and management of establishments (including the interests of good order or discipline, or of security) and the needs of individual inmates.

3. The guidance in this Instruction recognises the problems presented by disruptive or subversive inmates to the regulation and management of establishments (and especially training establishments). Where such problems cannot be resolved by other means, including charging the inmate with a disciplinary offence (an option which may not, in the event, be open to local management) or by re-locating the inmate within the establishment, local management may need to consider other options, ie:

control problems of a shorter-term nature:

(a) segregation under Prison Rule 43 or YOI Rule 46—see CI 26/1990; or

(b) temporary transfer to another establishment; or

(c) a permanent transfer to another establishment; or

control problems of a longer-term nature:

(d) transfer to a CRC Special Unit.

4. As to (b), CI 10/1974 set out arrangements for the temporary transfer of disruptive prisoners from dispersal prisons to local prisons for a period of up to 28 days. CI 10/1974 provided for such prisoners to be held on Rule 43 at the local prison—initially on the authority of the dispersal prison governor and thereafter subject to the further authority of the Regional Director on behalf of the Secretary of State.

5. Alongside those set out in CI 10/1974, arrangements have been in force in all regions to provide for the temporary transfer, on normal location, of disruptive prisoners (other than Category A) from dispersal prisons and certain other Category B prisons designated by Regional Directors. Such prisoners have mainly been transferred to local prisons, but in one region the option of temporary transfer to other training prisons has been available.

6. Similar temporary transfer arrangements have been available for Category A prisoners. These have been administered on a national rather than regional basis and subject to specific approval by the Category A Section which has also determined, in consultation with the governors of local prisons, the prisoner's destination, and provided movement instructions.

7. The general principles set out in section II of this Instruction reflect judgements in a judicial review case. Although this was specifically concerned with a transfer under the terms of CI 10/1974, the principles which were established there are of wider application.

8. The specific guidance contained in section III reflects the recommendations of a Working Group which considered the operation of CI 10/1974, and of the judicial review referred to above. This Instruction replaces CI 10/1974 (which is hereby cancelled) and the P5 memorandum of 20 July 1990 to dispersal and local prison governors; and it rescinds such instructions as may be in force governing the arrangements described in paragraph 5 above.

II. General principles applying to all transfers in the interests of good order or discipline

9. A transfer (whether on a permanent or temporary basis) inevitably entails disruption to the inmate and, if, in the process, it creates travelling difficulties for visitors, it may adversely affect the frequency of visits which the inmate receives. In addition, a transfer in the interests of good order or discipline can create a strong feeling of grievance in the individual concerned. Such implications may be an unavoidable consequence of a decision that a transfer is necessary. But, in the light of such implications, special care must be exercised to ensure that no inmate is transferred as a form of punishment, and that—eg, in the choice of other establishment or in the frequency of transfers—the arrangements are not so designed that they amount to a form of punishment or may reasonably be seen to do so.

A decision that an inmate should be transferred, and the associated arrangements, must therefore be based on reasoned grounds and must be defensible; and such reasoned grounds must be recorded on the inmate's record by the governor*.

10. A decision to transfer an inmate may be open to criticism if it cannot be shown that the problem which gave rise to that decision could not have been resolved by moving the inmate, on normal location, within the establishment—although it is recognised that this option may not provide a sufficient answer to the problem of disruption presented by the inmate.

11. Where the inmate's conduct may amount to a disciplinary offence, the governor has discretion to decide, on the basis of the facts which are before him or her, not to have a disciplinary charge brought against the inmate but instead to decide, on the basis of the same facts, that the inmate should be transferred. Governors may conclude that disciplinary proceedings would not be conducive to the good order or discipline of the establishment. They may also decide that the available evidence would not support a disciplinary charge. In such a circumstance, before deciding to transfer the governor should consider whether there might be other more readily obtainable evidence which would increase the available options. On no account must the discretion not to proceed with a disciplinary charge, but to transfer the inmate instead, be exercised with the motive of denying the inmate an opportunity to hear and answer an allegation made against the inmate. It must also be born in mind that reasons for the transfer will still have to be given to the inmate—see below—and may be challenged. In any event, it follows that the reasons for a decision to transfer rather than charge must be recorded carefully on the inmate's record.

12. Conversely, if the transfer of an inmate to another establishment on normal location will resolve the problem, this option may be preferable to segregation within the original establishment.

---

* In this Instruction, references to the governor include the most senior governor grade available who, in the governor's absence, is acting governor. It goes without saying that the same principles apply to a decision which is taken by Headquarters. Such a decision must also be reasoned and recorded either on the inmate's record, or on other retrievable case papers relating to the individual inmate.

The giving of reasons

13. Inmates must be told the reasons for their transfer as far as is practicable, and as soon as possible. In general, this should be done before, or at the time of, transfer. But an inmate does not have an absolute right to be informed of the reasons for a decision to transfer him or her *before* such a decision is taken, or *before or at the time that* the transfer takes place. If it is judged that the giving of reasons may result in additional control problems during the transfer, it will be sensible for the giving of reasons to be deferred until the transfer has been completed.

14. Similarly, an inmate who is segregated at the receiving establishment must be told the reasons as far as is practicable, and as soon as possible. In general, this should be done before, or at the time that, the inmate is placed in conditions of segregation. But again an inmate does not have an absolute right to be informed of the reasons for segregation *before* the decision is taken at the receiving establishment or *before or at the time that* the inmate is placed in conditions of segregation, if the interests of good order or discipline dictate otherwise.

15. An inmate who is not given the reasons for the decision as to transfer or (where this applies) segregation before or at the time must be given reasons as soon as possible thereafter (and at the latest within 48 hours of the transfer (or segregation) taking place unless, in a most exceptional case, the inmate's demeanour necessitates delay—in which case at the earliest opportune time).

16. The giving of reasons must be undertaken by a member of staff not below governor V level who is familiar with the facts of the case. Reasons should be given in writing if the inmate so requests. They should be as full as is reasonably possible. But the governor may decide that the reasons given to the inmate shall not include any material which should be withheld in the interests of good order or discipline (in particular, any material which identifies, or may tend to identify, and thereby put at risk, another inmate), or in the interests of security.

17. Where an inmate who is subject to segregation has not yet been given reasons, exceptionally authority may nevertheless be given for continued segregation if the inmate's demeanour makes the giving of reasons impossible and, where appropriate, such authority may be for a period extending beyond the time by which the inmate should be given reasons.

Written complaints by inmates

18. If an inmate makes a written complaint about a transfer within the terms of this instruction, it should be logged in the normal way by the Requests/Complaints clerk at the establishment where the inmate is held. But the substantive reply should be provided and signed by the governor of the establishment where the decision to transfer the inmate was taken. The normal time limit for reply of 7 days applies.

The risk of manipulation

19. In making a decision to transfer an inmate, due regard must be paid to the risk that the inmate may be seeking to manipulate the system, for example, as part of a pre-planned escape with outside assistance. This risk is greatest where an inmate who is knowledgeable about the system may be able to predict with some precision the timing and destination of a transfer resulting from his or her own misconduct.

III. *Temporary transfer of a disruptive prisoner from a training prison to a local prison*

20. This section is concerned with two specific types of arrangements which may be made for the temporary transfer of a prisoner from a training prison to a local prison, viz:

(*a*) short-term transfer with the possibility of immediate segregation at the local prison; and

(*b*) medium-term transfer on normal location.

It will be recognised that (a) supersedes the arrangement which has hitherto been available under CI 10/1974, while (b) replaces arrangements which have hitherto been operated by Regional Offices (or, in the case of Category A prisoners, by Headquarters). It should be understood that this Instruction does not preclude the transfer of an inmate—whether on a temporary or permanent basis—under other arrangements within the general discretion afforded by the Prison Act (see paragraph 2 above).

(a) Short-term transfer with the possibility of segregation

21. This option is available to dispersal prisons and in relation to prisoners held in a Special Security Unit or a CRC Special Unit located in an establishment other than a dispersal prison. References hereafter to dispersal prisons should be interpreted as including such units.

22. The purpose of this facility is to provide a brief 'cooling off' period for a seriously disruptive or subversive prisoner

(*a*) who needs to be removed from association in the dispersal prison because of an imminently explosive situation resulting from his actual or impending disruptive behaviour; and

(*b*) who is therefore considered by the governor as requiring segregation within the dispersal prison were the prisoner to remain there; but

(*c*) for whom location in the dispersal prison's segregation unit is considered not to be appropriate or practical because either

(i) the prisoner would still be able to exercise a seriously disruptive influence from the segregation unit (eg, because of inadequate separation of the segregation unit from the main prison), or

(ii) the extent to which the prisoner constitutes a focal point for prisoner unrest would mean that the mere act of location in the segregation unit could have a provocative and explosive effect on the rest of the establishment.

It will be clear from the above that such a transfer is to be regarded as exceptional.

23. A transfer under this arrangement will be at the discretion of the dispersal prison governor. It shall be for a maximum of one calendar month, whether the prisoner is held at the local prison under segregation or on normal location.

24. Before transferring a prisoner under this arrangement, the dispersal prison governor must telephone the local prison governor to confirm that there are no exceptional problems which make it impossible for the prisoner to be accommo-

dated, and to ensure that his transfer will not bring the prisoner into contact with any other prisoners from whom he should be kept apart. Appendix A indicates the location of the primary cells available to each dispersal prison (see, further, paragraph 28 below). In normal circumstances, establishments shown as having the primary cells should be contacted in the first instance. In the event that, because of the temporary non-availability of cells, or because of difficulties at the relevant local prisons such as are described above, the dispersal prison governor is unable to find a place for the prisoner in any of the primary cells identified for the dispersal prison in appendix A; or if, for some particular reason, location in a different local prison is considered more suitable; the Tactical Management Planning Unit (TMPU) in DOC 1 Division should be contacted and asked to nominate or arrange a location.

25. As under CI 10.1974, such a transfer may entail the segregation of the prisoner at the local prison. By definition, any prisoner who is being considered for transfer under this arrangement is likely, in the view of the dispersal prison governor, to require segregation.

26. But, while the dispersal prison governor may authorise segregation at the dispersal prison as an interim step before transfer takes place, a decision as to segregation at the local prison must rest entirely with the local prison governor. In accordance with Rule 43 (2) (as amended), a prisoner may be segregated on the authority of the local prison governor for a maximum of 3 days. If, in the judgement of the local prison governor, the prisoner requires segregation beyond the end of that period, the Board of Visitors must be asked to consider authorising continued segregation. Such cases must be processed and documented in accordance with the normal requirements laid down for Rule 43 cases—see CI 26/1990. *It is stressed that the need for segregation must be kept under review by the local prison governor.* If, at any time while the prisoner is detained at the local prison, the local prison governor concludes that segregation is no longer necessary, he or she must forthwith arrange for that prisoner to resume association.

27. The local prison governor must be fully advised by the dispersal prison governor of the prisoner's behaviour, and of other relevant background information, leading to the need for the transfer, in order that he or she can reach a fully-informed judgement of the need for the prisoner to be segregated on arrival. This information should be given by telephone in the first instance. Normally, the prisoner's record containing all the relevant background to the transfer should accompany the prisoner to the local prison. If, exceptionally, this is not possible, the prisoner's record must be sent to the local prison within 24 hours. Quite apart from the requirement for the local prison governor to have information on which to address the question of the need for segregation, for obvious reasons the local prison governor should have full information about a difficult prisoner who is to be accommodated at his or her establishment, albeit for only a relatively short time.

28. The cells identified in appendix A should normally be available to receive a prisoner transferred under this arrangement, provided the local prison governor authorises the prisoner's segregation. However, provided that such cells can be vacated at short notice if necessary, the cells need not be kept empty when not required to accommodate a prisoner on temporary transfer from a dispersal prison under this arrangement. If the local prison governor concludes that the prisoner need not be segregated (or, on reviewing the case at a later stage, that the prisoner

need not remain segregated), it will be the responsibility of the local prison governor to arrange other suitable accommodation at the establishment.

29. The dispersal prison governor will be responsible for getting the prisoner to the local prison. The reception of the prisoner will be reported to the TMPU by the local prison governor by means of the form at appendix B. The Board of Visitors must also be given a copy of the form, whether or not the prisoner is placed on Rule 43.

30. Except in the case of a prisoner in a Special Security Unit or a Category A prisoner in a CRC Special Unit (where the prior authority of the Category A Section must always be obtained), the transfer of a Category A prisoner to a local prison under this arrangement will not require clearance with the Category A Section of DOC 1 Division. But, insofar as this is practicable, the Category A Section will be notified by the dispersal prison governor in advance of the decision to transfer the prisoner and will be consulted about the need to provide a police escort. If such a transfer needs to be arranged outside normal office hours, a police escort will always be arranged by the dispersal prison governor if the last movement of the prisoner was similarly covered; and the Category A Section will be notified of the transfer by the dispersal prison governor as soon as possible afterwards, by telephone. The Category A Section must always be sent a copy of the form in appendix B if the prisoner is in Category A.

31. In the case of a non-Category A life sentence prisoner, a copy of the form in appendix B should also be sent to the Lifer Management Unit in DOC 2 Division.

32. If got any reason it is not considered feasible or desirable for the prisoner to be returned to the dispersal prison at the end of the period of temporary transfer, the dispersal prison governor must explore the possibility of a transfer to another prison. In a case where the prisoner's recent behaviour and past record indicate that he is likely to present persistent and long-term control problems, consideration should be given to the possibility of referral to the Special Unit Selection Committee (SUSC) for a place in a CRC Special Unit. In other cases, transfers should be considered in consultation with the Category A Section if the prisoner is in Category A, with the Lifer Management Unit in DOC 2 Division in the case of a non-Category A life sentence prisoner, or with the TMPU in other cases. Any such action must be initiated immediately. It must be remembered that it will take time to make alternative arrangements; and that prompt action is imperative if the maximum period of transfer of one month is to be adhered to.

33. If, during the period spent at the local prison, the local prison governor considers that the prisoner no longer needs to be held at the local prison, the dispersal prison governor should be contacted with a view to the prisoner's early return to the dispersal system. The decision will normally rest with the dispersal prison governor. It must be reasoned and recorded. In the event of any difficulty, the Category A Section (if the prisoner is in Category A) or the TMPU (in any other case) should be contacted.

34. In the interests of security (especially in the case of a Category A prisoner), a prisoner should be returned to the dispersal prison, or (if this is not possible) transferred to another prison, up to a few days before the expiry of the maximum period of one calendar month, so that the actual date of the movement is not predictable.

35. The dispersal prison governor will be responsible for getting the prisoner back

to the dispersal prison. If, exceptionally, the prisoner is not to return to the dispersal prison, the governor of that prison will nevertheless be responsible for the transfer arrangements. The Category A Section must be consulted about the escorting arrangements for a Category A prisoner; and, in the case of a non-Category A prisoner who is not to be returned to the dispersal prison, the dispersal prison governor will consult the TMPU about the transfer arrangements.

36. The return of the prisoner to the dispersal prison (or transfer to another prison) must be reported to the TMPU by the local prison governor by means of the form in appendix C. A copy must be given to the Board of Visitors. If the prisoner is in Category A, a copy must be sent to the Category A Section in DOC 1 Division. In the case of a non-Category A life sentence prisoner, a copy must be sent to the Lifer Management Unit in DOC 2 Division.

(b) Medium-term transfer on normal location

37. This type of arrangement is available to all Category B training prisons (including dispersal prisons) and corresponds to that which has hitherto been available under the regional arrangements described in paragraphs 5 and 6. The detailed arrangements with regard to the medium-term transfer of a Category A prisoner differ from those in relation to others—see paragraphs 47–48.

38. The purpose of the arrangement is to provide a facility whereby a prisoner

   (a) who, because of his actual or impending disruptive behaviour, poses a significant problem for the regulation and management of the training prison; and

   (b) who needs to be removed from the training prison for a period of time; but

   (c) for whom an immediate permanent transfer to another training prison is considered not to be a desirable or practical solution

may be temporarily transferred to a local prison.

39. This arrangement may be helpful to governors of training prisons in dealing with a difficult situation (for example, where it is necessary to separate a prisoner from associates involved with him in disruptive or subversive activities, such as drug trafficking); and, where it may take time to sort out a satisfactory longer-term solution, may afford valuable breathing-space while options are being explored. But this facility must be used sparingly, especially having regard to other pressures on local prisons.

40. A training prison governor who considers that such a transfer should be made will seek in writing the approval of the Tactical Management Planning Unit in DOC 1 Division. The training prison governor must provide the TMPU with full information about the prisoner's behaviour, and other relevant background, leading to the request for transfer. The maximum period of temporary transfer in any individual case will be determined in advance by the TMPU. Such period shall be no longer than is necessary in the interests of good order or discipline and shall in no case exceed 6 months. It is stressed that the period of transfer must be determined strictly in the light of the circumstances of the case, and the maximum period of 6 months shall not be regarded as the automatic norm. The TMPU will also determine the local prison to which the prisoner is to be transferred. The reasons for

decisions as to transfer, and as to the duration and location of such transfer, must be recorded. In the case of a life sentence prisoner, the Lifer Management Unit in DOC 2 Division must be informed.

41. It will be the responsibility of the training prison governor to get the prisoner to the local prison. The local prison governor will need to be fully advised of the background to the case, over and above the information supplied by the TMPU in the first instance. Normally, the prisoner's record containing all the relevant background to the case should accompany the prisoner to the local prison. If, exceptionally, this is not possible, the prisoner's record must be sent to the local prison within 24 hours.

42. The normal presumptions in such a case are that the prisoner

(*a*) will remain at the same local prison during the period of the transfer;

(*b*) will be on normal location (unless, in the light of the prisoner's subsequent behaviour, the local prison governor decides that segregation is necessary); and

(*c*) will be returned to the parent training prison.

43. If, while the prisoner is on temporary transfer under this arrangement, it is considered for any reason that it will not after all be feasible or desirable for the prisoner to be returned to the training prison, the training prison governor must explore the possibility of a transfer to another prison, through the TMPU and, in the case of a life sentence prisoner, the Lifer Management Unit in DOC 2 Division. It is imperative that action to make alternative arrangements is initiated as soon as possible.

44. During the period at the local prison, the local prison governor must regularly review, against the considerations which led to the decision to transfer the prisoner, the necessity for him to remain at the local prison. If the local prison governor considers that the prisoner no longer needs to be held at the local prison, the TMPU (and, in the case of a life sentence prisoner, the Lifer Management Unit in DOC 2 Division) should be contacted with a view to the prisoner's early return to the training prison. The decision will rest with the TMPU and the reasons must be recorded.

45. In any event, in the interests of security the prisoner should be returned to the training prison, or (if this is not possible) transferred to another prison, up to a few days before the expiry of the maximum period of temporary transfer originally determined by TMPU, so that the actual date of the movement is not predictable.

46. The training prison governor will be responsible for getting the prisoner back to the training prison. If, exceptionally, the prisoner is not to return to the parent prison, the training prison governor will nevertheless be responsible for the transfer arrangements, in consultation with the TMPU. Again, as indicated at paragraph 32, if a long-term prisoner's recent behaviour and record indicate that he could be a candidate for a CRC Special Unit, the training prison governor should make a referral to SUSC. The criteria for the CRC Special Units are circulated to dispersal and local prison governors at regular intervals.

47. The operational basis for medium-term transfers, as set out in paragraphs 38–39 above, applies equally to Category A prisoners; but the administrative arrangements are different. In particular, if the dispersal prison governor considers

that such a transfer should be made, he or she must seek the specific approval of Category A Section in DOC 1 Division. As in the case of non-Category A prisoners, full information must be provided as to the prisoner's behaviour and other relevant background. If approval is given, it will be for the Category A Section, in consultation with governors, to determine to which local prison the transfer is to be made. The Category A Section will notify the Lifer Management Unit of DOC 2 Division where appropriate.

48. The Category A Section will determine the appropriate arrangements for the prisoner's movement and issue instructions to the governors involved. The duration of the transfer will be determined, in consultation with the governor of the local prison, by the Category A Section which will also determine subsequent movements, ie a return to the original training prison (dispersal); or a further transfer to another local prison; or (and most likely) reallocation to another dispersal prison.

Further points on temporary transfers under this section

49. One of the purposes of the arrangements described in this section is to take the heat out of a situation before matters reach a point where breaches of discipline occur. But there will be occasions when a disciplinary charge is necessary. In that event, so far as practicable the charge should be preferred and the hearing held before the transfer takes place. Unless it is necessary to transfer a prisoner at once, any period of punishment should be completed prior to transfer.

50. While in a local prison, prisoners will be treated in the same way as any other convicted prisoner who is in the same security category and who is subject to the same conditions (ie, on normal location or under segregation as appropriate). In particular, they will do the same work (if available), receive pay on the same basis, and be subject to the same regime and privileges. Prisoners transferred under these arrangements will be permitted to take with them their possessions, excluding any which are over and above those normally allowed to other convicted prisoners at the local prison, which may be retained by the prisoner in his cell or with his property as decided by the governor.

51. For the avoidance of any doubt, instructions governing the treatment of Category A inmates, including the handling of casework by DOC 1 Division, continue to apply wherever a Category A prisoner is located.

IV. Other matters

Board of Visitors

52. A copy of this Instruction should be passed to the chairman of the Board of Visitors.

Operative date

53. Revised arrangements for the authorisation of segregation in cases of prisoners transferred under the terms of CI 10/1974 should have already been introduced immediately following the P5 memorandum of 20 July 1990 to governors of dispersal and local prisons. Otherwise, this Instruction will take effect from 25 September.

Prison Service Headquarters
Cleland House
13 September 1990
PDG/90 8/3/1

Appendix 5
The European Convention on Human Rights

The Governments signatory hereto, being Members of the Council of Europe,

Considering the Universal Declaration of Human Rights proclaimed by the General Assembly of the United Nations on 10 December 1948;

Considering that this Declaration aims at securing the universal and effective recognition and observance of the Rights therein declared;

Considering that the aim of the Council of Europe is the achievement of greater unity between its Members and that one of the methods by which the aim is to be pursued is the maintenance and further realisation of Human Rights and Fundamental Freedoms;

Reaffirming their profound belief in those Fundamental Freedoms which are the foundation of justice and peace in the world and are best maintained on the one hand by an effective political democracy and on the other by a common understanding and observance of the Human Rights upon which they depend;

Being resolved, as the Governments of European countries which are like-minded and have a common heritage of political traditions, ideals, freedom and the rule of law to take the first steps for the collective enforcement of certain of the Rights stated in the Universal Declaration;

Have agreed as follows:

Article 1

The High Contracting Parties shall secure to everyone within their jurisdiction the rights and freedoms defined in Section 1 of this Convention

Section I

Article 2

1. Everyone's right to life shall be protected by law. No one shall be deprived of his life intentionally save in the execution of a sentence of a court following his conviction of a crime for which this penalty is provided by law.

2. Deprivation of life shall not be regarded as inflicted in contravention of this Article when it results from the use of force which is no more than absolutely necessary:

   (*a*) in defence of any person from unlawful violence;
   (*b*) in order to effect a lawful arrest or to prevent the escape of a person lawfully detained;
   (*c*) in action lawfully taken for the purpose of quelling a riot or insurrection.

Article 3

No one shall be subjected to torture or inhuman or degrading treatment or punishment.

Article 4

1. No one shall be held in slavery or servitude.

2. No one shall be required to perform forced or compulsory labour.

3. For the purpose of this Article the term 'forced or compulsory labour' shall not include:

(a) any work required to be done in the ordinary course of detention imposed according to the provisions of Article 5 of this Convention or during conditional release from such detention;

(b) any service of a military character or, in the case of conscientious objectors in countries where they are recognized, service exacted instead of compulsory military service;

(c) any service exacted in case of an emergency or calamity threatening the life or well-being of the community;

(d) any work of service which forms part of normal civic obligations.

Article 5

1. Everyone has the right to liberty and security of person. No one shall be deprived of his liberty save in the following cases and in accordance with a procedure prescribed by law:

(a) the lawful detention of a person after conviction by a competent court;

(b) the lawful arrest or detention of a person for non-compliance with the lawful order of a court or in order to secure the fulfilment of any obligation prescribed by law;

(c) the lawful arrest or detention of a person effected for the purpose of bringing him before the competent legal authority on reasonable suspicion of having committed an offence or when it is reasonably considered necessary to prevent his committing an offence or fleeing after having done so;

(d) the detention of a minor by lawful order for the purpose of educational supervision or his lawful detention for the purpose of bringing him before the competent legal authority;

(e) the lawful detention of persons for the prevention of the spreading of infectious diseases, of persons of unsound mind, alcoholics or drug addicts, or vagrants;

(f) the lawful arrest or detention of a person to prevent his effecting an unauthorized entry into the country or of a person against whom action is being taken with a view to deportation or extradition.

2. Everyone who is arrested shall be informed promptly, in a language which he understands, of the reasons for his arrest and of any charge against him.

3. Everyone arrested or detained in accordance with the provisions of paragraph 1 (c) of this Article shall be brought promptly before a judge or other officer authorised by law to exercise judicial power and shall be entitled to trial within a reason-

able time or to release pending trial. Release may be conditioned by guarantees to appear for trial.

4. Everyone who is deprived of his liberty by arrest or detention shall be entitled to take proceedings by which the lawfulness of his detention shall be decided speedily by a court and his release ordered if the detention is not lawful.

5. Everyone who has been the victim of arrest or detention in contravention of the provisions of this article shall have an enforceable right to compensation.

Article 6

1. In the determination of his civil rights and obligations or of any criminal charge against him, everyone is entitled to a fair and public hearing within a reasonable time by an independent and impartial tribunal established by law. Judgment shall be pronounced publicly but the press and public may be excluded from all or part of the trial in the interest of morals, public order or national security in a democratic society, where the interests of juveniles or the protection of the private life of the parties so require, or to the extent strictly necessary in the opinion of the court in special circumstances where publicity would prejudice the interests of justice.

2. Everyone charged with a criminal offence shall be presumed innocent until proved guilty according to law.

3. Everyone charged with a criminal offence has the following minimum rights:

(a) to be informed promptly, in a language which he understands and in detail, of the nature and cause of the accusation against him;

(b) to have adequate time and facilities for the preparation of his defence;

(c) to defend himself in person or through legal assistance of his own choosing or, if he has not sufficient means to pay for legal assistance, to be given it free when the interests of justice so require;

(d) to examine or have examined witnesses against him and to obtain the attendance and examination of witnesses on his behalf under the same conditions as witnesses against him;

(e) to have the free assistance of an interpreter if he cannot understand or speak the language used in court.

Article 7

1. No one shall be held guilty of any criminal offence on account of any act or omission which did not constitute a criminal offence under national or international law at the time when it was committed. Nor shall a heavier penalty be imposed than the one that was applicable at the time the criminal offence was committed.

2. This article shall not prejudice the trial and punishment of any person for any act or omission which, at the time when it was committed, was criminal according to the general principles of law recognized by civilized nations.

Article 8

1. Everyone has the right to respect for his private and family life, his home and his correspondence.

2. There shall be no interference by a public authority with the exercise of this right except such as in accordance with the law and is necessary in a democratic

society in the interests of national security, public safety or the economic well-being of the country, for the prevention of disorder or crime, for the protection of health or morals, or for the protection of the rights and freedoms of others.

Article 9

1. Everyone has the right to freedom of thought, conscience and religion; this right includes freedom to change his religion or belief, and freedom, either alone or in community with others and in public or private, to manifest his religion or belief, in worship, teaching, practice and observance.

2. Freedom to manifest one's religion or beliefs shall be subject only to such limitations as are prescribed by law and are necessary in a democratic society in the interests of public safety, for the protection of public order, health or morals, or for the protection of the rights and freedoms of others.

Article 10

1. Everyone has the right to freedom of expression. This right shall include freedom to hold opinions and to receive and impart information and ideas without interference by public authority and regardless of frontiers. This article shall not prevent States from requiring the licensing of broadcasting, television or cinema enterprises.

2. The exercise of these freedoms, since it carries with it duties and responsibilities, may be subject to such formalities, conditions, restrictions or penalties as are prescribed by law and are necessary in a democratic society, in the interests of national security, territorial integrity or public safety, for the prevention of disorder or crime, for the protection of health or morals, for the protection of the reputation or rights of others, for preventing the disclosure of information received in confidence, or for maintaining the authority and impartiality of the judiciary.

Article 11

1. Everyone has the right to freedom of peaceful assembly and to freedom of association with others, including the right to form and to join trade unions for the protection of his interests.

2. No restrictions shall be placed on the exercise of these rights other than such as are prescribed by law and are necessary in a democratic society in the interests of national security or public safety, for the prevention of disorder or crime, for the protection of health or morals or for the protection of the rights and freedoms of others. This Article shall not prevent the imposition of lawful restrictions on the exercise of these rights by members of the armed forces, of the police or of the administration of the State.

Article 12

Men and women of marriageable age have the right to marry and to found a family, according to the national laws governing the exercise of this right.

Article 13

Everyone whose rights and freedoms as set forth in this Convention are violated shall have an effective remedy before a national authority notwithstanding that the violation has been committed by persons acting in an official capacity.

Article 14

The enjoyment of the rights and freedoms set forth in this Convention shall be secured without discrimination on any ground such as sex, race, colour, language, religion, political or other opinions, national or social origin, association with a national minority, property, birth or other status.

Article 15

1. In time of war of other public emergency threatening the life of the nation and High Contracting Party may take measures derogating from its obligations under this Convention to the extent strictly required by the exigencies of the situation, provided that such measures are not inconsistent with its other obligations under international law.

2. No derogation from Article 2, except in respect of deaths resulting from lawful acts of war, or from Articles 3, 4 (paragraph 1) and 7 shall be made under this provision.

3. Any High Contacting Party availing itself of this right of derogation shall keep the Secretary-General of the Council of Europe fully informed of the measures which it has taken and the reasons therefor. It shall also inform the Secretary-General of the Council of Europe when such measures have ceased to operate and the provisions of the Convention are again being fully executed.

Article 16

Nothing in Articles 10, 11 and 14 shall be regarded as preventing the High Contracting Parties from imposing restrictions on the political activity of aliens.

Article 17

Nothing in this Convention may be interpreted as implying for any State, group or person the right to engage in any activity or perform any act aimed at the destruction of any of the rights and freedoms set forth herein or at their limitation to a greater extent than is provided for in the Convention.

Article 18

The restrictions permitted under this Convention to the said rights and freedoms shall not be applied for any purpose other than those for which they have been prescribed.

Section II

Article 19

To ensure the observance of the engagements undertaken by the High Contracting Parties in the present Convention, there shall be set up:

1. A European Commission of Human Rights hereinafter referred to as 'the Commission';

2. A European Court of Human Rights, hereinafter referred to as 'the Court'.

Section III

Article 20

The Commission shall consist of a number of members equal to that of the High Contracting Parties. No two members of the Commission may be nationals of the same State.

Article 21

1. The members of the Commission shall be elected by the Committee of Ministers by an absolute majority of votes, from a list of names drawn up by the Bureau of the Consultative Assembly; each group of the Representatives of the High Contracting Parties in the Consultative Assembly shall put forward three candidates, of whom two at least shall be its nationals.

2. As far as applicable, the same procedure shall be followed to complete the Commission in the event of other States subsequently becoming Parties to this Convention, and in filling casual vacancies.

Article 22

1. The members of the Commission shall be elected for a period of six years. They may be re-elected. However, of the members elected at the first election, the terms of seven members shall expire at the end of three years.

2. The members whose terms are to expire at the end of the initial period of three years shall be chosen by lot by the Secretary-General of the Council of Europe immediately after the first election has been completed.

3. A member of the Commission elected to replace a member whose term of office has not expired shall hold office for the remainder of his predecessor's term.

4. The members of the Commission shall hold office until replaced. After having been replaced, they shall continue to deal with such cases as they already have under consideration.

Article 23

The members of the Commission shall sit on the Commission in their individual capacity.

Article 24

Any High Contracting Party may refer to the Commission, through the Secretary-General of the Council of Europe, any alleged breach of the provisions of the Convention by another High Contracting Party.

Article 25

1. The Commission may receive petitions addressed to the Secretary-General of the

Council of Europe from any person, non-governmental organization or group of individuals claiming to be the victim of a violation by one of the High Contracting Parties of the rights set forth in this Convention, provided that the High Contracting Party against which the complaint has been lodged has declared that it recognizes the competence of the Commission to receive such petitions. Those of the High Contracting Parties who have made such a declaration undertake not to hinder in any way the effective exercise of this right.

2. Such declarations may be made for a specific period.

3. The declarations shall be deposited with the Secretary-General of the Council of Europe who shall transmit copies thereof to the High Contracting Parties and publish them.

4. The Commission shall only exercise the powers provided for in this Article when at least six High Contracting Parties are bound by declarations made in accordance with the preceding paragraphs.

Article 26

The Commission may only deal with the matter after all domestic remedies have been exhausted, according to the generally recognized rules of international law, and within a period of six months from the date on which the final decision was taken.

Article 27

1. The Commission shall not deal with any petition submitted under Article 25 which:

(*a*) is anonymous, or

(*b*) is substantially the same as a matter which has already been examined by the Commission or has already been submitted to another procedure or international investigation or settlement and if it contains no relevant new information.

2. The Commission shall consider inadmissible any petition submitted under Article 25 which it considers incompatible with the provisions of the present Convention, manifestly ill-founded, or an abuse of the right of petition.

3. The Commission shall reject any petition referred to it which it considers inadmissible under Article 26.

Article 28

In the event of the Commission accepting a petition referred to it:

(*a*) it shall, with a view to ascertaining the facts undertake together with the representatives of the parties an examination of the petition and, if need be, an investigation, for the effective conduct of which the States concerned shall furnish all necessary facilities, after an exchange of views with the Commission:

(*b*) it shall place itself at the disposal of the parties concerned with a view to securing a friendly settlement of the matter on the basis of respect for Human Rights as defined in this Convention.

Article 29

1. The Commission shall perform the functions set out in Article 28 by means of a Sub-Commission consisting of seven members of the Commission.

2. Each of the parties concerned may appoint as members of this Sub-Commission a person of its choice.

3. The remaining members shall be chosen by lot in accordance with arrangements prescribed in the Rules of Procedure of the Commission.

Article 30

If the Sub-Commission succeeds in effecting a friendly settlement in accordance with Article 28, it shall draw up a Report which shall be sent to the States concerned, to the Committee of Ministers and to the Secretary-General of the Council of Europe for publication. This Report shall be confined to a brief statement of the facts and of the solution reached.

Article 31

1. If a solution is not reached, the Commission shall draw up a Report on the facts and state its opinion as to whether the facts found disclose a breach by the State concerned of its obligations under the Convention. The opinions of all the members of the Commission on this point may be stated in the Report.

2. The Report shall be transmitted to the Committee of Ministers. It shall also be transmitted to the States concerned, who shall not be at liberty to publish it.

3. In transmitting the Report to the Committee of Ministers the Commission may make such proposals as it thinks fit.

Article 32

1. If the question is not referred to the Court in accordance with Article 48 of this Convention within a period of three months from the date of the transmission of the Report to the Committee of Ministers, the Committee of Ministers shall decide by a majority of two-thirds of the members entitled to sit on the Committee whether there has been a violation of the Convention.

2. In the affirmative case the Committee of Ministers shall prescribe a period during which the Contracting Party concerned must take the measures required by the decision of the Committee of Ministers.

3. If the High Contracting Party concerned has not taken satisfactory measures within the prescribed period, the Committee of Ministers shall decide by the majority provided for in paragraph 1 above what effect shall be given to its original decision and shall publish the Report.

4. The High Contracting Parties undertake to regard as binding on them any decision which the Committee of Ministers may take in application of the preceding paragraphs.

Article 33

The Commission shall meet *in camera.*

Article 34

The Commission shall take its decision by a majority of the Members present and voting; the Sub-Commission shall take its decisions by a majority of its members.

Article 35

The Commission shall meet as the circumstances require. The meetings shall be convened by the Secretary-General of the Council of Europe.

Article 36

The Commission shall draw up its own rules of procedure.

Article 37

The secretariat of the Commission shall be provided by the Secretary-General of the Council of Europe.

Section IV

Article 38

The European Court of Human Rights shall consist of a number of judges equal to that of the Members of the Council of Europe. No two judges may be nationals of the same State.

Article 39

1. The members of the Court shall be elected by the Consultative Assembly by a majority of the votes cast from a list of persons nominated by the Members of the Council of Europe; each Member shall nominate three candidates, of whom two at least shall be its nationals.

2. As far as applicable, the same procedure shall be followed to complete the Court in the event of the admission of new members of the Council of Europe, and in filling casual vacancies.

3. The candidates shall be of high moral character and must either possess the qualifications required for appointment to high judicial office or be jurisconsults of recognized competence.

Article 40

1. The members of the Court shall be elected for a period of nine years. They may be re-elected. However, of the members elected at the first election the terms of four members shall expire at the end of three years, and the terms of four more members shall expire at the end of six years.

2. The members whose terms are to expire at the end of the initial periods of three and six years shall be chosen by lot by the Secretary-General immediately after the first election has been completed.

3. A member of the Court elected to replace a member whose term of office has not expired shall hold office for the remainder of his predecessor's term.

4. The members of the Court shall hold office until replaced. After having been replaced, they shall continue to deal with such cases as they already have under consideration.

Article 41

The Court shall elect its President and Vice-President for a period of three years. They may be re-elected.

Article 42

The members of the Court shall receive for each day of duty a compensation to be determined by the Committee of Ministers.

Article 43

For the consideration of each case brought before it the Court shall consist of a Chamber composed of seven judges. There shall sit as an *ex officio* member of the Chamber the judge who is a national of any State party concerned, or, if there is none, a person of its choice who shall sit in the capacity of judge; the names of the other judges shall be chosen by lot by the President before the opening of the case.

Article 44

Only the High Contracting Parties and the Commission shall have the right to bring a case before the Court.

Article 45

The jurisdiction of the Court shall extent to all cases concerning the interpretation and application of the present Convention which the High Contracting Parties or the Commission shall refer to it in accordance with Article 48.

Article 46

1. Any of the High Contracting Parties may at any time declare that it recognizes as compulsory *ipso facto* and without any special agreement the jurisdiction of the Court in all matters concerning the interpretation and application of the present Convention.

2. The declarations referred to above may be made unconditionally or on condition of reciprocity on the part of several or certain other High Contracting Parties or for a specified period.

3. These declarations shall be deposited with the Secretary-General of the Council of Europe who shall transmit copies thereof to the High Contracting Parties.

Article 47

The Court may only deal with a case after the Commission has acknowledged the failure of efforts for a friendly settlement and within the period of three months provided for in Article 32.

Article 48

The following may bring a case before the Court, provided that the High Contracting Party concerned, if there is only one, or the High Contracting Parties concerned, if there is more than one, are subject to the compulsory jurisdiction of the Court or, failing that, with the consent of the High Contracting Party concerned, if there is only one, or of the High Contracting Parties concerned if there is more than one:

(*a*) the Commission;

(*b*) a High Contracting Party whose national is alleged to be a victim;

(*c*) a High Contracting Party which referred the case to the Commission;

(*d*) a High Contracting Party against whom the complaint has been lodged.

Article 49

In the event of dispute as to whether the Court has jurisdiction, the matter shall be settled by the decision of the Court.

Article 50

If the Court finds that a decision or a measure taken by a legal authority or any other authority of a High Contracting Party, is completely or partially in conflict with the obligations arising from the present Convention, and if the internal law of the said Party allows only partial reparation to be made for the consequences of this decision or measure, the decision of the Court shall, if necessary, afford just satisfaction to the injured party.

Article 51

1. Reasons shall be given for the judgment of the Court.

2. If the judgment does not represent in whole or in part the unanimous opinion of the judges, any judge shall be entitled to deliver a separate opinion.

Article 52

The judgment of the Court shall be final.

Article 53

The High Contracting Parties undertake to abide by the decision of the Court in any case to which they are parties.

Article 54

The judgment of the Court shall be transmitted to the Committee of Ministers which shall supervise its execution.

Article 55

The Court shall draw up its own rules and shall determine its own procedure,

Article 56

1. The first election of the members of the Court shall take place after the declarations by the High Contracting Parties mentioned in Article 46 have reached a total of eight.

　　2. No case can be brought before the Court before this election.

European Convention for the Prevention of Torture and Inhuman or Degrading Treatment or Punishment

The member States of the Council of Europe, signatory hereto,

　　Having regard to the provisions of the Convention for the Protection of Human Rights and Fundamental Freedoms,

　　Recalling that, under Article 3 of the same Convention, 'no one shall be subjected to inhuman or degrading treatment or punishment',

　　Noting that the machinery provided for in that Convention operates in relation to persons who alleged that they are victims of violations of Article 3;

　　Convinced that the protection of persons deprived of their liberty against torture and inhuman or degrading treatment or punishment could be strengthened by non-judicial means of a preventive character based on visits;

　　Have agreed as follows:

Chapter I

Article 1

There shall be established a European Committee for the Prevention of Torture and Inhuman or Degrading Treatment or Punishment (hereinafter referred to as 'the Committee'). The Committee shall, by means of visits, examine the treatment of persons deprived of their liberty with a view to strengthening, if necessary, the protection of such persons from torture and from inhuman or degrading treatment or punishment.

Article 2

Each Party shall permit visits, in accordance with this Convention, to any place within its jurisdiction where persons are deprived of their liberty by a public authority.

Article 3

In the application of this Convention, the Committee and the competent national authorities of the Party concerned shall co-operate with each other.

Chapter II

Article 4

1. The Committee shall consist of a number of members equal to that of the Parties.

2. The members of the Committee shall be chosen from among persons of high moral character, known for their competence in the field of human rights or having professional experience in the areas covered by this Convention.

3. No two members of the Committee may be nationals of the same State.

4. The members shall serve in their individual capacity, shall be independent and impartial, and shall be available to serve the Committee effectively.

Article 5

1. The members of the Committee shall be elected by the Committee of Ministers of the Council of Europe by an absolute majority of votes, from a list of names drawn up by the Bureau of the Consultative Assembly of the Council of Europe; each national delegation of the Parties in the Consultative Assembly shall put forward three candidates, of whom two at least shall be its nationals.

2. The same procedure shall be followed in filling casual vacancies.

3. The members of the Committee shall be elected for a period of four years. They may only be re-elected once. However, among the members elected at the first election, the terms of three members shall expire at the end of two years. The members whose terms are to expire at the end of the initial period of two years shall be chosen by lot by the Secretary General of the Council of Europe immediately after the first election has been completed.

Article 6

1. The Committee shall meet in camera. A quorum shall be equal to the majority of its members. The decisions of the Committee shall be taken by a majority of the members present, subject to the provisions of Article 10, paragraph 2.

2. The Committee shall draw up its own rules of procedure.

3. The Secretariat of the Committee shall be provided by the Secretary General of the Council of Europe.

Chapter III

Article 7

1. The Committee shall organise visits to places referred to in Article 2. Apart from periodic visits, the Committee may organise such other visits as appear to it to be required in the circumstances.

2. As a general rule, the visits shall be carried out by at least two members of the Committee. The Committee may, if it considers it necessary, be assisted by experts and interpreters.

Article 8

1. The Committee shall notify the Government of the Party concerned of its intention to carry out a visit. After such notification, it may at any time visit any place referred to in Article 2.

    2. A Party shall provide the Committee with the following facilities to carry out its task:

    a. access to its territory and the right to travel without restriction;
    b. full information on the places where persons deprived of their liberty are being held;
    c. unlimited access to any place where persons are deprived of their liberty, including the right to move inside such places without restriction; '
    d. other information available to the Party which is necessary for the Committee to carry out its task. In seeking such information, the Committee shall have regard to applicable rules of national law and professional ethics.

    3. The Committee may interview in private persons deprived of their liberty.

    4. The Committee may communicate freely with any person whom it believes can supply relevant information.

    5. If necessary, the Committee may immediately communicate observation to the competent authorities of the Party concerned.

Article 9

1. In exceptional circumstances, the competent authorities of the Party concerned may make representations to the Committee against a visit at the time or to the particular place proposed by the Committee. Such representations may only be made on grounds of national defence, public safety, serious disorder in places where persons are deprived of their liberty, the medical condition of a person or that an urgent interrogation relating to a serious crime is in progress.

    2. Following such representations, the Committee and the Party shall immediately enter into consultations in order to clarify the situation and seek agreement on arrangements to enable the Committee to exercise its functions expeditiously. Such arrangements may include the transfer to another place of any person whom the Committee proposed to visit. Until the visit takes place, the Party shall provide information to the Committee about any person concerned.

Article 10

1. After each visit, the Committee shall draw up a report on the facts found during the visit, taking account of any observations which may have been submitted by the Party concerned. It shall transmit to the latter its report containing any recommendations it considers necessary. The Committee may consult with the Party with a view to suggesting, if necessary, improvements in the protection of persons deprived of their liberty.

    2. If the Party fails to co-operate or refuses to improve the situation in the light of the Committee's recommendations, the Committee may decide, after the Party has had an opportunity to make known its views, by a majority of two-thirds of its members to make a public statement on the matter.

Article 11

1. The information gathered by the Committee in relation to a visit, its report and its consultations with the Party concerned shall be confidential.

2. The Committee shall publish its report, together with any comments of the Party concerned, whenever requested to do so by that Party.

3. However, no personal data shall be published without the express consent of the person concerned.

Article 12

Subject to the rules of confidentiality in Article 11, the Committee shall every year submit to the Committee of Ministers a general report on its activities which shall be transmitted to the Consultative Assembly and made public.

Article 13

The members of the Committee, experts and other persons assisting the Committee are required, during and after their terms of office, to maintain the confidentiality of the facts or information of which they have become aware during the discharge of their functions.

Article 14

1. The names of persons assisting the Committee shall be specified in the notification under Article 8, paragraph 1.

2. Experts shall act on the instructions and under the authority of the Committee. They shall have particular knowledge and experience in the areas covered by this Convention and shall be bound by the same duties of independence, impartiality and availability as the members of the Committee.

3. A Party may exceptionally declare that an expert or other person assisting the Committee may not be allowed to take part in a visit to a place within its jurisdiction.

Chapter IV

Article 15

Each Party shall inform the Committee of the name and address of the authority competent to receive notifications to its Government, and of any liaison officer it may appoint.

Article 16

The Committee, its members and experts referred to in Article 7, paragraph 2 shall enjoy the privileges and immunities set out in the Annex to this Convention.

Article 17

1. This Convention shall not prejudice the provisions of domestic law or any international agreement which provide greater protection for persons deprived of their liberty.

2. Nothing in this Convention shall be construed as limiting or derogating from the competence of the organs of the European Convention on Human Rights or from the obligations assumed by the Parties under that Convention.

3. The Committee shall not visit places which representatives or delegates of Protecting Powers or the International Committee of the Red Cross effectively visit on a regular basis by virtue of the Geneva Conventions of 12 August 1949 and the Additional Protocols of 8 June 1977 thereto.

Chapter V

Article 18

This Convention shall be open for signature by the member States of the Council of Europe. It is subject to ratification, acceptance or approval. Instruments of ratification, acceptance or approval shall be deposited with the Secretary General of the Council of Europe.

Article 19

1. This Convention shall enter into force on the first day of the month following the expiration of a period of three months after the date on which seven member States of the Council of Europe have expressed their consent to be bound by the Convention in accordance with the provisions of Article 18.

2. In respect of any member State which subsequently expresses its consent to be bound by it, the Convention shall enter into force on the first day of the month following the expiration of a period of three months after the date of the deposit of the instrument of ratification, acceptance or approval.

Article 20

1. Any State may at the time of signature or when depositing its instrument of ratification, acceptance or approval, specify the territory or territories to which this Convention shall apply.

2. Any State may at any later date, by a declaration addressed to the Secretary General of the Council of Europe, extend the application of this Convention to any other territory specified in the declaration. In respect of such territory the Convention shall enter into force on the first day of the month following the expiration of a period of three months after the date of receipt of such declaration by the Secretary General.

3. Any declaration made under the two preceding paragraphs may, in respect of any territory specified in such declaration, be withdrawn by a notification addressed to the Secretary General. The withdrawal shall become effective on the first day of the month following the expiration of a period of three months after the date of receipt of such notification by the Secretary General.

Article 21

No reservation may be made in respect of the provisions of this Convention.

Article 22

1. Any Party may, at any time, denounce this Convention by means of a notification addressed to the Secretary General of the Council of Europe.

2. Such denunciation shall become effective on the first day of the month following the expiration of a period of twelve months after the date of receipt of the notification by the Secretary General.

Article 23

The Secretary General of the Council of Europe shall notify the member States of the Council of Europe of:

    a. any signature;

    b. the deposit of any instrument of ratification, acceptance or approval;

    c. any date of entry into force of this Convention in accordance with Article 19 and 20;

    d. any other act, notification or communication relating to this Convention, except for action taken in pursuance of Articles 8 and 10.

Annex

Privileges and immunities

Article 16

1. For the purpose of this annex, references to members of the Committee shall be deemed to include references to experts mentioned in Article 7, paragraph 2.

2. The members of the Committee shall, while exercising their functions and during journeys made in the exercise of their functions, enjoy the following privileges and immunities:

    a. immunity from personal arrest or detention and from seizure of their personal baggage and, in respect of words spoken or written and all acts done by them in their official capacity, immunity from legal process of every kind;

    b. exemption from any restrictions on their freedom of movement on exist from and return to their country of residence, and entry into and exit from the country in which they exercise their functions, and from alien registration in the country which they are visiting or through which they are passing in the exercise of their functions.

3. In the course of journeys undertaken in the exercise of their functions, the members of the Committee shall, in the matter of customs and exchange control, be accorded.

    a. by their own Government, the same facilities as those accorded to senior officials travelling abroad on temporary official duty;

    b. by the Governments of other Parties, the same facilities as those accorded to representatives of foreign Governments on temporary official duty.

4. Documents and papers of the Committee, in so far as they related to the business of the Committee, shall be inviolable.

The official correspondence and other official communications of the Committee may not be held up or subjected to censorship.

5. In order to secure for the members of the Committee complete freedom of speech and complete independence on the discharge of their duties, the immunity from legal process in respect of words spoken or written and all acts done by them in discharging their duties shall continue to be accorded, notwithstanding that the persons concerned are no longer engaged in the discharge of such duties.

6. Privileges and immunities are accorded to the members of the Committee, not for the personal benefit of the individuals themselves but in order to safeguard the independent exercise of their functions. The Committee alone shall be competent to waive the immunity of its members; it has not only the right, but is under a duty, to waive the immunity of one of its members in any case where, in its opinion, the immunity would impede the course of justice, and where it can be waived without prejudice to the purpose for which the immunity is accorded.

Appendix 6
Criminal Justice Act 1991, Part II

Preliminary

The Parole Board

32.—(1) There shall continue to be a body to be known as the Parole Board ('the Board') which shall discharge the functions conferred on it by this Part.

(2) It shall be the duty of the Board to advise the Secretary of State with respect to any matter referred to it by him which is connected with the early release of recall or prisoners.

(3) The Board shall deal with cases as respects which it makes recommendations under this Part on consideration of—

(*a*) any documents given to it by the Secretary of State; and

(*b*) any other oral or written information obtained by it

and if in any particular case the Board thinks it necessary to interview the person to whom the case relates before reaching a decision, the Board may authorise one of its members to interview him and shall consider the report of the interview made by that member.

(4) The Board shall deal with cases as respects which it gives directions under this Part on consideration of all such evidence as may be adduced before it.

(5) Without prejudice to subsections (3) and (4) above, the Secretary of State may make rules with respect to the proceedings of the Board, including provision authorising cases to be dealt with by a prescribed number of its members or requiring cases to be dealt with at prescribed times.

(6) The Secretary of State may also give to the Board directions as to the matters to be taken into account by it in discharging any functions under this Part; and in giving any such directions the Secretary of State shall in particular have regard to—

(*a*) the need to protect the public from serious harm from offenders; and

(*b*) the desirability of preventing the commission of them of further offences and of securing their rehabilitation.

(7) Schedule 5) to this Act shall have effect with respect to the Board.

New arrangements for early release

Duty to release short-term and long-term prisoners

33.—(1) As soon as a short-term prisoner has served one-half of his sentence, it shall be the duty of the Secretary of State—

(*a*) to release him unconditionally if that sentence is for a term of less than twelve months; and (*b*) to release him on licence if that sentence is for a term of twelve months or more.

(2) As soon as a long-term prisoner has served two-thirds of his sentence, it shall be the duty of the Secretary of State to release him on licence.

(3) As soon as a short-term or long-term prisoner who—

(*a*) has been released on licence under subsection (1) (*b*) or (2) above or section 35 or 36 (1) below; and

(*b*) has been recalled to prison under section 38 (2) or 39 (1) below,

would (but for his release) have served three-quarters of his sentence, it shall be the duty of the Secretary of State to release him unconditionally.

(4) Where a prisoner whose sentence is for a term of less than twelve months h been released on licence under section 36 (1) below and recalled to prison unde4 section 38 (2) below, subsection (3) above shall have effect as if for the reference to three-quarters of his sentence there were substituted a reference to one-half of that sentence.

(5) In this Part—

'long-term prisoner' means a person serving a sentence of imprisonment for a term of four years or more;

'short-term prisoner' means a person serving a sentence of imprisonment for a term of less than four years.

Duty to release discretionary life prisoners

34.—(1) A life prisoner is a discretionary life prisoner for the purposes of this Part if—

(*a*) his sentence was imposed for a violent or sexual offence the sentence for which is not fixed by law; and

(*b*) the court by which he was sentenced for that offence ordered that this section should apply to him as soon as he had served a part of his sentence specified in the order.

(2) A part of a sentence so specified shall be such part as the court considers appropriate taking into account—

(*a*) the seriousness of the offence, or the combination of the offence and other offences associated with it; and

(*b*) the provisions of this section as compared with those of section 33 (2) above and section 35 (1) below.

(3) As soon as, in the case of a discretionary life prisoner—

(*a*) he has served the part of his sentence specified in the order ('the relevant part'); and

(*b*) the Board has directed his release under this section,

it shall be the duty of the Secretary of State to release him on licence.

(4) The Board shall not give a direction under subsection (3) above with respect to a discretionary life prisoner unless—

(*a*) the Secretary of State has referred the prisoner's case to the Board; and (*b*) the Board is satisfied that it is no longer necessary for the protection of the public that the prisoner should be confined.

(5) A discretionary life prisoner may require the Secretary of State to refer his case to the Board at any time—

(*a*) after he has served the relevant part of his sentence; and

(*b*) where there has been a previous reference of his case to the Board, after the end of the period of two years beginning with the disposal of that reference; and

(*c*) where he is also serving a sentence of imprisonment for a term, after he has served one-half of that sentence;

and in this subsection 'previous reference' means a reference under subsection (4) above or section 39 (4) below made after the prisoner had served the relevant part of his sentence.

(6) In determining for the purpose of subsection (3) or (5) above whether a discretionary life prisoner has served the relevant part of his sentence, no account shall be taken of any time during which he was unlawfully at large within the meaning of section 49 of the Prison Act 1952 ('the 1952 Act').

(7) In this Part 'life prisoner' means a person serving one or more sentences of life imprisonment; but—

(*a*) a person serving two or more such sentences shall not be treated as a discretionary life prisoner for the purpose of this Part unless the requirements of subsection (1) above are satisfied as respects each of those sentences; and

(*b*) subsections (3) and (5) above shall not apply in relation to such a person until after he has served the relevant part of each of those sentences.

Power to release long-term and life prisoners

35.—(1) After a long-term prisoner has served one-half of his sentence, the Secretary of State may, if recommended to do so by the Board, release him on licence.

(2) If recommended to do so by the Board, the Secretary of State may, after consultation with the Lord Chief Justice together with the trial judge if available, release on licence a life prisoner who is not a discretionary life prisoner,

(3) The Board shall not make a recommendation under subsection (2) above unless the Secretary of State has referred the particular case, or the class of case to which that case belongs, to the Board for its advice.

Power to release prisoners on compassionate grounds

36.—(1) The Secretary of State may at any time release a prisoner on licence if he is satisfied that exceptional circumstances exist which justify the prisoner's release on compassionate grounds.

(2) Before releasing a long-term or life prisoner under subsection (1) above, the Secretary of State shall consult the Board, unless the circumstances are such as to render such consultation impracticable.

Duration and conditions of licences

37.—(1) Subject to subsection (2) below, where a short-term or long-term prisoner is released on licence, the licence shall, subject to any suspension under section 38 (2) below or, as the case may be, any revocation under section 39 (1) or (2) below, remain in force until the date on which he would (but for his release) have served three-quarters of his sentence.

(2) Where a prisoner whose sentence is for a term of less than twelve months is released on licence under section 36 (1) above, subsection (1) above shall have effect as if for the reference to three-quarters of his sentence there were substituted a reference to one-half of that sentence.

(3) Where a life prisoner is released on licence, the licence shall, unless previously revoked under section 39 (1) or (2) below, remain in force until his death.

(4) A person subject to a licence shall comply with such conditions (which shall include on his release conditions as to his supervision by a probation officer) as may for the time being be specified in the licence; and the Secretary of State may make rules for regulating the supervision of any description of such persons.

(5) The Secretary of State shall not include on release, or subsequently insert, a condition in the licence of a long-term or life prisoner, or vary or cancel any such condition, except—

(*a*) in the case of the inclusion of a condition in the licence of a discretionary life prisoner, in accordance with recommendations of the Board; and

(*b*) in any other case, after consultation with the Board.

(6) For the purposes of subsection (5) above, the Secretary of State shall be treated as having consulted the Board about a proposal to include, insert, vary or cancel a condition in any case if he has consulted the Board about the implementation of proposals of that description generally or in that class of case.

(7) The power to make rules under this section shall be exerciseable by statutory instrument which shall be subject to annulment in pursuance of a resolution of either House of Parliament.

Misbehaviour after release

Breach of licence conditions by short-term prisoners

38.—(1) A short-term prisoner—

(*a*) who is released on licence under this Part; and

(*b*) who fails to comply with such conditions as may for the time being be specified in the licence,

shall be liable on summary conviction to a fine not exceeding level 3 on the standard scale.

(2) The magistrates' court by which a person is convicted of an offence under subsection (1) above may, whether or not it passes any other sentence on him—

(*a*) suspend the licence for a period not exceeding six months; and

(*b*) order him to be recalled to prison for the period during which the licence is suspended.

(3) On the suspension of the licence of any person under this section, he shall be liable to be detained in pursuance of his sentence and, if at large, shall be deemed to be unlawfully at large.

Recall of long-term and life prisoners while on licence

39.—(1) If recommended to do so by the Board in the case of a long-term or life prisoner who has been released on licence under this Part, the Secretary of State may revoke his licence and recall him to prison.

(2) The Secretary of State may revoke the licence of any such person and recall him to prison without a recommendation by the Board, where it appears to him that it is expedient in the public interest to recall that person before such a recommendation is practicable.

(3) A person recalled to prison under subsection (1) or (2) above—

(a) may make representations in writing with respect to his recall; and

(b) on his return to prison, shall be informed of the reasons for his recall and of his right to make representations.

(4) The Secretary of State shall refer to the Board—

(a) the case of a person recalled under subsection (1) above who makes representations under subsection (3) above; and

(b) the case of a person recalled under subsection (2) above.

(5) Where on a reference under subsection (4) above the Board—

(a) directs in the case of discretionary life prisoner; or

(b) recommends in the case of any other person,

his immediate release on licence under this section, the Secretary of State shall give effect to the direction or recommendation.

(6) On the revocation of the licence of any person under this section, he shall be liable to be detained in pursuance of his sentence and, if at large, shall be deemed to be unlawfully at large.

Convictions during currency of original sentences

40.—(1) This section applies to a short-term or long-term prisoner who is released under this Part if—

(a) before the date on which he would (but for his release) have served his sentence in full, he commits an offence punishable with imprisonment; and

(b) whether before or after that date, he is convicted of that offence ('the new offence').

(2) Subject to subsection (3) below, the court by or before which a person to whom this section applies is convicted of the new offence may, whether or not it passes any other sentence on him, order him to be returned to prison for the whole or any part of the period which—

(a) begins with the date of the order; and

(b) is equal in length to the period between the date on which the new offence was committed and the date mentioned in subsection (1) above.

(3) A magistrates' court—

(a) shall not have power to order a person to whom this section applies to be returned to prison for a period of more than six months; but

(b) may commit him in custody or on bail to the Crown Court for sentence in accordance with section 42 of the 1973 Act (power of Crown Court to sentence persons convicted by magistrates' courts of indictable offences).

(4) The period for which a person to whom this section applies is ordered under subsection (2) above to be returned to prison—

(a) shall be taken to be a sentence of imprisonment for the purposes of this Part;

(b) shall, as the court may direct, either be served before and be followed by, or be served concurrently with, the sentence imposed for the new offence; and

(c) in either case, shall be disregarded in determining the appropriate length of that sentence.

Remand time and additional days

Remand time to count towards time served

41.—(1) This section applies to any person whose sentence falls to be reduced under section 67 of the Criminal Justice Act 1967 ('the 1967 Act') by any relevant period within the meaning of that section ('the relevant period').

(2) For the purpose of determining for the purposes of this Part—

(a) whether a person to whom this section applies has served one-half or two-thirds of his sentence; or

(b) whether such a person would (but for his release) have served three-quarters of that sentence,

the relevant period shall, subject to subsection (3) below, be treated as having been served by him as part of that sentence.

(3) Nothing in subsection (2) above shall have the effect of reducing the period for which a licence granted under this Part to a short-term or long-term prisoner remains in force to a period which is less than—

(a) one-quarter of his sentence in the case of a short-term prisoner; or

(b) one-twelfth of his sentence in the case of a long-term prisoner.

Additional days for disciplinary offences

42.—(1) Prison rules, that is to say, rules made under section 47 of the 1952 Act, may include provision for the award of additional days—

(a) to short-term or long-term prisoners; or

(b) conditionally on their subsequently becoming such prisoners, to persons on remand,

who (in either case) are guilty of disciplinary offences.

(2) Where additional days are awarded to a short-term or long-term prisoner, or to a person on remand who subsequently becomes such a prisoner, and are not remitted in accordance with prison rules—

(a) any period which he must serve before becoming entitled to or eligible for release under this Part; and

(b) any period for which a licence granted to him under this Part remains in force,

shall be extended by the aggregate of those additional days.

Special cases

Young offenders

43.—(1) Subject to subsections (4) and (5) below, this Part applies to persons serving sentences of detention in a young offender institution, or determinate sentences of detention under section 53 of the 1933 Act, as it applies to persons serving equivalent sentences of imprisonment.

(2) Subject to subsection (5) below, this Part applies to persons serving—

(*a*) sentences of detention during Her Majesty's pleasure or for life under section 53 of the 1933 Act; or

(*b*) sentences of custody for life under section 8 of the 1982 Act,

as it applies to persons serving sentences of imprisonment for life.

(3) References in this Part to prisoners (whether short-term, long-term or life prisoners), or to prison or imprisonment, shall be construed in accordance with subsections (1) and (2) above.

(4) In relation to a short-term prisoner under the age of 18 years to whom subsection (1) of section 33 applies, that subsection shall have effect as if it required the Secretary of State—

(*a*) to release him unconditionally if his sentence is for a term of twelve months or less; and

(*b*) to release him on licence if that sentence is for a term of more than twelve months.

(5) In relation to a person under the age of 22 years who is released on licence under this Part, section 37 (4) above shall have effect as if the reference to supervision by a probation officer included a reference to supervision by a social worker of a local authority social services department.

Sexual offenders

44. Where, in the case of a long-term or short-term prisoner—

(*a*) the whole or any part of his sentence was imposed for a sexual offence; and

(*b*) the court by which he was sentenced for that offence, having had regard to the matters mentioned in section 32 (6) (*a*) and (*b*) above, ordered that this section should apply,

sections 33 (3) and 37 (1) above shall each have effect as if for the reference to three-quarters of his sentence there were substituted a reference to the whole of that sentence.

Fine defaulters and contemnors

45.—(1) Subject to subsection (2) below, this Part (except sections 35 and 40 above) applies to persons committed to prison or to be detained under section 9 of the 1982 Act—

(*a*) in default of a payment of a sum adjudged to be paid by a conviction; or

(*b*) for contempt of court or any kindred offence,

as it applies to persons serving equivalent sentences of imprisonment; and references in this Part to short-term or long-term prisoners, or to prison or imprisonment, shall be construed accordingly.

(2) In relation to persons committed as mentioned in subsection (1) above, the provisions specified in subsections (3) and (4) below shall have effect subject to the modifications so specified.

(3) In section 33 above, for subsections (1) to (4) there shall be substituted the following subsections—

'(1) As soon as a person committed as mentioned in section 45 (1) below has served the appropriate proportion of his term, that is to say—

    (*a*) one-half, in the case of a person committed for a term of less than twelve months;

    (*b*) two-thirds, in the case of a person committed for a term of twelve months or more,

it shall be the duty of the Secretary of State to release him unconditionally.

(2) As soon as a person so committed who—

    (*a*) has been released on licence under section 36 (1) below; and

    (*b*) has been recalled under section 38 (2) or 39 (1) below,

would (but for his release) have served the appropriate proportion of his term, it shall be the duty of the Secretary of State to release him unconditionally.'

(4) In section 37 above, for subsections (1) to (3) there shall be substituted the following subsection—

'(1) Where a person committed as mentioned in section 45 (1) below is released on licence under section 36 (1) above, the licence shall, subject to—

    (*a*) any suspension under section 38 (2) below; or

    (*b*) any revocation under section 39 (1) below,

continue in force until the date on which he would (but for his release) have served the appropriate proportion of his term; and in this subsection "appropriate portion" has the meaning given by section 33 (1) above.'

Persons liable to removal from the United Kingdom

46.—(1) In relation to a long-term prisoner who is liable to removal from the United Kingdom, section 35 above shall have effect as if the words 'if recommended to do so by the Board' were omitted.

(2) In relation to a person who is liable to removal from the United Kingdom, section 37 (4) above shall have effect as if the words in parentheses were omitted.

(3) A person is liable to removal from the United Kingdom for the purposes of this section if—

    (*a*) he is liable to deportation under section 3 (5) of the Immigration Act 1971 and has been notified of a decision to make a deportation order against him;

    (*b*) he is liable to deportation under section 3 (6) of that Act;

    (*c*) he has been notified of a decision to refuse him leave to enter the United Kingdom; or

    (*d*) he is an illegal entrant within the meaning of section 33 (1) of that Act.

Persons extradited to the United Kingdom

47.—(1) A short-term or long-term prisoner is an extradited prisoner for the purposes of this section if—

(*a*) he was tried for the offence in respect of which his sentence was imposed—

    (i) after having been extradited to the United Kingdom; and

    (ii) without having first been restored or had an opportunity of leaving the United Kingdom; and

(*b*) he was for any period kept in custody while awaiting his extradition to the United Kingdom as mentioned in paragraph (*a*) above.

(2) If, in the case of an extradited prisoner, the court by which he was sentenced so ordered, section 67 of the 1967 Act (computation of sentences of imprisonment) shall have effect in relation to him as if a period specified in the order were a relevant period for the purposes of that section.

(3) The period that may be so specified is such period as in the opinion of the court is just in all the circumstances and does not exceed the period of custody mentioned in subsection (1) (b) above.

(4) In this section—

'extradited to the United Kingdom' means returned to the United Kingdom—

    (i) in pursuance of extradition arrangements;

    (ii) under any law of a designated Commonwealth county corresponding to the Extradition Act 1989;

    (iii) under that Act as extended to a colony or under any corresponding law of a colony; or

    (iv) in pursuance of a warrant of arrest endorsed in the Republic of Ireland under the law of that country corresponding to the Backing of Warrants (Republic of Ireland) Act 1965;

'extradition arrangements' has the meaning given by section 3 of the Extradition Act 1989;

'designated Commonwealth country' has the meaning given by section 5 (1) of that Act.

Life prisoners transferred to England and Wales

48.—(1) This section applies where, in the case of a transferred life prisoner, the Secretary of State, after consultation with the Lord Chief Justice, certifies his opinion, that if—

(*a*) he had been sentenced for his offence in England and Wales after the commencement of section 34 above; and

(*b*) the reference in subsection (1) (*a*) of that section to a violent or sexual offence the sentence for which is not fixed by law were a reference to any offence the sentence for which is not so fixed,

the court by which he was so sentenced would have ordered that that section should apply to him as soon as he had served a part of his sentence specified in the certificate.

(2) In a case to which this section applies, this Part except section 35 (2) above shall apply as if—

(*a*) the transferred life prisoner were a discretionary life prisoner for the purposes of this Part; and

(*b*) the relevant part of his sentence within the meaning of section 34 of this Act were the part specified in the certificate.

(3) In this section 'transferred life prisoner' means a person—

(*a*) on whom a court in a country or territory outside England and Wales has imposed one or more sentences of imprisonment or detention for an indeterminate period; and

(*b*) who has been transferred to England and Wales, in pursuance of—

(i) an order made by the Secretary of State under section 26 of the Criminal Justice Act 1961 or section 2 of the Colonial Prisoners Removal Act 1884; or

(ii) a warrant issued by the Secretary of State under the Repatriation of Prisoners Act 1984,

there to serve his sentence or sentences or the remainder of his sentence or sentences.

(4) A person who is required so to serve the whole or part of two or more such sentences shall not be treated as a discretionary life prisoner for the purposes of this Part unless the requirements of subsection (1) above are satisfied as respects each of those sentences; and subsections (3) and (5) of section 34 above shall not apply in relation to such a person until after he has served the relevant part of each of those sentences.

Supplemental

Alteration by order of relevant proportions of sentences

49.—(1) The Secretary of State may by order made by statutory instrument provide—

(*a*) that the references in section 33 (5) above to four years shall be construed as references to such other period as may be specified in the order;

(*b*) that any reference in this Part to a particular proportion of a prisoner's sentence shall be construed as a reference to such other proportion of a prisoner's sentence as may be so specified.

(2) An order under this section may make such transitional provisions as appear to the Secretary of State necessary or expedient in connection with any provisions made by the order.

(3) No order shall be made under this section unless a draft of the order has been laid before and approved by resolution of each House of Parliament.

Transfer by order of certain functions to Board

50.—(1) The Secretary of State, after consultation with the Board, may by order made by statutory instrument provide that, in relation to such class of case as may be specified in the order, the provisions of this Part specified in subsections (2) to (4) below shall have effect subject to the modifications so specified.

(2) In section 35 above, in subsection (1) for the word 'may' there shall be substituted the word 'shall'; but nothing in this subsection shall affect the operation of that subsection as it has effect in relation to a long-term prisoner who is liable to removal from the United Kingdom (within the meaning of section 46 above).

(3) In section 37 above, in subsection (5) (*a*) after the words 'in the case of' there shall be inserted the words 'the licence of a long-term prisoner or,' and subsection (6) shall be omitted.

(4) In section 39 above, in subsection (1) for the word 'may' there shall be substituted the word 'shall,' and subsection (2) shall be omitted.

(5) No order shall be made under this section unless a draft of the order has been laid before and approved by resolution of each House of Parliament.

Interpretation of Part II

51.—(1) In this Part—

'the Board' means the Parole Board;

'discretionary life prisoner' has the meaning given by section 34 above (as extended by section 43 (2) above);

'life prisoner; has the meaning given by section 34 (7) above (as extended by section 43 (2) above);

'long-term prisoner' and 'short-term prisoner' have the meanings given by section 33 (5) above (as extended by sections 43 (1) and 45 (1) above);

'sentence of imprisonment' does not include a committal in default of payment of any sum of money, or for want of sufficient distress to satisfy any sum of money, or for failure to do or abstain from doing anything required to be done or left undone.

'sexual offence' and 'violent offence' have the same meanings as in Part I of this Act.

(2) For the purposes of any reference in this Part, however expressed, to the term of imprisonment to which a person has been sentenced or which, or part of which, he has served, consecutive terms and terms which are wholly or partly concurrent shall be treated as a single term.

(3) Nothing in this Part shall require the Secretary of State to release a person who is serving—

(*a*) a sentence of imprisonment for a term; and

(*b*) one or more sentences of imprisonment for life,

unless and until he is entitled under this Part to be released in respect of each of those sentences.

(4) Subsections (2) and (3) of section 31 above shall apply for the purposes of this Part as they apply for the purposes of Part I of this Act.

SCHEDULE 12
TRANSITIONAL PROVISIONS AND SAVINGS

Early release: general

8.—(1) In this paragraph and paragraphs 9 to 11 below—

'existing licensee' means any person who, before the commencement of Part II of this Act, has been released on licence under section 60 of the 1967 Act and whose licence under that section is in force at that commencement;

'existing prisoner' means any person who, at that commencement, is serving a custodial sentence;

and sub-paragraphs (2) to (7) below shall have effect subject to those paragraphs.

(2) Subject to sub-paragraphs (3) to (7) below, Part II of this Act shall apply in relation to an existing licensee as it applies in relation to a person who is released on licence under that Part; and in its application to an existing prisoner, or to an existing licensee who is recalled under section 39 of this Act, that Part shall apply with the modifications made by those sub-paragraphs.

(3) Section 40 of this Act shall not apply in relation to an existing prisoner or licensee.

(4) In relation to an existing prisoner whose sentence is for a term of twelve months, section 33 (1) of this Act shall apply as if that sentence were for a term of less than twelve months.

(5) In relation to an existing prisoner or licensee whose sentence is for a term of—

(*a*) more than twelve months; and

(*b*) less than four years or, as the case may require, such other period as may for the time being be referred to in section 33 (5) of this Act,

Part II of this Act shall apply as if he were or had been a long-term rather than a short-term prisoner.

(6) In relation to an existing prisoner or licensee whose sentence is for a term of more than twelve months—

(*a*) section 35 (1) of this Act shall apply as if the reference to one half of his sentence were a reference to one-third of that sentence or six months, whichever is the longer; and

(*b*) sections 33 (3) and 37 (1) of this Act shall apply as if the reference to three-quarters of his sentence were a reference to two-thirds of that sentence.

(7) In relation to an existing prisoner or licensee—

(*a*) whose sentence is for a term of more than twelve months; and

(*b*) whose case falls within such class of cases as the Secretary of State may determine after consultation with the Parole Board,

section 35 (1) of this Act shall apply as if the reference to a recommendation by the Board included a reference to a recommendation by a local review committee established under section 59 (6) of the 1967 Act.

(8) In this paragraph 'custodial sentence' means—

(*a*) a sentence of imprisonment;

(*b*) a sentence of detention in a young offender institution;

(*c*) a sentence of detention (whether during Her Majesty's pleasure, for life or for a determinate term) under section 53 of the 1933 Act; or

(*d*) a sentence of custody for life under section 8 of the 1982 Act.

9.—(1) This paragraph applies where, in the case of an existing life prisoner, the Secretary of State certifies his opinion that, if—

(*a*) section 34 of this Act had been in force at the time when he was sentenced; and

(*b*) the reference in subsection (1) (*a*) of that section to a violent or sexual offence the sentence for which is not fixed by law were a reference to any offence the sentence of which is not so fixed,

the court by which he was sentenced would have ordered that that section should apply to him as soon as he had served a part of his sentence specified in the certificate.

(2) In a case to which this paragraph applies, Part II of this Act except section 35 (2) shall apply as if—

(*a*) the existing life prisoner were a discretionary life prisoner for the purposes of that Part; and

(*b*) the relevant part of his sentence within the meaning of section 34 of this Act were the part specified in the certificate.

(3) In this paragraph 'existing life prisoner' means a person who, at the commencement of Part II of this Act, is serving one or more of the following sentences, namely—

(*a*) a sentence of life imprisonment;

(*b*) a sentence of detention during Her Majesty's pleasure or for life under section 53 of the 1933 Act; or

(*c*) a sentence of custody for life under section 8 of the 1982 Act.

(4) A person serving two or more such sentences shall not be treated as a discretionary life prisoner for the purposes of Part II of this Act unless the requirements of sub-paragraph (1) above are satisfied as respects each of those sentences; and subsections (3) and (5) of section 34 of this Act shall not apply in relation to such a person until after he has served the relevant part of each of those sentences.

10. Prison rules made by virtue of section 42 of this Act, in relation to any existing prisoner or licensee who has forfeited any remission of his sentence, as if he had been awarded such number of additional days as may be determined by or under the rules.

Early release of young persons detained under 1933 Act

11. In relation to an existing prisoner or licensee whose sentence is a determinate sentence of detention under section 53 of the 1933 Act—

(*a*) Part II of this Act shall apply as if he were or had been a life rather than a long-term or short-term prisoner;

(*b*) section 35 (2) of this Act shall apply as if the requirement as to consultation were omitted; and

(*c*) section 37 (3) of this Act shall apply as if the reference to his death were a reference to the date on which he would (but for his release) have served the whole of his sentence.

Early release of prisoners serving extended sentences

12.—(1) In relation to an existing prisoner or licensee on the passing of whose sentence an extended sentence certificate was issued—

(*a*) section 33 (3) of this Act shall apply as if the duty to release him unconditionally were a duty to release him on licence; and

(*b*) section 37 (1) of this Act shall apply as if the reference to three-quarters of his sentence were a reference to the whole of that sentence.

(2) In this paragraph 'extended sentence certificate' means a certificate issued under section 28 of the 1973 Act stating that an extended term of imprisonment was imposed on an offender under that section.

Early release of fine defaulters and contemnors

13. Part II of this Act shall apply in relation to any person who, before the commencement of that Part, has been committed to prison or to be detained under section 9 of the 1982 Act—

(*a*) in default of payment of a sum adjudged to be paid by a conviction; or

(*b*) for contempt of court or any kindred offence,

as it applies in relation to any person who is so committed after that commencement.

The Parole Board Rules 1992

Made 1992
Coming into force.........................1992

ARRANGEMENT OF RULES

Part I
Introductory

1. Title and commencement
2. Application and interpretation

Part II
General

3. Appointment of panel
4. Listing the case for hearing
5. Information and reports by the Secretary of State

In exercise of the powers conferred upon me by section 32 (5) of the Criminal Justice Act 1991ᵃ, I hereby make the following Rules:—

Part 1
Introductory

Title and commencement

1. These Rules may be cited as the Parole Board Rules 1992 and shall come into force on 1st October 1992.

Application and interpretation

2. (1) Subject to rule 19, these Rules apply where a prisoner's case is referred to the Board by the Secretary of State under section 34 (4) (a), section 34 (5) or section 39 (4) of the Act.

(2) In these Rules, unless a contrary intention appears—

'Board' means the Parole Board, continued by section 32 (1) of the Act,

a 1991 c. 53.

'Chairman' means the chairman of the board appointed under paragraph 1 of Schedule 5 to the Act,

'panel' means those members of the Board constituted in accordance with rule 3,

'parties' means the prisoner and the Secretary of State,

'prisoner' means a discretionary life prisoner within the meaning of section 34 (1) of the Act,

'the Act' means the Criminal Justice Act 1991.

Part II
General

Appointment of panel

3. (1) The Chairman shall appoint three members of the Board to form a panel for the purpose of conducting proceedings in relation to a prisoner's case.

   (2) The members of the panel appointed under paragraph (1) shall include a person who hold judicial office and who shall act as chairman of the panel.

Listing the case for hearing

4. The Board shall list the case for hearing and, as soon as practicable thereafter, notify the parties of the date when the case was so listed.

Information and reports by the Secretary of State

5. (1) Within eight weeks of the case being listed, the Secretary of State shall serve on the Board and, subject to paragraph (2), the prisoner or his representative—

   (a) the information specified in Part A of Schedule 1 to these Rules,
   (b) the reports specified in Part B of that Schedule, and
   (c) such further information that the Secretary of State considers to be relevant to the case.

   (2) Any part of the information or reports referred to in paragraph (1) which, in the opinion of the Secretary of State, should be withheld from the prisoner on the ground that its disclosure would adversely affect the health or welfare of the prisoner or others, shall be recorded in a separate document and served only on the Board together with the reasons for believing that its disclosure would have that effect.

   (3) Where a document is withheld from the prisoner in accordance with paragraph (2), it shall nevertheless be served as soon as practicable on the prisoner's representative if he is—

   (a) a barrister or solicitor,
   (b) a registered medical practitioner, or
   (c) a person whom the chairman of the panel directs is suitable by virtue of his experience of professional qualification;

provided that no information disclosed in accordance with this paragraph shall be disclosed either directly or indirectly to the prisoner or to any other person without the authority of the chairman of the panel.

Representation, etc.

6. (1) Subject to paragraph (2), a party may be represented by any person who he has authorised for that purpose.

(2) The following are ineligible to act as a representative before the Board—

(a) any person liable to be detained under the Mental Health Act 1983,

(b) any person serving a sentence of imprisonment;

(c) any person who is on licence having been released under Part III of the Criminal Justice Act 1967 or under Part II of the Act,

(d) any person with a previous conviction for an imprisonable offence which remains unspent under the Rehabilitation of Offenders Act 1974[a].

(3) Within three weeks of the case being listed, a party shall notify the Board and the other party of the name, address and occupation of any person authorised in accordance with paragraph (1).

(4) Where a prisoner does not authorise a person to act as his representative, the Board may, with his agreement, appoint someone to act on his behalf.

(5) A party may apply, in accordance with the procedure set out in rule 7 (1) and (2), to be accompanied at the hearing by such other person or persons as he wishes, in addition to any representative he may have authorised; but before granting any such application the Board shall obtain the agreement of—

(a) in the case where the hearing is to be held at a prison, the governor, and

(b) in any other case, the person in whom is vested the authority to agree.

Witnesses

7. (1) Where a party wishes to call witnesses at the hearing, he shall make a written application to the Board, a copy of which he shall serve on the other party, within 12 weeks of the case being listed, giving the name, address and occupation of the witness he wishes to call and the substance of the evidence he proposes to adduce.

(2) The chairman of the panel may grant or refuse an application under paragraph (1) and shall communicate his decision to both parties, giving reasons in writing, in the case of a refusal, for his decision.

Evidence of the prisoner

8. (1) Where the prisoner wishes to make representations about his case, he shall serve them on the Board and the Secretary of State within 15 weeks of the case being listed.

(2) Any other documentary evidence that the prisoner wishes to adduce shall

a 1974 c. 53.

be served on the Board and the Secretary of State at least 14 days before the date of the hearing.

Directions

9. (1) Subject to paragraph (3), the chairman of the panel may give, vary or revoke directions for the conduct of the case, including directions in respect of—

   (*a*) the timetable for the proceedings,

   (*b*) the varying of the time within which or by which an act is required, by these Rules, to be done,

   (*c*) the service of documents,

   (*d*) as regards any documents which have been received by the Board but which have been withheld from the prisoner in accordance with rule 5 (2), whether the disclosure of such documents would adversely affect the health or welfare of the prisoner or others, and

   (*e*) the submission of evidence;

   and following his appointment under rule 3, the chairman of the panel shall consider whether such directions need to be given at any time.

   (2) Within 14 days of being notified of a direction under paragraph (1) (d), either party may appeal against it to the Chairman, who shall notify the other party of the appeal; the other party may make representations on the appeal to the Chairman whose decision shall be final.

   (3) Directions under paragraph (1) may be given, varied or revoked either—

   (*a*) of the chairman of the panel's own motion, or

   (*b*) on the written application of a party to the Board which has been served on the other party and which specifies the direction which is sought;

   but in either case, both parties shall be given an opportunity to make written representations or, where the chairman of the panel thinks it necessary, and subject to paragraph (6) (b), to make oral submissions at a preliminary hearing fixed in accordance with paragraph (4).

   (4) Where the chairman of the panel decides to hold a preliminary hearing, he shall give the parties at least 14 days' notice of the date, time and place which has been fixed in respect thereof.

   (5) A preliminary hearing shall be held in private and information about the proceedings and the names of any persons concerned in the proceedings shall not be made public.

   (6) Except in so far as the chairman of the panel otherwise directs, at a preliminary hearing—

   (*a*) the chairman of the panel shall sit alone, and

   (*b*) the prisoner shall not attend save where he is unrepresented.

   (7) The chairman of the panel shall take a note of the giving, variation or revocation of a direction under this rule and serve a copy on the parties as soon as practicable thereafter.

Part III
The hearing

Oral hearing

10. (1) Except in so far as both parties and the chairman of the panel agree otherwise, there shall be an oral hearing of the prisoner's case.
    (2) The prisoner shall, within five weeks of the case being listed, notify the
Board and the Secretary of State whether he wishes to attend the hearing.

Notice of hearing

11. (1) When fixing the date of the hearing the Board shall consult the parties.
    (2) The Board shall give the parties at least three weeks notice of the date, time
and place scheduled for the hearing or such shorter notice to which the parties may consent.

Location, privacy of proceedings

12. (1) The hearing shall be held at the prison or other institution where the prisoner is detained.
    (2) The hearing shall be held in private and, except in so far as the chairman of
the panel otherwise directions, information about the proceedings and the
names of any persons concerned in the proceedings shall not be made public.
    (3) The chairman of the panel may admit to the hearing such persons on such
terms and conditions as he considers appropriate.

Hearing procedure

13. (1) At the beginning of the hearing the chairman of the panel shall explain the
order of proceeding which the panel proposes to adopt.
    (2) Subject to this rule, the panel shall conduct the hearing in such manner as it
considers most suitable to the clarification of the issues before it and generally to the just handling of the proceedings; it shall so far as appears to it
appropriate, seek to avoid formality in the proceedings.
    (3) The parties shall be entitled to appear and be heard at the hearing and take
such part in the proceedings as the panel thinks proper; and the parties may
hear each others' evidence, put questions to each other, call any witnesses
who the Board has authorised to give evidence in accordance with rule 7,
and put questions to any witness or other person appearing before the
panel.
    (4) The chairman of the panel may require any person present at the hearing
who is, in his opinion, behaving in a disruptive manner to leave and may
permit him to return, if at all, only on such conditions as he may specify.
    (5) The panel may receive in evidence any document or information notwith-
standing that such document or information would be inadmissible in a
court of law but no person shall be compelled to give any evidence or produce any document which he could not be compelled to give or produce on
the trial of an action.

(6) The chairman of the panel may require the prisoner, or any witness appearing for the prisoner, to leave the hearing where evidence is being examined which the chairman of the panel, in accordance with rule 9 (1) (d) (subject to any successful appeal under rule 9 (2)), previously directed should be withheld from the prisoner as being injurious to the health or welfare of the prisoner or another person.

(7) After all the evidence has been given, the prisoner shall be given a further opportunity to address the panel.

Adjournment

14. (1) The panel may at any time adjourn a hearing for the purpose of obtaining further information or for such other purposes as it may think appropriate.

(2) Before adjourning any hearing, the panel may give such directions as it thinks fit for ensuring the prompt consideration of the application at a resumed hearing.

(3) Before the panel resumes any hearing which was adjourned without a further hearing date being fixed it shall give the parties not less than 14 days' notice, or such shorter notice to which all parties may consent, of the date, time and place of the resumed hearing.

The decision

15. (1) Any decision of the majority of the members of the panel shall be the decision of the panel.

(2) The decision by which the panel determines a case shall be recorded in writing with reasons, signed by the chairman of the panel, and communicated in writing to the parties not more than seven days after the end of the hearing.

Part IV
Miscellaneous

Time

16. Where the time prescribed by or under these Rules for doing any act expires on a Saturday, Sunday or public holiday, the act shall be in time if done on the next working day.

Transmission of documents etc.

17. Any document required or authorised by these Rules to be served or otherwise transmitted to any person may be sent by pre-paid post or delivered,—

(a) in the case of a document directed to the Board or the chairman of the panel, to the office of the Board;

(b) in any other case, to the last known address of the person to whom the document is directed.

Irregularities

18. Any irregularity resulting from failure to comply with these Rules before the panel has determined a case shall not of itself render the proceedings void, but the panel may, and shall, if it considers that the person may have been prejudiced, take such steps as it thinks fit, before determining the case, to cure the irregularity, whether by the amendment of any document, the giving of any notice, the taking of any step or otherwise.

References to the Board following recall

19. Where the Secretary of State refers a prisoner's case to the Board under section 39 (4) of the Act, and the prisoner has made representations under section 39 (3) of the Act, these Rules shall apply subject to the following modifications—

   (*a*) rules 5 (1), 6 (3), 7 (1), 8 (1) and (2), 9 (2) and (4), 10 (2), 11 (2), 14 (3) and 15 (2) shall apply as if for references to the periods of time specified therein there were substituted a reference to such period of time as the chairman of the panel shall in each case determine, taking account of both the desirability of the Board reaching an early decision in the prisoners' case and the need to ensure fairness to the prisoner;

   (*b*) rule 5 shall apply as if for the references in paragraph (1) (a) and (b) of that rule to the information and reports specified in Schedule 1 there were substituted a reference to the information and reports specified in Schedule 2.

One of Her Majesty's Principal
Secretaries of State

Home Office
August 1992

SCHEDULE 1   Rule 5 (1)
INFORMATION AND REPORTS FOR SUBMISSION TO THE BOARD BY THE SECRETARY OF STATE ON A REFERENCE TO THE BOARD UNDER SECTION 34 (4) (A) OR 34 (5) OF THE ACT

Part A
Information relating to the prisoner.

1. The full name of the prisoner.

  2. The age of the prisoner.

  3. The prison in which the prisoner is detained and details of other prisons in which the prisoner has been detained, the date and reasons for any transfer.

  4. The date the prisoner was sentenced and the details of the offence.

  5. The previous convictions and parole history, if any, of the prisoner.

  6. The comments, if available, of the trial judge in passing sentence.

  7. Where applicable, the conclusions of the Court of Appeal in respect of any appeal by the prisoner against conviction or sentence.

8. The details of any life sentence plan prepared for the prisoner which have previously been disclosed to him.

Part B
Reports relating to the prisoner

1. Any pre-trial and pre-sentence reports examined by the sentencing court and any post-trial police report on the circumstances of the offence(s).

2. Any report on a prisoner while he was subject to a transfer direction under section 47 of the Mental Health Act 1983.

3. Any current reports on the prisoner's performance and behaviour in prison and, where relevant, on his health including any opinions on his suitability for release on licence (reports previously examined by the Board need only be summarised) as well as his compliance with any sentence plan.

4. An up-to-date home circumstances report prepared for the Board by a probation officer, including reports on the following—

(a) details of the home address, family circumstances, and family attitudes towards the prisoner;

(b) alternative options if the offender cannot return home;

(c) the opportunity for employment on release;

(d) the local community's attitude towards the prisoner (if known), including the attitudes and concerns of the victim(s) of the offence(s);

(e) the prisoner's response to previous periods of supervision;

(f) the prisoner's behaviour during any temporary leave during the current sentence;

(g) the prisoner's response to discussions of the objectives of supervision where applicable;

(h) an assessment of the risk of re-offending;

(i) a programme of supervision;

(j) a recommendation for release; and

(k) recommendations regarding any special licence conditions.

SCHEDULE 2　　Rules 5 (1) and 19 (b)
INFORMATION AND REPORTS FOR SUBMISSION TO THE BOARD BY THE SECRETARY OF STATE ON A REFERENCE TO THE BOARD UNDER SECTION 39 (4) OF THE ACT

Part A
Information relating to the prisoner

1. The full name of the prisoner.

2. The age of the prisoner.

3. The prison in which the prisoner is detained and details of other prisons in

which the prisoner has been detained, the date and reasons for any transfer.

4. The date the prisoner was sentenced and the details of the offence.

5. The previous convictions and parole history, if any, of the prisoner.

6. The details of any life sentence plan prepared for the prisoner which have previously been disclosed to him.

7. The details of any previous recalls of the prisoner including the reasons for such recalls and subsequent re-release on licence.

8. The statement of reasons for the most recent recall which was given to the prisoner under section 39 (3) (*b*) of the Act.

9. The details of any memorandum which the Board considered prior to making its recommendation for recall under section 39 (1) of the Act or confirming the Secretary of State's decision to recall under section 39 (2) of the Act, including the reasons why the Secretary of State considered it expedient in the public interest to recall that person before it was practicable to obtain a recommendation from the Board.

Part B
Reports relating to the prisoner

1. The reports considered by the Board prior to making its recommendation for recall under section 39 (1) of the Act or its confirmation of the Secretary of State's decision to recall under section 39 (2) of the Act.

2. Any other relevant reports.

Extract from Standing Order 3C*

4. Calculating Release Dates

4.1 *Remand Time* Once the length of the single term has been established, and the correct release scheme confirmed, the next step is to identify the amount of remand time which under section 67 of the Criminal Justice Act 1967 may be regarded as reducing the length of each sentence making up the single term. In general, the 1967 Act provides that remand time may only reduce the sentence given for the offence for which the prisoner was on remand. Where there is more than one sentence to be considered, it may be that not all remand time will attract to all sentences. In the case of concurrent *and* overlapping sentences, apportionment of relevant remand and custody time to each sentence will identify which sentences provides the latest Sentence Expiry Date (SED) for the purposes of calculating release dates. Only remand time applicable to the dominant sentence will then be calculated as reducing the sentence length.

* Standing Order 3C is currently undergoing revision. A revised version is expected to come into effect sometime after this work is published. Details can be obtained from the Prison Service.

4.2   The following principles should be followed in calculating remand time relevant to a sentence. Any sentence imposed for offences in respect of which the prisoner was before a court on or before the date of receiving the first sentence will be treated as being reduced by all the time spent in custody before the sentence was imposed subject to the following criteria:

    a) That the proceedings out of which the sentence arises must have been before the court on or before the date of first sentence.

    b) That the offences for which the sentence was passed must have been committed on or before the date of first remand into custody.

4.3   If the sentence under consideration satisfies both these criteria then it falls to be reduced by all the time spent in custody before first sentence was imposed. However, if the sentence does not meet criteria at (*a*) then it is not necessary to give further consideration to any remand time, for the sentence will not fall to be reduced.

4.4   Should the offence be committed after the date of first remand ie on bail, then the subsequent sentence may be reduced only by periods of remand after the actual commission of the offence.

4.5   Should the offence be committed in custody then only remand time applicable after the first court appearance on that charge will count towards the subsequent sentence.

4.6   One day appearances at court to surrender to bail do not count as custody.

4.7   Periods of certified policy custody apply only to the sentence given for the offence for which the prisoner was detained by the police. These periods are not transferrable and cannot reduce the release dates of other sentences.

4.8   For the counting of remand time any court disposal may be treated as a sentence but custodial remand preceding a 'not guilty' verdict cannot be counted against any subsequent sentence in respect of matters not before the court at the verdict.

4.9   Periods of custody to be counted as remand time are given at Annex C.

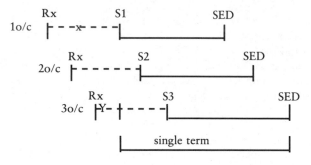

Example 4

4.10 In this example, only sentences 1 and 2 would be reduced by all the remand period (X). The offence for which sentence 3 was imposed was committed after the date of the first remand, whilst in custody. Only the portion of remand time arising after the commission of this offence (period Y) will reduce this sentence. The single term runs from day 1 of sentence number 1 for the full term ordered to be served by the court on sentence number 3 (**before** deduction of any remand). Since it also attracts the least amount of relevant remand time it will also clearly provide the latest actual Sentence Expiry Date for the purpose of calculating release dates. It should however be noted that in cases such as this where there are overlapping sentences, and these are of varying length and attract varying remand times, the sentence providing the longest term for assessing the length of the single term may not always provide the SED for calculating release dates once remand time has been deducted.

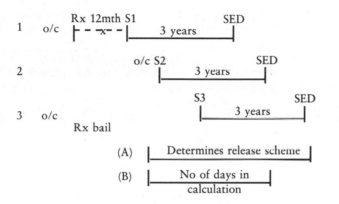

Example 5

In example 5 period (A) represents the single term for deciding under which of the release schemes the sentence falls to be calculated. This period takes no account of remand time.

When deciding the actual single term for calculation purposes the remand time should be apportioned to each of the sentences to determine which sentence provides the latest SED. This sentence becomes the dominant sentence and the single term will be calculated from the date of first sentence to the expiry of the full term of the dominant sentence, period (B) above. Sentence 2 does not attract any remand time because it was committed after the date of first sentence. Sentence 3 will be reduced by all of period X. Sentence 2 provides the latest SED and becomes the dominant sentence.

4.11 *Additional Days Awarded* If a prisoner is found guilty of a breach of prison rules the governor may order that the prisoner serve additional days. Additional Days Awarded (ADAs) will be added to all dates except the SED. A prisoner on remand may be awarded additional days prospectively. Should the prisoner subsequently receive a custodial sentence than the release dates will be deferred by the number of additional days awarded. (special provisions exist for existing prisoners see CI 29/1992)

4.12 Where a prisoner serving a sentence is given additional days awarded and subsequently receives an over lapping concurrent sentence the single term will be calculated in the normal way. The resultant release dates will be deferred by the additional days awarded.

4.13 A prisoner may have additional days awarded remitted in which case all release dates affected by ADAs will be advanced by the number of days remitted.

4.14 Each of a prisoner's release dates may be advanced by the addition of additional days until it reaches the SED.

4.15 *Appeals* Where the Crown Court, on appeal from a decision of a Magistrates' Court, substitutes a fresh sentence for that imposed by the lower court (or reimposes the same sentence) the release dates will be calculated as follows:

  a) Determine the number of days in the sentence imposed or reimposed by the Crown Court, reckoned from the date of the determination of the appeal;

  b) deduct any remand time applicable to the original sentence:

  c) unless the Court orders otherwise, deduct the term already served.

  d) calculate the appropriate release dates on the balance of the sentence.

4.16 Where the Court of Appeal Criminal Division varies a sentence on appeal, the new sentence will be calculated as commencing from the date of the original sentence, and will be treated as being reduced by any period which was treated as reducing the original unless otherwise ordered by the court.

4.17 Where a prisoner serving concurrent or consecutive sentences appeals against one of the sentences and it is quashed on appeal, it will be treated as a nullity and the remaining sentence(s) will be recalculated as though the quashed sentence had never been imposed.

4.18 It should be remembered that on appeal the Court of Appeal can increase as well as reduce sentences.

4.19 The following steps provide the basis for considering *single sentence* calculation under the three release schemes. All dates will be calculated from the date of sentence (for concurrent sentences see section 5 and consecutive sentences section 6 of this instruction):

All prisoners

  i. Identify the appropriate release scheme by reference to the total length of the sentence from the date of imposition to the expiry of the full term (the 'single term').

  ii. Determine the amount of remand time and police custody, if any, that is applicable to the sentence.

  iii. Deduct relevant remand and police custody time from the total number of days in the sentence. This will provide the number of days to be used to calculate the **Sentence Expiry Date (SED)**.

Prisoners serving under 12 months

These prisoners will be given an **Automatic Release Date (ARD)** and a **Sentence Expiry Date (SED)**.

iv. Steps i–iii above should be followed in calculating the **Sentence Expiry Date**.

v. To calculate the **Automatic Release Date** divide the total number of days in the single term by 2, rounding down half days and deduct the resultant figure from the number of days used to calculate the SED.

Prisoners serving between 12 months and less than 4 years

A prisoner serving 12 months and less than 4 years will be given a **Sentence Expiry Date (SED)**, a **Licence Expiry Date (LED)** and a **Conditional Release Date (CRD)**.

vi. Steps i–iii above should be followed in calculating the **Sentence Expiry Date**.

vii. To calculate the **Conditional Release Date** divide the total number of days in the single term by 2, rounding down half days and deduct the resultant figure from the number of days used to calculate the SED.

viii. To calculate the **Licence Expiry Date** divide the total number of days in the sentence by 4, rounding down fractions, and deduct the resultant figure from the number of days to the SED. If there are less days than this between the CRD and the LED, the LED should be altered accordingly. The minimum time to be served on licence is the total number of days in the sentence divided by 4, rounded down, or to the SED whichever comes first.

Prisoners serving 4 years and over

A prisoner serving 4 years and over will be given a **Sentence Expiry Date (SED)**, a **Parole Eligibility Date (PED)**, a **Non Parole Release Date (NPD)** and a **Licence Expiry Date (LED)**.

ix. Steps i–iii above should be used in calculating the **Sentence Expiry Date**.

x. To calculate the **Parole Eligibility Date** divide the total number of days in the sentence by 2, rounding down, and deduct the resultant figure from the number of days used to calculate the SED.

xi. To calculate the **Non Parole Release Date** divided the total number of days in the sentence by 3, rounding down, and deduct the resultant figure from the number of days used to calculate the SED.

xii. The **Licence Expiry Date** will be calculated by dividing the total number of days in the calculation period by 4 and deducting the resultant figure from the number of days to the SED. If there are less days than this between the NPD and the LED then the LED should be altered accordingly. The minimum time to be served on licence is the total number of days in the sentence divided by 12 (rounded down), or the SED whichever comes first.

N.B. Where the whole or part of a prisoner's sentence was imposed for a sexual offence the court may order that the licence shall run to the very end of that sentence. eg. A prisoner receives 12 months for a sexual offence and 12 months consecutive for another offence. The court may order that the licence shall run to the full two year point.

Example 6

Prisoner sentences to 9 month on 1 October 1992.

S	ARD	SED

A	Total length of sentence.	273 days		
B	Time in custody to count	—		
C	Number of days to SED (A-B)	273 days	SED = 30/6/93	
D	Number of days to ARD (C-$\underline{A}$)	137 days	ARD = 14/2/93	
	2			

The prisoner will be discharged on his Automatic Release Date.

Example 7

Prisoner sentenced to 2 years on 1 October 1992 with 30 days remand time.

RX	S	CRD	LED	SED

A	Total length of sentence	730 days		
B	Time in custody to count	30 days		
C	Sentence Expiry Date (A-B)	700 days	SED = 31/8/94	
D	Conditional Release Date C-($\underline{A}$)	335 days	CRD = 31/8/93	
	2			
E	Licence Expiry Date C-($\underline{A}$)	518 days	LED = 2/3/94	
	4			

The prisoner will be discharged under supervision on his Conditional Release Date.

Example 8

Prisoner sentences to 4 years on 1 October 1992 with 3 months on remand.

RX	S	PED	NPD	LED	SED

A	Total length of sentence.	1461 days		
B	Time in custody to count	92 days		
C	Sentence Expiry Date (A-B)	1369 days	SED = 30/6/96	
D	Parole Eligibility Date C-($\underline{A}$)	639 days	PED = 1/7/94	
	2			
E	Non Parole Release Date C-($\underline{A}$)	882 days	NPD = 1/3/95	
	3			
F	Licence Expiry Date C-($\underline{A}$)	1004 days	LED = 1/7/95	
	4			

5. Concurrent sentences

5.1 Particular care needs to be taken where two or more sentences of imprisonment are being served concurrently. The full term of each sentence must be considered in determining the length of the single term and the relevant release scheme. Deduction of remand and any policy custody time relevant to each sentence will then determine which sentence gives the latest SED (the dominant sentence). This will then provide the basis for setting the appropriate release arrangements.

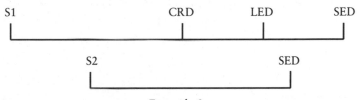

Example 9

5.2 In example 9, the second sentence, passed after the prisoner has started to serve the first sentence, fails to extend either the single term or the latest SED. As such, it has no effect on the prisoner's release dates as originally calculated.

5.3 If a prisoner receives a concurrent sentence which overlaps the SED of the original sentence then the first step is to identify the appropriate release scheme. This is done by taking the total number of days from the date of first sentence to the end of the sentence that expires latest (before the deduction of remand time). The total sentence length will then identify under which release scheme the sentence falls to be calculated.

5.4 Once the release scheme has been identified the next step is to identify the sentence that provides the latest SED after the deduction of remand time. This sentence is the dominant sentence and only remand time applicable to it will then be used to reduce the sentence. The number of days to be used in the actual calculation of release dates will be the number of days from the date of first reception into custody as a sentenced prisoner to the SED of the dominant sentence excluding remand time. The release dates will then be calculated as described in paragraph 4.18 of this instruction and described in example 5.

6. Consecutive sentences

6.1 Sentences or terms which by order of a court are to be served consecutively will, for the purpose of calculation of dates of release be aggregated and treated as a single term equal to the combined total of such sentences (see Example 2). This will also apply where such sentences or terms are imposed by different courts on different days providing the warrant or court order clearly specifies this. Remand time applicable to any of the consecutive sentences will be aggregated to reduce the total sentence.

6.2 Care is necessary in examining the order of the court or warrant of commitment in respect of a consecutive sentence or term of imprisonment imposed on a prisoner who at the date of the new sentence is already subject to two or more consecutive sentences or terms of imprisonment. Normally, where the intention of the court is that the new sentence or term should run from the end of the last of the consecutive sentences or terms to which the prisoner is already subject, the court will clearly indicate that the further sentence is consecutive to the total period of imprisonment to which the prisoner is already subject.

There may be instances, however, where the precise intention of the court is not clear and the intention of the court should be confirmed in writing and, where appropriate, a new warrant obtained. In case of doubt the Area Manager's Office should be contacted.

7. Fines and civil sentences

7.1   Under the Criminal Justice Act 1991, terms of imprisonment imposed for non payment of fines, including confiscation orders, or for civil offences are treated separately. Such terms cannot be considered with other terms of imprisonment as forming part of the single term as they attract distinct automatic release arrangements and do not fall to be reduced by custody time prior to sentence. Default terms for fines and civil sentences (which previously attracted remission) of 12 months or less will have an ARD set at the halfway point of their sentence; those serving over 12 months will have an ARD set at the two thirds point of their sentence. Annex D lists those civil offences which attract early release and those which must be served in full. It should also be noted that fine default terms and sentences for civil offences do not attract the Act's supervision or at risk protection.

7.2   Where such terms are ordered to run consecutive to a period of determinate imprisonment, they will commence from the day after the prisoner would otherwise have been released on their ARD, CRD, APD or NPD (which ever is appropriate). Where they are ordered to run concurrently, then care must be taken to determine whether the separately calculated release date acts to defer the prisoner's release from the determinate sentence.

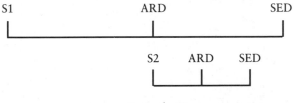

Example 10

In example 10, S2 is a default term ordered to run consecutive to S1, a term of imprisonment. Separately calculated, it commences the day after the ARD on S1.

ANNEX A

CRIMINAL JUSTICE ACT: SENTENCE ABBREVIATIONS

Release schemes and existing prisoners

Automatic Unconditional Release Scheme (AUR)

This scheme applies to all adult prisoners serving under 12 months. Prisoners are released unconditionally at the 1/2 point. They are not subject to licences but are subject to a period at risk. This scheme also includes young offenders aged under 18 years serving sentences up to and *including* 12 months. They will continue to attract the current young offender supervision requirements.

Automatic Conditional Release Scheme (ACR)

This scheme applies to all prisoners serving 12 months and less than 4 years (Young Offenders serving *over* 12 months and under 4 years). Prisoners will be released automatically on licence at the 1/2 point. They will be on licence up to the 3/4 point and thereafter at risk of being returned to prison if they re offend to complete their sentence. Sex offenders may have their licence extended to the end of their sentence on the order of the court.

Discretionary Release Scheme (DCR)

This scheme applies to all prisoners serving 4 years and over who may be granted early release any time between the PED and the NPD. Prisoners under this scheme must be released on the NPD if they are not granted early release. They will be on licence from the day they are released until the LED at the 3/4 point. They will be at risk of being returned to prison between the 3/4 point and the completion of the sentence. Sex offenders may (subject to the order of the sentencing Judge) be on licence until the completion of the sentence.

Existing Prisoners (EP)

Those prisoners who received sentences prior to 1 October 1992 who will maintain their existing release entitlements. They will not be subject to the licensing and at risk provisions of the Act. They will remain existing prisoners throughout their term even if they receive a subsequent sentence of appeal after 1 October 1992. EDR will equal ARD or NPD. LDR will equal SED and PED will remain the same (existing prisoners' actual parole release dates will reflect any loss of remission converted to ADAs).

Release Dates

Approved Parole Date (APD)

The day on which a prisoner granted early release on licence is actually released from prison. It will apply to prisoners serving 4 years and over and existing prisoners serving 12 months and over.

Automatic Release Date (ARD)

The day on which a prisoner must be released unconditionally. It will apply to:

a) All prisoners serving a sentence of less than 12 months—set at the 1/2 point.
b) All fine defaulters and contemnors. For those sentenced to under 12 months it will be set at the 1/2 point. For those prisoners serving 12 months and over it will be set at the 2/3 point.
c) Prisoners serving 12 months and less than 4 years who have been conditionally released and whilst on licence are returned to custody by order of a court having had their licence suspended. In such circumstances a court may order a prisoner's return to custody for any time up to the expiry of the total licence period. If the period of suspension ordered exceeds the total remaining licence period available on return to custody they will be released on what would have been the original licence expiry date and this would be known as the ARD.
d) Existing prisoners serving under 12 months. Will replace the EDR.

Conditional Release Date (CRD)

a) The day on which a prisoner serving 12 months and less than 4 years will be released on licence—set at the 1/2 point.
b) Prisoners in breach of their licence whose licence is suspended by the court, where the period of suspension ends before the original LED.

Parole Eligibility Date (PED)

The day on which a prisoner serving 4 years and over becomes eligible for consideration for release on licence. The PED is set at the 1/2 point. (Existing prisoners serving 12 months and over will maintain their PEDs at 1/3 point of the sentence.)

Non Parole Release Date (NPD)

The day on which a prisoner serving 4 years and over who has not previously been granted parole must be released. It will be set at the 2/3 point. The prisoner will be released on licence to the 3/4 point.

Existing prisoners serving over 12 months set at the two thirds point of sentence. Will replace the EDR.

Licence Expiry Date (LED)

The day on which compulsory supervision ends and will apply to prisoners serving 12 months and over. It will be set at the 3/4 point but may be extended to the completion of the sentence for sex offenders if ordered by the sentencing court.

Post Recall Release Date (PRRD)

The day on which prisoners serving 4 years or more and who have been released on licence either on their APD or their NPD and whose licence has been revoked by the Secretary of State, are released after serving a period of imprisonment for that revocation. The PRRD will be set at the prisoner's original LED or earlier if allowed by the Parole Board.

Sentence Expiry Date (SED)

The day on which the sentence is completed. The prisoner has no further liability. It will apply to all prisoners and will mark the completion of the sentence.

Unlawfully at Large (UAL)

Time spent by a prisoner not in the lawful custody of the authorities. It applies to prisoners who have escaped or absconded or failed to return to prison after home leave or time spent between the date of revocation/suspension of licence and reception into custody.

Disciplinary terms

Additional Days Awarded (ADA)

Days added to the prisoner's release date for breach of prison discipline. ADAs will be added to all dates with the exception of the Sentence Expiry Date.

Additional Days Awarded Prospectively (ADAP)

Days added to a prisoner's release date for a breach of prison discipline whilst a prisoner is on remand, including prisoners who are convicted and awaiting sentence. ADA(P) will only be activated in the event of a prisoner receiving a sentence of imprisonment.

Additional Days Remitted (ADAR)

Days added to a prisoner's sentence, either prospectively or immediate, for the breach of prison rules and subsequently remitted by the prison Governor.

Appendix 7
Framework Document for the Prison Service

1. Introduction

1.1 HM Prison Service for England and Wales is an executive agency of the Home Office established on 1 April 1993.

1.2 The Agency's title is 'HM Prison Service'. It is referred to as 'the Prison Service' in this Framework Document.

1.3 The Prison Service is responsible for providing prison services in England and Wales, both directly and through contractors. Its main statutory duties are set out in the Prison Act 1952 and rules made under that Act.

1.4 The Home Secretary is the Minister responsible for the Prison Service.

1.5 On 1 March 1993 the Prison Service had 38,233 staff in post, and there were 42,870 prisoners in custody at 128 establishments in England and Wales.

1.6 The Prison Service's financial allocation for 1993–94 is £1,525m.

2. Role and task

2.1 The Prison Service's Statement of Purpose, Vision, Goals and Values is:

Statement of Purpose

Her Majesty's Prison Service serves the public by keeping in custody those committed by the courts.

Our duty is to look after them with humanity and help them lead law-abiding and useful lives in custody and after release.

Vision

Our vision is to provide a service, through both directly managed and contracted prisons, of which the public can be proud and which will be regarded as a standard of excellence around the world.

Goals

Our principal goals are to:

- keep prisoners in custody
- maintain order, control, discipline and a safe environment
- provide decent conditions for prisoners and meet their needs, including health care
- provide positive regimes which help prisoners address their offending behaviour and allow them as full and responsible a life as possible
- help prisoners prepare for their return to the community
- deliver prison services using the resources provided by Parliament with maximum efficiency.

In meeting these goals, we will co-operate closely with other criminal justice agencies and contribute to the effectiveness and development of the criminal justice system as a whole.

Values

In seeking to realise our vision and meet our goals, we will adhere to the following values:

- *Integrity* is fundamental to everything we do. We will meet our legal obligations, act with honesty and openness, and exercise effective stewardship of public money and assets
- *Commitment* by our staff and to our staff. Staff are the most important asset of the Prison Service. They will be empowered to develop and use their skills and abilities to the full, while being held accountable for their performance. Teamwork will be encouraged. They will be treated with fairness, respect and openness. Their safety and well-being will be a prime concern
- *Care* for prisoners. Prisoners will be treated with fairness, justice and respect as individuals. Their punishment is deprivation of liberty and they are entitled to certain recognised standards while in prison. They will be given reasons for decisions and, where possible, involved in discussions about matters affecting them. In working with prisoners, we will involve their families and others in the community as fully as possible
- *Equality of opportunity*. We are committed to equality of opportunity and the elimination of discrimination on improper grounds
- *Innovation and improvement* are essential to the success of the Service, requiring the acceptance of change and the delivery of continuing improvements in quality and efficiency.

Key performance indicators

2.2 The Prison Service's performance is monitored against a wide range of measures and indicators which derive from its Statement of Purpose, Vision, Goals and Values. The Prison Service's key performance indicators[1] are shown in the table below.

Relevant goal	Key performance indicator[1]
Keep prisoners in custody	the number of escapes from prison establishments and from escorts
Maintain order, control, discipline and a safe environment	the number of assaults on staff, prisoners and others
Provide decent conditions for prisoners and meet their needs, including health care	the proportion of prisoners held in units of accommodation intended for fewer numbers
	the number of prisoners with 24 hour access to sanitation
Provide positive regimes which help prisoners address their offending behaviour and allow them as full and responsible a life as possible	the number of hours a week which, on average, prisoners spend in purposeful activity
	the proportion of prisoners held in establishments where prisoners are unlocked on weekdays for a total of at least 12 hours
Help prisoners prepare for their return to the community	the proportion of prisoners held in establishments where prisoners have the opportunity to exceed the minimum visiting entitlement
Deliver prison services using the resources provided by Parliament with maximum efficiency	the average cost per prisoner place

[1] A full description of these key performance indicators is given in the Prison Service's Business Plan.

2.3 The Prison Service will review and develop its key performance indicators as necessary. In particular, it will examine in 1993–94 the feasibility of developing performance measures or indicators relating to reconviction rates and health care.

3. Accountability

Home Secretary

3.1 The Home Secretary is accountable to Parliament for the Prison Service. The Home Secretary allocates resources to the Prison Service and approves its Corporate and Business Plans, including its key targets. The Home Secretary will not normally become involved in the day-to-day management of the Prison Service but will expect to be consulted by the Director General on the handling of operational matters which could give rise to grave public or Parliamentary concern.

3.2 The Home Secretary will receive reports from the Director General on the following matters:

   (i) escape of a Category A prisoner
   (ii) apparent suicide of a prisoner
   (iii) serious disturbance involving a number of prisoners and damage to person or property or other matter which is likely to arouse Parliamentary or public concern
   (iv) any incident, issue
   (v) national or particularly serious local industrial action or dispute
   (vi) major change in an establishment's functions or the proposed permanent closure of an establishment.

The Home Secretary may also request reports from the Director General on other matters.

Permanent Secretary

3.3 The Permanent Under-Secretary of State for the Home Office ('the Permanent Secretary') is the principal adviser to the Home Secretary on matters affecting the Home Office ('the Department') as a whole, including expenditure allocation and finance and the co-ordination of the Department's contribution to the criminal justice system. He or she is responsible for advising the Home Secretary on the Prison Service's corporate and business plans, proposed key targets and performance.

Director General

3.4 The Director General is the Chief Executive of the Prison Service. He or she is appointed for a fixed period by the Home Secretary, with the approval of the Prime Minister, normally following open competition. The appointment may be renewed.

3.5 The Director General is responsible for the day-to-day management of the Prison Service and is also the Home Secretary's principal policy adviser on matters relating to the Prison Service. The Director General is directly accountable to the Home Secretary for the Prison Service's performance and operations. In particular, the Director General is responsible for:

   • preparing the Prison Service's draft corporate and business plans, including key targets, and submitting them to the Home Secretary for approval

- achieving the Prison Service's key targets
- managing the Prison Service's resources efficiently, effectively and economically
- submitting quarterly reports as directed by the Home Secretary on the Prison Service's performance
- submitting an Annual Report to the Home Secretary.

3.6 The Director General may appoint a management board, including non-executive directors.

Accounting Officer responsibilities

3.7 The Permanent Secretary is the principal Accounting Officer and as such is responsible for ensuring a high standard of financial management in the Department as a whole.

3.8 The Director General is appointed by the Treasury as an additional Accounting Officer in the Home Office with responsibility for the Prisons Vote. As such the Director General is responsible for ensuring that proper procedures are followed for securing the regularity and propriety of expenditure on the Vote, that the public funds for which he or she is responsible are properly and well managed and that the requirements of Government Accounting are met, that the Prison Service observes any general guidance issued by the Treasury or the Cabinet Office, and for putting into effect any recommendations accepted by Government of the Public Accounts Committee, other Parliamentary Select Committees or other Parliamentary authority. The respective responsibilities of the principal Accounting Officer and the additional Accounting Officer are set out in more detail in a financial memorandum.[2]

3.9 The Director General and the Permanent Secretary are both liable to be summoned before the Public Accounts Committee in connection with their respective responsibilities. It is for the Home Secretary to decide who should appear at Departmental Select Committee hearings. In practice, where a Committee's interest lies in the policies or day-to-day operations of the Prison Service, the Director General will normally appear.

Members of Parliament

3.10 Members of Parliament will be encouraged to write direct to the Director General on matters for which the Director General has delegated responsibility. The Home Secretary will normally ask the Director General to reply to correspondence from Members of Parliament on such matters. When a Member of Parliament asks a Parliamentary Question on a delegated matter, the Home Secretary will normally reply to the effect that the Director General will write direct to the Member. Replies from the Director General to Members of Parliament in response to Parliamentary Questions will be published in *Hansard*. Parliamentary Questions on non-delegated matters will normally be answered by Ministers, advised by the Director General.

Parliamentary Commissioner for Administration

3.11 The Prison Service is subject to the jurisdiction of the Parliamentary Commissioner for Administration. The Permanent Secretary is the Principal Officer of the Department for this purpose but will delegate to the Director General responsibility for replying on any matters concerning the Prison Service.

---

[2] Copies of this memorandum can be obtained from the Head of DPF 4 Division, Prison Service Headquarters, Abell House, John Islip Street, London SW1P 4LH.

Her Majesty's Chief Inspector of Prisons

3.12 The Home Secretary receives reports from Her Majesty's Chief Inspector of Prisons on inspections of prison establishments. The Home Secretary will ask the Director General to respond to recommendations on delegated matters.

Prisons Ombudsman

3.13 The Home Secretary intends to appoint a Prisons Ombudsman. The Home Secretary will receive an annual report from the Prisons Ombudsman. The Director General will respond to recommendations from the Prisons Ombudsman in respect of particular complaints.

Boards of Visitors

3.14 The Home Secretary receives annual reports from Boards of Visitors on the state of prison premises, the administration of prisons and the treatment of prisoners. The Home Secretary will ask the Director General to respond on delegated matters.

4. Planning, finance and support services

Planning framework

4.1 Each year the Prison Service will prepare a corporate plan which sets out the Service's objectives and strategy over the following three years. A more detailed plan, the business plan, will be produced each year to cover the first year of the planning period. The corporate and business plans will be prepared to a timetable consistent with the Departmental Public Expenditure Survey and published.

4.2 The corporate plan will include:

- the goals and key performance indicators of the Prison Service
- the strategies which it intends to follow to achieve its goals
- operating assumptions about the prison population, the accommodation and the financial and manpower resources which are expected to be available.
- information about its sensitivity to variations in those assumptions
- plans for improving efficiency, including plans for market testing.

4.3 The business plan will include the Prison Service's key targets.

4.4 If there are major unforeseen changes in the Prison Service's operating assumptions, the corporate or business plan may need to be revised in the course of the year. Proposals to this effect made by the Director General require the approval of the Home Secretary.

Financial provision

4.5 The Prison Service is financed by supply estimates voted by Parliament. There is a separate Prisons Vote. The Agency operates within gross running costs control. As part of the annual planning process, the Prison Service puts forward proposals to the Home Secretary for current and capital expenditure to meet its projected needs, taking into account planned efficiency improvements.

Accounting responsibilities

4.6 Each year the Director General produces and signs the Appropriation Account for the Prisons Vote, which is prepared in accordance with Treasury requirements. The Annual Report and accounts are submitted by the Director

General to the Home Secretary and are laid before Parliament, published and placed in the libraries of both Houses of Parliament. The accounts for 1993–94 and 1994–95 will be published in October 1994 and October 1995 respectively. Thereafter the Prison Service will plan to lay the account before the summer recess. It is intended that from 1994–95 accruals accounts will be produced in accordance with a direction issued by the Treasury under section 5 of the Exchequer and Audit Departments Act 1921 and will be audited by the Comptroller and Auditor General. These will reflect the requirements of Parliamentary accountability and the nature and management needs of the business. The Director General will provide a reconciliation of the Appropriation Account and the accruals accounts. The Vote accounts are subject to external auditing by the Comptroller and Auditor General.

Internal audit and other services

4.7 The Director General is responsible for making arrangements for the provision of internal audit and consultancy, inspection and review services. The Director General may use either Home Office Internal Audit or other auditors to conduct the work in accordance with best commercial practice and the standards and objectives of the Government Internal Audit Manual. Home Office Internal Audit will have access to the Prison Service as necessary to perform any work required by the principal Accounting Officer in order to fulfil the duties set out at paragraph 3.7.

Financial delegations

4.8 The Director General has authority to approve all voted expenditure which is consistent with the corporate and business plans subject to the exceptions set out in Annex A, which may be reviewed from time to time. The Director General will delegate responsibility for expenditure to the lowest level he or she considers possible, consistent with the needs of financial control and propriety.

Assets register

4.9 The Director General will develop arrangements for recording and publishing the value of the Prison Service's assets.

Support services

4.10 Annex B lists support services which are presently performed by the Prison Service for the Department, and by the Department for the Prison Service. Any changes in these arrangements shall be made in consultation with the other party. The Prison Service and the Department will each seek, where it is practicable and economic to do so, to introduce charging for services provided for the other.

5. Personnel matters

General

5.1 Prison Service staff will continue to be civil servants employed in the Home Office on Civil Service terms, conditions and pension arrangements. Variations may be made subject to the approval of the Treasury as necessary.

5.2 The Home Office Staff Handbook will continue to apply to Prison Service staff until replaced by a Prison Service Staff Handbook.

5.3 The Director General will consult staff and their representatives on matters affecting their pay and conditions of service, including any amendments to the Staff Handbook.

Personnel management

5.4 The Director General has authority on all personnel management matters in relation to Prison Service staff up to and including Grade 4, with the exception of ATs and HEODs allocated to the Department. Postings of staff at Grades 3–5 into, out of and within the Prison Service require the agreement of the Director General and the Department's Principal Establishment Officer.

5.5 In exercising delegated personnel management responsibilities the Director General will have regard to any relevant central and Departmental guidance and policy statements.

5.6 The Prison Service is committed to the training and development of its staff. Its training strategy aims to promote the development of its staff in order to meet the needs of the Prison Service.

Pay and grading arrangements

5.7 The Director General will review the pay and grading arrangements of the Prison Service to ensure that they are suited to its needs and circumstances. Any changes in pay and grading arrangements must be agreed by the Treasury. The Director General will be responsible for pay bargaining from 1 April 1994.

5.8 The Director General is responsible for the creation, number and grading of posts up to and including Grade 6. The creation of posts at Grade 5 requires the agreement of the Permanent Secretary. The creation of posts at Grade 4 and above requires the agreement of the Permanent Secretary and the Treasury.

Industrial relations

5.9 The Director General is responsible for ensuring good industrial relations in the Prison Service. The Prison Service will continue to operate Whitley arrangements.

6. Review and variation of framework

6.1 The Prison Service's Framework Document will be reviewed jointly by the Home Secretary and the Director General at intervals of not more than three years. The Home Secretary, the Permanent Secretary or the Director General may propose amendments to this framework at any time. Any amendments are subject to agreement by the Home Secretary, the Director General, the Treasury and the Office of Public Service and Science.

6.2 Copies of this Framework Document and any subsequent amendments will be published and placed in the libraries of both Houses of Parliament and in prison libraries.

Annexes

Annex A—Financial memorandum

Annex B—Exceptions to financial delegation

Annex C—Support services provided by the Prison Service

Annex D—Support services provided by the Department

Annex E—Staff mobility: Memorandum of Understanding between the Prison Service and Personnel Management Division (PMD).

Criminal Justice Act 1991 Sections 84–92 and Schedule 10
Contracted out prisons

Contracting out of certain prisons

84.—(1) The Secretary of State may enter into a contract with another person for the running by him of any prison which—

    (*a*) is established after the commencement of this section; and

    (*b*) is for the confinement of remand prisoners, that is to say, persons charged with offences who are remanded in or committed to custody pending their trial, or persons committed to custody on their conviction who have not been sentenced for their offences;

and while such a contract is in force, the prison to which it relates shall be run subject to and in accordance with sections 85 and 86 below, the 1952 Act (as modified by section 87 below) and prison rules.

(2) In this Part—

'contracted out prison' means a prison as respects which such a contract is for the time being in force;

'the contractor,' in relation to such a prison, means the person who has contracted to run it.

(3) The Secretary of State may by order made by statutory instrument provide that this section shall have effect as if there were omitted from subsection (1) above either—

    (*a*) paragraph (*a*) and the word 'and' immediately following that paragraph; or

    (*b*) paragraph (*b*) and the said word 'and'; or

    (*c*) the words from 'which,' in the first place where it occurs, to the end of paragraph (*b*).

(4) An order under subsection (3) (*b*) or (*c*) above shall provide that section 87 below shall have effect as if subsection (5) were omitted.

(5) No order shall be made under subsection (3) above unless a draft of the order has been laid before and approved by resolution of each House of Parliament.

Officers of contracted out prisons

85.—(1) Instead of a governor, every contracted out prison shall have—

(*a*) a director, who shall be a prisoner custody officer appointed by the contractor and specially approved for the purposes of this section by the Secretary of State; and

(*b*) a controller, who shall be a Crown servant appointed by the Secretary of State;

and every officer of such a prison who performs custodial duties shall be a prisoner custody officer who is authorised to perform such duties.

(2) Subject to subsection (3) below, the director shall have such functions as are conferred on him by the 1952 Act (as modified by section 87 below) or as may be conferred on him by prison rules.

(3) The director shall not—

(*a*) inquire into a disciplinary charge laid against a prisoner, conduct the hearing of such a charge or make, remit or mitigate an award in respect of such a charge; or

(*b*) except in cases of urgency, order the removal of a prisoner from association with other prisoners, the temporary confinement of a prisoner in a special cell or the application to a prisoner of any other special control or restraint.

(4) The controller shall have such functions as may be conferred on him by prison rules and shall be under a duty—

(*a*) to keep under review, and report to the Secretary of State on, the running of the prison by or on behalf of the director; and

(*b*) to investigate, and report to the Secretary of State on, any allegations made against prisoner custody officers performing custodial duties at the prison.

(5) The contractor shall be under a duty to do all that he reasonably can (whether by giving directions to the officers of the prison or otherwise) to facilitate the exercise by the controller of all such functions as are mentioned in or conferred by subsection (4) above.

Powers and duties of prisoner custody officers employed at contracted out prisons

86.—(1) A prisoner custody officer performing custodial duties at a contracted out prison shall have the following powers, namely—

(*a*) to search in accordance with prison rules any prisoner who is confined in the prison; and

(*b*) to search any other person who is in or is seeking to enter the prison, and any article in the possession of such a person.

(2) The powers conferred by subsection (1) (*b*) above to search a person shall not be construed as authorising a prisoner custody officer to require a person to remove any of his clothing other than an outer coat, jacket or gloves.

(3) A prisoner custody officer performing custodial duties at a contracted out prison shall have the following duties as respects prisoners confined in the prison, namely—

(a) to prevent their escape from lawful custody;

(*b*) to prevent, or detect and report on, the commission or attempted commission by them of other unlawful acts;

(*c*) to ensure good order and discipline on their part; and

(*d*) to attend to their wellbeing.

(4) The powers conferred by subsection (1) above, and the powers arising by virtue of subsection (3) above, shall include power to use reasonable force where necessary.

Consequential modifications of 1952 Act

87.—(1) In relation to a contracted out prison, the provisions of the 1952 Act specified in subsections (2) to (8) below shall have effect subject to the modifications so specified.

(2) In section 7 (1) (prison officers), the reference to a governor shall be construed as a reference to a director and a controller.

(3) Section 8 (powers of prison officers) and section 11 (ejectment of prison officers and their families refusing to quit) shall not apply.

(4) In sections 10 (5), 12 (3), 13 (1) and 19 (1) and (3) (various functions of the governor of a prison), references to the governor shall be construed as references to the director.

(5) In section 12 (1) and (2) (place of confinement of prisoners), and reference to a prisoner or prisoners shall be construed as a reference to a remand prisoner or prisoners.

(6) In section 13 (2) (legal custody of prisoners), the reference to an officer of the prison shall be construed as a reference to a prisoner custody officer performing custodial duties at the prison.

(7) In section 14 (2) (cells), the reference to a prison officer shall be construed as a reference to a prisoner custody officer performing custodial duties at the prison.

(8) Section 35 (vesting of prison property in the Secretary of State) shall have effect subject to the provisions of the contract entered into under section 84 (1) above.

Intervention by the Secretary of State

88.—(1) This section applies where, in the case of a contracted out prison, it appears to the Secretary of State—

(*a*) that the director has lost, or is likely to lose, effective control of the prison or any part of it; and

(*b*) that the making of an appointment under subsection (2) below is necessary in the interests of preserving the safety of any person, or of preventing serious damage to any property.

(2) The Secretary of State may appoint a Crown servant to act as governor of the prison for the period—

(*a*) beginning with the time specified in the appointment; and

(*b*) ending with the time specified in the notice of termination under subsection (4) below.

(3) During that period—

(*a*) all the functions which would otherwise be exercisable by the director or the controller shall be exercisable by the governor;

(*b*) the contractor shall do all that he reasonably can to facilitate the exercise by the governor of those functions; and

(*c*) the officers of the prison shall comply with any directions given by the governor of those functions.

(4) Where the Secretary of State is satisfied—

(*a*) that the governor has secured effective control of the prison or, as the case may be, the relevant part of it; and

(*b*) that the governor's appointment is no longer necessary as mentioned in subsection (1) (b) above,

he shall, by a notice to the governor, terminate the appointment at a time specified in the notice.

(5) As soon as practicable after making or terminating an appointment under this section, the Secretary of State shall give a notice of the appointment, or a copy of the notice of termination, to the contractor, the director and the controller.

Supplemental

Certification of prisoner custody officers

89.—(1) In this Part 'prisoner custody officer' means a person in respect of whom a certificate is for the time being in force certifying—

(*a*) that he has been approved by the Secretary of State for the purpose of performing escort functions or custodial duties or both; and

(*b*) that he is accordingly authorised to perform them.

(2) The provisions of Schedule 10 to this Act shall have effect with respect to the certification of prisoner custody officers.

(3) In this section and Schedule 10 to this Act—

'custodial duties' means custodial duties at a contracted out prison;

'escort functions' means the functions specified in section 80 (1) above,

Protection of prisoner custody officers

90.—(1) Any person who assaults a prisoner custody officer acting in pursuance of prisoner escort arrangements, or performing custodial duties at a contracted out prison, shall be liable on summary conviction to a fine not exceeding level 5 on the standard scale or to imprisonment for a term not exceeding six months or to both.

(2) Section 17 (2) of the Firearms Act 1968 (additional penalty for possession of firearms when committing certain offences) shall apply to offences under subsection (1) above.

(3) Any person who resists or wilfully obstructs a prisoner custody officer acting in pursuance of prisoner escort arrangements, or performing custodial duties at a contracted out prison, shall be liable on summary conviction to a fine not exceeding

level 3 on the standard scale.

(4) For the purposes of this section, a prisoner custody officer shall not be regarded as acting in pursuance of prisoner escort arrangements at any time when he is not readily identifiable as such an officer (whether by means of a uniform or badge which he is wearing or otherwise).

Wrongful disclosure of information

91.—(1) A person who is or has been employed (whether as a prisoner custody officer or otherwise) in pursuance of prisoner escort arrangements, or at a contracted out prison, shall be guilty of an offence if he discloses, otherwise than in the course of his duty or as authorised by the Secretary of State, any information which he acquired in the course of his employment and which relates to a particular prisoner.

(2) A person guilty of an offence under subsection (1) above shall be liable—

(a) on conviction of indictment, to imprisonment for a term not exceeding two years or a fine or both;

(b) on summary conviction, to imprisonment for a term not exceeding six months or a fine not exceeding the statutory maximum or both.

Interpretation of Part IV

92.—(1) In this Part—

'contracted out prison' and 'the contractor' have the meanings given by section 84 (2) above;

'court-house' means a petty sessional court-house within the meaning of the 1980 Act or an occasional court-house appointed under section 147 of that Act;

'court security officer' has the meaning given by section 76 (1) above;

'prison' includes a young offender institution or remand centre;

'prisoner' means any person who—

(a) is held in custody in a prison;

(b) is kept in police detention after being charged with an offence;

(c) has been committed to detention at a police station under section 128 (7) of the 1980 Act; or

(d) is in the custody of a court;

'prisoner custody officer' has the meaning given by section 89 (1) above;

'prisoner escort arrangements' has the meaning given by section 80 (2) above.

(2) Unless the contrary intention appears, expressions used in sections 76 to 79 above which are also used in the 1979 Act have the same meanings as in that Act.

(3) Section 80, 81 (1) and (2) (a), 82 and 89 to 91 above, subsection (1) above and Schedule 10 to this Act shall have effect as if—

(a) any reference in section 80 (1), 81 (1), 82 or 91 above to prisoners included a reference to persons kept in secure accommodation by virtue of a security requirement imposed under section 23 (4) of the 1969 Act (remands and committals to local authority accommodation); and

(b) any reference in section 80 (1) (c) to (e) above to a prison included a reference to such accommodation.

Section 89   SCHEDULE 10
CERTIFICATION OF PRISONER CUSTODY OFFICERS

Preliminary

1. In this Schedule—

'certificate' means a certificate under section 89 of this Act;
'the relevant functions', in relation to a certificate, means the escort functions or
custodial duties authorised by the certificate.

Issue of certificates

2.—(1) Any person may apply to the Secretary of State for the issue of a certificate
in respect of him.

(2) The Secretary of State shall not issue a certificate on any such application
unless he is satisfied that the applicant—

(a) is a fit and proper person to perform the relevant functions; and
(b) has received training to such standard as he may consider appropriate for the
performance of those functions.

(3) Where the Secretary of State issues a certificate, then, subject to any suspension under paragraph 3 or revocation under paragraph 4 below, it shall continue in
force until such date or the occurrence of such event as may be specified in the certificate.

(4) A certificate authorising the performance of both escort functions and custodial duties may specify different dates or events as respects those functions and
duties respectively.

Suspension of certificate

3.—(1) This paragraph applies where at any time it appears—

(a) in the case of a prisoner custody officer acting in pursuance of prisoner escort
arrangements, to the prisoner escort monitor for the area concerned; or
(b) in the case of such an officer performing custodial duties at a contracted out
prison, to the controller of that prison,

that the officer is not a fit and proper person to perform the escort functions or, as
the case may be, custodial duties.

(2) The prisoner escort monitor or controller may—

(a) refer the matter to the Secretary of State for a decision under paragraph 4
below; and
(b) in such circumstances as may be prescribed by regulations made by the
Secretary of State, suspend the officer's certificate so far as it authorises the
performance of escort functions or, as the case may be, custodial duties pending that decision.

(3) The power to make regulations under this paragraph shall be exercisable by
statutory instrument which shall be subject to annulment in pursuance of a resolution of either House of Parliament.

Revocation of certificate

4. Where at any time it appears to the Secretary of State that a prisoner custody officer is not a fit and proper person to perform escort functions or custodial duties, he may revoke that officer's certificate so far as it authorises the performance of those functions or duties.

False statements

5. If any person, for the purpose of obtaining a certificate for himself or for any other person—

    (*a*) makes a statement which he knows to be false in a material particular, or

    (*b*) recklessly makes a statement which is false in a material particular,

he shall be liable on summary conviction to a fine not exceeding level 4 on the standard scale.

Index